Function Spaces, Differential Operators and Nonlinear Analysis

The Hans Triebel Anniversary Volume

Dorothee Haroske
Thomas Runst
Hans-Jürgen Schmeisser
Editors

Springer Basel AG

Editors' address:

Institute of Mathematics
Friedrich-Schiller-University
07740 Jena
Germany
e-mail: mhj@minet.uni-jena.de

2000 Mathematics Subject Classification 35-02, 42-02, 46-02, 46E35, 47-02

A CIP catalogue record for this book is available from the Library of Congress, Washington D.C., USA

Bibliographic information published by Die Deutsche Bibliothek
Die Deutsche Bibliothek lists this publication in the Deutsche Nationalbibliografie; detailed bibliographic data is available in the Internet at <http://dnb.ddb.de>.

ISBN 978-3-0348-9414-2 ISBN 978-3-0348-8035-0 (eBook)
DOI 10.1007/978-3-0348-8035-0

Originally published by Birkhäuser Verlag in 2003
Softcover reprint of the hardcover 1st edition 2003
Member of the BertelsmannSpringer Publishing Group
Printed on acid-free paper produced from chlorine-free pulp. TCF ∞
Cover design: Micha Lotrovsky, CH-4106 Therwil, Switzerland

ISBN 978-3-0348-9414-2

9 8 7 6 5 4 3 2 1 www.birkhäuser-science.com

This collection of research papers is dedicated to
Hans Triebel
on the occasion of his 65th birthday

Contents

Preface

This volume is dedicated to our teacher and friend Hans Triebel. The core of the book is based on lectures given at the International Conference "Function Spaces, Differential Operators and Nonlinear Analysis" (FSDONA–01) held in Teistungen, Thuringia / Germany, from June 28 to July 4, 2001, in honour of his 65th birthday. This was the fifth in a series of meetings organised under the same name by scientists from Finland (*Helsinki, Oulu*), the Czech Republic (*Prague, Plzeň*) and Germany (*Jena*) promoting the collaboration of specialists in East and West, working in these fields.

This conference was a very special event because it celebrated Hans Triebel's extraordinary impact on mathematical analysis. The development of the modern theory of function spaces in the last 30 years and its application to various branches in both pure and applied mathematics is deeply influenced by his lasting contributions. In a series of books Hans Triebel has given systematic treatments of the theory of function spaces from different points of view, thus revealing its interdependence with interpolation theory, harmonic analysis, partial differential equations, nonlinear operators, entropy, spectral theory and, most recently, analysis on fractals.

The presented collection of papers is a tribute to Hans Triebel's distinguished work. The book is subdivided into three parts:

- Part I contains the two invited lectures by O.V. Besov (*Moscow*) and D.E. Edmunds (*Sussex*) having a survey character and honouring Hans Triebel's contributions.
- The papers in Part II reflect seven recent developments in the theory of function spaces, linear and nonlinear partial differential equations presented by outstanding experts in the field.
- Shorter communications related to the topics of the conference and Hans Triebel's research are collected in Part III.

Hans Triebel's personal qualities leave a lasting impression on his colleagues, students and friends. Many of us have benefited from his extensive knowledge, his ideas and hospitality. We are glad to have the opportunity to express our deep gratitude to him.

We acknowledge with gratitude financial support by the DFG, the Graduiertenkolleg *"Analytic and stochastic structures and systems"* and the Friedrich-Schiller-University Jena. It is a pleasure for us to give our special thanks to our colleagues Michele Bricchi, Serguei Dachkovski, Hans-Gerd Leopold and Winfried

Sickel for many helpful suggestions concerning the present book and the great amount of work they did to make things run smoothly. We are very much indebted to Birkhäuser Verlag for their important support during the conference and for this project. Finally, we would like to thank all the speakers and all participants in the conference FSDONA-01 for their energy and enthusiasm which made its success possible.

Jena, in April 2002

Dorothee D. Haroske
Thomas Runst
Hans-Jürgen Schmeißer

Part I

D. Haroske, T. Runst, H.-J. Schmeisser (eds.): Function Spaces, Differential Operators and Nonlinear Analysis. The Hans Triebel Anniversary Volume.

Spaces of Differentiable Functions

Oleg Besov[1] and Gennadiy Kalyabin[2]

Dedicated to Prof. Hans Triebel on the occasion of his 65th birthday

Introduction

A brilliant exposition of numerous aspects of the theory of function spaces (embeddings and equivalent norms, description in terms of smoothness properties; decompositions and approximations; interpolation via real and complex methods; trace problems; extension operators for regular and irregular domains; applications to PDO and ΨDO etc.) is given in the series of famous books [T-78], [T-83], [T-92], [T-01] by Professor Hans Triebel. Some other approaches one may find in fundamental monographs [S-88], [N-75], [St-70], [M-85], [dVLo-93], [BIN-96] and survey papers [KL-87], [BKuLN-90] where the detailed references are given.

The goal of this paper is to supplement these texts (avoiding the reproduction of passages from them) with some selected topics which touch mainly the historically significant points and with more recent results obtained by mathematicians of Russia in the theory of function spaces.

In Section 1 a brief history is expounded of the most known types of function spaces: Sobolev spaces W_p^l, Nikol'skiy spaces H_p^s, Lipschitz type spaces B_{pq}^s and the Lizorkin-Triebel spaces L_{pq}^s (often referred to also as Triebel-Lizorkin spaces F_{pq}^s). The spaces of functions of generalized smoothness are surveyed in Section 2 where the results on capacities are of major interest. The embedding theory for spaces on domains (Section 3) contains many quite fresh results related to regular and irregular domains. Function spaces on domains which allow the extension onto the whole Euclidean space R^n and corresponding extension operators are described in Section 4. In the final Section 5 diverse additional questions are briefly touched upon.

The definitions, theorems, remarks are given not as separate parts but are simply set forth in the text with necessary references.

Authors would like to express their deep gratitude to A. Kufner, V. Maz'ya, S. Pohozaev for valuable remarks which are taken into account.

---

[1] This work was supported by grants of RFBR-99-01-00868, LSSRF-00-15-96047 and INTAS-99-01080

[2] This work was supported by the grant of RFBR-99-01-00868

It is important to emphasize that the active cooperation between Russian and German mathematicians (more widely: between scientists of the former Soviet Union – fSU – and those of Western countries) initiated by Professor Hans Triebel and the Mathematical Institute of Friedrich-Schiller-University (FSU), Jena, has played a significant and helpful role in the achievements in function spaces theory during the last decades.

1. Brief history

In [T-83: 2.2.1] three periods in function spaces theory are marked out: the classical period (Lebesgue spaces L_p, the spaces C, C^m, Hölder spaces $C^s, s > 0$, Hardy classes H_p of analytic functions); the constructive period (1930–75) and the period of systematization which is continuing until now. In the sequel we deal with the two last periods.

1.1. Sobolev spaces

The theory of spaces of differentiable functions of several real variables originates from the paper by S.L. Sobolev [S-38]. He proposed the notion of generalized derivative and introduced the Banach spaces $W_p^l(\Omega)$ of functions f, defined on a domain $\Omega \subset R^n$ with the norm

$$\| f \mid W_p^l(\Omega)\| := \sum_{|\alpha|\le l} \| D^\alpha f \mid L_p(\Omega)\|, \tag{1.1}$$

where $l \in \mathbf{N}$, $1 \le p < \infty$, $L_p(\Omega)$ is the Lebesgue space of functions with the norm

$$\| f \mid L_p(\Omega)\| := \left(\int_\Omega |f(x)|^p \, dx\right)^{1/p}. \tag{1.2}$$

These Sobolev spaces and their generalizations play until now an important role in the theory of function spaces and in applications to mathematical physics. It was Sobolev who first obtained the embedding theorems which are related to a domain satisfying the cone condition. For a function $f \in W_p^l(\Omega)$ they assert its summability to the power $q > p$ on the same domain $\Omega = \Omega^{(n)} \subset R^n$ or on sufficiently smooth manifolds $\Omega^{(m)}$ of the dimension $m < n$, belonging to Ω:

$$W_p^l(\Omega) \subset L_q(\Omega^{(m)}) \quad \text{for } \delta := l - \frac{n}{p} + \frac{m}{q} > 0, \tag{1.3}$$

where $1 \le p < q \le \infty$, $1 \le m \le n$; in the case $m = n$, $q = \infty$ one may replace $L_\infty(\Omega)$ in (1.3) by the space $C(\Omega)$ of all functions continuous and bounded on Ω.

For $q < \infty$ the embedding (1.3) holds also if $\delta \ge 0$. These supplements are contained in results by S.L. Sobolev ($p > 1$, $m = n$), V.P. Il'in [I-54] ($p > 1$), E. Gagliardo ($p = 1$) (cf. [BIN-96: Ch. 3]). V.I. Kondrashov (1945) has established that the embedding (1.3) is *compact* provided the domain is bounded and $\delta > 0$. The first method used by Sobolev to prove the embeddings was the integral representation method in which the function is decomposed into certain sums of potential type integrals applied to the function itself and to its derivatives.

The critical case $m = n = lp$ was studied by S.I. Yudovich (1961) who proved that for $1 < p < \infty$ the space $W^l_p(R^n)$ is embedded into Orlicz space $L^\star_M$ with M being the Young function $M(u) := \exp(u^{p'})$ (see [BIN-96: 10.6] for more historical comments).

1.2. Nikol'skiy spaces

The next important step was made in the 1950-ies by S.M. Nikol'skiy who introduced the function spaces $H^s_p(R^n)$[1], $1 \le p \le \infty$, $s > 0$, and established the theory of these spaces. A function $f \in H^s_p(R^n)$ if f is defined on the Euclidean space R^n and has the finite norm

$$\| f \mid H^s_p(R^n)\| := \| f \mid L_p(R^n)\| + \sum_{j=1}^{n} \sup_{h>0} h^{-s+k} \| \Delta^M_{j,h} D^k_j f \mid L_p(R^n)\|, \quad (1.4)$$

where $M > 0, k \ge 0$ are integers such that $k < s$, $M + k > s$, $D^k_j f(x) := \partial^k f / \partial x^k_j$ are the generalized derivatives and

$$\Delta^M_{j,h} f(x) := \sum_{l=0}^{M} (-1)^{(M+l)} C^l_M \; f(x + lhe_i) \quad (1.5)$$

are the coordinate-wise differences with the step $h > 0$ of the M-th order.

For different admissible pairs (M, k) the norms (1.4) are equivalent to each other and to the following *equivalent norming in terms of approximation via entire functions of exponential type* [N-51]:

$$\| f \mid H^s_p(R^n)\|^{(A)} := \| f \mid L_p(R^n)\| + \sup_{N>1} N^s \inf_{g_N \in \mathrm{EFET}(N)} \|f - g_N \mid L_p(R^n)\|, \quad (1.6)$$

Here EFET(N) stands for a set of all entire functions of the exponential type N which in 2π-periodical case coincides with the set of all trigonometric polynomials of the degree $\le N$ (cf. [N-75: Ch. 3]). Of great significance are the Bernstein-Nikol'skiy inequalities for the class EFET(N)

$$\| D^\alpha g_N \mid L_q(R^m)\| \le c_{p,q} N^{|\alpha| + \frac{n}{p} - \frac{m}{q}} \| g_N \mid L_p(R^n)\|, \quad 1 \le p \le q \le \infty. \quad (1.7)$$

Having based on these relationships S.M. Nikol'skiy obtained the embedding theorem for H^s_p-spaces:

$$H^s_p(R^n) \subset H^\rho_q(R^m), \; 1 \le p \le q \le \infty, \; 1 \le m \le n, \; \rho = s - \frac{n}{p} + \frac{m}{q} > 0 \quad (1.8)$$

which establishes for $m < n$ the integral-differential properties of the *trace* of the function $f \in H^s_p(R^n)$ on the m-dimensional subspace R^m. It is important that for $q = p$ this theorem is reversible, i.e. any function $\varphi \in H^\rho_p(R^m)$ is a trace of some function $f \in H^s_p(R^n)$, thus the sharp description of the trace space is given.

[1] Here H^s_p does not denote the Bessel-potential (or Lebesgue or Liouville) spaces (defined in Section 1.4).

1.3. Lipschitz spaces of functions

The statement of the complete results on the trace problem of Sobolev spaces required the consideration of Lipschitz spaces of functions. The first results were obtained by N. Aronszajn for $p = 2$ [A-55]. L.N. Slobodeckiy built the theory of the anisotropic Sobolev spaces $W_2^l(R^n)$, $l = (l_1, \dots, l_n)$ with integer and fractional smoothness parameters [Sl-58]. E. Gagliardo characterized the traces of functions of Sobolev space $W_p^1(R^n)$ on the hyperplanes in R^n [Ga-57].

In the 1960-ies O.V. Besov has studied the 4-parametric family scale of function spaces $B_{p,q}^s(R^n)$, $s > 0$, $1 \le p$, $q \le \infty$, $n \in \mathbf{N}$ defined by the finiteness of the norm

$$\| f \mid B_{p,q}^s(R^n)\| := \| f \mid L_p(R^n)\| + \sum_{j=1}^n \left\{ \int_0^1 \left(\frac{\| \Delta_{j,h}^M D_j^k f \mid L_p(R^n)\|}{h^{s-k}} \right)^q \frac{dh}{h} \right\}^{1/q} . \tag{1.9}$$

It is clear that $B_{p,\infty}^s(R^n) = H_p^s(R^n)$. The embeddings analogous to (1.8) are valid for B-spaces as well as the description of traces on R^m. The spaces $B_{p,q}^s(R^n)$ are closely related to Sobolev spaces: for s integer and $p = q = 2$ one has $B_{2,2}^s(R^n) = W_2^s(R^n)$, and for $1 < p < \infty$ in terms of the B-spaces the traces on R^m for Sobolev spaces are described:

$$W_p^s(R^n)\Big|_{R^m} = B_{p,p}^{s-\frac{n-m}{p}}(R^m) \quad \text{for} \quad s - \frac{n-m}{p} > 0 . \tag{1.10}$$

The latter result for $m = n - 1$, $s = 1$, $p = 2$ was established by N. Aronszajn and for $1 < p < \infty$ by E. Gagliardo. The general case $1 \le m < n - 1$ was studied by O.V. Besov [B-61] (see details in [N-75: Introduction]).

1.4. Lizorkin-Triebel spaces

P.I. Lizorkin [L-72] has introduced the spaces $L_{p,q}^s(R^n)$ $(1 < p, q < \infty, s > 0)$ with the norms

$$\| f \mid L_{p,q}^s(R^n)\| := \| \|\{2^{ks} f_k(x)\}|l_q\| \mid L_p(R^n) \| \tag{1.11}$$

where $f = \sum_{k=1}^{\infty} f_k$ is the decomposition of the function f:

$$f_1(x) := S_1 f(x), \quad f_k(x) := S_{2^k} f(x) - S_{2^{k-1}} f(x),\ k \ge 1 \tag{1.12}$$

and S_N in Fourier images correspond to the multiplication by the characteristic function of the cube $Q_N := \{\xi \in R^n : \max|\xi_j| \le N\}$. The paper [L-67] in which the $L_p(R^n, l_q)$-norms first appear in function spaces theory also should be mentioned.

A little bit later H. Triebel (cf. [T-83: 2.3.5]) suggested to use instead of the cutting operators S_N the smooth decompositions of the Fourier transform

$$\varphi_1(\xi) := \varphi(\xi) \in C_0^\infty(R^n); \quad \varphi(\xi) = 1,\ |\xi| \le 1; \quad \varphi(\xi) = 0,\ |\xi| \ge 2$$

$$\varphi_k(\xi) := \varphi(2^{-k}\xi) - \varphi(2^{-k+1}\xi),\ k > 1 \tag{1.13}$$

so that $\sum_{k=1}^\infty \varphi_k(\xi) \equiv 1$.

For $0 < p < \infty$, $0 < q \leq \infty$, $-\infty < s < \infty$ the space $F^s_{p,q}(R^n)$ consists by definition of all functions f which belong to the Schwartz space of distributions $S'(R^n)$ and possess the finite (quasi)-norms:

$$\| f \mid F^s_{p,q}(R^n)\|^{(\varphi)} := \| \ \|\{2^{ks} F^{-1} \ \varphi_k(\xi) Ff(x)\}| \ l_q\| \mid L_p(R^n)\| \tag{1.14}$$

where F, F^{-1} are the direct and the inverse Fourier transforms (for different $\varphi(\xi)$, satisfying (1.13) one obtains equivalent norms). It is important that for the range of parameters considered by Lizorkin (i.e. for $1 < p, q < \infty$, $s > 0$) these two definitions yield the same space: $L^s_{p,q}(R^n) = F^s_{p,q}(R^n)$. With this case we shall deal in the sequel.

For $q = 2$ Lizorkin-Triebel spaces coincide with Liouville spaces $L^s_p(R^n)$ (= Bessel potentials spaces) [T-83: 2.2.1]:

$$\| f \mid L^s_p(R^n)\| := \|F^{-1}(\ (1 + |\xi|^2)^{s/2} \ Ff(\xi) \ )(x)| \ L_p(R^n)\|. \tag{1.15}$$

Note that the norm in the space $B^s_{p,q}(R^n)$ differs from the norm in $L^s_{p,q}(R^n)$ only by interchanging the order of taking $L_p(R^n)$- and l_q-norms in (1.13). One of Triebel's observations (sometimes called "the thumb rule") is that almost all propositions established for L-spaces are valid (and can be proved much easier) for B-spaces as well. So further in this paper we shall concentrate our attention mostly on Lizorkin-Triebel spaces.

In [T-83: 2.5], [T-92: Chs 2,4] one may find many results related to the equivalent norms in terms of differences, local averaged oscillations and other smoothness characteristics, to the trace problem (on hyperplanes R^m and Lipschitz domains) for the whole scale $F^s_{p,q}(R^n)$. For the spaces $L^s_{p,q}(R^n), 1 < p, q < \infty$, $s > 0$ these results have been first established by Kalyabin (1977, 79, 81, 83). It turned out (rather unexpectedly) that the trace spaces on R^m, the criteria of embedding into the space $C(R^n)$, the multiplicativity property of Lizorkin-Triebel spaces are *not at all influenced by the second summability parameter* $q \in (1, \infty)$ in full contrast to B-spaces (for more details see further Section 2 and [KL-87]).

1.5. Other types of function spaces

The study of anisotropic spaces $H^{\overline{s}}_p$, $\overline{s} := (s_1, s_2, \dots s_n)$ with various differential properties in various coordinate directions has been initiated by S.M. Nikol'skiy. Later he also proposed another type of anisotropy, namely the mixed $L_{\overline{p}}$-norms, $\overline{p} := (p_1, \dots, p_n)$

$$\| f \mid L_{\overline{p}}(R^n)\| := \Big(\int \Big(\dots \Big(\int |f(x_1, \dots, x_n)|^{p_1} dx_1 \Big)^{\frac{p_2}{p_1}} \dots \Big)^{\frac{p_n}{p_{n-1}}} dx_n \Big)^{\frac{1}{p_n}} \tag{1.16}$$

which appeared to be of great importance (detailed exposition and precise references may be found in [BIN-96: Ch. 3]).

S.M. Nikol'skiy was also the first who investigated the spaces $SH^{\overline{s}}_p$ defined in terms of mixed differences properties. This approach was further developed by T. Amanov (1965) (the spaces $SB^{\overline{s}}_{p,q}$). H.-J. Schmeisser (1980) has constructed a similar theory for the more complicated case of F-type spaces. The idea of these spaces may be well understood from their important particular case of the so-called Liouville-type spaces. By definition the space $SL_p^{\overline{s}}$, $\overline{s} := (s_1, s_2, \dots s_n)$, $s_j > 0$ is the set of all functions $f \in L_p(R^n)$ with the finite norm

$$\| f|SL_p^{\overline{s}} \| := \|F^{-1}(1 + \prod_{j=1}^{n} |\xi_j|^{2s_j})^{1/2} \; Ff| \; L_p(R^n) \| \, . \tag{1.17}$$

For all integer s_j the norm in $SL_p^{\overline{s}}$ is equivalent to the sum of L_p-norms of f and its mixed derivative $D_{x_1}^{s_1} D_{x_2}^{s_2} \dots D_{x_n}^{s_n} f$. The complete system of embeddings and equivalent norms for the scales $L_p^{\overline{s}}$, $SB_{p,q}^{\overline{s}}$, $SF_{p,q}^{\overline{s}}$ (known as the *spaces with dominated mixed derivatives*) one can find in the book [SchT-87].

2. Spaces of generalized smoothness

Many problems in analysis demand tools which would allow to measure the smoothness not by a single number (or several numbers) but by a comparison with certain calibre functions or calibre sequences. The idea of such spaces is ascending to H. Weyl. P.L. Ul'yanov (1967) has considered the spaces H_p^{ω} in the one-dimensional case. More general B-type spaces have been studied by M.L. Gol'dman (1972) and G.A. Kalyabin (1976). All details may be found in [KL-87] and [BKuLN-90] where the complete system of embeddings, equivalent normings in various terms, criteria of multiplicativity and some applications of spaces of generalized smoothness are expounded. Below only a small portion of these results are presented.

2.1. Generalized spaces of L- and B-type

Let us introduce two parameters-sequences of positive numbers $\{\alpha_k\}$ and $\{N_k\}$ such that for some $c_0 > 1$, $c_1 > 1$ and all k

$$\alpha_{k+1} \le c_1 \alpha_k; \;\; \{\alpha_k^{-1}\} \in l_{q'} \text{ or resp. } \; l_{p'}; \quad N_{k+1} \ge c_0 N_k \tag{2.1}$$

and define the space $B_{p,q}^{(\alpha,N)}(R^n)$ (or resp. $L_{p,q}^{(\alpha,N)}(R^n)$) as the set of all functions $f \in L_p(R^n)$ for which there exist sequences $\{f_k(x)\} \in \mathrm{EFET}(N_k)$ such that

$$\| \; \|\{\alpha_k\|f_k - f|L_p\|\}|l_q\| < \infty, \text{ resp.} \| \; \|\{\alpha_k(f_k(x) - f(x))\}|l_q\| \; |L_p\| < \infty. \tag{2.2}$$

The norm in the corresponding space is equal to the infimum of the quantity (2.2) over all sequences $\{f_k(x)\} \in \mathrm{EFET}(N_k)$.

Restrictions (2.1) imposed on $\{\alpha_k\}$, $\{N_k\}$ are motivated by two requirements:

(i) the space $B(L)_{p,q}^{(\alpha,N)}(R^n)$ does not coincide with the whole L_p,
(ii) the operators of dilation: $f(x) \to f(tx)$, $t > 0$ are bounded in the corresponding space.

Thus the spaces of analytic functions (or Gevrey type spaces) are not embraced by these scales. Recall that the ordinary (power-scaled) spaces $L_{p,q}^s$, $B_{p,q}^s$ correspond to the case $\alpha_k := 2^k$, $N_k := 2^{k/s}$.

2.2. Traces on hyperplanes and embeddings

Let $\{\beta_k\}$ stand for a sequence of multipliers which obeys the bilateral growth condition:

$$c_2\beta_k \leq \beta_{k+1} \leq c_1\beta_k, \; 1 < c_2 \leq c_1 \tag{2.3}$$

(e.g. one may choose $\beta_k := 2^k$). Denote for given $\{\alpha_k\}$ and $\{N_k\}$

$$n_k := \min\{n : \sum_{m\geq n} \alpha_m^{-p'} \leq \beta_k^{-p'}\}; \quad \hat{N}_k := N_{n_k} . \tag{2.4}$$

Then the spaces $L_{p,q}^{(\alpha,N)}(R^n)$ and $L_{p,q}^{(\beta,\hat{N})}(R^n)$ coincide (with equivalence of the norms) and an analogous assertion (which may be called the *standardization theorem*) holds for B-spaces if one replaces p' in (2.4) by q'.

Let $m < n$, put $\alpha_k := \beta_k N_k^{-\frac{n-m}{p}}$. Then *independently of* $q \in (1,\infty)$ every function belonging to $L_{p,q}^{(\beta,N)}(R^n)$ has a trace on R^m if, and only if, $\{\alpha_k^{-1}\} \in l_{p'}$ and in this case the reversible embeddings hold

$$L_{p,q}^{(\beta,N)}(R^n)\Big|_{R^m} = L_{p,q}^{(\alpha,N)}(R^m) . \tag{2.5}$$

The embedding

$$L_{p,q}^{(\beta,N)}(R^n) \hookrightarrow C(R^n) \quad \text{holds if, and only if,} \quad \{\beta_k^{-1}N_k^{\frac{n}{p}}\} \in l_{p'} \tag{2.6}$$

and the same condition is sufficient and necessary for the space $L_{p,q}^{(\beta,N)}(R^n)$ to be a Banach algebra with respect to point-wise multiplication.

For generalized B-spaces more results on embeddings have been obtained by M.L. Gol'dman [G-85] one of which is as follows:

Let $0 < q < \infty; 1 \leq p < r < \infty$. Denote $\rho := qr(q-r)^{-1}; q > r;\ \rho := \infty; q \leq r$. The embedding

$$B_{p,q}^{(\beta,N)}(R^n) \hookrightarrow L_r(R^n) \iff \{\beta_k^{-1}N_k^{n(\frac{1}{p}-\frac{1}{r})}\} \in l_\rho . \tag{2.7}$$

2.3. Capacity estimates

The capacity of the *compact* set $E \subset R^n$ with respect to the space $L_{p,q}^{(\beta,N)}(R^n)$ is by the definition (cf. [M-85: 7.2.1], [Net-89]) the quantity

$$\text{Cap}\,(E, L_{p,q}^{(\beta,N)}(R^n)) := \inf\{\| f \mid L_{p,q}^{(\beta,N)}(R^n)\|^p : \ f(x) \geq 1, \quad x \in E\} . \tag{2.8}$$

In [K-94] the following universal upper estimate for Cap $(E, L_{p,q}^{(\beta,N)}(R^n))$ was obtained which is valid for all $q \in (1,\infty)$:

$$\text{Cap}\,(E, L_{p,q}^{(\beta,N)}(R^n)) \leq c_0\Big(\sum(\beta_k(\text{ meas}(E+\Delta^{(k)}))^{1/p})^{-p'}\Big)^{-p/p'} \tag{2.9}$$

where $\Delta^{(k)} := \{x : 0 < N_k x_j < 1\}$ is an open cube of rank k, "meas" denotes the Lebesgue measure and $+$ stands for the arithmetic sum of sets. The inverse estimate (of course with a different constant ε_0 instead of c_0) also holds provided the set E is convex or is a lattice-wise structured union of such sets.

Yu.V. Netrusov introduced a remarkable quantity of combinatorial nature: for $m = (m_1, \ldots, m_n) \in Z^n$ and $k > 0$ let us consider the "elementary brick" of rank k $\Delta_m^{(k)} := \Delta^{(k)} + N_k^{-1}m$, let further PWCF_k^+ be the set of all functions which are non-negative and piece-wise constant, i.e. have the same value on the interior of each brick $\Delta_m^{(k)}$. For convenience let us assume that the values at any boundary point of the brick $\Delta_m^{(k)}$ equal the greatest of the values inside all bricks for which this point is boundary. Now define

$$\nu_p(E) := \inf\Big\{\int\sum|\beta_k g_k(x)|^p dx : g_k(x) \in \mathrm{PWCF}_k^+; \sum g_k(x) \geq 1\ \forall x \in E\Big\}. \quad (2.10)$$

In [Net-89: th.2.2] it is proved that *for any compact $E \subset R^n$ and any* $q \in (1, \infty)$ holds $\nu_p(E) \asymp \mathrm{Cap}\,(E; L_{p,q}^{(\beta,N)}(R^n))$.

Let us impose the additional technical assumption $N_{k+1} = l_k\, N_k$, l_k-integers, which means that every brick $\Delta_m^{(k)}$ of rank k is the union of all bricks $\Delta_{\tilde{m}}^{(k+1)}$ of rank $(k+1)$ having a non-empty intersection with $\Delta_m^{(k)}$. This requirement holds automatically for all power-scaled spaces $L_{p,q}^s(R^n)$, $\beta_k := 2^{k/s}$, $N_k := 2^k$, $l_k = 2$ and it can be fulfilled *always* at the expense of changing of $\{\beta_k\}, \{N_k\}$ with norms equivalence.

Introduce now two auxiliary sequences:

$$\zeta_k := N_k^{np'/p} \sum_{s \leq k} \beta_s^{-p'}; \quad \eta_k := \sum_{s>k} \beta_s^{-p'}\, N_s^{np'/p}\,. \quad (2.11)$$

Then the Netrusov capacity of a single cube of rank k is given by the formula: $\nu_p(\Delta_m^{(k)}) = (\zeta_k + \eta_k)^{-1/p'}$. Besides the "sophisticated additivity" property takes place which can be described as follows. Let the sets E^i be situated in different cubes $\Delta_{m^i}^{(k)}$ of rank k which are in turn the sub-cubes of one cube $\Delta_m^{(k-1)}$ of rank $k-1$; then

$$\Big(\nu_p\Big(\bigcup_i E_i\Big) - \eta_k \Big)^{-p'} = \sum_i \Big(\nu_p\Big(\bigcup_i E_i\Big) - \eta_k \Big)^{-p'}. \quad (2.12)$$

These relationships allow to compute the capacity of an arbitrary union of T elementary cubes using only $O(T)$ arithmetic operations of the type $a + b, ab, a^b$. Thus, the computational complexity is the same in order as for computation of the Lebesgue measure of the same set though the capacity is often regarded as a quantity much more complicated than the measure.

2.4. Spaces of variable smoothness

H.-G. Leopold has introduced and studied the spaces $B_{p,q}^a(R^n)$ with $1 < p, q < \infty$, and $a := a(x, \xi)$ being the symbol of a hypoelliptic pseudodifferential operator belonging to a certain class $S(m, m', \delta)$ (see all definitions and further details in [Le-91]). In particular the symbol $a(x, \xi) := \langle\xi\rangle^{\sigma(x)}$, $\langle\xi\rangle := (1 + |\xi|^2)^{1/2}$ where the real-valued exponent $\sigma(x) = d + \psi(x)$, $\psi \in S(R^n)$ for any $\delta \in (0,1)$ belongs to $S(m, m', \delta)$, $m := \sup\sigma$, $m' := \inf\sigma > 0$.

The norm of $f \in S'(R^n)$ in $B^a_{p,q}(R^n)$ is defined as

$$\| f \mid B^a_{p,q}(R^n) \| := \Big(\sum_{k=0}^{\infty} 2^{kq} \, \| \varphi_k(x,D) f \mid L_p \|^q \Big)^{\frac{1}{q}} \tag{2.13}$$

where the system of functions $\{\varphi_k(x,\xi)\}$, $k \in N_0$, associated with the symbol $a(x,\xi)$ satisfies for some natural J, N the conditions:

$$\varphi_k(x,\xi) \in C^\infty(R^n \times R^n), \quad \varphi_k(x,\xi) \geq 0;$$

$$\operatorname{supp} \varphi_k \subset \{(x,\xi) : |a(x,\xi)| < 2^{J+N+k}\}, \; 0 \leq k \leq N, \tag{2.14}$$

$$\operatorname{supp} \varphi_k \subset \{(x,\xi) : 2^{J-N+k} < |a(x,\xi)| < 2^{J+N+k}\}, \; k > N;$$

$$\sup_k \sup_x |D^\alpha_\xi D^\beta_x \varphi_k(x,\xi)| \leq c_{\alpha,\beta} \langle \xi \rangle^{-|\alpha|+\delta|\beta|}; \quad \sum_{k=0}^{\infty} \varphi_k(x,\xi) \equiv N-1. \tag{2.15}$$

(Such systems may be constructed for any symbol $a \in S(m,m',\delta)$.)

For $a \in S(m,m',\delta)$ with certain conditions of almost increasing and almost decreasing Leopold has found the characterization of $B^a_{pq}(R^n)$ in terms of differences with variable step.

Recently [B-99] the characterization of $B^a_{p,q}(R^n)$ for the same class of symbols was obtained in terms of differences with constant and piece-wise constant steps, e.g. for $q = p$ one has the equivalent norming (cp. (1.9)):

$$\| f \mid B^a_{p,p} \|_\Delta := \Big(\sum_{k=0}^{\infty} \sum_{j=1}^{n} \| a(x, 2^k e_1) \Delta^M_{j,\varepsilon 2^{-k}} f(x) \mid L_p \|^p \Big)^{\frac{1}{p}} + \| f \mid L_p \| \tag{2.16}$$

where ε is fixed: $0 < \varepsilon < \varepsilon^\star(a) \in (0,1]$ and the order of difference $M > m + \delta$.

3. Spaces of functions given on domains

3.1. Embedding theorems for regular domains

As it was mentioned in **1.1** the first embedding theorems have been established by S.L. Sobolev [S-38] for spaces $W^l_p(G)$ of functions given on the domain G with the cone condition. These theorems are now generalized to the case of domains which satisfy the "flexible horn condition" (FHC). These are called domains for which there exist $\varepsilon > 0, T > 0$ and for every $x \in G$ there exists a piecewise smooth path $\gamma = \gamma_x : [0,T] \to G$, $\gamma(0) = x$ such that the flexible horn

$$\bigcup_{0 \leq t \leq T} \big(\gamma(x,t) + B(0,\varepsilon t)\big) \subset G. \tag{3.1}$$

Note that the bounded domain with flexible horn condition is so-called John domain. The boundary of such a domain may be of fractal structure, e.g. a domain surrounded by the snowflake satisfies FHC. For a domain G with FHC the spaces $B^s_{p,q}(G)$ and $L^s_{p,q}(G)$ may be defined by difference ratios analogous to (1.9) (cf. also [BIN-96: § 18, § 29]). One should keep in mind that for a function given on the domain its difference $\Delta_h^{(m)} f(x)$ with vector step h is defined to be 0 unless the whole segment $[x, x+mh] \subset G$.

For all these spaces the embedding theorems and theorems on interpolation are established which are formulated exactly in the same manner as for the case $G = R^n$ (cf. [BIN-96: Chs. 3,4], [T-78], [Mu-74]).

To prove his embedding theorems S.L. Sobolev has proposed the method of *integral representations* according to which the given function is represented as a sum of certain potential type integrals applied to this function and to the derivatives thereof. For embeddings of the spaces $B^s_{p,q}(G)$, $L^s_{p,q}(G)$ the integral representations via differences should be used (cf. details and history [BIN-96: § 7]).

As an example let us describe the integral representation of a function $f \in L_{loc}(G)$ via its coordinate-wise differences over the flexible cone (3.1):

$$f(x) = f_T(x) + \sum_{j=1}^{n} \int_0^T t^{n-2} \Psi_j f(x,t)\, dt; \quad \text{where}$$

$$f_T(x) := T^{-n} \int_{R^n} K_0\Big(\frac{y}{T}, \frac{\gamma(x,T)}{T}\Big) f(x+y) dy; \tag{3.2}$$

$$\Psi_j f(x,t) := \int_{R^n} \int_{-\infty}^{+\infty} K_{j,m}\Big(\frac{y}{t}, \frac{u}{t}, \frac{\gamma(x,t)}{t}, \frac{\gamma'_t(x,t)}{t}\Big)\, \Delta^{(m)}_{\delta u e_j} f(x+y+ue_j) du dy$$

where m is a natural number, e_j are the vectors of the standard basis in R^n and where the specially constructed C_0^∞-kernels $K_0, K_{j,m}$ equal zero provided their arguments are outside the corresponding flexible horn.

3.2. Function spaces on irregular domains

V.G. Maz'ya has found the necessary and sufficient conditions on the domain G for the embedding $W_p^1(G) \subset L_q(G)$ which are formulated in terms of capacities and isoperimetric inequalities. For domains with a single cusp (say $G := \{(x_1^2 + x_2^2 + \ldots x_{n-1}^2) < x_n^{2\sigma};\ 0 < x_n < 1\} \subset R^n$) and $\sigma(n-1)+1 > pl$ he has established (cf. [M-85: §4.9]) that the embedding $W_p^l(G) \subset L_q(G)$ holds if, and only if, $l - [\sigma(n-1)+1]/p + [\sigma(n-1)+1]/q \geq 0$.

Different sharp exponent appears when the domain has a massive set of singular points on the boundary. For $\sigma \geq 1$ a domain $G \subset R^n$ is called a domain with the flexible σ-cone condition ($G \in \sigma$-FHC) if, for certain $t^*, \kappa > 0$ and any $x \in G$, there exists a piecewise smooth path

$$\gamma : [0, t^*] \to G,\ \gamma(0) = x,\ \Big|\frac{d}{dt}\gamma(t)\Big| \leq 1 \text{ a.e., } \operatorname{dist}(\gamma(t), \partial G) \geq \kappa t^\sigma. \tag{3.3}$$

When G is bounded the condition $G \in \sigma$-FHC is equivalent to the σ-John condition for G.

For $\sigma > 1$ the boundary of $G \in \sigma$-FHC may contain a set of points unreachable from inside the domain. Recently (cf. [KiMa-00] in the case $l = 1$, $p \geq 1$,

[B-00] for all natural $l,\ p>1$) it was established that for the σ-John domain G

$$l-\frac{\sigma(n-1)+1}{p}+\frac{n}{q}\geq 0 \quad\Rightarrow\quad W_p^l(G)\subset L_q(G). \tag{3.4}$$

For $\sigma=1$ one obtains again the classical Sobolev embedding (1.3) for $G\in$FHC. For a narrower class of domains (domains with *continuous* flexible σ-cone condition) the embedding $W_p^l(G)\subset L_{q^*}(G)$ holds with the bigger exponent q^* [La-97].

It is important to note that for the case $l=1$ P. Hajlasz and P. Koskela [HKo-98] using the Nikodym example have constructed such a domain $G\in\sigma$-FHC that $W_p^l(G)\not\subset L_q(G)$ if $l-(\sigma(n-1)+1)/p+n/q<0$. D.A. Labutin [La-01] noticed that the very same domain allows to establish the sharpness of the result (3.4) for any $l\geq 1$.

3.3. Embedding of weighted function spaces

The problems of embedding, compactness, traces, etc., for weighted generalizations of Sobolev spaces have been studied by many authors (cf. detailed survey in [BKuLN-90]).

Here we give sufficient conditions (close to necessary ones) for the three-weights (u,v,w) embedding of the type $W^l_{p,v;r,u}(G)\in L_{q,w}(G)$ investigated in [B-00], which is characterized by the inequality

$$\| f\mid L_{q,w}(G)\|\leq c\Big(\sum_{|\alpha|=l}\| D^\alpha f\mid L_{p,v}(G)\|+\| f\mid L_{r,u}(G)\|\Big) \tag{3.5}$$

where $1<p<q<\infty, 1\leq r<q,\ l-n/p+n/q\geq 0,\quad \| f\mid L_{p,v}(G)\|:=\| fv^{1/p}\mid L_p(G)\|$.

Let G be a bounded domain and let for some $\delta_0\in(0,1), c_0\geq 1$ and every $x\in G$ there exist a piece-wise smooth path ($t^*=t^*(x)$)

$$\gamma(x,t):[0,t^*]\to G,\ \gamma(x,0)=x;\quad |\gamma_t'(x,t)|\leq c_0\quad \text{for a.e. } t\in(0,t^*), \tag{3.6}$$

and a positive continuous piece-wise smooth function $r(x,t)$ such that

$$\begin{aligned} &r(x,t)\leq\delta_0\mathrm{dist}(\gamma(x,t),\partial G),\ r(x,t^*)\geq\delta_0^2,\ |r_t'(x,t)|\leq c_0,\ \text{for a.e. } t\in(0,t^*),\\ &\delta_0 r(x,t')\leq r(x,t'')\ \text{ if }\ |\gamma(x,t')-\gamma(x,t'')|\leq\delta_0(r(x,t')+r(x,t''));\\ &\qquad\qquad\qquad\qquad\text{for }\ t',t''\in[0,t^*]. \end{aligned} \tag{3.7}$$

Introduce the quantity

$$V_R(x,t):=r(x,t)^{-\frac{n}{p'}}\Big(\int_{E_{x,R,t}} v(y)^{-\frac{p}{p'}}dy\Big)^{\frac{1}{p'}}, \tag{3.8}$$

where $E_{x,R,t}:=B(\gamma(x,t),r(x,t))\setminus B(x,R),\quad E_{x,R}:=\bigcup_{0<t<t^*}E_{x,R,t}$.

Then the main condition sufficient for (3.5) is

$$\sup_{x\in G}\sup_{R>0}\left(\int_0^{t^*}(t+r(x,0))^{(l-1)p'}r(x,t)^{\frac{(1-n)p'}{p}}V_R(x,t)^{p'}dt\right)^{\frac{1}{p'}}|G\cap B(x,R)|_w^{\frac{1}{q}}<\infty,$$

where $|E|_w := \int_E w(x)dx,\ E\subset R^n$.

In similar terms the conditions of compactness of the corresponding embeddings for weighted Sobolev spaces are also obtained.

4. Extension onto the whole Euclidean space

The history of the main ideas related to the problem of extension of differentiable functions given on a domain $G\subset R^n$ onto the whole Euclidean space with preservation (or with minimal loss) of smoothness properties may be found in [St-70: Ch 6], [M-85: §1.1], [KL-87: § 7], [BKuLN-90], [BIN-96: § 9.8]. In this section we set forth several new questions of this old topic.

4.1. Norm estimates for extension operators

One of the important problems is to find bilateral estimates of the minimal norm of extension operators in terms of geometric characteristics of the domain. For the particular case when $n=2$, G is a *convex* domain on the plane R^2, $1<p<\infty$ in [K-99] it has been established that the least possible norm of extension operators $T_G : W_p^1(G)\to W_p^1(R^2)$ can be bilaterally estimated via the diameter $\delta=\delta(G)$ and the area $S=S(G)$ of the given convex domain. Let us introduce the quantity

$$\tau_p(G) := \sup_f\{\ \inf_F\{\ \|\ F\mid W_p^1(R^2)\ \|: F=f \text{ on } G\}:\ \|\ f|\ W_p^1(G)\|=1\} \qquad (4.1)$$

Then for $\delta(G)\le 1$ the inequalities

$$c_p^{-1}(S^{-1}\gamma_p(\delta))^{1/p}\le\tau_p(G)\le c_p(S^{-1}\gamma_p(\delta))^{1/p} \qquad (4.2)$$

hold where the notation is used

$$\gamma_p(\delta) := \begin{cases}\delta^{2-p} & \text{if}\quad 1<p<2;\\ \ln^{-1}(2/\delta) & \text{if}\quad p=2;\\ 1 & \text{if}\quad 2<p<\infty\,.\end{cases} \qquad (4.3)$$

For the case when $\delta(G)>1$ (or if G is unbounded) the usual partition of unity technique (together with the theorem) shows that $\tau_p(G)\asymp S_*^{-1/p}(G)$ where $S_*(G)$ means the infimum of areas of intersections of G with disks $B_1(x)$ with radius 1 centered at some $x\in G$.

In all the cases described in the above assertion there exists a *linear* extension operator $T: W_p^1(G)\to W_p^1(R^2)$ whose norm does not exceed $\tilde c_p\tau_p(G)$.

Note that both restrictions posed in this result (namely that **only** *2D-domains* and the *first* order derivatives have been considered) are essential. In particular there is nothing similar to (4.2) for n-dimensional bricks ($n>2$) with arbitrary (*independent*) edges or for the spaces W_p^m, $m>1$ though in [MP-96] the asymptotic behaviour (as $\varepsilon\to+0$) has been studied for $\tau_p((\varepsilon\ \Omega)\times[0.1]\ )$ where Ω is a sufficiently "good" domain in R^n. In the same paper one can find more information concerning extension operators from many other special domains.

4.2. Extension operators for Sobolev spaces in the halfline

Even for the one-dimensional case new interesting results were obtained recently related to the Sobolev spaces of unboundedly growing order. Recall that $W_2^m(I)$ is the Sobolev space of all functions $f(x)$ defined on the interval $I := (a, b) \subset R^1$, having absolutely continuous derivative $f^{(m-1)}(x)$ and such that the norm

$$\|f\|_{W_2^m(I)} := \Big(\int_I (|f(x)|^2 + |f^{(m)}(x)|^2)\, dx\Big)^{1/2} < \infty . \tag{4.4}$$

Burenkov [Bu-98] has proved that for an interval of finite length the least norm $\tau_m(I)$ of the extension operator $T_m : W_2^m(I) \to W_2^m(R^1)$ satisfies the estimate:

$$\ln \tau_m(I) = m(\ln m - \ln(b-a) + \delta_m); \quad |\delta_m| \le c_0 \tag{4.5}$$

The case of the extension from the halfline turned out to be more difficult. In [BuG-97] extension operators $T_m : W_2^m(R_-^1) \to W_2^m(R^1)$ of Hestenes type have been constructed whose norms do not exceed 8^m. On the other hand in [BuK-98] it was shown that $W_2^m(R_-^1)$ contains such a function $f_m(x)$ that any of its extensions onto the whole line has a norm in $W_2^m(R^1)$ greater than $0.08\, m^{-1/4} 2^m \|f_m\|_{W_2^m(R_-^1)}$.

Denote by $y(x) := y(x; f), x > 0$ the solution of the equation $(-1)^m y^{(2m)} + y = 0$ tending to 0 as $x \to +\infty$ and satisfying the initial conditions $y^{(s)}(+0) = a_s = f^{(s)}(-0), s \in \{0, 1, \dots, m-1\}$. Then the extension of f onto the positive halfline by y (which is a linear operator) provides the minimal possible norm in $W_2^m(R^1)$.

In [K-02] the asymptotics as $m \to \infty$ which is sharp in the logarithmic scale has been established for the norm of this operator of best extension:

$$\frac{\ln\, \tau_m(R_-^1)}{m} \to K_0 := \frac{4}{\pi} \int_0^{\pi/4} \ln \operatorname{ctg} x\, dx = 1.1662\cdots = \ln 3.2099\dots \tag{4.6}$$

4.3. Extrapolation problems

The latter result is closely related to the problem posed by L.D. Kudryavtzev of extrapolation (i.e. extension from a point set) with minimal norm in $W_2^m(R_+^1)$. For a given set of boundary values $a := \{a_s\}$ denote by $\psi_m(a)$ the minimum of $\| u \mid W_2^m(R_+^1) \|^2$ taken over all function u with these initial data (this quantity equals to $\| y \mid W_2^m(R_+^1) \|^2$). Let further Ω_m and ω_m stand for the maximum and minimum of $\psi_m(a)$ when a is running over the unit sphere $\sum_{s=0}^{m-1} |a_s|^2 = 1$. It can be proved that $\ln \Omega_m = -\ln \omega_m \approx K_0 m/2,\ m \to \infty$.

In [K-99a] a similar problem for extrapolation in the Bernstein classes B_σ of entire functions of exponential type $\sigma > 0$ has been studied which can be reduced to the study of the Hilbert type matrices $H_m(\beta, \theta),\ \beta > 0, \theta > 0$ with elements

$h_{j,k} = h_{j,k}^{(m)}(\beta,\theta) := \beta^{j+k}(j+k+\theta)^{-1}$, $j,k \in \{0,\dots,m-1\}$. An asymptotic formula for minimal eigenvalues of these matrices was established in [K-00]:

$$\lambda_{\min}\big(H_m(\beta,\theta)\big) = K\sqrt{m}\,\rho^{-4m}(1+o(1)),\ m\to\infty; \tag{4.7}$$

$$\rho = \rho(\beta) := \sqrt{\beta^{-1}} + \sqrt{1+\beta^{-1}};\quad K = K(\beta,\theta) := \frac{8\pi\sqrt{2\pi}\,(1+\beta)^{1/4}}{\beta^{\theta+1/2}\rho^{2\theta+2}(\beta)}.$$

Introduce the quantity $\varphi_m(a,\sigma) = \min\{\ \| f \mid L_2(R^1)\| : f \in B_\sigma,\ f^{(s)}(0) = a_s,$ $s \in \{0,\dots,m-1\}\}$. Having based on (4.7) one can prove that there exist two positive numbers $\gamma^{(o)}(\sigma), \gamma^{(e)}(\sigma)$ such that

$$\Phi_m(\sigma) := \max\{\varphi_m(a,\sigma)\ : \sum |a_s|^2 = 1\} \approx \gamma_m^{(o,e)}(\sigma)\ m^{-1/4}\rho^m(\sigma^2) \tag{4.8}$$

where $\gamma_m^{(o,e)}(\sigma) = \gamma^{(e)}(\sigma)$ for m even and $\gamma_m^{(o,e)}(\sigma) = \gamma_m^{(o)}(\sigma)$ for m odd.

4.4. The internal normings of retraction spaces

Another problem studied during the last decade is to find for a given space $V(R^n)$ of functions defined on the whole Euclidean space and a domain (or more generally an arbitrary open set) $G \subset R^n$ internal characteristics which use only the values of the function inside G equivalent to the retraction norm

$$\| f \mid V(G) \|^{(e)} := \inf\{\ \| F \mid V(R^n)\ \| :\ F = f\ \text{ on }\ G\ \}. \tag{4.9}$$

Here we give two examples. Let us consider the plain domain G, described by the inequalities $\{0 < x < 1,\ 0 < y < \varphi(x)\}$ where φ is a positive function satisfying the Lipschitz condition. If $\varphi(x)/x \to 0,\ x \to 0$ then at the point $(0,0)$ the upper part of the boundary of G is tangent to the lower one and thus one has here the point of sharpness (i.e. *the cusp*) at which the Lipschitzness of the whole boundary is violated.

Let us denote by $Tg(x)$ the trace of a function $g(x,y),\ (x,y) \in G$, on the interval $\{y = 0,\ 0 < x < 1\}$. Then the norm $\| f \mid W_p^l(G) \|^{(e)}$ is equivalent ([K-95]) to another norm expressed in terms of internal properties of f:

$$\| f \mid W_p^l(G) \|^{(i)} := \sum_{|\alpha|\le l} \| D^\alpha f \mid L_p(G) \| + \sum_{k=0}^{l-1} \| T\partial^k f/\partial y^k \mid B_{p,p}^{l-k-1/p}(0,1) \| \tag{4.10}$$

It would be interesting to extend this rather simple result to the multidimensional case and to more sophisticated spaces.

The next problem is related to the *family* of retraction spaces. The function $g(x)$ given on the interval $(0,b),\ b > 0$ belongs to the retraction space $L_{p,q}^{(\beta,N)}(0,b)$ if there exists a function $f(x) \in L_{p,q}^{(\beta,N)}$ (cf. Section 2) which coincides with $g(x)$ on $(0,b)$. The behaviour of the determining sequences $\{\beta_k\}$, $\{N_k\}$ will be reflected by the specific function introduced in [K-97]

$$\gamma(b) := \Big(\sum_k (\beta_k (N_k^{-1} + b)^{1/p})^{-p'}\Big)^{-p/p'};\quad (1/p + 1/p' = 1). \tag{4.11}$$

It is easy to calculate that for the power-scaled spaces $L^s_{p,q}$ and $0 < b < 1$ the function

$$\gamma(b) \quad \text{is equivalent to} \quad \begin{cases} b^{1-ps} & \text{if} \quad 0 < s < 1/p; \\ (\ln 2/b)^{1-p} & \text{if} \quad s = 1/p; \\ 1 & \text{if} \quad s > 1/p\,. \end{cases}$$

Let us for $t \in R$ denote by $\Delta_t g(x)$ the difference $g(x+h) - g(x)$ provided both points $x, x+h$ belong to $(0,b)$ (otherwise put $\Delta_h g(x) = 0$). Introduce the averaged local oscillation (of first order) defined for $h > 0$ by the formula

$$\Omega_h(g,x) := \int_{-1}^{1} |\Delta_{ht} g(x)|\, dt. \tag{4.12}$$

Suppose that the constant c_0 in restriction (2.1) on the sequence $\{N_k\}$ is greater than c_2 in condition (2.3) restricting the sequence $\{\beta_k\}$ (this implies $s < 1$ for power-scaled spaces). Then the norm $\| \; g \mid L^{(\beta,N)}_{p,q}(0,b) \; \|^{(e)}$ is equivalent to

$$\| \; g \; | L^{(\beta,N)}_{p,q}(0,b) \; \|^{(i)} =: \frac{\gamma(b)^{1/p}}{b} \int_0^b \; | \; g(x) \; | \; dx + \| \{\beta_k \Omega_{N_k^{-1}}(g,x)\} | L_p(l_q,(0,b)) \|$$

and the ratio of these two quantities is bilaterally bounded for $b > 0$.

5. Diverse results on function spaces

5.1. Pointwise multipliers

For a given function space $V(R^n)$ the *space of multipliers* $MV(R^n)$ is a set of all such functions g for which the product gf belongs $V(R^n)$ for any $f \in V(R^n)$ equipped with the norm

$$\| \; g \mid MV(R^n) \; \| := \sup\{ \; \| \; gf \mid V(R^n) \; \| : \; \| \; f \mid V(R^n) \; \| \le 1\} \,. \tag{5.1}$$

The problem of explicit description for $MV(R^n)$ was studied for different $V(R^n)$ by V.G. Mazya, T.O. Shaposhnikova and many other authors, cf. [MSh-85] where detailed references are given. In particular, those cases turned out to be easier (in the sense of description of multiplier spaces) for which the initial space $V(R^n)$ is embedded into $L_\infty(R^n)$.

Here we consider the modified Sobolev space $\tilde{W}^l_p(R^n)$ defined by the finiteness of the norm

$$\| \; f \mid \tilde{W}^l_p(R^n) \; \| := \| \; f \mid L_p(B^n_1) \; \| + \| \; \nabla_l f \mid L_p(R^n) \; \|;$$

$$|\nabla_l f(x)| := \Big(\sum_{|\gamma|=l} |D^\gamma f(x)|^2 \Big)^{1/2} \tag{5.2}$$

which differs from $W^l_p(R^n)$ by taking the L_p-norm of the function over the unit ball B^n_1 only and not over the whole R^n.

Further only the case $p > n$ is considered which provides the embedding $\tilde{W}_p^l(R^n) \hookrightarrow C_{loc}^{l-1}$. However in contrast to $W_p^l(R^n)$ the class $\tilde{W}_p^l(R^n)$ contains unbounded functions, e.g.

$$f(x) := (1+|x|^2)^{\alpha/2} \in \tilde{W}_p^l(R^n); \quad 0 < \alpha < \beta := l - n/p. \tag{5.3}$$

In fact infinity is the only singular point for functions from the space $\tilde{W}_p^l(R^n)$. The following statement yields an effective characterization of the space $M\tilde{W}_p^l(R^n)$ [K-93].

For $p > n$ the norm $\| g \mid M\tilde{W}_p^l(R^n) \|$ (cf. (5.1)) is equivalent to

$$\sup_{x\in R^n} | g(x)| + \| \nabla_l g \mid L_p(B_1^n)\| + \sup_{r>1} r^\beta \Big(\int_{r\le|x|\le 2r} |\nabla_l g(x)|^p dx \Big)^{1/p}. \tag{5.4}$$

5.2. The K-functional for certain pairs of function spaces

Let A_0, A_1 be two Banach spaces, $a \in A_0 + A_1$, $t > 0$,

$$K(t,a) = K(t,a; A_0, A_1) := \inf \{ \| a_0 \mid A_0 \| + t \| a_1 \mid A_1 \| : \ a_0 + a_1 = a \}. \tag{5.5}$$

According to the classical result by J. Peetre $K(t, f; L_p, W_p^1) \asymp \omega(f,t)_p$, $1 \le p \le \infty$, R. DeVore (cf. [dVLo-93]) has posed the question: to what effective quantity is $K(t, f; L_p, W_q^l)$ equivalent for $q \ne p$? Further we describe the results of [K-95a] related to the case $1 < p < q < \infty$, $n = 1$, $l \ge 1$.

Any function $g \in L_p = L_p(R^1)$ may be represented as the series of its Fourier sums $f = \sum f_k$, each summand f_k being an entire function of exponential type 2^k belonging to L_p. The Shannon-Kotel'nikov theorem claims that

$$g \in \mathrm{EFET}(N) \cap L_p \Rightarrow g(x) = \sum_{m\in \mathbf{Z}} g\Big(\frac{\pi m}{N}\Big) \ \mathrm{sinc}(Nx - \pi m); \ \ \mathrm{sinc}\, z := \frac{\sin z}{z}. \tag{5.6}$$

Thus for any $f \in L_p$ one obtains the double series

$$f(x) = \sum_{k=1}^{\infty} \sum_{m\in\mathbf{Z}} \varphi_{k,m} \ \mathrm{sinc}(2^k x - \pi m) \tag{5.7}$$

where the coefficients $\varphi_{k,m} = \varphi_{k,m}(f)$ are constructively defined. Consider the dyadic intervals $\Delta_{k,m} := (2^{-k} m\pi, 2^{-k}(m+1)\pi)$ and calculate the numbers $\delta_{k,m}(f) := \sum |2^{sl} \varphi_{s,\tilde{m}}(f)|^2$ where the (finite) sum is taken over all such pairs $(s, \tilde{m})$ that the given interval $\Delta_{k,m} \subset \Delta_{s,\tilde{m}}$.

Now let us for given $\lambda > 0$ associate to the set $\Psi_\lambda^{(0)}$ (and resp. to $\Psi_\lambda^{(1)}$) all such pairs (k, m) that

$$2^{kl} (\delta_{k,m}(f))^{\frac{q}{p}-1} > \lambda, \quad \text{or resp.} \quad 2^{kl}(\delta_{k,m}(f))^{\frac{q}{p}-1} \le \lambda. \tag{5.8}$$

Decompose the function f given by the series (5.7) into two sums:

$$f_\lambda^{(i)}(x) := \sum \sum\nolimits_{(k,m)\in\Psi_\lambda^{(i)}} \varphi_{k,m} \ \mathrm{sinc}(2^k x - \pi m); \quad i \in \{0,1\} \tag{5.9}$$

and introduce the (non-increasing) function

$$\tau(\lambda) = \tau(\lambda; f) := \frac{\| \left(\sum \sum_{(k,m)\in\Psi_\lambda^{(0)}} |\varphi_{k,m}\chi_{k,m}(x)|^2 \right)^{1/2} | \; L_p \, \|}{\| \left(\sum \sum_{(k,m)\in\Psi_\lambda^{(1)}} |2^{kl}\varphi_{k,m}\chi_{k,m}(x)|^2 \right)^{1/2} | \; L_q \, \|} \tag{5.10}$$

where $\chi_{k,m}(x)$ stands for the characteristic function of the interval $\Delta_{k,m}$. Then the decomposition $f = f_\lambda^{(0)} + f_\lambda^{(1)}$ is quasi-optimal for $t = \tau(\lambda)$, i.e.

$$\| \, f_\lambda^{(0)} \mid L_p \, \| + \tau(\lambda) \, \| \, f_\lambda^{(1)} \mid W_q^l \, \| \leq \; c \, K(\tau(\lambda), f; \; L_p, W_q^l) \tag{5.11}$$

with a constant c depending only on p, q, l.

References

[A-55] N. Aronszajn. *Boundary values of functions with finite Dirichlet integral.* Tech. Report N 14, 1955, Univ. of Kansas.

[B-61] O.V. Besov. *Investigation of one family of function spaces in connection with imbedding and extension theorems.* Proc. Steklov Inst. Math. 1961, v. 60, pp. 42–81.

[B-67] O.V. Besov. *Extension of functions from L_p^l and W_p^l.* Proc. Steklov Inst. Math. 1967, v. 89, pp. 5–17.

[B-99] O.V. Besov. *On the spaces of functions of variable smoothness defined by pseudodifferential operators.* Proc. Steklov Inst. Math. 1999, v. 227, pp. 56–74.

[B-00] O.V. Besov. *The Sobolev Imbedding Theorem for a Domain with Irregular Boundary.* Russ. Acad. Math. Dokl. 2000, v. 373, No 2, pp. 151–154.

[B-01] O.V. Besov. *On the Compactness of Embeddings of Weighted Sobolev Spaces on a Domain with Irregular Boundary.* Proc. Steklov Inst. Math. 2001, v. 232, pp. 66–87.

[BIN-96] O.V. Besov, V.P. Il'in and S.M. Nikol'skij. *Integral Representation of Functions and Imbedding Theorems.* Fizmatlit, Moscow, 1996.

[BKuLN-90] O.V. Besov , L.D. Kudryavtzev, P.I. Lizorkin and S.M. Nikol'skij. *Studies on the theory of spaces of differentiable functions of several variables.* Proc. Steklov Inst. Math. 1990, v. 182, pp. 73–140.

[Bu-98] V.I. Burenkov. *Sobolev Spaces on Domains.* Teubner, Stuttgart, Leipzig, 1998.

[BuG-97] V.I. Burenkov, A.L. Gorbunov. *Sharp estimates of the least norms of extension operators for Sobolev spaces.* Russ. Math. Izvest. 1997, v. 61, No 1, pp. 3–44 .

[BuK-98] V.I. Burenkov, G.A. Kalyabin. *Lower estimates of the norms of extension operators for Sobolev spaces on the halfline.* Math. Nachr. 2000, v. 218, pp. 19–23.

[dVLo-93] R. DeVore and G.G. Lorentz. *Constructive Approximation.* Springer, Berlin, 1993.

[G-85] M.L. Gol'dman. *On Imbedding of Generalized Nikol'skii-Besov Spaces into Lorentz Spaces.* Proc. Steklov Inst. Math. 1985, v. 172, pp 128–139.

[Ga-57] E.L. Gagliardo. *Caratterizzationi delle trace sulla frontiera relative adalcune classi di functioni in n variabili.* Rend. Sem. Mat. Padova, 1957, f.27, pp. 284–305.

[HKo-98] P. Hajlasz and P. Koskela. *Isoperimetric inequalities and imbedding theorems in irregular domains.* J. London Math. Soc. 1998, v. 58, pp. 425–450.

[I-54] V.P. Il'in. *On the imbedding theorem for the limit exponent.* Sov. Math. Dokl, 1954, v. 96, No 5, pp. 905–908.

[K-93] G.A. Kalyabin. *Pointwise multipliers in certain Sobolev spaces containing unbounded functions.* Proc. Steklov Inst. Math. 1993, v. 204, pp. 160–165.

[K-94] G.A. Kalyabin. *On exact values and bilateral estimates of certain capacities.* Math. Nachrichten, 1994, v. 169, pp. 149–159.

[K-95] G.A. Kalyabin. *The internal norming of retractions of Sobolev spaces onto the plain domains with points of sharpness.* Abst. Int. Conf. dedicated to the 90-th anniversary of S.M. Nikol'skiy, Moscow, RUSSIA, April 27–May 4, 1995.

[K-95a] G.A. Kalyabin. *The effective description for the sum of certain pairs of function spaces.* Proc. Steklov Inst. Math. 1995, v. 210, pp. 120–128.

[K-97] G.A. Kalyabin. *The uniform norming of retractions on short intervals for certain function spaces.* Georgian Math. J. 1997, v. 4, No. 5, pp. 443–450.

[K-99] G.A. Kalyabin. *Bilateral estimates for the least norm of extension operators from the convex plane domains for Sobolev classes.* Proc. Steklov Inst. Math. 1999, v. 227, pp. 140–145.

[K-99a] G.A. Kalyabin. *On extrapolations with minimal norms in Bernstein classes.* Proc. Razmadze Math. Inst. 1999, v. 119, Tbilisi, pp. 85–92.

[K-00] G.A. Kalyabin. *Asymptotic formula for minimal eigenvalues of Hilbert type matrices.* Proc. of the 2nd ISAAC Congress, Kluwer Academic Publishers, Dordrecht/Boston/London, 2000, v. 2, pp. 1155–1162.

[K-02] G.A. Kalyabin. *The best extension operators for Sobolev classes on the halfline.* Funct. Anal. Appl. 2002 (to appear).

[KL-87] G.A. Kalyabin and P.I. Lizorkin. *Spaces of functions of generalized smoothness.* Math. Nachr. 1987, v. 133, pp. 7–32.

[KiMa-00] T. Kilpeläinen, and J. Malý. *Sobolev Inequalities on Sets with Irregular Boundaries.* Z. Anal. Anwend. 2000, v. 19, No. 2, pp. 369–380.

[L-67] P.I. Lizorkin. *On Fourier integral multipliers in spaces* $L_{p,\theta}$. Proc. Steklov Inst. Math. 1967, v. 89, pp. 231–248.

[L-72] P.I. Lizorkin. *Operators related to the fractional differentiation and classes of differentiable functions.* Proc. Steklov Inst. Math. 1972, v. 117, pp. 172–177.

[La-97] D.A. Labutin. *Integral representations of functions and embeddings of Sobolev spaces on cuspidal domains.* Math. Notes 1997, v. 61, pp. 164–179.

[La-01] D.A. Labutin. *Sharpness of Sobolev Inequalities for a Class of Irregular Domains.* Proc. Steklov Inst. Math. 2001, v. 232, pp. 211–215.

[Le-91] H.-G. Leopold. *On function spaces of variable order of differentiation.* Forum Math. 1991, v. 3, pp. 1–21.

[M-85] V.G. Maz'ya. *Sobolev spaces.* Springer, Berlin, 1985.

[MP-96] V.G. Maz'ya, and S.V. Poborchi. *On extension of functions in Sobolev spaces on parameter dependent domains.* Math. Nachr. 1996, v. 178, pp. 5–41.

[MSh-85] V.G. Maz'ya, and T.O. Shaposhnikova. *The theory of multipliers in spaces of differentiable functions.* Pitman, Boston, 1985.

[Mu-74] T. Muramatu. *On Besov Spaces and Sobolev Spaces of Generalized Functions Defined on General Region.* Publ. Res. Inst. Math. Sci. Kyoto Univ., 1974, v. 9, No. 2, pp. 325–396.

[N-51] S.M. Nikol'skij. *Inequalities for entire functions of the finite degree and their application to the theory of differentiable functions of several variables.* Proc. Steklov Inst. Math. 1951, v. 38, pp. 244–278.

[N-75] S.M. Nikol'skij. *Approximation of Functions of several variables and Imbedding Theorems.* Springer, Berlin, 1975.

[Net-89] Yu.V. Netrusov. *Metric estimates of the capacity of sets in Besov spaces.* Proc. Steklov Inst. Math. 1989, v. 190, pp. 159–185.

[S-38] S.L. Sobolev. *On one theorem of functional analysis.* Math. Sb. 1938, No 4, pp. 471–497.

[S-88] S.L. Sobolev. *Certain Applications of Functional Analysis in Mathematical Physics.* Nauka, Moscow, 1988.

[SchT-87] H.-J. Schmeiszer and H. Triebel. *Topics in Fourier Analysis and Function Spaces.* Geest-Portig, Leipzig, 1987, and Wiley, Chichester, 1987.

[Sl-58] L.N. Slobodeckiy. *Fractional Sobolev spaces and their applications to the boundary value problems for partial differential equations.* USSR Math. Doklady, 1958, v. 118, No 2, pp. 243–246.

[St-70] E.M. Stein. *Singular Integrals and Differentiability Properties of Functions.* Princeton Univ. Press, 1970.

[T-78] H. Triebel. *Interpolation Theory, Function Spaces, Differential Operators.* North-Holland, Amsterdam, 1978.

[T-83] H. Triebel. *Theory of Function Spaces.* Birkhäuser, Basel, 1983.

[T-92] H. Triebel. *Theory of Function Spaces II.* Birkhäuser, Basel, 1992.

[T-01] H. Triebel. *The structure of functions.* Birkhäuser, Basel, 2001.

Oleg V. Besov
Function Theory Department
Steklov Institute of Mathematics
Gubkina 8
117966 GSP-1 Moscow, Russia
E-mail address: besov@mi.ras.ru

Gennadiy A. Kalyabin
Chair of Mathematics and Natural Sciences
Samara Academy of Humanities
8 Radialnaya 31
443011 Samara, Russia
E-mail address: kalyabin@lycos.com

D. Haroske, T. Runst, H.-J. Schmeisser (eds.): Function Spaces, Differential Operators and Nonlinear Analysis. The Hans Triebel Anniversary Volume.

Entropy, Embeddings and Equations

David E. Edmunds

Dedicated to Prof. Hans Triebel on the occasion of his 65th birthday

1. Introduction

At first sight the number 1978 may not seem to be very interesting. It appears to lack the star quality of its more glamorous neighbour 1729, for example, which (as everyone knows, thanks to the celebrated story of Hardy's visit to the sick Ramanujan in a taxi with that number) is the smallest natural number representable as the sum of two cubes in two different ways:

$$1729 = 10^3 + 9^3 = 12^3 + 1^3.$$

Indeed, 1978 cannot be written as the sum of two cubes at all, and even if we allow ourselves the luxury of a third cube, there is only the rather sad representation

$$1978 = 12^3 + 5^3 + 5^3.$$

One might therefore doubt whether either Hardy or Ramanujan would have been fascinated by it, although with Ramanujan I suppose one could never be quite certain. Nevertheless, for me it is one of the most important numbers, as it was in the year 1978 that I first met Hans Triebel. This was at the Spring School on 'Nonlinear analysis, function spaces and applications' held in Horni Bradlo in what was then Czechoslovakia. I am doubly indebted to our Czech colleagues for organising this conference, since not only did they enable me to meet Hans but also my long-standing and very fruitful collaboration with the Function Spaces group in Prague has its roots in this same meeting. The circumstances were idyllic: excellent weather (so important for someone from my country!), the beautiful countryside away from other distractions, an inexhaustible supply of jokes and above all, stimulating companions. My recollection of some of the details of that remarkable meeting may have dimmed with the passage of time, but what is still fresh in my mind is the immediate impact which Hans made on me. Since then it has been my great good fortune to have worked with him on a more or less regular basis. His visits to my department have made him part of the Sussex scene: he is often to be found walking along the path at the foot of the cliffs in Brighton, a productive place for theorems, it seems. It is an enormous pleasure for me to have this opportunity to pay tribute to his outstanding mathematical ability and exceptional personal qualities.

In this lecture I propose to speak about some of Hans Triebel's contributions to Analysis. Since his work is both broad and deep, limitations of time and my own knowledge oblige me to be quite selective in my choice of material.

2. Entropy

We begin with the notion of the metric entropy of a set, an idea which goes back to Pontryagin and Schnirelmann in 1932. Given any compact subset K of a metric space E and any $\varepsilon > 0$, let $N_\varepsilon(K)$ be the least $N \in \mathbf{N}$ such that K can be covered by N balls in E of radius ε. Then

$$H_\varepsilon(K) := \log_2 N_\varepsilon(K)$$

is called the metric entropy, or ε-entropy, of K. Of particular importance is the case arising from the study of a compact linear map T acting from a Banach space A to another such space B, with $E = B$ and K taken to be the image under T of the closed unit ball U_A in A; we then write

$$H(\varepsilon, T) = H(\varepsilon, T; A, B) = H_\varepsilon(T(U_A)).$$

Great attention has been paid over the years to the behaviour of $H(\varepsilon, T)$ when T is an embedding between function spaces. For example, Kolmogorov and Tikhomirov [22] found its asymptotic behaviour as $\varepsilon \to 0$ when T is the natural embedding of $C^s([0,1]^n)$ in $C([0,1]^n)$, s being a natural number. Vitushkin and Henkin [43] used ideas like this in their work on the superposition of functions related to Hilbert's thirteenth problem. Birman and Solomyak [2], [3] considered the embedding map

$$id : W_p^s(\Omega) \to L_q(\Omega),$$

where Ω is a bounded domain in $\mathbf{R}^n$ with smooth boundary, $s \in \mathbf{N}$, $1 < p < \infty$, $1 < q < \infty$, $s > n(1/p - 1/q)_+$ and $W_p^s(\Omega)$ is the usual Sobolev space of order s, based on $L_p(\Omega)$. Using the method of piecewise polynomial approximations which has now become a standard technique they obtained two-sided estimates for $H(\varepsilon, id)$. Further work in this last direction is contained in Triebel's 1970 paper [33], which is also remarkable for its study of the interpolation properties of the ideals

$$E_\alpha(A,B) = \left\{ T \in K(A,B) : \sup_{\varepsilon>0} \varepsilon H^\alpha(\varepsilon, T) < \infty \right\}, \quad 0 < \alpha < \infty,$$

where $K(A, B)$ is the space of all compact linear maps from A to B.

Corresponding to $H(\varepsilon, T)$ there are the entropy numbers of the map T. The theory of these was introduced by Pietsch in his book 'Operator ideals' [29]; certain functions inverse to the ε-entropy may be found in [33] and in a paper by Mityagin and Pelczynski [24]. The formal definition of the entropy numbers is as follows:

Let A and B be quasi-Banach spaces with closed unit balls U_A and U_B respectively, and let $T \in L(A, B)$, the space of all bounded linear maps from A to

B. Then for all $k \in \mathbf{N}$, the kth entropy number $e_k(T)$ of T is defined by

$$e_k(T) = \inf\left\{\varepsilon > 0 : T(U_A) \subset \bigcup_{j=1}^{2^{k-1}} (b_j + \varepsilon U_B) \text{ for some } b_1, \ldots, b_{2^{k-1}} \in B\right\}.$$

Some of the elementary properties of these numbers are given in the following statements, in which A, B, C are quasi-Banach spaces and $S, T \in L(A, B)$, while $R \in L(B, C)$.

(i) $\|T\| \geq e_1(T) \geq e_2(T) \geq \ldots \geq 0$; $e_1(T) = \|T\|$ if B is a Banach space.

(ii) For all $k, l \in \mathbf{N}$,

$$e_{k+l-1}(R \circ S) \leq e_k(R) e_l(S).$$

(iii) If B is a p-Banach space ($0 < p \leq 1$), then for all $k, l \in \mathbf{N}$,

$$e^p_{k+l-1}(S + T) \leq e^p_k(S) + e^p_l(T).$$

It is plain that T is compact if, and only if, $e_k(T) \longrightarrow 0$ as $k \to \infty$; when T is compact, the entropy numbers provide a convenient way of assessing 'how compact' a map it is. Of course, several other sequences of numbers have been devised with the same purpose at least partly in mind. Of these we shall mention here only the approximation numbers $a_k(T)$, which are defined by

$$a_k(T) = \inf\left\{\|T - L\| : L \in L(A, B), \text{rank } L < k\right\}.$$

These numbers resemble the entropy numbers in that they have properties exactly similar to (i)–(iii) above, but in some other respects are quite different. Thus if $T \in L(A, B)$, then $a_k(T) = 0$ if, and only if, rank $T < k$; while for all $k \in \mathbf{N}, e_k(T) \neq 0$ unless $T = 0$. Moreover, the interpolation properties of the entropy numbers (see [29] for the Banach space case and [19] for quasi-Banach spaces; in both cases one end-point has to be fixed) are much better than those of the approximation numbers (but see [7] for an interesting result about the approximation numbers). How the entropy numbers behave under interpolation when both end-points are allowed to be different is not known: it seems to offer a considerable challenge. If A and B are Hilbert spaces, the approximation numbers of $T \in L(A, B)$ are just the singular values of T, that is, the eigenvalues of the positive square root of T^*T. However, outside the comfortable world of Hilbert spaces there is no particularly useful connection between the approximation numbers of a map $T \in L(A, A) := L(A)$ and its eigenvalues. It is here that the entropy numbers come into their own.

To justify this last claim, let A be a (complex) quasi-Banach space and let $T \in L(A)$ be compact. The spectrum of T, apart from the point 0, consists solely of isolated eigenvalues of finite algebraic multiplicity. We denote by $\lambda_k(T)$ ($k \in \mathbf{N}$) the non-zero eigenvalues of T, repeated according to algebraic multiplicity and ordered by non-increasing modulus. If T has only finitely many distinct such eigenvalues and M is the sum of their algebraic multiplicities, we set $\lambda_k(T) = 0$ for all $k > M$. Then it turns out that for all $k \in \mathbf{N}$,

$$|\lambda_k(T)| \leq \sqrt{2} e_k(T).$$

This beautiful inequality was first proved by Carl [5] when A is a Banach space. Later he and Triebel [6] gave a more geometric proof, obtaining the inequality

$$\left(\prod_{m=1}^{k} |\lambda_m(T)|\right)^{1/k} \leq \inf 2^{n/(2k)} e_n(T), \quad k \in \mathbf{N},$$

again in Banach spaces, and this in turn was established in the setting of quasi-Banach spaces in [12]. Thus if one wishes to estimate the eigenvalues of an operator, a possible strategy is to try to estimate its entropy numbers. When dealing with integral, differential or pseudodifferential operators, the map induced by them can often be represented as the composition of various maps, one such factor being an embedding map between spaces of Sobolev type. This means that knowledge of the precise behaviour of the entropy numbers of such embeddings and of the norms of the remaining factors, together with the submultiplicative property (ii) above, would lead to the required estimate. For this reason we now turn to function spaces.

3. Function spaces

To begin with, we consider function spaces defined on the whole of $\mathbf{R}^n$. Let $\mathcal{S}$ denote the Schwartz space of all complex-valued, rapidly decreasing C^∞ functions on $\mathbf{R}^n$ and let $\mathcal{S}'$ be its dual. Let $\phi \in \mathcal{S}$ with

$$\phi(x) = 1 \quad \text{if} \quad |x| \leq 1 \quad \text{and} \quad \phi(x) = 0 \quad \text{if} \quad |x| \geq 3/2,$$

and put $\phi_0 = \phi$, $\phi_1(x) = \phi(x/2) - \phi(x)$ and

$$\phi_k(x) = \phi_1(2^{-k+1}x), \quad x \in \mathbf{R}^n,\ k \in \mathbf{N}.$$

Since

$$\sum_{k=0}^{\infty} \phi_k(x) = 1 \quad \text{for all} \quad x \in \mathbf{R}^n,$$

the ϕ_k form a dyadic resolution of unity. Given any $f \in \mathcal{S}$,

$$\widehat{f}(\xi) = (Ff)(\xi) = (2\pi)^{-n/2} \int_{\mathbf{R}^n} e^{-i\xi \cdot x} f(x) dx, \quad \xi \in \mathbf{R}^n,$$

denotes the Fourier transform of f; by $F^{-1}f$ or $\check{f}$ we shall mean the inverse Fourier transform of f. We now define the Besov spaces $B^s_{p,q}(\mathbf{R}^n)$ and the Lizorkin-Triebel spaces $F^s_{p,q}(\mathbf{R}^n)$.

Definition 1. (i) *Let $s \in \mathbf{R}$ and $p, q \in (0, \infty]$. Then $B^s_{p,q}(\mathbf{R}^n)$ is the collection of all $f \in \mathcal{S}'$ such that*

$$\left\| f \mid B^s_{p,q}(\mathbf{R}^n) \right\| = \left(\sum_{j=0}^{\infty} 2^{jsq} \left\| \left(\phi_j \widehat{f}\right)^{\vee} \mid L_p(\mathbf{R}^n) \right\|^q \right)^{1/q}$$

(with the usual modification if $q = \infty$) is finite.

(ii) *Let $s \in \mathbf{R}$, $0 < p < \infty$ and $0 < q \leq \infty$. Then $F^s_{p,q}(\mathbf{R}^n)$ is the collection of all $f \in \mathcal{S}'$ such that*

$$\left\| f \mid F^s_{p,q}(\mathbf{R}^n) \right\| = \left\| \left(\sum_{j=0}^{\infty} 2^{jsq} \left| \left(\phi_j \widehat{f} \right)^{\vee} (\cdot) \right|^q \right)^{1/q} \mid L_p(\mathbf{R}^n) \right\|$$

(with the usual modification if $q = \infty$) is finite.

To the extensive theory of these spaces Triebel has made notable contributions: see his books [34], [35], [36], [38], [39] as well as [12]. Here we cannot go into this in detail and so we content ourselves with some basic observations. First, the particular choice of the function ϕ does not really influence the spaces, as change of ϕ merely gives equivalent quasi-norms. Moreover, if $p \geq 1$ and $q \geq 1$, then both $B^s_{p,q}(\mathbf{R}^n)$ and $F^s_{p,q}(\mathbf{R}^n)$ are Banach spaces. In addition, there is the remarkable fact that even though at first sight the definitions above may seem hard to understand and remote from common experience, these two scales of spaces actually include astonishingly many of the well-loved spaces which are in common use in analysis. For example:

(i) If $1 < p < \infty$, then $F^0_{p,2}(\mathbf{R}^n) = L_p(\mathbf{R}^n)$.

(ii) If $1 < p < \infty$ and $s \in \mathbf{N}_0 = \mathbf{N} \cup \{0\}$, then $F^s_{p,2}(\mathbf{R}^n) = W^s_p(\mathbf{R}^n)$, the classical Sobolev space often normed by

$$\left\| f \mid W^s_p(\mathbf{R}^n) \right\| = \sum_{|\alpha| \leq s} \left\| D^\alpha f \mid L_p(\mathbf{R}^n) \right\|.$$

(iii) If $1 < p < \infty$ and $s \in \mathbf{R}$, then $F^s_{p,2}(\mathbf{R}^n) = H^s_p(\mathbf{R}^n)$, the (fractional) Sobolev space usually normed by

$$\left\| f \mid H^s_p(\mathbf{R}^n) \right\| = \left\| F^{-1} \left(\left(1 + |x|^2 \right)^{s/2} (Fu)(x) \right) \mid L_p(\mathbf{R}^n) \right\|.$$

(iv) Define the iterated differences Δ^m_h by

$$\left(\Delta^1_h f \right)(x) = f(x+h) - f(x), \quad \left(\Delta^{l+1}_h f \right)(x) = \Delta^1_h \left(\Delta^l_h f \right)(x),$$

with $x, h \in \mathbf{R}^n$ and $l \in \mathbf{N}$. Then $B^s_{\infty,\infty}(\mathbf{R}^n) = \mathcal{C}^s(\mathbf{R}^n)$, the Hölder-Zygmund space usually normed by

$$\left\| f \mid \mathcal{C}^s(\mathbf{R}^n) \right\| = \sup_{x \in \mathbf{R}^n} |f(x)| + \sup_{x,h \in \mathbf{R}^n, 0 < |h| \leq 1} |h|^{-s} \left| \Delta^m_h f(x) \right|,$$

where $0 < s < m \in \mathbf{N}$.

Atomic decompositions play an important part of the theory of function spaces. The basic idea is to show that elements of the spaces $B^s_{p,q}(\mathbf{R}^n)$ and $F^s_{p,q}(\mathbf{R}^n)$ can be represented as infinite sums of smooth functions. It turns out that $f \in \mathcal{S}'$

belongs to $B^s_{p,q}(\mathbf{R}^n)$ $(0 < p \le \infty,\ 0 < q \le \infty,\ s \in \mathbf{R})$ if, and only if, it can be represented in the form

$$f = \sum_{\nu=0}^{\infty} \sum_{m \in \mathbf{Z}^n} \lambda_{\nu m} a_{\nu m},$$

where the $\lambda_{\nu m}$ are complex numbers such that

$$\|\lambda \mid b_{p,q}\| := \left(\sum_{\nu=0}^{\infty} \left(\sum_{m \in \mathbf{Z}^n} |\lambda_{\nu m}|^p \right)^{q/p} \right)^{1/q} < \infty$$

(with natural modifications if p or q is infinite) and the $a_{\nu m}$ are certain smooth, compactly supported functions, depending on s, p and n, called atoms. Moreover,

$$\inf \|\lambda \mid b_{p,q}\|,$$

where the infimum is taken over all admissible representations of f in this form, is a quasi-norm on $B^s_{p,q}(\mathbf{R}^n)$ equivalent to that given earlier. A similar statement holds for $F^s_{p,q}(\mathbf{R}^n)$. Results in this direction go back to Frazier and Jawerth ([14], [15], [16]), with significant contributions by Netrusov ([25], [26], [27]); Triebel gave new complete proofs of these decompositions in his book "Fractals and Spectra" [39].

A technical point about the atomic decompositions described above is that one usually constructs an optimal atomic decomposition for the particular function f, and so the atoms used may well depend upon that function: one does not have a universal set of atoms. Influenced by this and other considerations, Triebel introduced in [36] subatomic decompositions: the atoms were split, the new elementary particles being called quarks. To explain the basic idea involved in this, let $\psi \in C_0^\infty(\mathbf{R}^n)$ be non-negative, with

$$\operatorname{supp} \psi \subset \{y \in \mathbf{R}^n \mid \quad |y| < 2^r\}$$

for some $r \ge 0$, and

$$\sum_{m \in \mathbf{Z}^n} \psi(x - m) = 1 \quad \text{if} \quad x \in \mathbf{R}^n.$$

Let $s \in \mathbf{R}$, $0 < p \le \infty$, $\beta \in \mathbf{N}_0^n$, $\nu \in \mathbf{N}_0$, $m \in \mathbf{Z}^n$ and $\psi^\beta(x) = x^\beta \psi(x)$. Then

$$(\beta qu)_{\nu m}(x) = 2^{-\nu(s-n/p)} \psi^\beta(2^\nu x - m), \quad x \in \mathbf{R}^n,$$

is called an $(s,p) - \beta - quark$ related to the cube $Q_{\nu m}$ in $\mathbf{R}^n$ with sides parallel to the axes of coordinates, centred at $2^{-\nu} m$ and with side length $2^{-\nu}$. In terms of these quarks the fundamental decomposition theorem for $B^s_{p,q}(\mathbf{R}^n)$, when $0 < p \le \infty$, $0 < q \le \infty$ and $s > n\left(\frac{1}{p} - 1\right)_+$, reads as follows. Let $\rho > r$ and put

$$\lambda = \left\{\lambda^\beta : \beta \in \mathbf{N}_0^n\right\} \quad \text{with} \quad \lambda^\beta = \left\{\lambda^\beta_{\nu m} \in \mathbf{C} : \nu \in \mathbf{N}_0, m \in \mathbf{Z}^n\right\};$$

let $b^\rho_{p,q}$ be the space of all such λ with

$$\left\|\lambda \mid b^\rho_{p,q}\right\| := \sup_{\beta \in \mathbf{N}_0^n} 2^{\rho|\beta|} \left\|\lambda^\beta \mid b_{p,q}\right\| < \infty.$$

Then $B^s_{p,q}(\mathbf{R}^n)$ is the family of all $f \in \mathcal{S}'$ which can be represented as

$$f = \sum_{\beta \in \mathbf{N}_0^n} \sum_{\nu=0}^{\infty} \sum_{m \in \mathbf{Z}^n} \lambda^{\beta}_{\nu m} (\beta q u)_{\nu m}$$

for some $\lambda \in b^{\rho}_{p,q}$. Moreover,

$$\inf \left\| \lambda \mid b^{\rho}_{p,q} \right\|$$

is a quasi-norm on $B^s_{p,q}(\mathbf{R}^n)$ equivalent to that given earlier. Even better, to each $f \in B^s_{p,q}(\mathbf{R}^n)$ there are optimal coefficients $\lambda^{\beta}_{\nu m}(f)$, determined by a constructive procedure, which can be used in the representation above and for which

$$\left\| \lambda(f) \mid b^{\rho}_{p,q} \right\| \sim \left\| f \mid B^s_{p,q}(\mathbf{R}^n) \right\|,$$

where the equivalence constants are independent of f. This means that questions about $B^s_{p,q}(\mathbf{R}^n)$ can be reduced to matters concerning sequence spaces.

All this may be found in the forthcoming book [39] by Triebel (see also [38]), together with the corresponding result for $F^s_{p,q}(\mathbf{R}^n)$ and details of how the condition that s should be sufficiently large may be removed at the cost of some additional complications.

So far all the function spaces we have considered have been over the whole of $\mathbf{R}^n$. When it comes to spaces on domains, one has the choice of defining these by restriction or in some intrinsic way. The first of these procedures is as follows. Let Ω be a domain (a connected open set) in $\mathbf{R}^n$, let $s \in \mathbf{R}$ and $0 < q \leq \infty$. The restriction of $g \in \mathcal{S}'$ to Ω is denoted by $g \mid \Omega$ and is considered as an element of $\mathcal{D}'(\Omega)$. If $0 < p \leq \infty$, then $B^s_{p,q}(\Omega)$ is the restriction of $B^s_{p,q}(\mathbf{R}^n)$ to Ω, quasi-normed by

$$\left\| f \mid B^s_{p,q}(\Omega) \right\| = \inf \left\| g \mid B^s_{p,q}(\mathbf{R}^n) \right\|,$$

where the infimum is taken over all $g \in B^s_{p,q}(\mathbf{R}^n)$ with $g \mid \Omega = f$. The space $F^s_{p,q}(\Omega)$ is defined in a similar manner, under the restriction $0 < p < \infty$. If Ω is a bounded domain with C^∞ boundary, then there are bounded linear extension operators from these spaces to the corresponding spaces on $\mathbf{R}^n$ and, if $s > n\left(\frac{1}{p} - 1\right)_+$, some intrinsic characterisations are known (see [36]). For the substantial contributions of Triebel to intrinsic characterisations of these spaces on non-smooth domains and to atomic and sub-atomic decompositions of spaces on domains we refer to [40], [41], [12], [36] and [38].

We conclude this section on spaces by describing some recent work of Triebel related to the distance function and Hardy's inequality. Let Ω be a bounded C^∞ domain in $\mathbf{R}^n$, let $1 < p < \infty$ and suppose that $s \in \mathbf{N}_0$. The classical Sobolev space $W^s_p(\Omega)$ is defined by restriction, as above, and one may distinguish two subspaces of it: $\overset{\circ}{W}{}^s_p(\Omega)$, the closure of $C_0^\infty(\Omega)$ in it; and

$$\widetilde{W}^s_p(\Omega) = \left\{ f \in W^s_p(\mathbf{R}^n) : \text{ supp } f \subset \overline{\Omega} \right\}.$$

The first subspace is given the norm inherited from $W^s_p(\Omega)$, while the second is given that of $W^s_p(\mathbf{R}^n)$. It is well known that these two subspaces coincide, in the sense of equivalent norms. Let

$$d(x) = \text{ dist } (x, \partial\Omega), \quad x \in \mathbf{R}^n,$$

be the distance of a point $x \in \mathbf{R}^n$ from the boundary $\partial\Omega$ of Ω. The Hardy inequality asserts that there is a positive constant c such that for all $f \in \overset{\circ}{W}{}^s_p(\Omega)$,

$$\int_\Omega d^{-sp}(x)\,|f(x)|^p\,dx \le c^p \left\|f \mid W^s_p(\Omega)\right\|^p,$$

which we may write as

$$\left\|f \mid L_p(\Omega, d^{-s})\right\| \le c\left\|f \mid W^s_p(\Omega)\right\|.$$

Here $L_p(\Omega, d^{-s})$ is the weighted Lebesgue space with weight d^{-s} and the natural norm. It then emerges that

$$\overset{\circ}{W}{}^s_p(\Omega) = W^s_p(\Omega) \cap L_p(\Omega, d^{-s}).$$

Triebel has recently (see [39]) extended this in the following natural way. Letting $A^s_{p,q}$ stand for either $B^s_{p,q}$ or $F^s_{p,q}$, we proceed as in the case of the Sobolev spaces above and denote by $\overset{\circ}{A}{}^s_{p,q}(\Omega)$ the closure of $C_0^\infty(\Omega)$ in $A^s_{p,q}(\Omega)$ and give it the quasi-norm inherited from $A^s_{p,q}(\Omega)$, while we write

$$\widetilde{A}^s_{p,q}(\Omega) = \left\{f \in \mathcal{D}'(\Omega) : f = g \mid \Omega \quad \text{for some } g \in A^s_{p,q}(\mathbf{R}^n) \text{ with supp } g \subset \overline{\Omega}\right\}$$

and give it the norm

$$\left\|f \mid \widetilde{A}^s_{p,q}(\Omega)\right\| = \inf \left\|g \mid A^s_{p,q}(\mathbf{R}^n)\right\|,$$

where the infimum is taken over all g with $g \mid \Omega = f$ and supp $g \subset \overline{\Omega}$. His result is that if

$$0 < p < \infty,\ 0 < q \le \infty,\ s > n\left(\frac{1}{\min(p,q)} - 1\right)_+,$$

then

$$\widetilde{F}^s_{p,q}(\Omega) = F^s_{p,q}(\Omega) \cap L_p(\Omega, d^{-s})$$

in the sense of equivalent quasi-norms. Moreover, if

$$0 < p < \infty,\ 0 < q < \infty,\ s > n\left(\frac{1}{p} - 1\right)_+ \quad \text{and} \quad s - 1/p \notin \mathbf{N}_0,$$

then

$$\widetilde{F}^s_{p,q}(\Omega) = \overset{\circ}{F}{}^s_{p,q}(\Omega) \quad \text{and} \quad \widetilde{B}^s_{p,q}(\Omega) = \overset{\circ}{B}{}^s_{p,q}(\Omega).$$

For further information about results of this type see [39].

4. Entropy numbers of embeddings

Let Ω be a bounded domain in $\mathbf{R}^n$ with C^∞ boundary and let the spaces $B^s_{p,q}(\Omega)$ and $F^s_{p,q}(\Omega)$ be defined by restriction, as explained above. Let

$$-\infty < s_2 < s_1 < \infty; \quad p_1, p_2, q_1, q_2 \in (0, \infty],$$

and suppose that

$$\delta_+ := s_1 - s_2 - n\left(\frac{1}{p_1} - \frac{1}{p_2}\right)_+ > 0.$$

Let $A^s_{p,q}(\Omega)$ stand for either $B^s_{p,q}(\Omega)$ or $F^s_{p,q}(\Omega)$, with the understanding that for the F-spaces we must have $p < \infty$. Under these conditions it is known that $A^{s_1}_{p_1,q_1}(\Omega)$ is compactly embedded in $A^{s_2}_{p_2,q_2}(\Omega)$; let id be the corresponding embedding map. The basic result concerning the entropy numbers of this embedding is the following, in which by $a_k \sim b_k$, for two sequences of positive numbers a_k and b_k, we shall mean that there are two positive constants c_1 and c_2 such that for all k, $c_1 a_k \le b_k \le c_2 a_k$.

Theorem 2. *Under the above hypotheses, the entropy numbers $e_k(id)$ satisfy*

$$e_k(id) \sim k^{-(s_1-s_2)/n}.$$

This contains many of the results of this kind to be found in the literature. For historical remarks and detailed references we refer to [12]. A forerunner of work of this kind is that of Birman and Solomyak [2], [3], to which we have already referred, on the embedding of the Sobolev space $W^s_p(\Omega)$ in $L_p(\Omega)$, where $1 < p < q < \infty$, $s \in \mathbf{N}$ and $s - n/p > -n/q$. They showed that the entropy numbers e_k of this embedding satisfied

$$e_k \sim k^{-s/n}.$$

Theorem 2 is given in the book [12] with Triebel and is based on our earlier papers [9] and [10]. The proof relies on the Fourier-analytical definition of the spaces together with a reduction of the problem to that of obtaining corresponding estimates for the entropy numbers of mappings between finite-dimensional sequence spaces. However, in his book 'Fractals and Spectra' [38] Triebel gave a new proof of Theorem 2 in which he used atomic and subatomic decompositions of Besov spaces together with a reduction of the problems to ones involving mappings between weighted finite-dimensional spaces. One of the advantages of this procedure is that it may be applied to similar problems involving function spaces on fractals and compact embeddings on spaces related to fractals. For details of this we refer to [38].

In certain limiting situations, different target spaces for embeddings arise naturally. To explain this, let Ω be a bounded domain in $\mathbf{R}^n$ with C^∞ boundary, let $1 < p < \infty$, $s \in \mathbf{R}$ and put $H^s_p(\Omega) = F^s_{p,2}(\Omega)$; $H^s_p(\Omega)$ is the (fractional) Sobolev space of order s. We introduce the Zygmund space $L_p(\log L)_a(\Omega)$. When $a \in \mathbf{R}$ this is defined to be the set of all measurable complex-valued functions f on Ω

such that

$$\int_\Omega |f(x)|^p \log^{ap}(2+|f(x)|)dx < \infty;$$

equivalently, $f \in L_p(\log L)_a(\Omega)$ if, and only if,

$$\left(\int_0^{|\Omega|} \{(1+|\log t|)^a f^*(t)\}^p \, dt\right)^{1/p} < \infty,$$

where $|\Omega|$ is the Lebesgue n-measure of Ω and f^* is the non-increasing rearrangement of f. This last displayed formula is, in general, only a quasi-norm, but an equivalent norm is obtained by replacing $f^*(t)$ by $f^{**}(t) = t^{-1}\int_0^t f^*(s)ds$. When $a < 0$, $L_\infty(\log L)_a(\Omega)$ is defined to be the set of all complex-valued measurable functions f on Ω for which there exists a constant $\lambda > 0$ such that

$$\int_\Omega \exp\left\{(\lambda\,|f(x)|)^{-1/a}\right\} dx < \infty.$$

This is an Orlicz space, given the Luxemburg norm corresponding to a Young function which behaves like $\exp(t^{-1/a})$ for large values of t; it may also be given the (equivalent) norm

$$\sup_{t>0} \{(1+|\log t|)^a f^*(t)\}.$$

It is known that $H_p^{n/p}(\Omega)$ is compactly embedded in $L_q(\Omega)$ for every q in $[1,\infty)$ and is not embedded at all in $L_\infty(\Omega)$. Moreover, it is embedded in $L_\infty(\log L)_a(\Omega)$ if, and only if, $a \le -1/p'$, the embedding being compact if, and only if, $a < -1/p'$; results of this type can be traced back to Yudovich [44], Peetre [28], Pohozaev [30], Trudinger [42] and Strichartz [32].They also hold with $H_p^{n/p}(\Omega)$ replaced by $B_{p,p}^{n/p}(\Omega)$. In a real 'tour de force' Triebel [37] succeeded in proving that if $a < -1-2/p$ and id_1 stands for the natural embedding of either $H_p^{n/p}(\Omega)$ or $B_{p,p}^{n/p}(\Omega)$ in $L_\infty(\log L)_a(\Omega)$, then

$$e_k(id_1) \sim k^{-1/p}, \quad k \in \mathbf{N}.$$

This remarkable result was obtained by refining the Fourier approach used in non-limiting situations in [9] and [10] and having good estimates for the various constants which turn up. He also obtained upper and lower estimates for $e_k(id_1)$ when $-1-2/p \le a < -1/p'$, but the upper and lower bounds involved different powers of k and so the results cannot be claimed to be optimal. Further progress in this direction is very desirable.

Moving from the $L_\infty(\log L)_a(\Omega)$ situation to one with $L_q(\log L)_a(\Omega)$, $1 < q < \infty$, it was shown in [11] that if $1/p = 1/q + s/n$ and $a < 0$, then the embedding

$$id_2 : H_p^s(\Omega) \to L_q(\log L)_a(\Omega)$$

is compact (if $a = 0$ it is continuous but not compact) and that

$$e_k(id_2) \sim k^{-s/n}, \quad k \in \mathbf{N},$$

if in addition $a < -2s/n$. In [8] this was shown to hold also when $a < -s/n$, while if $-s/n < a < 0$, then

$$e_k(id_2) \sim k^a, \quad k \in \mathbf{N}.$$

For additional results of this type, both for entropy numbers and approximation numbers, we refer to [11], [12] and [8].

We conclude this section by remarking that Triebel's other contributions to this area include estimates of entropy and approximation numbers of embeddings of weighted spaces of B and F type on the whole of $\mathbf{R}^n$ (work carried out with D. Haroske [19], [20]). An excellent survey of this and related topics is given by Haroske in [18].

5. Sharp inequalities

Our starting point here is the fractional Sobolev space $H_p^{n/p}(\mathbf{R}^n)$ with $1 < p < \infty$. Just as mentioned earlier in the context of spaces on bounded domains, this space is embedded in an Orlicz space, with Young function Φ such that $\Phi(t) \sim \exp(t^{p'})$ for large t. We reformulate this as an inequality: there exists $c > 0$ such that for all $f \in H_p^s(\mathbf{R}^n)$ and all choices of $t_j \in [2^{-j}, 2^{-j+1}]$,

$$\sup_{j\in\mathbf{N}} j^{-1/p'} f^*(t_j) \leq c \left\| f \mid H_p^{n/p}(\mathbf{R}^n) \right\|.$$

Moving outside the family of Orlicz spaces, Hansson [17] and, independently, Brézis and Wainger [4] showed that this inequality could be improved to

$$\int_0^1 \left(\frac{f^*(t)}{1 + |\log t|} \right)^p \frac{dt}{t} \leq c \left\| f \mid H_p^{n/p}(\mathbf{R}^n) \right\|.$$

Note that when $p = n$, this improved inequality can be obtained from certain capacity estimates of Maz'ya [23]. Now let κ be a positive, continuous function on the interval $(0, 1]$ and let A_p be either $H_p^{n/p}(\mathbf{R}^n)$ or $B_{p,p}^{n/p}(\mathbf{R}^n)$. In [13] it was shown, by means of atomic decomposition techniques and a little nonlinear interpolation, that the following two assertions are equivalent:

(i) There is a constant $c > 0$ such that

$$\int_0^1 \left(\frac{f^*(t)\kappa(t)}{1 + |\log t|} \right)^p \frac{dt}{t} \leq c \| f \mid A_p \|$$

for all $f \in A_p$.

(ii) The function κ is bounded.

Quite recently Triebel [39] has obtained a very general result, expressed in the language of envelope functions, which includes this result. We now give consequences of this work in a form which makes it possible to compare it with the theorem just stated. Let $0 < \varepsilon < 1$, let κ be a positive, monotonically decreasing function on $(0, \varepsilon]$ and suppose that $0 < u \leq \infty$.

Theorem 3. *Let $0 < p < \infty$ and $1 < q \le \infty$. Then there is a number $c > 0$ such that*

$$\left(\int_0^\varepsilon \left(\frac{f^*(t)\kappa(t)}{|\log t|^{1/q'}}\right)^u \frac{dt}{t\,|\log t|}\right)^{1/u} \le c \left\| f \mid B^{n/p}_{p,q}(\mathbf{R}^n)\right\|$$

for all $f \in B^{n/p}_{p,q}(\mathbf{R}^n)$ if, and only if, κ is bounded and $q \le u \le \infty$. If $1 < q < \infty$, then there are positive constants c_0 and c_1 such that

$$\sup_{0<t<\varepsilon} \frac{f^*(t)}{|\log t|^{1/q'}} \le c_0 \left(\int_0^\varepsilon \left(\frac{f^*(t)}{|\log t|}\right)^q \frac{dt}{t}\right)^{1/q} \le c_1 \left\| f \mid B^{n/p}_{p,q}(\mathbf{R}^n)\right\|$$

for all $f \in B^{n/p}_{p,q}(\mathbf{R}^n)$. If $q = \infty$ then

$$\sup_{0<t<\varepsilon} \frac{f^*(t)}{|\log t|} \le c \left\| f \mid B^{n/p}_{p,\infty}(\mathbf{R}^n)\right\|.$$

Without the hypothesis that κ is monotonic, there is a $c > 0$ such that

$$\sup_{0<t<\varepsilon} \frac{f^*(t)\kappa(t)}{|\log t|^{1/q'}} \le c \left\| f \mid B^{n/p}_{p,q}(\mathbf{R}^n)\right\|$$

if, and only if, κ is bounded.

The corresponding result for the F spaces reads as follows.

Theorem 4. *Let $1 < p < \infty$ and $0 < q \le \infty$. Then there is a number $c > 0$ such that*

$$\left(\int_0^\varepsilon \left(\frac{f^*(t)\kappa(t)}{|\log t|^{1/p'}}\right)^u \frac{dt}{t\,|\log t|}\right)^{1/u} \le c \left\| f \mid F^{n/p}_{p,q}(\mathbf{R}^n)\right\|$$

for all $f \in F^{n/p}_{p,q}(\mathbf{R}^n)$ if, and only if, κ is bounded and $p \le u \le \infty$. If $1 < q < \infty$, then there are positive constants c_0 and c_1 such that

$$\sup_{0<t<\varepsilon} \frac{f^*(t)}{|\log t|^{1/p'}} \le c_0 \left(\int_0^\varepsilon \left(\frac{f^*(t)}{|\log t|}\right)^p \frac{dt}{t}\right)^{1/p} \le c_1 \left\| f \mid F^{n/p}_{p,q}(\mathbf{R}^n)\right\|$$

for all $f \in F^{n/p}_{p,q}(\mathbf{R}^n)$. Without the hypothesis that κ is monotonic, there is a $c > 0$ such that

$$\sup_{0<t<\varepsilon} \frac{f^*(t)\kappa(t)}{|\log t|^{1/p'}} \le c \left\| f \mid F^{n/p}_{p,q}(\mathbf{R}^n)\right\|$$

if, and only if, κ is bounded.

The proofs of these results rely upon atomic decomposition techniques together with various other ideas. One is the important remark of Netrusov [27] that given any $f \in F^{n/p}_{p,\infty}(\mathbf{R}^n)$, there is a function $g \in B^{n/p}_{p,p}(\mathbf{R}^n)$ such that

$$|f(x)| \le g(x) \quad \text{a.e. in } \mathbf{R}^n, \quad \text{and} \quad \left\| g \mid B^{n/p}_{p,p}(\mathbf{R}^n)\right\| \le c \left\| f \mid F^{n/p}_{p,\infty}(\mathbf{R}^n)\right\|,$$

for some constant c which is independent of f and g. Another point is the construction of 'extremal' functions f which belong to $H_p^{n/p}(\mathbf{R}^n) \cap B_{p,p}^{n/p}(\mathbf{R}^n)$ and have the singularity behaviour

$$f(x) = |\log|x||^{1/p'} (\log|\log|x||)^{-\sigma}, \quad \text{where} \quad \sigma > 1/p,$$

near the origin.

Inequalities of Hardy type follow from the results described above. Some are contained in [13], but for the latest information in this direction we refer to [37], where the following is proved. Again let $0 < \varepsilon < 1$ and let κ be a positive monotonically decreasing function on $(0, \varepsilon]$; let $1 < p < \infty$ and $0 < q \leq \infty$. Then there is a constant $c > 0$ such that

$$\int_{|x| \leq \varepsilon} \left| \frac{\kappa(|x|) f(x)}{\log|x|} \right|^p \frac{dx}{|x|^n} \leq c \left\| g \mid F_{p,q}^{n/p}(\mathbf{R}^n) \right\|^p$$

for all $f \in F_{p,q}^{n/p}(\mathbf{R}^n)$ if, and only if, κ is bounded. A corresponding result holds for $B_{p,q}^{n/p}(\mathbf{R}^n)$ when $0 < p < \infty$ and $1 < q < \infty$.

In all of this we have concentrated on the 'critical' case in which the smoothness parameter in the spaces is of the form n/p. There are corresponding results, due to Triebel, for the 'sub-critical' and 'super-critical' situations, for a comprehensive account of which we refer to [39].

6. Elliptic equations and eigenvalues

Here we concentrate on three particular areas to which Triebel has made notable contributions and give very brief accounts.

6.1. Eigenvalues of degenerate elliptic equations

Let Ω be a bounded domain in $\mathbf{R}^n$ with C^∞ boundary, let A be a properly elliptic operator

$$Af = \sum_{|\alpha| \leq 2m} a_\alpha(x) D^\alpha f, \quad \text{where each} \quad a_\alpha \in C^\infty(\overline{\Omega}),$$

and suppose that there are boundary operators

$$B_j f = \sum_{|\alpha| \leq l_j} b_{j,\alpha}(x) D^\alpha f, \quad \text{where each} \quad b_{j,\alpha} \in C^\infty(\overline{\Omega}),$$

with $j = 1, \ldots, m$ and $0 \leq l_1 < \cdots < l_m \leq 2m - 1$, which form a normal system satisfying the complementing condition. We assume that the problem

$$Af = 0 \quad \text{in} \quad \Omega, \quad B_j f = 0 \quad \text{on} \quad \partial\Omega \quad \text{for } j = 1, \ldots, m,$$

has only the trivial C^∞ solution. It is known (see [12] for references) that if $0 < p < \infty$ and $s \geq n\left(\frac{1}{p} - 1\right)_+$, then A maps

$$\left\{ f \in H_p^{s+2m}(\Omega) : B_j f = 0 \quad \text{on} \quad \partial\Omega \quad \text{for } j = 1, \ldots, m \right\} \tag{1}$$

isomorphically onto $H_p^s(\Omega)$. Moreover, denoting by $A_{s,p}$ the operator A with domain of definition (1), 0 does not belong to the spectrum $\sigma(A_{s,p})$ of $A_{s,p}$, which consists of isolated eigenvalues of finite algebraic multiplicity. Now let A^{-1} be the inverse of $A_{s,p}$: we omit the subscripts as it will be clear from the context between which spaces A^{-1} acts. We consider the map B, where

$$Bf = b_2 A^{-1} b_1 f,$$

and b_1, b_2 belong to certain Lebesgue spaces so chosen that B is a compact map of some L_p space to itself. A typical result which can be found in [12] is the following:

Theorem 5. *Suppose that $p, r_1, r_2 \in [1, \infty]$ and that*

$$\frac{1}{p} + \frac{1}{r_1} < 1, \quad p < r_2, \quad \delta := \frac{2m}{n} - \left(\frac{1}{r_1} + \frac{1}{r_2}\right) > 0.$$

Let

$$b_1 \in L_{r_1}(\Omega), \quad b_2 \in L_{r_2}(\Omega).$$

Then $B : L_p(\Omega) \to L_p(\Omega)$ is compact and its kth eigenvalue μ_k (counted with respect to algebraic multiplicity and ordered by decreasing modulus) satisfies

$$|\mu_k| \le c\, \|b_1 \mid L_{r_1}(\Omega)\|\, \|b_2 \mid L_{r_2}(\Omega)\|\, k^{-2m/n},$$

where c is independent of b_1, b_2 and k.

The idea of the proof is to factorise B in the form

$$B = b_2 \circ id \circ A^{-1} \circ b_1,$$

where $b_1 : L_p(\Omega) \to L_q(\Omega)$ $(1/q = 1/r_1 + 1/p)$, $A^{-1} : L_q(\Omega) \to H_q^{2m}(\Omega)$, $id : H_q^{2m}(\Omega) \to L_t(\Omega)$ $(1/t = 1/p - 1/r_2)$ and $b_2 : L_t(\Omega) \to L_p(\Omega)$. Under the assumptions made, id is compact and

$$e_k(id) \sim k^{-2m/n}.$$

Together with Carl's inequality and the submultiplicativity of the entropy numbers, this gives the result. Note that if for almost all $x \in \Omega$, $b_1(x) \neq 0$ and $b_2(x) \neq 0$, then B is invertible in $L_p(\Omega)$ and, at least formally,

$$D := B^{-1} = b_1^{-1} A b_2^{-1}$$

is a degenerate elliptic operator, considered as an unbounded operator in $L_p(\Omega)$, with eigenvalues λ_k, counted according to algebraic multiplicity and ordered by increasing modulus. Since $\lambda_k = \mu_k^{-1}$, we have

$$|\lambda_k| \ge c^{-1}\, \|b_1 \mid L_{r_1}(\Omega)\|^{-1}\, \|b_2 \mid L_{r_2}(\Omega)\|^{-1}\, k^{2m/n}.$$

For elliptic operators of order $2m$ and with smooth coefficients it is well known that the modulus of the kth eigenvalue behaves like a multiple of $k^{2m/n}$, so that the result above is in accordance with this.

Various extensions of this idea can be found in [12] and [38]. The book [12] also contains results (some based on work of Haroske and Triebel [20]) concerning the negative spectrum of certain self-adjoint operators and using an 'entropy' version of the well-known Birman-Schwinger principle.

6.2. The fractal drum

Let Ω be a bounded C^∞ domain in $\mathbf{R}^n$ and let $0 < d < n$. A compact subset Γ of $\mathbf{R}^n$ is called a d-set if there is a Radon measure μ such that for some positive constants c_1 and c_2,

$$\operatorname{supp}\mu = \Gamma \quad \text{and} \quad c_1 s^d \leq \mu\left(B\left(\gamma, s\right)\right) \leq c_2 s^d \quad \text{for all } \gamma \in \Gamma \text{ and all } s \in (0,1),$$

where $B(\gamma, s)$ is the ball centred at $\gamma \in \Gamma$ and of radius s.

Let $\phi \in \mathcal{S}$ and let $tr_\Gamma \phi$ be the pointwise trace of ϕ on $\Gamma : tr_\Gamma \phi = \phi\,|_\Gamma$. The operator tr^Γ is defined by

$$\left(tr^\Gamma f\right)(\phi) = \int_\Gamma \left(tr_\Gamma f\right)(\gamma)\phi(\gamma)d\mu(\gamma), \quad \phi \in C_0^\infty(\Omega).$$

This maps $\mathcal{D}(\Omega) := C_0^\infty(\Omega)$ to $\mathcal{D}'(\Omega)$ and can be extended to a bounded linear map, still denoted by tr^Γ,

$$tr^\Gamma : \overset{\circ}{H}{}^1(\Omega) \to H^{-1}(\Omega).$$

Let $-\Delta$ be the Dirichlet Laplacian in Ω, set

$$B = (-\Delta)^{-1} \circ tr^\Gamma$$

and equip $\overset{\circ}{H}{}^1(\Omega)$ with the inner product

$$(f,g)_{\overset{\circ}{H}{}^1(\Omega)} = \sum_{j=1}^{n} \int_\Omega \frac{\partial f}{\partial x_j}(x)\frac{\partial \overline{g}}{\partial x_j}(x)dx.$$

Then it turns out that B is a non-negative, compact, self-adjoint operator acting in $\overset{\circ}{H}{}^1(\Omega)$, with

$$\left\|\sqrt{B}f \mid \overset{\circ}{H}{}^1(\Omega)\right\| = \|f \mid L_2(\Gamma)\|, \quad f \in \overset{\circ}{H}{}^1(\Omega),$$

and

$$\ker(B) = \left\{ \quad f \in \overset{\circ}{H}{}^1(\Omega) : tr_\Gamma f = 0 \right\}.$$

In fact, Triebel (see [39]) has proved the following striking result:

Theorem 6. *Let Ω be a bounded C^∞ domain in $\mathbf{R}^n$ and let $\Gamma \subset \Omega$ be a compact d-set for some $d \in (n-2, n)$ ($d \in (0,1)$ if $n = 1$). Then $B = (-\Delta)^{-1} \circ tr^\Gamma$ is a compact, non-negative self-adjoint operator acting in $\overset{\circ}{H}{}^1(\Omega)$, generated by the quadratic form*

$$(Bf,g)_{\overset{\circ}{H}{}^1(\Omega)} = \int_\Gamma f(\gamma)\overline{g(\gamma)}d\mu(\gamma), \quad f,g \in \overset{\circ}{H}{}^1(\Omega),$$

and with kernel $\overset{\circ}{H}{}^1(\Omega\backslash\Gamma)$. Let $\{\rho_k\}$ be the positive eigenvalues of B, each repeated according to multiplicity and ordered by decreasing magnitude, and let $\{u_k\}$ be the related eigenvalues: $Bu_k = \rho_k u_k$. Then:

(i) *The largest eigenvalue is simple:* $\rho_1 > \rho_2 \geq \rho_3 \geq \dots$. *Moreover,*

$$\rho_k \sim k^{-1+(n-2)/d}, \quad k \in \mathbf{N}.$$

(ii) *The eigenfunctions* u_k *are harmonic in* $\Omega \backslash \Gamma$:

$$\Delta u_k(x) = 0 \quad \textit{if} \quad x \in \Omega \setminus \Gamma.$$

(iii) *The first eigenfunction* u_1 *has no zeros in* Ω :

$$u_1(x) = cu(x) \quad \textit{with} \quad c \in \mathbf{C}, \quad \textit{and} \quad u(x) > 0 \quad \textit{if} \quad x \in \Omega.$$

For a proof and for extensive discussion of this and related matters we refer to [39]. Note that the characterisation of the kernel of B relies on density properties connected to the celebrated 'spectral synthesis' problem to which Hedberg and Netrusov have made notable contributions: see [1] for a full discussion of this topic. Some motivation for the theorem above comes from the fact that the $\eta_k := \rho_k^{-1/2}$ may be regarded as the eigenfrequencies of a vibrating membrane (interpreted as the bounded domain Ω) fixed at its boundary, while all its mass is concentrated on the fractal set Γ. We observe that if $n = 2$, $\eta_k \sim k^{1/2}$ and so the dependence on d is lost: one cannot hear the fractal!

6.3. Semi-linear equations

It is well known that if $1 \leq p < \infty$ and $f \in W_p^1(\mathbf{R}^n)$, then $|f| \in W_p^1(\mathbf{R}^n)$. This is a most useful property of the space $W_p^1(\mathbf{R}^n)$: for example, it is at the heart of the proof (by Kinnunen [21]) that the Hardy-Littlewood maximal operator maps $W_p^1(\mathbf{R}^n)$ boundedly to itself when $1 < p < \infty$. The book by Runst and Sickel [31] contains much material on related questions, in the context of the spaces $B_{p,q}^s(\mathbf{R}^n)$ and $F_{p,q}^s(\mathbf{R}^n)$, but here we draw attention to some recent work of Triebel which is contained in his book [39].

Let $\mathbf{B}_{p,q}^s(\mathbf{R}^n), \mathbf{F}_{p,q}^s(\mathbf{R}^n)$ be the real parts of $B_{p,q}^s(\mathbf{R}^n)$, $F_{p,q}^s(\mathbf{R}^n)$ respectively, and define maps T and T^+ by

$$Tf = |f|, \quad T^+ f = f_+, \quad f_+(x) = \max(f(x), 0)$$

for real functions f. If $A_{p,q}^s(\mathbf{R}^n)$ is either $B_{p,q}^s(\mathbf{R}^n)$ or $F_{p,q}^s(\mathbf{R}^n)$ and $\mathbf{A}_{p,q}^s(\mathbf{R}^n)$ is the real part of $A_{p,q}^s(\mathbf{R}^n)$, we say that $\mathbf{A}_{p,q}^s(\mathbf{R}^n)$ has the *truncation property* if T^+ maps $\mathbf{A}_{p,q}^s(\mathbf{R}^n)$ boundedly to itself. Given any $p \in (0, \infty]$ and any $s \in \mathbf{R}$, the pair $(1/p, s)$ is called a *truncation couple* if, for any $q \in (0, \infty]$, the space $\mathbf{B}_{p,q}^s(\mathbf{R}^n)$ has the truncation property; the family of all truncation couples is denoted by R_n. The following theorem (see [39]) is proved by means of atomic decomposition techniques.

Theorem 7. *Let*

$$\sigma_p = n\left(\frac{1}{p} - 1\right)_+ \quad \textit{and} \quad \sigma_{pq} = \left(\frac{1}{\min(p,q)} - 1\right)_+ .$$

Then:

(i) $R_n = \{(1/p, s) : 0 \leq 1/p < \infty, \sigma_p < s < 1 + 1/p\}$.

(ii) *If either*

$$1 < p < \infty, \quad 0 < q \leq \infty, \quad \sigma_{pq} < s < 1 + 1/p$$

or

$$0 < p \leq 1, \quad 0 < q \leq \infty, \quad \sigma_{pq} < s < 1 + 1/p, \quad s \neq 1/p,$$

then $\mathbf{F}^s_{p,q}(\mathbf{R}^n)$ *has the truncation property.*

From this Triebel is able to prove the following for the spaces $\mathbf{H}^s_p(\mathbf{R}^n)$, which are the real parts of $H^s_p(\mathbf{R}^n)$.

Corollary 8. *If either*

$$1 < p < \infty, \quad 0 \leq s < 1 + 1/p$$

or

$$0 < p \leq 1, \quad \sigma_p < s < 1 + 1/p, \quad s \neq 1/p,$$

then $\mathbf{H}^s_p(\mathbf{R}^n)$ *has the truncation property. If*

$$0 < p < \infty, \quad s \geq 1 + 1/p \quad \text{and} \quad s > \sigma_p,$$

then $\mathbf{H}^s_p(\mathbf{R}^n)$ *does not have the truncation property.*

In addition to these mapping properties of T and T^+, there is the question of what continuity properties these may have, a matter of some importance for the proof of existence of solutions of semilinear integral and differential equations.Valuable information is given by the following theorem (see [39]):

Theorem 9. *Suppose that*

$$0 \leq 1/p < \infty, \quad \sigma_p < s < 1 + 1/p, \quad 0 < q \leq \infty,$$

and let $\mathbf{A}^s_{p,q}(\mathbf{R}^n)$ *be either* $\mathbf{B}^s_{p,q}(\mathbf{R}^n)$ *or* $\mathbf{F}^s_{p,q}(\mathbf{R}^n)$ *(with* $p < \infty$ *in the* $\mathbf{F}$ *case). Then* T *(and hence also* T^+*) is not Lipschitz-continuous on* $\mathbf{A}^s_{p,q}(\mathbf{R}^n)$.

This negative answer to the quest for Lipschitz continuity led Triebel to look for another map with better properties. This is the map Q, defined as follows. Suppose that

$$0 < p \leq \infty, \quad 0 < q \leq \infty, \quad s > \sigma_p,$$

and let f belong to $B^s_{p,q}(\mathbf{R}^n)$ or $H^s_p(\mathbf{R}^n)$ and have the quarkonial decomposition

$$f(x) = \sum_{\beta,\nu,m} \lambda^\beta_{\nu m}(f)(\beta qu)_{\nu m}(x), \quad x \in \mathbf{R}^n,$$

where the $(\beta qu)_{\nu m}$ are $(s,p) - \beta$ quarks. Then Q is defined by

$$(Qf)(x) = \sum_{\beta,\nu,m} \left|\lambda^\beta_{\nu m}(f)\right| (\beta qu)_{\nu m}(x), \quad x \in \mathbf{R}^n.$$

Theorem 10. *Under the above conditions,*

(i) *there is a constant* $c > 0$ *such that for all* $f, g \in B^s_{p,q}(\mathbf{R}^n)$,

$$\|Qf - Qg \mid B^s_{p,q}(\mathbf{R}^n)\| \leq c \|f - g \mid B^s_{p,q}(\mathbf{R}^n)\|;$$

(ii) *if in addition* $p < \infty$, *then there is a constant* $c > 0$ *such that for all* $f, g \in H^s_p(\mathbf{R}^n)$,

$$\|Qf - Qg \mid H^s_p(\mathbf{R}^n)\| \leq c \|f - g \mid H^s_p(\mathbf{R}^n)\|.$$

This can be used to establish the existence of solutions of certain integral and differential equations. For example, Triebel [39] shows that the following theorem holds.

Theorem 11. *Let*

$$1 \leq p \leq \infty, \quad 1 \leq q \leq \infty, \quad 0 < s < 1 + 1/p, \quad \sigma > s$$

and let $K : \mathbf{R}^n \times \mathbf{R}^n \to [0, \infty)$ *be such that*

$$a := \int_{\mathbf{R}^n} \|K(\cdot, y) \mid B^\sigma_{\infty,\infty}(\mathbf{R}^n)\| \, dy < \infty.$$

Then there is a positive number ε *such that if* $a \leq \varepsilon$, *then for all* $h \in \mathbf{B}^s_{p,q}(\mathbf{R}^n)$ *there is a unique* $u \in \mathbf{B}^s_{p,q}(\mathbf{R}^n)$ *with*

$$u(x) = \int_{\mathbf{R}^n} K(x, y) u_+(x - y) dy + h(x), \quad x \in \mathbf{R}^n.$$

The proof uses the properties of Q, the contraction mapping theorem and supersolution techniques.

Semilinear differential equations of the form

$$(-\Delta + id)u(x) = \varepsilon u_+(x) + h(x)$$

can also be handled by means of the Q map: see [39].

I have tried to give an impression of the flood of new and extraordinarily interesting results which Hans is still producing. These represent creativity and dynamism at a level which would be the envy of much younger mathematicians. And there is much yet to come!

References

[1] Adams, D.R. and Hedberg, L.I., *Function spaces and potential theory*, Springer 1996.

[2] Birman, M.S. and Solomyak, M.Z., Piecewise polynomial approximations of functions of the classes W_p^α, Mat. Sb. **73** (115) (1967), 331–355 (Russian). Engl. transl. Math. USSR Sb. (1967), 295–317.

[3] Birman, M.S. and Solomyak, M.Z., Spectral asymptotics of non-smooth elliptic operators I, Trans. Moscow Math. Soc. **27** (1972), 1–52.

[4] Brézis, H. and Wainger, S., A note on limiting cases of Sobolev embeddings, Comm. Partial Diff. Equations **5** (1980), 773–789.

[5] Carl, B., Entropy numbers, s-numbers and eigenvalue problems, J. Functional Anal. **41** (1981), 290–306.

[6] Carl, B. and Triebel, H., Inequalities between eigenvalues, entropy numbers, and related quantities of compact operators in Banach spaces, Math. Ann. **251** (1980), 129–133.

[7] Cobos, F. and Signes, T., On a result of Peetre about interpolation of operator spaces, Publicacions Matemàtiques **44** (2000), 457–481.

[8] Edmunds, D.E. and Netrusov, Yu., Entropy numbers of embeddings of Sobolev spaces in Zygmund spaces, Studia Math. **128** (1998), 71–102.

[9] Edmunds, D.E. and Triebel, H., Entropy numbers and approximation numbers in function spaces, Proc. London Math. Soc. **58** (1989), 137–152.

[10] Edmunds, D.E. and Triebel, H., Entropy numbers and approximation numbers in function spaces, II, Proc. London Math. Soc. **64** (1992), 153–169.

[11] Edmunds, D.E. and Triebel, H., Logarithmic Sobolev spaces and their applications to spectral theory, Proc. London Math. Soc. **71** (1995), 333–371.

[12] Edmunds, D.E. and Triebel, H., *Function spaces, entropy numbers, differential operators*, Cambridge Univ. Press 1996.

[13] Edmunds, D.E. and Triebel, H., Sharp Sobolev embeddings and related Hardy inequalities, Math. Nachr. **207** (1999), 79–92.

[14] Frazier, M. and Jawerth, B., Decomposition of Besov spaces, Indiana Univ. Math. J. **34** (1985), 777–799.

[15] Frazier, M. and Jawerth, B., A discrete transform and decomposition of distribution spaces, J. Functional Anal. **93** (1990), 34–170.

[16] Frazier, M., Jawerth, B. and Weiss, G., *Littlewood-Paley theory and the study of function spaces*, CBMS-AMS Regional Conf. Ser. **79**, 1991.

[17] Hansson, K., Imbedding theorems of Sobolev type in potential theory, Math. Scand. **45** (1979), 77–102.

[18] Haroske, D., Embeddings of some weighted function spaces on $\mathbf{R}^n$; entropy and approximation numbers, An. Univ. Craiova, Ser. Mat. Inform. **24** (1997), 1–44.

[19] Haroske, D. and Triebel, H., Entropy numbers in weighted function spaces and eigenvalue distributions of some degenerate pseudodifferential operators I, Math. Nachr. **167** (1994), 131–156.

[20] Haroske, D. and Triebel, H., Entropy numbers in weighted function spaces and eigenvalue distributions of some degenerate pseudodifferential operators II, Math. Nachr. **168** (1994), 109–137.

[21] J. Kinnunen, The Hardy-Littlewood maximal function of a Sobolev function, Israel J. Math. **100** (1997), 117–124.

[22] Kolmogorov, A.N. and Tikhomirov, V.M., ε-entropy and ε-capacity of sets in functional spaces (Russian), Uspekhi Mat. Nauk **14** (2) (1959), 3-86. Eng. transl. Amer. Math. Soc. Transl. Ser. 2 **17** (1961), 277–364.

[23] Maz'ya, V., *Sobolev spaces*, Springer, Berlin, 1985.

[24] Mityagin, B.S. and Pelczyński, A., Nuclear operators and approximative dimension, Proc. I.C.M., Moscow 1966, 366–372.

[25] Netrusov, Yu., Imbedding theorems of Besov spaces into ideal spaces (Russian), Zap. Naučn. Sem. Leningrad. Otdel. Mat. Inst. Steklov (LOMI) **159** (1987), 69–82. Engl. transl. J. Soviet Math. **47** (1989), 2871–2881.

[26] Netrusov, Yu., Imbedding theorems for Lizorkin-Triebel spaces (Russian), Zap. Naučn. Sem. Leningrad. Otdel. Mat. Inst. Steklov (LOMI) **159** (1987), 103–112. Engl. transl. J. Soviet Math. **47** (1989), 2896–2903.

[27] Netrusov, Yu., Sets of singularities of functions of spaces of Besov and Lizorkin-Triebel type (Russian), Trudy Mat. Inst. Steklov **187** (1989), 162–177. Engl. transl. Proc. Steklov Inst. Math. **187** (1990), 185–203.

[28] Peetre, J., Espaces d'interpolation et théorème de Soboleff, Ann. Inst. Fourier **16** (1966), 279–317.

[29] Pietsch, A., *Operator ideals*, North-Holland, Amsterdam, 1980.

[30] Pohozaev, S., On eigenfunctions of the equation $\Delta u + \lambda f(u) = 0$ (Russian), Dokl. Akad. Nauk SSSR **165** (1965), 36–39.

[31] T. Runst and W. Sickel, *Sobolev spaces of fractional order, Nemytskij operators, and nonlinear partial differential equations,* W. de Gruyter, Berlin, 1996.

[32] R.S. Strichartz, A note on Trudinger's extension of Sobolev's inequality, Indiana Univ. Math. J. **21** (1972), 841–842.

[33] Triebel, H., Interpolationseigenschaften von Entropie- und Durchmesseridealen kompakter Operatoren, Studia Math. **34** (1970), 89–107.

[34] Triebel, H., *Interpolation theory, function spaces, differential operators*, North-Holland, Amsterdam, 1978. 2nd revised edition Leipzig: Barth 1995.

[35] Triebel, H., *Theory of function spaces*, Birkhäuser, Basel, 1983.

[36] Triebel, H., *Theory of function spaces II*, Birkhäuser, Basel, 1992.

[37] Triebel, H., Approximation numbers and entropy numbers of embeddings of fractional Besov-Sobolev spaces in Orlicz spaces, Proc. London Math. Soc. **66** (1993), 589–618.

[38] Triebel, H., *Fractals and spectra*, Birkhäuser, Basel, 1997.

[39] Triebel, H., *The structure of functions*, Birkhäuser, Basel, 2001.

[40] Triebel, H. and Winkelvoss, H., The dimension of a closed subset of $\mathbf{R}^n$ and related function spaces, Acta Math. Hungarica **68** (1995), 117–133.

[41] Triebel, H. and Winkelvoss, H., Intrinsic atomic characterisations of function spaces on domains, Math. Z. **221** (1996), 647–673.

[42] Trudinger, N., On imbeddings into Orlicz spaces and some applications, J. Math. Mech. **17** (1967), 473–483.

[43] Vitushkin, A.G. and Henkin, G.M., Linear superposition of functions (Russian), Uspekhi Mat. Nauk **22** (1) (1967), 77–124. Eng. transl. Russian Math. Surveys **22** (1967), 77–125.

[44] Yudovich, V.I., Some estimates connected with integral operators and with solutions of elliptic equations, Soviet Math. Doklady **2** (1961), 746–749.

D.E. Edmunds
CMAIA
University of Sussex at Brighton
Falmer
Brighton BN1 9QH, United Kingdom
E-mail address: D.E.Edmunds@sussex.ac.uk

Part II

D. Haroske, T. Runst, H.-J. Schmeisser (eds.): Function Spaces, Differential Operators and Nonlinear Analysis. The Hans Triebel Anniversary Volume.

Nonvariational Elliptic Systems via Galerkin Methods

Claudianor O. Alves[1] and Djairo G. de Figueiredo[1]

Dedicated to Prof. Hans Triebel on the occasion of his 65th birthday

Abstract. In this paper we show the existence of solutions for a class of non-variational elliptic systems, via the method of Galerkin, and the use of fixed point theorems in finite dimension.
key words and phrases: Galerkin's methods, lack of compactness, Sobolev embedding.
AMS 1991 Subject Classification: 35J20 35J10 35A15

1. Introduction

In this paper we study the existence of solution for elliptic systems of the form

$$\begin{cases} -\Delta u = au^{\alpha} + f(x,u,v) & \text{in } \Omega \\ -\Delta v = bv^{\beta} + g(x,u,v) & \text{in } \Omega \\ u, v > 0 & \text{in } \Omega \\ u = v = 0 & \text{on } \partial\Omega \end{cases} \tag{P}$$

where $\Omega \subset \mathbb{R}^N$ $(N \geq 2)$ is a bounded domain with smooth boundary, and a, b, α and β are positive constants with either $\alpha < 1$ or $\beta < 1$, or both less than 1. The nonlinear terms $f, g : \Omega \times \mathbb{R}^+ \times \mathbb{R}^+ \to \mathbb{R}^+$ are Lipschitz continuous functions satisfying the following conditions:

$$f(x,0,0) = g(x,0,0) = 0 \tag{H_0}$$

$$f(x,0,v) = 0 \Leftrightarrow v = 0 \text{ and } g(x,u,0) = 0 \Leftrightarrow u = 0 \tag{H_1}$$

We remark that, as it stands, system (P) is not variational, neither gradient nor Hamiltonian; for these concepts see the survey papers [5, 6], and extensive references appearing there.

Non-variational elliptic systems have been treated by topological arguments. And in such cases, one needs a priori bounds for the solutions of the system. We remark that our aim is to study system (P) under the above assumptions and rather mild conditions at infinity. So our main conditions are of a local character near zero. Under such conditions, it is improbable that a priori bounds will exist.

[1] Partially supported by CNPq and PRONEX

Since, under the present assumptions, neither variational nor topological methods apply to system (P), we have opted to using Galerkin approximations. The difficulties are then overcome, because we work in subspaces of finite-dimensional, and we can prove that the estimates on the approximate solutions found are independent of the dimensions of the subspaces.

We emphasize that in the literature there exist few results when α or β belongs to $(0,1)$ (see [12]), or when f and g have a supercritical growth. These facts yield serious difficulties because in general we cannot use variational or topological techniques.

This paper is organized as follows. In Sections 2 and 3 we consider the situations when f and g can be subcritical or critical nonlinearities. More precisely, we assume the following hypothesis:

$$|f(x,u,v)| \leq K_1\left(|v|^p+|u|^q\right) \text{ and } |g(x,u,v)| \leq K_2\left(|v|^p+|u|^q\right) \qquad (H_2)$$

with $p,q \in [1,\frac{N+2}{N-2}]$, for $N \geq 3$.

In Section 2, we prove a result on an approximate elliptic system related to (P) and in Section 3 we establish the following result:

Theorem 1. *Assume* (H_0)–(H_2). *Then,*

i) *If* $\alpha,\beta \in (0,1)$, *there exist positive numbers* a^* *and* b^* *such that* (P) *has a solution when* $(a,b) \in (0,a^*) \times (0,b^*)$.

ii) *If* $\alpha \in (0,1)$ *and* $\beta > 1$, *there exists* $a^* > 0$ *such that* (P) *has a solution when* $(a,b) \in (0,a^*) \times (0,\infty)$.

iii) *If* $\beta \in (0,1)$ *and* $\alpha > 1$, *there exists* $b^* > 0$ *such that* (P) *has a solution when* $(a,b) \in (0,\infty) \times (0,b^*)$.

In Section 4, we consider a situation when f and g have supercritical growth, always with $N \geq 3$. We consider there the special case when

$$f(x,u,v) = f(v) \text{ and } g(x,u,v) = g(u) \qquad (H_3)$$

and assume that

$$\liminf_{t\to+\infty} \frac{f(t)}{t^r} = +\infty \quad \text{and} \quad \liminf_{t\to+\infty} \frac{g(t)}{t^r} = +\infty \ \ \forall r \in (1,\frac{N+2}{N-2}). \qquad (H_4)$$

We also suppose that for some $\theta,\eta > 0$ one has

$$\limsup_{t\to+\infty} \frac{f(t)}{t^{2^*-1+\theta}} < +\infty \quad \text{and} \quad \limsup_{t\to+\infty} \frac{g(t)}{t^{2^*-1+\eta}} < +\infty \qquad (H_5)$$

and that there exist two sequences $(M_n),(T_n)$, with $M_n,T_n \to +\infty$, such that:

$$\frac{f(t)}{t^r} \leq \frac{f(M_n)}{(M_n)^r} \ \text{ if } t \in [0,M_n] \ \forall n \in \mathbb{N} \qquad (H_6)$$

and

$$\frac{g(t)}{t^s} \leq \frac{g(T_n)}{(T_n)^s} \ \text{ if } t \in [0,T_n] \ \forall n \in \mathbb{N} \qquad (H_7)$$

for all $r,s \in (0,\frac{N+2}{N-2})$.

Note that the function $f(t) = t^{2^*-1+\frac{\theta}{2}}$ satisfies the above hypotheses. However, there are more general functions that satisfy the hypotheses above, and moreover they are not necessarily increasing functions. This observation is important, because it shows that our results apply even when the methods of sub- and supersolution cannot be used to find solutions of system (P).

Using special truncations of f and g, as introduced in [16], we prove the following result on the supercritical case, for the system below.

$$\begin{cases} -\Delta u = au^\alpha + f(v) & \text{in } \Omega \\ -\Delta v = bv^\beta + g(u) & \text{in } \Omega \\ u, v > 0 & \text{in } \Omega \\ u = v = 0 & \text{on } \partial\Omega \end{cases} \tag{$P_{\alpha,\beta}$}$$

Theorem 2 *Assume (H_0)–(H_1), (H_3)–(H_7). Then exist a^*, b^*, γ_1 and γ_2 positive numbers such that $(P_{\alpha,\beta})$ has a solution when one of below conditions hold:*

i) *If $\alpha, \beta \in (0,1)$, then $(a,b) \in (0,a^*) \times (0,b^*)$ and $(\theta,\eta) \in (0,\gamma_1) \times (0,\gamma_2)$.*

ii) *If $\alpha \in (0,1)$ and $\beta > 1$, then $(a,b) \in (0,a^*) \times (0,\infty)$ and $(\theta,\eta) \in (0,\gamma_1) \times (0,\gamma_2)$.*

iii) *If $\beta \in (0,1)$ and $\alpha > 1$, then $(a,b) \in (0,\infty) \times (0,b^*)$ and $(\theta,\eta) \in (0,\gamma_1) \times (0,\gamma_2)$.*

2. An approximate elliptic system

In this section we will study the existence of a solution of the following system

$$\begin{cases} -\Delta u = au^\alpha + f(x,u,v) + \lambda\phi & \text{in } \Omega \\ -\Delta u = bv^\beta + g(x,u,v) + \lambda\phi & \text{in } \Omega \\ u, v > 0 & \text{in } \Omega \\ u = v = 0 & \text{on } \partial\Omega \end{cases} \tag{$(P)_A$}$$

where $\phi \in C_0^\infty(\Omega)$ is a fixed positive function and λ is a positive parameter. Here and in the sequel, if $\alpha \in (0,1)$ we denote by w_1 the unique solution of the sublinear problem below

$$\begin{cases} -\Delta u = u^\alpha & \text{in } \Omega \\ u > 0 & \text{in } \Omega \\ u = 0 & \text{on } \partial\Omega \end{cases} \tag{P_α}$$

and if $\beta \in (0,1)$, let us denote by w_2 the unique solution of problem

$$\begin{cases} -\Delta v = v^\beta & \text{in } \Omega \\ v > 0 & \text{in } \Omega \\ v = 0 & \text{on } \partial\Omega\,. \end{cases} \tag{P_β}$$

The existence and uniqueness of w_1 and w_2 was proved in [4] (see also [3]).

To obtain the solutions of $(P)_A$, we apply Galerkin methods together with the Fixed Point Theorem below (see, for instance, [13]).

Proposition *Let $F : \mathbb{R}^k \to \mathbb{R}^k$ be a continuous function such that $\langle F(\xi), \xi \rangle \geq 0$ on $|\xi| = r$. Then, there exists $z_0 \in \overline{B}_r(0)$ such that $F(z_0) = 0$.*

The main result in this section is the following theorem.

Theorem 3. *Assume (H_0)–(H_1). Then,*

i) *If $\alpha, \beta \in (0,1)$, there exist positive numbers a^*, b^* and λ^* such that $(P)_A$ has a solution when $(a, b, \lambda) \in (0, a^*) \times (0, b^*) \times (0, \lambda^*)$.*

ii) *If $\alpha \in (0,1)$ and $\beta > 1$, there exist $a^*, \lambda^* > 0$ such that $(P)_A$ has a solution when $(a, b, \lambda) \in (0, a^*) \times (0, \infty) \times (0, \lambda^*)$.*

iii) *If $\beta \in (0,1)$ and $\alpha > 1$, there exist $b^*, \lambda^* > 0$ such that $(P)_A$ has a solution when $(a, b, \lambda^*) \in (0, \infty) \times (0, b^*) \times (0, \lambda^*)$.*

Moreover, if (u_λ, v_λ) is the solution obtained, one of the inequalities below holds:

$$u_\lambda \geq a^{\frac{1}{1-\alpha}} w_1 \ \text{ or } \ v_\lambda \geq b^{\frac{1}{1-\beta}} w_2 \ \text{ on } \ \overline{\Omega}\,.$$

Proof of Theorem 3. We shall prove only item i) of the theorem. Similar arguments can be used to prove items ii) and iii).

Let $\Sigma = \{e_1, \ldots, e_n, \ldots\}$ be an orthonormal basis of the Hilbert space $H_0^1(\Omega)$. For each $m \in \mathbb{N}$ define the subspace

$$V_m = span\{e_1, \ldots, e_m\}.$$

It is well known that $(V_m, \|\ \|)$ and $(\mathbb{R}^m, |\ |)$ are isometrically isomorphic by the natural linear map $T : V_m \to \mathbb{R}^m$ given by

$$v = \sum_{i=1}^{m} \xi_i e_i \to T(v) = \xi = (\xi_1, \xi_2, \ldots, \xi_m).$$

So

$$\|v\| = |T(v)| = |\xi|$$

where $|\ |$ and $\|\ \|$ denote the usual norms in $\mathbb{R}^N$ and $H_0^1(\Omega)$, respectively.

For each $m \in \mathbb{N}$ we will show that there exist positive numbers a^*, b^* and λ_*, independent of m, such that for all $(a, b, \lambda) \in (0, a^*) \times (0, b^*) \times (0, \lambda_*)$, there exists $(u_m, v_m) \in V_m \times V_m$ verifying the equalities below

$$\int_\Omega \nabla u_m \nabla e_i dx = a \int_\Omega (u_{m+})^\alpha e_i dx + \int_\Omega f(x, u_{m+}, v_{m+}) e_i dx + \lambda \int_\Omega \phi e_i dx$$

and

$$\int_\Omega \nabla v_m \nabla e_i dx = b \int_\Omega (v_{m+})^\beta e_i dx + \int_\Omega g(x, u_{m+}, v_{m+}) e_i dx + \lambda \int_\Omega \phi e_i dx,$$

for $i = 1, \ldots, m$. Here $u_+ = \max\{u, 0\}$.

Consider the following function $F : \mathbb{R}^{2m} \to \mathbb{R}^{2m}$ given by

$$F(\eta, \xi) = (F_1(\eta, \xi), \ldots, F_m(\eta, \xi), G_1(\eta, \xi), \ldots, G_m(\eta, \xi))$$

where

$$F_i(\eta,\xi) = \int_\Omega \nabla u \nabla e_i dx - a\int_\Omega (u_+)^\alpha e_i dx - \int_\Omega f(x,u_+,v_+)e_i dx - \lambda \int_\Omega \phi e_i dx$$

and

$$G_i(\eta,\xi) = \int_\Omega \nabla v \nabla e_i dx - b\int_\Omega (v_+)^\beta e_i dx - \int_\Omega g(x,u_+,v_+)e_i dx - \lambda \int_\Omega \phi e_i dx$$

where in the above definitions we are using the identifications

$$\eta \mapsto u = \sum_{i=1}^m \eta_i e_i \quad \text{and} \quad \xi \mapsto \sum_{i=1}^m \xi_i e_i = v.$$

Note that

$$\begin{aligned}
&\langle F(\eta,\xi),(\eta,\xi)\rangle \\
&= \int_\Omega |\nabla u|^2 dx - a\int_\Omega (u_+)^{\alpha+1}dx - \int_\Omega f(x,u_+,v_+)udx - \lambda\int_\Omega \phi u dx \\
&\quad + \int_\Omega |\nabla v|^2 dx - b\int_\Omega (v_+)^{\beta+1}dx - \int_\Omega g(x,u_+,v_+)vdx - \lambda\int_\Omega \phi v dx
\end{aligned}$$

which implies

$$\begin{aligned}
\langle F(\eta,\xi),(\eta,\xi)\rangle \geq\ & \|u\|^2 - ac_1\|u\|^{\alpha+1} - c_2(\|u\|^p + \|v\|^q)\|u\| - \lambda c_3\|u\| + \|v\|^2 \\
& - c_4(\|u\|^p + \|v\|^q)\|v\| - bc_5\|v\|^{\beta+1} - \lambda c_6\|v\|.
\end{aligned}$$

Denoting by

$$\|(u,v)\|^2 = \|u\|^2 + \|v\|^2$$

we obtain

$$\begin{aligned}
\langle F(\eta,\xi),(\eta,\xi)\rangle \geq\ & \|(u,v)\|^2 - ac_6\|(u,v)\|^{\alpha+1} - c_7\|(u,v)\|^{p+1} - c_8\|(u,v)\|^{q+1} \\
& - bc_9\|(u,v)\|^{\beta+1} - \lambda c_{10}\|(u,v)\|.
\end{aligned}$$

Therefore, there exist $a^*, b^*, \lambda^*, \rho$ and $r > 0$, such that

$$\langle F(\eta,\xi),(\eta,\xi)\rangle \geq r > 0 \ \text{ on } \ \|(u,v)\| = \rho \ \ (\rho \text{ is independent of } m)$$

for all $(a,b,\lambda) \in (0,a^*) \times (0,b^*) \times (0,\lambda^*]$.

It follows from the proposition that, for each $m \in \mathbb{N}$ there exists $(u_m, v_m) \in V_m \times V_m$ verifying

$$F(u_m,v_m) = (0,0), \quad \|(u_m,v_m)\| \leq \rho.$$

This means

$$\int_\Omega \nabla u_m \nabla w dx = a\int_\Omega (u_{m_+})^\alpha w dx + \int_\Omega f(x,u_{m_+},v_{m_+})wdx + \lambda\int_\Omega \phi w dx \ \ \forall w \in V_m$$

$$\int_\Omega \nabla v_m \nabla w dx = b\int_\Omega (v_{m_+})^\beta w dx + \int_\Omega g(x,u_{m_+},v_{m_+})wdx + \lambda\int_\Omega \phi w dx \ \ \forall w \in V_m$$

with

$$\|u_m\|, \|v_m\| \leq \rho \ \ \forall m \in \mathbb{N}.$$

Let $u, v \in H_0^1(\Omega)$ be the weak limit of $\{u_m\}$ and $\{v_m\}$, respectively. So

$$u_m \rightharpoonup u \text{ and } v_m \rightharpoonup v \text{ in } H_0^1(\Omega)$$

and

$$u_m(x) \to u(x) \text{ and } v_m(x) \to v(x) \text{ a.e in } \Omega.$$

Considering $w \in V_k, \psi \in V_k$ and $m \geq k$ we have

$$\int_\Omega \nabla u_m \nabla w dx = a \int_\Omega (u_{m_+})^\alpha w dx + \int_\Omega f(x, u_{m_+}, v_{m_+}) w dx + \lambda \int_\Omega \phi w dx$$

and

$$\int_\Omega \nabla v_m \nabla \psi dx = b \int_\Omega (v_{m_+})^\beta \psi dx + \int_\Omega g(x, u_{m_+}, v_{m_+}) \psi dx + \lambda \int_\Omega \phi \psi dx.$$

Then taking the limits as $m \to \infty$, we obtain

$$\int_\Omega \nabla u \nabla w dx = a \int_\Omega (u_+)^\alpha w dx + \int_\Omega f(x, u_+, v_+) w dx + \lambda \int_\Omega \phi w dx \;\; \forall w \in V_k \quad (2)$$

and

$$\int_\Omega \nabla v \nabla \psi dx = b \int_\Omega (v_+)^\beta \psi dx + \int_\Omega g(x, u_+, v_+) \psi dx + \lambda \int_\Omega \phi \psi dx \quad \forall \psi \in V_k. \quad (3)$$

Since any $W \in H_0^1$ can be approximated by a sequence (w_k) with $w_k \in W_k$, we obtain from (2) and (3) that

$$\int_\Omega \nabla u \nabla W dx = a \int_\Omega (u_+)^\alpha W dx + \int_\Omega f(x, u_+, v_+) W dx + \lambda \int_\Omega \phi W dx$$

and by similar argument we have for $U \in H_0^1$

$$\int_\Omega \nabla v \nabla U dx = b \int_\Omega (v_+)^\beta U dx + \int_\Omega g(x, u_+, v_+) U dx + \lambda \int_\Omega \phi U dx.$$

Consequently (u, v) is a solution of the system

$$\begin{cases} -\Delta u = a u_+^\alpha + f(x, u_+, v_+) + \lambda \phi & \text{in } \Omega \\ -\Delta v = b v_+^\beta + g(x, u_+, v_+) + \lambda \phi & \text{in } \Omega \\ u = v = 0 & \text{on } \partial\Omega\,. \end{cases}$$

Recalling that $\lambda > 0$ and $\phi \geq 0$, it follows from the Maximum Principle that $u, v > 0$ in Ω. Moreover, we also have

$$-\Delta u \geq a u^\alpha \text{ and } -\Delta v \geq b v^\beta.$$

Using a result from [3], we conclude that $u \geq a^{\frac{1}{1-\alpha}} w_1$ and $v \geq b^{\frac{1}{1-\beta}} w_2$ in $\overline{\Omega}$. □

3. Proof of Theorem 1

For each $\lambda \in (0, \lambda_*)$, let us consider the solution (u_λ, v_λ) of $(P)_\lambda$ given by Theorem 3. So

$$\text{either } u_\lambda \geq a^{\frac{1}{1-\alpha}} w_1, \text{ or } v_\lambda \geq b^{\frac{1}{1-\beta}} w_2 \text{ in } \Omega.$$

Moreover, we also have

$$\|u_\lambda\|, \|v_\lambda\| \leq \rho \ \forall \lambda \in (0, \lambda_*).$$

If u and v are the weak limit of $\{u_\lambda\}$ and $\{v_\lambda\}$, respectively, we get

$$\text{either } u(x) \geq a^{\frac{1}{1-\alpha}} w_1(x) \text{ or } v(x) \geq b^{\frac{1}{1-\beta}} w_2(x) \quad \text{a.e in } \Omega$$

and thanks to the Sobolev embedding and the Lebesgue Theorem we can conclude by taking the limits as $\lambda \to 0$ that

$$\begin{cases} -\Delta u = au^\alpha + f(x, u, v) & \text{in } \Omega \\ -\Delta v = g(x, u, v) + bv^\beta & \text{in } \Omega \\ u, v > 0 & \text{in } \Omega \\ u = v = 0 & \text{on } \partial\Omega \,. \end{cases}$$

4. A supercritical case

In this section, we work with the situation when the functions f and g have supercritical growth. We study the existence of solution for the following elliptic system

$$\begin{cases} -\Delta u = au^\alpha + f(v) & \text{in } \Omega \\ -\Delta v = bv^\beta + g(u) & \text{in } \Omega \\ u, v > 0 & \text{in } \Omega \\ u = v = 0 & \text{on } \partial\Omega \end{cases} \qquad (P_{\alpha,\beta})$$

where either α or β belongs to $(0, 1)$ and f and g verify the hypotheses (H_0)–(H_1) and (H_3)–(H_7) given in the introduction.

Proof of Theorem 2. As in Section 2, we consider only the case $\alpha, \beta \in (0, 1)$. We begin our proof, considering an auxiliary system, obtained by using truncations of the nonlinear functions f and g. Let us denote by f_n and g_n the functions defined by

$$f_n(t) = \begin{cases} 0, & t \leq 0 \\ f(t), & 0 \leq t \leq M_n \\ \frac{f(M_n)}{(M_n)^r} t^r, & t \geq M_n \end{cases}$$

and

$$g_n(t) = \begin{cases} 0, & t \leq 0 \\ g(t), & 0 \leq t \leq T_n \\ \frac{g(T_n)}{(T_n)^s} t^s, & t \geq T_n \end{cases}$$

where $r, s \in (1, \frac{N+2}{N-2})$ verifying $(2^* - 1) - r < \theta$ and $(2^* - 1) - s < \theta$. The functions f_n and g_n were defined in [16] to study a scalar equation. Using the functions f_n and g_n we consider the system

$$\begin{cases} -\Delta u = au^\alpha + f_n(v) & \text{in} \quad \Omega \\ -\Delta v = bv^\beta + g_n(u) & \text{in} \quad \Omega \\ u, v > 0 & \text{in} \quad \Omega \\ u = v = 0 & \text{on} \quad \partial\Omega. \end{cases} \tag{P_n}$$

Our aim is to find a solution (u_n, v_n) of (P_n) with the property that, for large n,

$$\|u_n\|_{L^\infty(\Omega)} \leq M_n \quad \text{and} \quad \|v_n\|_{L^\infty(\Omega)} \leq M_n.$$

If this is done, we can conclude that (u_n, v_n) is a solution to $(P_{\alpha,\beta})$.

To obtain a solution of (P_n), we use the same arguments employed in Section 2. Namely, we consider the approximate system

$$\begin{cases} -\Delta u = au^\alpha + f_n(v) + \lambda\phi & \text{in} \quad \Omega \\ -\Delta v = bv^\beta + g_n(u) + \lambda\phi & \text{in} \quad \Omega \\ u, v > 0 & \text{in} \quad \Omega \\ u = v = 0 & \text{on} \quad \partial\Omega. \end{cases} \tag{$P_{\lambda,n}$}$$

Observing that f_n and g_n have the following growths

$$|f_n(v)| \leq M_n^{2\theta}|v|^r \quad \text{and} \quad |g_n(u)| \leq T_n^{2\eta}|u|^s$$

we obtain by direct computations, that there exist a_n^*, b_n^* and λ_n^* positive numbers depending of M_n and T_n such that $(P_{\lambda,n})$ has a solution $(u_{\lambda,n}, v_{\lambda,n})$ for $(a, b, \lambda) \in (0, a_n^*) \times (0, b_n^*) \times (0, \lambda_n^*)$. Moreover, we have the estimate

$$\|u_{\lambda,n}\|, \|v_{\lambda,n}\| \leq 1 \;\; \forall\, n \geq n_0 \quad \text{for some } n_0.$$

Taking the limits as $\lambda \to 0$ we conclude that the weak limit of $(u_{\lambda,n}, v_{\lambda,n})$, call it (u_n, v_n), satisfies

$$\begin{cases} -\Delta u_n = au_n^\alpha + f_n(v_n) & \text{in} \quad \Omega \\ -\Delta v_n = bv_n^\beta + g_n(u_n) & \text{in} \quad \Omega \\ u_n, v_n > 0 & \text{in} \quad \Omega \\ u_n = v_n = 0 & \text{on} \quad \partial\Omega\,. \end{cases}$$

Now, we will show that (u_n, v_n) is a solution of $(P_{\alpha,\beta})$ for large n. In order to simplify the notations, we denote the solution (u_n, v_n) by (u, v) and by $\widehat{f}$ and $\widehat{g}$ the following functions

$$\widehat{f}(u, v) = au^\alpha + f_n(v) \quad \text{and} \quad \widehat{g}(u, v) = bv^\beta + g_n(v).$$

It is obvious that $\widehat{f}$ and $\widehat{g}$ verify the following inequalities

$$|\widehat{f}(u, v)| \leq 1 + |u|^r + (M_n)^{2\theta}|v|^r$$

and

$$|\widehat{g}(u, v)| \leq 1 + |v|^s + (T_n)^{2\eta}|u|^s.$$

Note that $\widehat{f} \in L_{loc}^{\frac{2^*}{r}}(\Omega)$ and $\widehat{g} \in L_{loc}^{\frac{2^*}{s}}(\Omega)$, thus

$$u \in W_{loc}^{2,\frac{2^*}{r}}(\Omega) \text{ and } v \in W_{loc}^{2,\frac{2^*}{s}}(\Omega)$$

with

$$\|u\|_{W^{2,\frac{2^*}{r}}(\Omega)} \leq \widehat{C}_1\{|\widehat{f}|_{L^{\frac{2^*}{r}}(\Omega)} + |u|_{L^{\frac{2^*}{r}}(\Omega)}\}$$

and

$$\|v\|_{W^{2,\frac{2^*}{s}}(\Omega)} \leq \widehat{C}_1\{|\widehat{g}|_{L^{\frac{2^*}{s}}(\Omega)} + |v|_{L^{\frac{2^*}{s}}(\Omega)}\}.$$

Using the Sobolev embedding and the fact that the norms of u and v in $H_0^1(\Omega)$ are bounded, we find for large n

$$\|u\|_{W^{2,\frac{2^*}{r}}(\Omega)} \leq \widehat{C}_3(M_n)^{2\theta} \text{ and } \|v\|_{W^{2,\frac{2^*}{s}}(\Omega)} \leq \widehat{C}_3(T_n)^{2\eta}.$$

In what follows, we work only with the first equation in $(P_{\alpha,\beta})$ in order to find an estimate for the solution u. Similar arguments can be applied to the second equation in order to obtain an estimate of v.

Note that if $\frac{2^*}{r} > \frac{N}{2}$, we get

$$W^{2,\frac{2^*}{r}}(\Omega) \hookrightarrow L^\infty(\Omega)$$

and so,

$$\|u\|_{L^\infty(\Omega)} \leq \widehat{C}_4(M_n)^{2\theta} \quad \forall\, x \in \Omega.$$

If θ such that $2\theta < 1$, for large M_n we have $\widehat{C}_4(M_n)^{2\theta} \leq M_n$, hence

$$\|u\|_{L^\infty(\Omega)} \leq M_n.$$

Repeating the same arguments and choosing s such that $2\eta < 1$, we also have for large T_n

$$\|v\|_{L^\infty(\Omega)} \leq T_n$$

thus (u, v) is a solution for (P) with $\gamma_1 = \gamma_2 = \frac{1}{2}$.

If $\frac{2^*}{r} = \frac{N}{2}$, we get

$$W^{2,\frac{2^*}{r}}(\Omega) \hookrightarrow L^t(\Omega) \quad \forall t \in [1, \infty).$$

Fixing $t > \frac{N}{2}r$, it follows that $\widehat{f} \in L^{\frac{t}{r}}(\Omega)$, and consequently $u \in W^{2,\frac{t}{r}}(\Omega)$. Moreover,

$$\|u\|_{W^{2,\frac{t}{r}}(\Omega)} \leq \widehat{C}_5\{|\widehat{f}|_{L^{\frac{t}{r}}(\Omega)} + |u|_{L^{\frac{t}{r}}(\Omega)}\}.$$

Using the above estimates, we find

$$\|u\|_{W^{2,\frac{t}{r}}(\Omega)} \leq \widehat{C}_5\{(M_n)^{(2\theta)(1+r)} + (M_n)^{2\theta}\}.$$

If r and θ such that $2\theta < \frac{1}{1+r}$, we get for large M_n

$$\|u\|_{W^{2,\frac{t}{r}}(\Omega)} \leq M_n$$

so,

$$\|u\|_{L^\infty(\Omega)} \leq M_n.$$

Using the same type of arguments, if s and 2η such that $2\eta < \frac{1}{1+s}$, and we get

$$\|v\|_{L^\infty(\Omega)} \leq T_n$$

Thus (u, v) is a solution to $(P_{\alpha,\beta})$.

Note that in this case we have $(\theta, \eta) \in (0, \frac{1}{2(1+r)}) \times (0, \frac{1}{2(1+s)})$. If $\frac{2^*}{r} < \frac{N}{2}$, we get

$$W^{2,\frac{2^*}{r}}(\Omega) \hookrightarrow L^{p_1}(\Omega) \text{ with } \frac{1}{p_1} = \frac{r}{2^*} - \frac{2}{N},$$

thus $u, v \in L^{p_1}(\Omega)$ that implies $\widehat{f}(u, v), \widehat{g}(u, v) \in L^{p_1}(\Omega)$.

Consequently $u, v \in W^{2,p_1}(\Omega)$ with

$$\|u\|_{W^{2,p_1}(\Omega)} \leq C\left\{|\widehat{f}|_{L^{p_1}(\Omega)} + |u|_{L^{p_1}(\Omega)}\right\}$$

and

$$\|v\|_{W^{2,p_1}(\Omega)} \leq C\left\{|\widehat{g}|_{L^{p_1}(\Omega)} + |v|_{L^{p_1}(\Omega)}\right\}.$$

Repeating the above arguments, we study again the cases $\frac{p_1}{r} > \frac{N}{2}$, $\frac{p_1}{r} = \frac{N}{2}$ and $\frac{p_1}{r} < \frac{N}{2}$. After finite steps we will obtain numbers γ_1 and γ_2 depending of the dimension of Ω, that is, $\gamma_1 = \gamma_1(N)$ and $\gamma_2 = \gamma_2(N)$ such that $\|u\|_\infty \leq M_n$ and $\|v\|_\infty \leq T_n$ for large n. □

Acknowledgement. The first author would like to acknowledge the hospitality given to him during his visit to IMECC – UNICAMP.

References

[1] C.O. Alves and D.G. de Figueiredo, *Nonvariational Elliptic Systems* (to appear in Discrete and Continuous Dynamical System).

[2] C.O. Alves, D.C. de Morais Filho and M.A. Souto, *On systems of elliptic equations involving subcritical or critical Sobolev exponents*, Nonlinear Analysis 42 (2000), 771–787.

[3] A. Ambrosetti, H. Brezis and G. Cerami, *Combined effects of concave and convex nonlinearities in some elliptic problems*, J. Funct. Anal. 122 (1994), 519–543.

[4] H. Brezis, and L. Oswald, *Remarks on sublinear elliptic equations*, Nonlinear Analysis TMA. 10 (1986), 55–64.

[5] D.G. de Figueiredo, *Semilinear elliptic systems*, Proceedings of the Second School on Nonlinear Functional Analysis and Applications to Differential Equations, ICTP Trieste 1997, World Scientific Publishing Company (1998), Editors A. Ambrosetti, K.-C. Chang, I. Ekeland, 122–152.

[6] D.G. de Figueiredo, *Semilinear elliptic systems: A survey of superlinear problems.* Resenhas IME-USP, vol. 2, No.4 (1996), 373–391.

[7] D.G. de Figueiredo and P.L. Lions, *On pairs of positive solutions for a class of semilinear elliptic problems*, Indiana Univ. Math. J., 3 (1985), 591–606.

[8] D.G. de Figueiredo, P.L. Lions and R. Nussbaum, *A priori estimates and existence of positive solutions of semilinear elliptic equations.* J. Math. Pures et Appl. 61 (1982), 41–63.

[9] D.G. de Figueiredo and P. Felmer, *On superquadratic elliptic systems*, Trans. Amer. Math. Soc. 343 (1994), 99–116.

[10] D.G. de Figueiredo and J. Yang, *A priori bounds for positive solutions of a nonvariational elliptic system* (Preprint).

[11] D.C. de Morais Filho, *An Ambrosetti-Prodi type problem form an elliptic system of equations via Monotone Iteration methods and Leray-Schauder degree theory*, Abstract and Applied Analysis 01 (1996), 137–152.

[12] P. Felmer and S. Martínez, *Existence and uniqueness of positive solutions to certain differential systems* (Preprint).

[13] S. Kesavan, *Topics in functional analysis and applications*, John Wiley & Sons (1989).

[14] J. Moser, *A sharp form of an inequality by N. Trudinger*, Indiana Univ. Math. J., 20 (1971), 1077–1092.

[15] M.A. Souto, *A priori estimates and existence of positive solutions of nonlinear cooperative elliptic systems*, Diff. Int. Eq. 8 (1995), 1245–1258.

[16] P.H. Rabinowitz, *Variational methods for nonlinear elliptic eigenvalue problems*, Indiana Univ. Math. J. 23 (1974), 729–754.

[17] N.S. Trudinger, *On imbeddings into Orlicz spaces and some applications*, J. Math. Mech. 17 (1967), 473–484.

Claudianor O. Alves
Universidade Federal da Paraíba
Departamento de Matemática e Estatística
58109-970 Campina Grande-PB, Brazil
E-mail address: `coalves@dme.ufpb.br`

Djairo G. de Figueiredo
IMECC-UNICAMP
Caixa Postal 6065
13081-970 Campinas-SP, Brazil
E-mail address: `djairo@ime.unicamp.br`

D. Haroske, T. Runst, H.-J. Schmeisser (eds.): Function Spaces, Differential Operators and Nonlinear Analysis. The Hans Triebel Anniversary Volume.

Superposition Operators in Zygmund and BMO Spaces

Gérard Bourdaud

Dedicated to Prof. Hans Triebel on the occasion of his 65th birthday

Abstract. We characterize the functions f for which the superposition operator $T_f(g) := f \circ g$ acts in the Hölder-Zygmund spaces $B^s_{\infty,\infty}(\mathbb{R}^n)$ and in the spaces $BMO(\mathbb{R}^n)$, $VMO(\mathbb{R}^n)$, $CMO(\mathbb{R}^n)$. Further we study the continuity and the differentiability of T_f in those various function spaces.

1. Introduction

If f is a function defined on the real line or the complex plane, the *superposition* operator associated with f is defined by $T_f(g) := f \circ g$. The Superposition Operator Problem (S.O.P.) in a given function space E consists in finding the *acting condition* on the function f, which ensures that $T_f(E) \subseteq E$. If this condition is fulfilled, we can also study the regularity, i.e. the continuity and the differentiability of T_f.

Here our main subject will be the S.O.P. in the Hölder-Zygmund spaces $B^s_{\infty,\infty}(\mathbb{R}^n)$, for $s > 0$, and in $BMO(\mathbb{R}^n)$.

All what follows is the result of a collective work with Massimo Lanza de Cristoforis and Winfried Sickel. The detailed proofs will appear in forthcoming joint papers [5, 6]. In two cases, we give here proofs which seem to be better than the original ones (c.f. paragraphs 6.3 and 7.1). Proposition 1 is a still unpublished result.

2. Superposition in spaces of Sobolev type

First of all, it may be of interest to insert this problem into the more general framework of the S.O.P. in spaces of Sobolev type. Let us denote by $E^s_p(\mathbb{R}^n)$ the Besov space $B^s_{p,q}(\mathbb{R}^n)$ or the Lizorkin-Triebel space $F^s_{p,q}(\mathbb{R}^n)$. Here we assume that $s \in \mathbb{R}$ and $1 \leq p \leq \infty$. We omit the "microscopic regularity parameter" $q \in [1, \infty]$ and we overlook the distinction between B-spaces and F-spaces, since they have a minor influence on the general philosophy of the S.O.P.

It turns out that the main regularity parameter s takes *four* critical values with respect to the S.O.P.: $0,\ 1,\ 1+\frac{1}{p},\ \frac{n}{p}$. In the intervals between these critical values, the operator T_f behaves, roughly speaking, as follows.

(1) For $s < 0$, it is not even defined, except for linear functions f.

(2) For $0 < s < 1$, f acts in $E_p^s(\mathbb{R}^n)$ iff

- f is *locally* Lipschitz continuous in case $E_p^s(\mathbb{R}^n) \subset L^\infty(\mathbb{R}^n)$;
- f is *globally* Lipschitz continuous in case $E_p^s(\mathbb{R}^n) \not\subset L^\infty(\mathbb{R}^n)$.

(see [4] for the proof).

(3) For $0 < s < 1+\frac{1}{p}$, there exists a class $\mathcal{F}$ of "good functions" (including for instance the C^2 functions with integrable second derivative) for which the sublinear estimation

$$\|f \circ g\|_{E_p^s(\mathbb{R}^n)} \leq C(f)\, \|g\|_{E_p^s(\mathbb{R}^n)}$$

holds for all $f \in \mathcal{F}$ and $g \in E_p^s(\mathbb{R}^n)$.

(4) For $1+\frac{1}{p} < s < \frac{n}{p}$, the S.O.P. has only the trivial solution: f linear.

(5) For $s > \frac{n}{p}$, every C^∞ function acts on $E_p^s(\mathbb{R}^n)$.

For the statements (4),(5), we refer to our talk at FSDONA 92 [3] and to the book of Thomas Runst and Winfried Sickel [19].

As usual when a parameter takes a critical value, we can fall on one of both sidesdots or elsewhere. Let us illustrate this idea by several examples.

1. According to Sickel and Triebel [22], we have $B^0_{p,q}(\mathbb{R}^n) \subseteq L^1_{loc}$ iff $q \leq \min(p,2)$. If $q > \min(p,2)$, the operator T_f is not defined on $B^0_{p,q}(\mathbb{R}^n)$, except for linear f's. If $q \leq \min(p,2)$ and $p \neq 2$ (or if $q < 2 = p$), there is no elementary expression for the norm in $B^0_{p,q}(\mathbb{R}^n)$. For this reason, the characterization of acting functions seems to be a difficult problem. Finally $B^0_{2,2}(\mathbb{R}^n)$ coincides with $L^2(\mathbb{R}^n)$, for which the acting condition is

$$\sup_{t\neq 0} \frac{|f(t)|}{|t|} < +\infty\,.$$

We can observe that the above condition does not imply continuity nor, a fortiori, local Lipschitz continuity.

2. The classical Sobolev space $W^{1,p}(\mathbb{R}^n)$ coincides with $F^1_{p,2}(\mathbb{R}^n)$ for $1 < p < +\infty$. It is well known that $W^{1,p}$ behaves as the spaces E^s_p $(0 < s < 1)$ with respect to the S.O.P. [18]. In the Besov spaces $B^1_{p,q}(\mathbb{R}^n)$ (for $p \neq 2$ or $q \neq 2$) and the Lizorkin-Triebel spaces $F^1_{p,q}(\mathbb{R}^n)$ (for $q \neq 2$), the S.O.P. is essentially open. In the following lines, we solve it only for $B^1_{\infty,\infty}(\mathbb{R}^n)$ and we give some partial results for $B^1_{\infty,q}(\mathbb{R}^n)$ $(q < \infty)$ and $F^1_{\infty,2}(\mathbb{R}^n)$.

3. For $1+\frac{1}{p} < \frac{n}{p}$ and $q > 1$, the linear functions are the only which act in $B^{1+\frac{1}{p}}_{p,q}(\mathbb{R}^n)$. The existence of a nonlinear function acting in $B^{1+\frac{1}{p}}_{p,1}(\mathbb{R}^n)$, for $1+\frac{1}{p} < \frac{n}{p}$, is still an open question.

4. Since $E_p^{n/p}(\mathbb{R}^n)$ is not an algebra (except for $q = 1$ in the Besov case and for $p = 1$ in the Lizorkin-Triebel case), we see that the function $f(t) := t^2$ does not act in $E_p^{n/p}(\mathbb{R}^n)$. Nevertheless every C^∞ function, with *bounded derivatives* of all orders, acts on $E_p^{n/p}(\mathbb{R}^n)$.

We shall end this discussion with two conjectures.

Conjecture 1. *Assume that* $s > 1 + \frac{1}{p}$ *and that* $E_p^s(\mathbb{R}^n) \subseteq L^\infty$. *Then a function* f *acts in* $E_p^s(\mathbb{R}^n)$ *iff it belongs locally to* $E_p^s(\mathbb{R})$.

Conjecture 2. *Assume that* $\frac{n}{p} > 1 + \frac{1}{p}$ *and that* $E_p^{n/p}(\mathbb{R}^n) \nsubseteq L^\infty$. *Then a function* f *acts in* $E_p^s(\mathbb{R}^n)$ *iff it belongs locally-uniformly to* $E_p^{n/p}(\mathbb{R})$.

These conjectures are satisfied by the classical Sobolev spaces [2] and by the Hölder-Zygmund spaces $B^s_{\infty,q}(\mathbb{R}^n)$, as we are going to see.

3. Acting conditions

Theorem 1. *Let* $s > 0$, $s \neq 1$, $1 \leq q \leq \infty$. *Then* f *acts in* $B^s_{\infty,q}(\mathbb{R}^n)$ *iff the following condition holds, respectively:*

(i) f *is locally Lipschitz continuous, for* $0 < s < 1$.

(ii) f *belongs locally to* $B^s_{\infty,q}(\mathbb{R})$, *for* $s > 1$.

For $s = 1$, our characterization is less complete.

Theorem 2. *The function* f *acts in* $B^1_{\infty,1}(\mathbb{R}^n)$ *iff it belongs locally to* $B^1_{\infty,1}(\mathbb{R})$.

Theorem 3. *The function* f *acts in* $B^1_{\infty,\infty}(\mathbb{R}^n)$ *iff it is locally Lipschitz continuous and satisfies the condition*

$$f(x+t) + f(x-t) - 2f(x) = O\left(\frac{t}{|\log t|}\right),$$

as $t \to 0+$, *uniformly on each compact subset of* $\mathbb{R}$.

Remark. We can prove that any f which acts in $B^1_{\infty,\infty}(\mathbb{R}^n)$ acts in $B^1_{\infty,q}(\mathbb{R}^n)$ $(q > 1)$ too. But we do not know if the reciprocal is true.

In contrast with the case of Hölder-Zygmund spaces, very few is known on the S.O.P. for the spaces $F^s_{\infty,q}(\mathbb{R}^n)$.

Theorem 4. *Let* $0 < s < 1$, $1 \leq q \leq \infty$. *Then the function* f *acts in* $F^s_{\infty,q}(\mathbb{R}^n)$ *iff it is locally Lipschitz continuous.*

Proposition 1. *Let* f *be a differentiable function on* $\mathbb{R}$ *such that*

$$f'(x+t) - f'(x) = O\left(\frac{1}{|\log|t||}\right),$$

as $t \to 0$, *uniformly on each compact subset of* $\mathbb{R}$. *Then* f *acts in* $F^1_{\infty,2}(\mathbb{R}^n)$.

We now turn to the spaces $F^0_{\infty,q}(\mathbb{R}^n)$. If $q > 2$, then $F^0_{\infty,q}(\mathbb{R}^n)$ is a space of "true" distributions, on which T_f has no reasonable sense. For $q < 2$, $F^0_{\infty,q}(\mathbb{R}^n)$ is a subspace of $L^2_{loc}(\mathbb{R}^n)$ (in actual fact a subspace of $bmo(\mathbb{R}^n)$, see hereafter), but its norm is not easily manageable. On the other hand $F^0_{\infty,2}(\mathbb{R}^n)$ is the local version of $BMO(\mathbb{R}^n)$, introduced by David Goldberg [14] under the name of $bmo(\mathbb{R}^n)$. Then we have a chance to yield a solution of the S.O.P. for $F^0_{\infty,2}(\mathbb{R}^n)$. Indeed we shall characterize acting functions not only in $bmo(\mathbb{R}^n)$ but also in $BMO(\mathbb{R}^n)$ and related spaces.

Let us recall the basic definitions. The classical space $BMO(\mathbb{R}^n)$ is endowed with the norm

$$\|g\|_* := \|g\|_{BMO} + \fint_{Q_0}|g|\,, \quad \text{where} \quad \|g\|_{BMO} := \sup_Q \fint_Q \left|g - \fint_Q g\right| .$$

Here the supremum is taken on all cubes Q with sides parallel to the coordinate axes and we denote by $\fint_Q g$ the mean value of the function g on Q, and by Q_0 the unit cube $[-1/2, +1/2]^n$. The additional mean on Q_0 is needed, since we want $BMO(\mathbb{R}^n)$ to be a Banach space of functions, *not functions modulo constants* as in the usual setting for BMO. We denote by $bmo(\mathbb{R}^n)$ the space of functions $g \in BMO(\mathbb{R}^n)$ which satisfy also the condition

$$\sup_{|Q|=1} \fint_Q |g| < +\infty\,,$$

endowed with the norm

$$\|g\|_{bmo} := \|g\|_{BMO} + \sup_{|Q|=1} \fint_Q |g|\,.$$

We denote by $cmo(\mathbb{R}^n)$ and $CMO(\mathbb{R}^n)$ the respective closures of $\mathcal{D}(\mathbb{R}^n)$ in $bmo(\mathbb{R}^n)$ and $BMO(\mathbb{R}^n)$. Finally we denote by $VMO(\mathbb{R}^n)$ the closure, in $BMO(\mathbb{R}^n)$, of the set of C^∞ functions which belong to BMO with all their derivatives. The subspace $vmo(\mathbb{R}^n)$ is defined in the same way from $bmo(\mathbb{R}^n)$. The spaces VMO and CMO were introduced respectively by Sarason [20] and Coifman-Weiss [9]. (The reader should be aware of the fact that the symbols VMO and CMO are used with different meanings at different places in the literature.)

According to Sarason, a function $g \in BMO(\mathbb{R}^n)$ (resp. $bmo(\mathbb{R}^n)$) belongs to $VMO(\mathbb{R}^n)$ (resp. $vmo(\mathbb{R}^n)$) iff it fulfils the condition

$$\lim_{a\to 0}\left(\sup_{|Q|\le a} \fint_Q \left|g - \fint_Q g\right|\right) = 0\,. \tag{1}$$

Now we are ready to solve the S.O.P. for BMO and its subspaces. We assume, without further reference, that f is a Borel measurable function of $\mathbb{C}$ to itself.

Theorem 5. *The following properties are equivalent.*

(i) $\sup_{x,y\in\mathbb{C}} (1 + |x-y|)^{-1}|f(x) - f(y)| < +\infty$.

(ii) $T_f\,(BMO(\mathbb{R}^n)) \subseteq BMO(\mathbb{R}^n)$.

(iii) $T_f\,(bmo(\mathbb{R}^n)) \subseteq bmo(\mathbb{R}^n)$.
(iv) $T_f\,(cmo(\mathbb{R}^n)) \subseteq BMO(\mathbb{R}^n)$.

Theorem 6. *The following properties are equivalent.*
(i) *f is uniformly continuous.*
(ii) $T_f\,(VMO(\mathbb{R}^n)) \subseteq VMO(\mathbb{R}^n)$.
(iii) $T_f\,(vmo(\mathbb{R}^n)) \subseteq vmo(\mathbb{R}^n)$.
(iv) $T_f\,(cmo(\mathbb{R}^n)) \subseteq VMO(\mathbb{R}^n)$.

Theorem 7. *The following two statements hold.*
(i) *We have $T_f\,(cmo(\mathbb{R}^n)) \subseteq cmo(\mathbb{R}^n)$ iff f is uniformly continuous and $f(0) = 0$.*
(ii) *We have $T_f\,(CMO(\mathbb{R}^n)) \subseteq CMO(\mathbb{R}^n)$ iff f is uniformly continuous.*

4. Regularity of the superposition operator

4.1. Regularity in Hölder-Zygmund spaces

We begin with an important complement to Theorems 1-3. Let Φ_s denote the set of functions which act in $B^s_{\infty,\infty}(\mathbb{R}^n)$. If $f \in \Phi_s$, then T_f is bounded in the following sense: for any $R > 0$, we have

$$N_R(f) := \sup\{\|f \circ g\|_{B^s_{\infty,\infty}(\mathbb{R}^n)} \,:\, \|g\|_{B^s_{\infty,\infty}(\mathbb{R}^n)} \leq R\} < \infty\,.$$

Moreover Φ_s, endowed with the family of seminorms (N_R), becomes a Fréchet space. Of course, for each value of s, we can replace the seminorms N_R by an equivalent set of more concrete seminorms. For example, by Theorem 3, the following seminorms define as well the topology of Φ_1:

$$\gamma_R(f) := \|f'\|_{L^\infty([-R,R])} + \sup_{|x|\leq R, 0<t\leq 1/e} t^{-1}|\log t||f(x+t) + f(x-t) - 2f(x)|\,.$$

We denote by $W^r(\Phi_s)$ the Sobolev space based on Φ_s, i.e. the set of f such that $f^{(j)} \in \Phi_s$ for $j = 0, \ldots, r$. We endow $W^r(\Phi_s)$ with the natural Fréchet structure inherited from Φ_s. Finally we remark that $C^\infty(\mathbb{R})$ is embedded in $W^r(\Phi_s)$ for all $r \in \mathbb{N}$.

Theorem 8. *Let $r \in \mathbb{N}$ and $s > 0$. T_f is a mapping of class C^r from $B^s_{\infty,\infty}(\mathbb{R}^n)$ to itself iff f belongs to the closure of C^∞ into $W^r(\Phi_s)$.*

Thus we obtain the following explicit characterization of regular superposition operators in Hölder-Zygmund spaces, which is well known in case s noninteger and seems to be novel for Zygmund classes.

Corollary 1. *Let $s > 0$ and $r \in \mathbb{N}$. Then the following list provides, for the various values of s, a necessary and sufficient condition for the operator T_f to be of class C^r from $B^s_{\infty,\infty}(\mathbb{R}^n)$ to itself.*

(i) *f is of class C^{r+1}, if $0 < s < 1$.*

(ii) *f is of class C^{r+1} and*

$$f^{(r)}(x+t) + f^{(r)}(x-t) - 2f^{(r)}(x) = o\left(\frac{t}{|\log t|}\right),$$

as $t \to 0+$, uniformly on each compact subset of $\mathbb{R}$, if $s = 1$.

(iii) *f is of class C^{r+m} and*

$$f^{(r+m)}(x+t) - f^{(r+m)}(x) = o\left(|t|^{s-m}\right),$$

as $t \to 0$, uniformly on each compact subset of $\mathbb{R}$, if $m < s < m+1$ for some $m \in \mathbb{N}^*$.

(iv) *f is of class C^{r+s-1} and*

$$f^{(r+s-1)}(x+t) + f^{(r+s-1)}(x-t) - 2f^{(r+s-1)}(x) = o(t),$$

as $t \to 0+$, uniformly on each compact subset of $\mathbb{R}$, if s is integer and $s > 1$.

4.2. The case of BMO and related spaces

The continuity of T_f in the framework of BMO norms depends strongly on the subspace of BMO we consider. Indeed we have the following alternative: either the acting condition implies continuity or there is no nonlinear continuous superposition operator.

Theorem 9. *If f is a uniformly continuous function, then*

(i) *the mapping $T_f : bmo(\mathbb{R}^n) \to bmo(\mathbb{R}^n)$ is continuous at every element of $vmo(\mathbb{R}^n)$,*

(ii) *the mapping $T_f : BMO(\mathbb{R}^n) \to BMO(\mathbb{R}^n)$ is continuous at every element of $CMO(\mathbb{R}^n)$.*

Theorem 10. *T_f is continuous from $bmo(\mathbb{R}^n)$ or $VMO(\mathbb{R}^n)$ to $BMO(\mathbb{R}^n)$ iff f is $\mathbb{R}$-affine.*

We end with the following result, which shows that there is no nonlinear differentiable superposition operator in the BMO framework.

Theorem 11. *T_f is $\mathbb{R}$-differentiable from $\mathcal{D}(\mathbb{R}^n)$, endowed with the norm of $bmo(\mathbb{R}^n)$, to $BMO(\mathbb{R}^n)$ iff f is $\mathbb{R}$-affine.*

5. A fundamental lemma

In the proofs of necessity of acting conditions, we use, most of the time, the following general result.

Definition 1. *Let Ω be an open subset of $\mathbb{R}^n$. Let E be a vector subspace of $L^1_{loc}(\Omega)$. We say that E is a* Banach $\mathcal{D}(\Omega)$-module *if the two following properties hold.*

(i) *E is endowed with a complete norm such that the canonical injection $E \to L^1_{loc}(\Omega)$ is continuous.*

(ii) *For all $g \in E$ and all $\varphi \in \mathcal{D}(\Omega)$, we have $\varphi g \in E$.*

Lemma 1. *Let Ω be an open subset of $\mathbb{R}^n$. Let E, F be two Banach $\mathcal{D}(\Omega)$-modules. Let f be a function such that $f(0) = 0$ and $T_f(E) \subseteq F$. Then there exist a ball $B \subset \Omega$ and two constants $c_1, c_2 > 0$ such that $\|f \circ g\|_F \leq c_2$ for all g with support in B and satisfying $\|g\|_E \leq c_1$.*

The proof is essentially given in [3, Lemma 3] and [4, Lemma 1]. The idea is classical in Harmonic Analysis (cf. *e.g.*, Katznelson [15, ch. VIII, § 8.3]).

6. Sketch of the proofs (Acting conditions)

Let us underline that parts of the above theorems are more or less classical. In case s not integer, $q = \infty$ — i.e. ordinary Hölder spaces —, Theorems 1 and 8 are due to Drábek [11]. Theorem 4 is a straightforward consequence of the characterization of $F^s_{\infty,q}(\mathbb{R}^n)$ by differences in case $0 < s < 1$ (cf. Sickel [21]) and of our previous work [4]. The equivalence between (i) and (ii) in Theorem 5 is due to Fominykh [13] in dimension one and to Chevalier [8] in the general case. In the context of compact manifolds, Brezis and Nirenberg proved the following properties: if f is uniformly continuous, then f acts in VMO and T_f is continuous on CMO [7, Lem. A.7-8, p. 238].

In the sequel, we denote by $\| - \|$ the norm in E^s_p if the context is clear.

6.1. Sufficient conditions in Hölder-Zygmund spaces

The Besov spaces can be easily described by the means of the first and second order moduli of continuity

$$\omega(g;t) := \sup_{|h| \leq t\,,\, x \in \mathbb{R}^n} |g(x+h) - g(x)|\,,$$

$$\eta(g;t) := \sup_{|h| \leq t\,,\, x \in \mathbb{R}^n} |g(x+h) + g(x-h) - 2g(x)|\,.$$

For $0 < s < 2$ and $1 \leq q \leq \infty$, the function g belongs to $B^s_{\infty,q}(\mathbb{R}^n)$ if, for some (and then for all) $a \in]0, +\infty]$, one has

$$\|g\|_\infty + \left(\int_0^a \left(\frac{\eta(g;t)}{t^s} \right)^q \frac{dt}{t} \right)^{1/q} < +\infty\,.$$

For $0 < s < 1$, it is well known that $g \in B^s_{\infty,q}(\mathbb{R}^n)$ iff

$$\|g\|_\infty + \left(\int_0^a \left(\frac{\omega(g;t)}{t^s} \right)^q \frac{dt}{t} \right)^{1/q} < +\infty\,.$$

Of course any of the above expressions delivers an equivalent norm in the corresponding Besov space.

Our main tool is the following.

Proposition 2. *For all functions $f : \mathbb{R} \to \mathbb{R}$ and $g : \mathbb{R}^n \to \mathbb{R}$, the following holds:*

$$\eta(f \circ g; t) \leq \omega(f; \eta(g;t)) + \eta(f; \omega(g;t))\,, \quad \forall t > 0\,.$$

Using the above proposition and standard embeddings between Lipschitz and Besov spaces yields the proofs of Theorems 1 and 2.

Now we consider the case $s = 1$, $1 < q \leq \infty$. Without loss of generality, we assume that f is Lipschitz continuous and satisfies

$$\eta(f;t) = O\left(\frac{t}{|\log t|}\right), \quad \text{as} \quad t \to 0+ . \tag{2}$$

Let us take $g \in B^1_{\infty,q}(\mathbb{R}^n)$ and define

$$\kappa(t) := \frac{\omega(g;t)}{t|\log t|} .$$

Then Marchaud and Hardy inequalities yield

$$\left(\int_0^{1/e} \kappa(t)^q \frac{dt}{t}\right)^{1/q} < +\infty \tag{3}$$

(cf. [1, Theorem 4.4] and [12, Corollary 6.21].) By assumption (2) and by an easy computation, we obtain

$$\frac{1}{t}\,\eta(f;\omega(g;t)) = O\left(\kappa(t)\right) \quad \text{as} \quad t \to 0+ . \tag{4}$$

The Lipschitz continuity of f gives us

$$\frac{1}{t}\omega(f;\eta(g;t)) = O\left(\frac{1}{t}\eta(g;t)\right) \quad \text{as} \quad t \to 0+ . \tag{5}$$

Then we take the $L^q\left(\frac{dt}{t}\right)$-norms of (4) and (5) and we apply the property (3) and Proposition 2. We conclude that $f \circ g \in B^1_{\infty,q}(\mathbb{R}^n)$.

6.2. Sufficient condition in $F^1_{\infty,2}(\mathbb{R}^n)$

Without loss of generality, we assume that f is a differentiable function, with bounded derivative and that

$$\omega(f';t) = O\left(\frac{1}{|\log t|}\right), \quad \text{as} \quad t \to 0+ . \tag{6}$$

Let $g \in F^1_{\infty,2}(\mathbb{R}^n)$. By the embedding $F^1_{\infty,2}(\mathbb{R}^n) \subset B^1_{\infty,\infty}(\mathbb{R}^n)$ and by (3), we have

$$\omega(g;t) = O\left(t\,|\log t|\right), \quad \text{as} \quad t \to 0+ .$$

Then by (6) we obtain

$$\omega(f' \circ g;t) = O\left(\frac{1}{|\log t|}\right), \quad \text{as} \quad t \to 0+ .$$

From the above estimation and from Stegenga's theorem (cf. [24] and [19, 4.9.1, Theorem 2]), we deduce that $f' \circ g$ is a multiplier in $bmo(\mathbb{R}^n)$. Since $F^1_{\infty,2}(\mathbb{R}^n)$ is the Sobolev space $W^1\left(bmo(\mathbb{R}^n)\right)$, we conclude that $f \circ g$ belongs to $F^1_{\infty,2}(\mathbb{R})$.

6.3. Necessary condition in $B^1_{\infty,\infty}(\mathbb{R}^n)$

Let f be a function which acts in $B^1_{\infty,\infty}(\mathbb{R}^n)$. We fix a number a and consider the function $f_a(x) := f(a+x) - f(a)$. We apply Lemma 1 to f_a and $E = F := B^1_{\infty,\infty}(\mathbb{R}^n)$. By dilation and translation invariance of $B^1_{\infty,\infty}(\mathbb{R}^n)$, we can assume that the ball B in Lemma 1 is the unit ball. Let $\varphi \in \mathcal{D}(\mathbb{R}^n)$ with support in B such that $\varphi(x) = 1$ for $|x| \le 1/2$. We observe that the function

$$v(x) := x_1 \log|x|\,\varphi(x) \quad , \forall x = (x_1, x_2, \dots, x_n) \in \mathbb{R}^n$$

belongs to $B^1_{\infty,\infty}(\mathbb{R}^n)$ (Indeed v belongs to the smaller space $F^1_{\infty,2}(\mathbb{R}^n)$.) Let b be a real number such that $|b-a| \le \dfrac{c_1}{2\|\varphi\|}$ and $\varepsilon := \dfrac{c_1}{2\|v\|}$. Then the function

$$g(x) := (b-a)\varphi(x) + \varepsilon v(x)$$

is supported by B and satisfies $\|g\| \le c_1$. From $\|f_a \circ g\| \le c_2$, we deduce

$$|f_a(g(x)) + f_a(g(-x)) - 2f_a(g(0)| \le c_2|x|\,. \tag{7}$$

Let $0 < t < \varepsilon e^{-1}$ and define $s \in]0, e^{-1}[$ by the equality $t = \varepsilon s|\log s|$. If we take $x := (s, 0, \dots, 0)$ in the estimation (7), we obtain

$$|f(b+t) + f(b-t) - 2f(b)| \le c_3 \frac{t}{|\log t|}\,,$$

for some constant c_3.

Remark. The above proof differs slightly from that given in [5]. It has the advantage to be valid as well for spaces in domains.

6.4. Sufficient conditions in BMO and related spaces

A function f satisfies the condition (i) of Theorem 5 iff it is the sum of a bounded function and of a Lipschitz continuous function (see [6, Proposition 1]). Since the properties (ii) and (iii) of Theorem 5 are immediate if f is bounded, we can assume that f is Lipschitz continuous. Denote by C the Lipschitz constant of f. Then we have

$$⨍_Q \left| f\circ g - f\left(⨍_Q g\right)\right| \le C\,\|g\|_{BMO}\,,$$

hence

$$\|f\circ g\|_{BMO} \le 2C\,\|g\|_{BMO}\,, \tag{8}$$

and

$$⨍_Q |f\circ g| \le C(1 + ⨍_Q |g|) + |f(0)|\,.$$

Thus f acts in $BMO(\mathbb{R}^n)$ and in $bmo(\mathbb{R}^n)$.

Now let f be a uniformly continuous function. As it is well known, there exists a concave increasing function ω of $[0,\infty[$ to itself such that

$$|f(x) - f(y)| \le \omega(|x-y|) \qquad \forall x, y \in \mathbb{C}, \qquad \lim_{t\to 0} \omega(t) = 0 \tag{9}$$

(cf. *e.g.*, DeVore and Lorentz [10, Lemma 6.1, p. 43].) Thus by Jensen's inequality we have

$$⨍_Q \left| f \circ g - f\left(⨍_Q g\right)\right| \leq \omega\left(⨍_Q \left|g - \left(⨍_Q g\right)\right|\right).$$

By the characterization of VMO given by (1), we conclude that f acts in $VMO(\mathbb{R}^n)$ and in $vmo(\mathbb{R}^n)$. Moreover, by the continuity property of T_f (Theorem 9), we see that f acts in $CMO(\mathbb{R}^n)$ and — if $f(0) = 0$ — in $cmo(\mathbb{R}^n)$.

6.5. Necessary condition in BMO

First of all we need some specific test functions.

Lemma 2. *There exist two sequences $(\theta_j)_{j\geq 1}$ and $(\psi_j)_{j\geq 1}$ of functions of $\mathcal{D}(\mathbb{R}^n)$ such that*

- $\theta_j(x) = 1$ *for* $|x| \leq 2^{-j}$, $\theta_j(x) = 0$ *for* $|x| \geq 1$, $0 \leq \theta_j \leq 1$ *and*
$$\lim_{j\to\infty} \|\theta_j\|_{bmo} = 0.$$
- $\psi_j(x) = 1$ *for* $|x| \leq 2^j$, $\psi_j(x) = 0$ *for* $|x| \geq 4^j$, $0 \leq \psi_j \leq 1$, *and*
$$\lim_{j\to\infty} \|\psi_j\|_{BMO} = 0.$$

Proof. Let u be a C^∞ function on $\mathbb{R}$ such that $0 \leq u \leq 1$, and

$$u(t) = 1 \quad \text{for} \quad t \leq -1 \quad , \quad u(t) = 0 \quad \text{for} \quad t \geq 0.$$

We define θ_j and ψ_j as follows.

$$\theta_j(x) = u\left(\frac{\log_2 |x|}{j}\right) \quad , \quad \psi_j(x) = u\left(\frac{\log_2 |x|}{j} - 2\right).$$

Then by (8) the sequences (θ_j) and (ψ_j) have the required properties. This ends up the proof of Lemma 2.

Assume $T_f(cmo(\mathbb{R}^n)) \subseteq BMO(\mathbb{R}^n)$ and $f(0) = 0$. Then we can use Lemma 1. Let $\phi \in \mathcal{D}(\mathbb{R}^n)$ such that $\phi = 1$ on $\frac{1}{2}Q_0$, $\operatorname{supp}\phi \subseteq Q_0$, $0 \leq \phi \leq 1$. Let a, b be two complex numbers. According to Lemma 2, there exist a function $\theta \in \mathcal{D}(\mathbb{R}^n)$ and an integer $j \geq 1$ such that $\operatorname{supp}\theta \subseteq Q_0$, $\theta = 1$ on the cube $2^{-j}Q_0$, and

$$|a| \; \|\theta\|_{bmo} \leq \frac{c_1}{2}.$$

Now we set

$$g(x) := (b - a)\phi(2^{j+1}x) + a\theta(x)$$

If $|a - b|$ is sufficiently small, we have $\|g\|_{bmo} \leq c_1$. By exploiting the inequality $\|f \circ g\|_{BMO} \leq c_2$, we obtain easily $|f(b) - f(a)| \leq c_3$. Thus f satisfies the condition (i) of Theorem 5.

6.6. Necessary conditions in VMO and CMO

We use the following variant of Lemma 1 (see [6] for the proof).

Lemma 3. *If conditions $T_f\,(cmo(\mathbb{R}^n)) \subseteq VMO(\mathbb{R}^n)$ and $f(0) = 0$ hold, then for every $\varepsilon > 0$, there exist a cube K contained in the cube Q_0, and two constants $c_1 > 0$, $c_2 > 0$ such that*

$$⨍_Q \left| f \circ g - \left(⨍_Q f \circ g \right) \right| \le \varepsilon,$$

for all $g \in cmo(\mathbb{R}^n)$ with $\operatorname{supp} g \subseteq K$, $\|g\|_{bmo} \le c_1$, and for all cubes Q with $|Q| \le c_2$.

Now let ε be an arbitrary positive number. By the same argument as in the last section and by the above lemma, we prove that $|f(a) - f(b)| \le \varepsilon$ if $|a - b|$ is sufficiently small, hence the uniform continuity of f. Assume moreover that $T_f\,(cmo(\mathbb{R}^n)) \subseteq cmo(\mathbb{R}^n)$; since 0 is the only constant function in $cmo(\mathbb{R}^n)$, we obtain $f(0) = 0$.

7. Sketch of the proofs (Regularity)

7.1. Sufficient conditions in Hölder-Zygmund spaces

For simplicity, we set $E := B^s_{\infty,\infty}(\mathbb{R}^n)$, $\Phi := \Phi_s$. Since E is a Banach algebra, we can introduce the following continuous linear mapping

$$\begin{array}{rccc} M: & E & \to & \mathcal{L}(E,E) \\ & g & \mapsto & \{h \mapsto gh\} \end{array}.$$

Let us assume that f belongs to the closure of $C^\infty(\mathbb{R})$ into $W^r(\Phi)$. We are going to prove, by induction on r, that T_f is a mapping of class C^r on E.

Step 1: the case $r = 0$. We assume first that $f' \in \Phi$. Let us fix a function $g \in E$ and take any $h \in E$, with $\|h\| \le 1$. From the identity

$$T_f(g+h) - T_f(g) = h \int_0^1 T_{f'}(g+th)\,dt\,, \tag{10}$$

we deduce

$$\|T_f(g+h) - T_f(g)\| \le \|h\|\, N_{\|g\|+1}(f')\,, \tag{11}$$

hence the continuity of T_f at g.

In the general case, we approximate f by a sequence (f_j) of C^∞ functions. From (11), we obtain

$$\|T_f(g+h) - T_f(g)\| \le \|h\|\, N_{\|g\|+1}(f_j') + 2N_{\|g\|+1}(f - f_j)\,.$$

A standard argument yields the continuity of T_f at g.

Step 2: the induction. Assume f belongs to the closure of $C^\infty(\mathbb{R})$ into $W^{r+1}(\Phi)$. From (10), we deduce

$$T_f(g+h) - T_f(g) - (f' \circ g)h = h \int_0^1 (T_{f'}(g+th) - T_{f'}(g))\,dt\,. \tag{12}$$

By *Step 1* we know that $T_{f'}$ is continuous on E. The formula (12) yields the differentiability of T_f and the equality $dT_f = M \circ T_{f'}$. By inductive assumption, $T_{f'}$ is of class C^r. Then dT_f is of class C^r as a mapping from E to $\mathcal{L}(E,E)$, which means that T_f is a mapping of class C^{r+1}.

Remark. In the paper [5], we prove the regularity of T_f by a different way, using an abstract regularity theorem of Lanza [17].

7.2. Necessary conditions in Hölder-Zygmund spaces

Step 1. Assume that T_f is continuous on E. By exploiting ideas of Drábek [11, p. 52], Sobolevskij [23] and Lanza [16, Proposition 4.21], we prove the following properties.

(i) f is of class C^1, if $0 < s < 1$.

(ii) f is of class C^1 and

$$f(x+t) + f(x-t) - 2f(x) = o\left(\frac{t}{|\log t|}\right),$$

as $t \to 0+$, uniformly on each compact subset of $\mathbb{R}$, if $s = 1$.

(iii) f is of class C^m and

$$f^{(m)}(x+t) - f^{(m)}(x) = o\left(|t|^{s-m}\right),$$

as $t \to 0$, uniformly on each compact subset of $\mathbb{R}$, if $m < s < m+1$ for some $m \in \mathbb{N}^*$.

(iv) f is of class C^{s-1} and

$$f^{(s-1)}(x+t) + f^{(s-1)}(x-t) - 2f^{(s-1)}(x) = o(t),$$

as $t \to 0+$, uniformly on each compact subset of $\mathbb{R}$, if s is integer and $s > 1$.

Step 2. Assume that T_f is of class C^{r+1} on E. From the formula $T_{f'}(g) = dT_f(g)(1)$, we deduce that $T_{f'}$ is of class C^r in E. Then the various conditions of Corollary 1 follow by induction and by Step 1.

We refer to [5] for the details.

7.3. Continuity in vmo and CMO

To prove Theorem 9, we follow essentially Brezis and Nirenberg [7, Lemma A.8, p. 238].

Assume that f and ω satisfy the condition (9). Let us fix $g \in BMO(\mathbb{R}^n)$ and define

$$I_Q(v) := \fint_Q \left| f \circ (g+v) - f \circ g - \fint_Q (f \circ (g+v) - f \circ g) \right|.$$

An easy computation yields the following estimation.

$$I_Q(v) \le \min\left(2\omega(2\fint_Q |g - \fint_Q g|) + \omega(2\fint_Q |v - \fint_Q v|),\, 2\omega(\fint_Q |v|)\right). \qquad (13)$$

Let $g \in vmo(\mathbb{R}^n)$ and $\varepsilon > 0$. By the condition (1), we can find a small positive number c such that

$$\omega(2 f_Q |g - f_Q g|) \leq \varepsilon \,,$$

for $|Q| \leq c$. If $c < |Q| \leq 1$, we have

$$f_Q |v| \leq c^{-1} \|v\|_{bmo} \,.$$

From (13), we deduce that $\sup_{|Q| \leq 1} I_Q \leq 3\varepsilon$, if $\|v\|_{bmo}$ is sufficiently small, hence the continuity of T_f at g. If $g \in CMO(\mathbb{R}^n)$, we have to complete the above proof by considering also big cubes. We refer to [6] for the details.

7.4. Degeneracy result for continuity in bmo or VMO

Now we turn to Theorem 10. Without loss of generality, we assume that $f(0) = 0$. Let α, β be arbitrary complex numbers.

Let assume that T_f is continuous from $bmo(\mathbb{R}^n)$ to $BMO(\mathbb{R}^n)$. By Lemma 2 there exists a sequence $(\theta_j)_{j \geq 1}$ of functions such that $\theta_j(x) = 1$ on the cube $K_j = [-j^{-1}, j^{-1}]^n$, and $\lim_{j\to\infty} \|\theta_j\|_{bmo} = 0$. Let γ denote the characteristic function of $[0,1]^n$. By an easy computation, we obtain

$$\|f \circ (\beta\gamma + \alpha\theta_j) - f \circ (\beta\gamma)\|_{BMO} \geq 2^{-n}(1 - 2^{-n}) \, |f(\beta + \alpha) - f(\beta) - f(\alpha)| \,.$$

By taking the limit as j tends to infinity, we obtain $f(\alpha + \beta) = f(\alpha) + f(\beta)$, that is the $\mathbb{R}$-linearity of f.

In case T_f is continuous from $VMO(\mathbb{R}^n)$ to $BMO(\mathbb{R}^n)$, we proceed in a similar way, using the sequence (ψ_j) of Lemma 2. Let M be a large positive constant. Let K_j, K_j', K_j'' be the cubes of center $a_j := (2M4^j, 0, \ldots, 0)$ and halfsidelength $2^j, 2^j + 1$, and 2^{j+1}, respectively. We note that

$$|K_j' \setminus K_j| = O(2^{j(n-1)}) \quad \text{as} \quad j \to +\infty \,. \tag{14}$$

Let $(\phi_j)_{j\geq 1}$ be a sequence of functions of $\mathcal{D}$ such that

$$\phi_j(x) = 1 \quad \text{for} \quad x \in [-1,1]^n \,, \quad \phi_j(x) = 0 \quad \text{for} \quad x \notin [-1 - 2^{-j}, 1 + 2^{-j}]^n$$

and

$$|\phi_j| \leq 2, \qquad \sup_{j \geq 1} 2^{-j} \|\nabla \phi_j\|_\infty < +\infty \,.$$

We define the function $g \in VMO(\mathbb{R}^n)$ by setting

$$g(x) = \phi_j \left(\frac{x - a_j}{2^j} \right) \qquad \text{if } x \in K_j'' \text{ for some } j \geq 1,$$

and $g(x) = 0$ elsewhere. Let $u_j(x) := \psi_j \left(M^{-1}(x - a_j) \right)$. Then we have

$$\lim_{j\to+\infty} \|u_j\|_* = 0$$

and $u_j(x) = 1$ on the cube K_j''. We now set

$$c_j := f_{K_j''} (f \circ (\beta g + \alpha u_j) - f \circ (\beta g)) \,.$$

By (14) and by the uniform continuity of f, we deduce that

$$c_j = 2^{-n}(f(\beta+\alpha) - f(\beta)) + (1-2^{-n})f(\alpha) + \varepsilon_j\,,$$

with $\lim_{j\to+\infty}\varepsilon_j = 0$. Then we have

$$\begin{aligned}\|f\circ(\beta g+\alpha u_j) - f\circ(\beta g)\|_{BMO} &\geq \frac{1}{|K_j''|}\int_{K_j} |f\circ(\beta g+\alpha u_j) - f\circ(\beta g) - c_j| \\ &= 2^{-n}\left|(1-2^{-n})(f(\beta+\alpha) - f(\beta) - f(\alpha)) - \varepsilon_j\right|.\end{aligned}$$

Thus by taking the limit as $j\to+\infty$, we obtain $f(\beta+\alpha) = f(\beta) + f(\alpha)$.

7.5. Degeneracy result for differentiability in BMO

Let f be a function of $\mathbb{C}$ to itself, viewed as a function of two real variables, say x, y. Assume $T_f : \mathcal{D}(\mathbb{R}^n) \to BMO(\mathbb{R}^n)$ to be differentiable in the bmo-norm. A standard argument gives the existence of $\frac{\partial f}{\partial x}$ and $\frac{\partial f}{\partial y}$ and the formula

$$dT_f[u](v) = \left(\frac{\partial f}{\partial x}\circ u\right)(\Re v) + \left(\frac{\partial f}{\partial y}\circ u\right)(\Im v)\,,$$

for any $u, v\in\mathcal{D}(\mathbb{R}^n)$. Since the mapping

$$u \mapsto f\circ u - (\Re u)\frac{\partial f}{\partial x}(0) - (\Im u)\frac{\partial f}{\partial y}(0) - f(0)$$

is also differentiable, there is no loss of generality in assuming that

$$f(0) = \frac{\partial f}{\partial x}(0) = \frac{\partial f}{\partial y}(0) = 0\,.$$

Now we set

$$\sigma(t) := \sup\left\{\frac{\|f\circ u\|_{BMO}}{\|u\|_{bmo}} : u\in\mathcal{D}(\mathbb{R}^n),\ 0<\|u\|_{bmo}\leq t\right\} \qquad ,\ \forall t>0.$$

By assumptions, we have

$$\lim_{t\to 0}\sigma(t) = 0. \tag{15}$$

From the inequality

$$\|f\circ u\|_{BMO} \leq t\sigma(t) \quad \text{for} \quad \|u\|_{bmo}\leq t\,,$$

we deduce, as in the proof of Theorem 5, that

$$|f(a) - f(b)| \leq C|a-b|\sigma(c|a-b|)\,,$$

for two constants $C, c>0$ and for all a, b with $|a-b|$ sufficiently small. By (15), we conclude that f is differentiable, and that its differential is identically zero.

References

[1] C. Bennett and R. Sharpley. *Interpolation of Operators.* Academic Press, Boston, 1988.

[2] G. Bourdaud. *Le calcul fonctionnel dans les espaces de Sobolev.* Invent. Math. 104 (1991), pp. 435–446.

[3] G. Bourdaud. The Functional Calculus in Sobolev Spaces. Function Spaces, Differential Operators and Nonlinear Analysis, Schmeisser, Triebel (eds.), Teubner-Texte zur Math. Vol. 133, pp. 127–142, Leipzig 1993.

[4] G. Bourdaud. *Fonctions qui opèrent sur les espaces de Besov et de Triebel.* Ann. Inst. Henri Poincaré, Analyse non linéaire, 10 (1993), pp. 413–422.

[5] G. Bourdaud and M. Lanza de Cristoforis. *Functional calculus in Hölder-Zygmund spaces* (submitted).Inst. de Math. Jussieu, Prépub. 260 (sept. 2000).

[6] G. Bourdaud, M. Lanza de Cristoforis and W. Sickel. *Functional calculus on BMO and related spaces.* J. Funct. Anal. (to appear). Inst. de Math. Jussieu, Prépub. 296 (june 2001).

[7] H. Brezis and L. Nirenberg. *Degree theory and BMO; Part I: Compact manifolds without boundaries.* Selecta Mathematica, New Series, 1 (1995), pp. 197–263.

[8] L. Chevalier. *Quelles sont les fonctions qui opèrent de BMO dans BMO ou de BMO dans $\overline{L^\infty}$?* Bull. London Math. Soc. 27 (1995), pp. 590–594.

[9] R. Coifman and G. Weiss. *Extension of Hardy spaces and their use in analysis.* Bull. Amer. Math. Soc. 83 (1977), pp. 569–645.

[10] R.A. DeVore and G.G. Lorentz. *Constructive Approximations.* Springer Verlag, Berlin, 1993.

[11] P. Drábek. *Continuity of Nemyckii's operator in Hölder spaces.* Comm. Math. Univ. Carolinae 16 (1975), pp. 37–57.

[12] G.B. Folland. *Real Analysis, modern Techniques and their applications.* John Wiley & Sons, 1984.

[13] M.A. Fominykh. *Transformation of BMO functions* (Russian). Vestnik Moskov. Univ. Ser. I Mat. Mekh. 94, 2 (1985), pp. 20–24.

[14] D. Goldberg. *A local version of real Hardy space.* Duke Math. J. 46 (1979), pp. 27–42.

[15] Y. Katznelson. *An Introduction to Harmonic Analysis.* Dover, New York, 1976.

[16] M. Lanza de Cristoforis. *Higher order differentiability properties of the composition and of the inversion operator.* Indag. Matem. N.S. 5 (1994), pp. 457–482.

[17] M. Lanza de Cristoforis. *Differentiability properties of an abstract autonomous composition operator.* J. London Math. Soc. 61 (2000), pp. 923–936.

[18] M. Marcus and V.J. Mizel. *Complete characterization of functions which act, via superposition, on Sobolev Spaces.* Trans. Amer. Math. Soc. 251 (1979), pp. 187–218.

[19] T. Runst and W. Sickel. *Sobolev Spaces of Fractional Order, Nemytskij Operators, and Nonlinear Partial Differential Equations.* De Gruyter, Berlin, 1996.

[20] D. Sarason. *Functions of vanishing mean oscillation.* Trans. Amer. Math. Soc. 20 (1975), pp. 391–405.

[21] W. Sickel. *Spaces around bmo-Sobolev classes.* Preprint, FSU Jena, 2001.

[22] W. Sickel and H. Triebel. *Hölder inequalities and sharp embeddings in function spaces of $B^s_{p,q}$ and $F^s_{p,q}$ type*. Z. Anal. Anwendungen 14 (1995), pp. 105–140.

[23] Je. P. Sobolevskij. *The superposition operator in Hölder spaces* (Russian). Voronezh, VINITI No. 3765-84, 1984.

[24] D.A. Stegenga. *Bounded Toeplitz operators on H^1 and applications of duality between H^1 and the functions of bounded mean oscillation*. Amer. J. Math. 98 (1976), pp. 573–589.

Gérard Bourdaud
Institut de Mathématiques de Jussieu
Equipe d'Analyse Fonctionnelle
Case 186, 4 place Jussieu
F-75252 Paris Cedex 05
France
E-mail address: bourdaud@ccr.jussieu.fr

D. Haroske, T. Runst, H.-J. Schmeisser (eds.): Function Spaces, Differential Operators and Nonlinear Analysis. The Hans Triebel Anniversary Volume.

Asymptotics of a Singular Solution to the Dirichlet Problem for an Elliptic Equation with Discontinuous Coefficients Near the Boundary

Vladimir Kozlov[1] and Vladimir Maz'ya[1]

Dedicated to Prof. Hans Triebel on the occasion of his 65th birthday

Abstract. We consider the Dirichlet problem for elliptic equations of arbitrary order and prove an asymptotic formula for a singular solution near a boundary point. The only a priori assumption on the coefficients of the principal part of the equation is the smallness of the local oscillation near the point.

1. Introduction

In this article, we are interested in the behaviour of solutions to the Dirichlet problem for arbitrary even order $2m$ strongly elliptic equations in divergence form near a point $\mathcal{O}$ at the smooth boundary. We require only that the coefficients of the principal part of the operator have small oscillation near this point and the coefficients in lower order terms are allowed to have singularities at the boundary. In our recent paper [4], we derived an explicit asymptotic formula near $\mathcal{O}$ for solutions with finite energy integral. Here, our objective is to obtain an analogous asymptotic representation for solutions with infinite energy integral which have the least possible singularity. Since this representation is new even in the case of the second order equations, we start with describing it for this particular case.

Let us consider the uniformly elliptic equation

$$-\text{div}\,(A(x)\,\text{grad}\,u(x)) = f(x) \quad \text{in } G \tag{1}$$

complemented by the Dirichlet condition

$$u = 0 \quad \text{on } \partial G \setminus \{\mathcal{O}\}, \tag{2}$$

where G is a domain in $\mathbb{R}^n$ with smooth boundary. We assume that the elements of the $n \times n$-matrix $A(x)$ are measurable and bounded complex-valued functions. We deal with a solution u having a finite Dirichlet integral outside any neighborhood of $\mathcal{O}$ and require, for simplicity, that $f = 0$ in a certain δ-neighborhood $G_\delta =$

[1] The authors were supported by the Swedish Natural Science Research Council (NFR)

$\{x \in G : |x| < \delta\}$ of the origin. We suppose that there exists a constant symmetric matrix A with positive definite real part such that the function

$$\varkappa(r) := \sup_{G_r} ||A(x) - A||$$

is sufficiently small for $r < \delta$. We introduce the function

$$\mathcal{R}(x) = \frac{\langle (A(x) - A)\nu, \nu\rangle - n\langle \nu, (A(x) - A)A^{-1}x\rangle\langle \nu, x\rangle\langle A^{-1}x, x\rangle^{-1}}{|S^{n-1}|(\det A)^{1/2}\langle A^{-1}x, x\rangle^{n/2}}, \tag{3}$$

where $\langle z, \zeta\rangle = z_1\zeta_1 + \ldots + z_n\zeta_n$ and ν is the interior unit normal at $\mathcal{O}$. (For the notation $(\det A)^{1/2}$ and $\langle A^{-1}x, x\rangle^{n/2}$ see [2], Section 6.2.)

The following asymptotic formula is a corollary of our main Theorem 1

$$u(x) = \exp\Big(\int_{G_\delta \setminus G_{|x|}} \mathcal{R}(y)dy + O\Big(\int_{|x|}^{\delta} \varkappa(\rho)^2 \frac{d\rho}{\rho}\Big)\Big)$$
$$\times \Big(C\Big(\frac{\operatorname{dist}(x, \partial G)}{\langle A^{-1}x, x\rangle^{n/2}} + O\Big(|x|^{2-n-\varepsilon} \int_{|x|}^{\delta} \varkappa(\rho) \frac{d\rho}{\rho^{2-\varepsilon}}\Big)\Big) + O\Big(|x|^{1-\varepsilon}\Big)\Big), \tag{4}$$

where $C = \text{const}$ and ε is a small positive number.

In Theorem 1 we obtain a general asymptotic formula similar to (4) for solutions of the Dirichlet problem for the uniformly strongly elliptic equation with complex-valued measurable coefficients

$$\sum_{0 \le |\alpha|, |\beta| \le m} (-\partial_x)^\alpha (\mathcal{L}_{\alpha\beta}(x)\partial_x^\beta u(x)) = f(x) \quad \text{on } B_\delta^+, \tag{5}$$

where $B_\delta^+ = \mathbb{R}_+^n \cap B_\delta$, $\mathbb{R}_+^n = \{x = (x', x_n) \in \mathbb{R}^n : x_n > 0\}$ and $B_\delta = \{x \in \mathbb{R}^n : |x| < \delta\}$. Here and elsewhere by ∂_x we mean the vector of partial derivatives $(\partial_{x_1}, \ldots, \partial_{x_n})$. The only a priori assumption on the coefficients $\mathcal{L}_{\alpha\beta}$ is smallness of the function

$$\sum_{|\alpha| = |\beta| = m} |\mathcal{L}_{\alpha\beta}(x) - L_{\alpha\beta}| + \sum_{|\alpha + \beta| < 2m} x_n^{2m - |\alpha + \beta|} |\mathcal{L}_{\alpha\beta}(x)|,$$

where $x \in B_\delta^+$ and $L_{\alpha\beta}$ are constants.

The proof of Theorem 1 follows the same lines as that of the main result in [4]. In order to make our exposition self-contained, we give complete formulations of intermediate results and their detailed proofs instead of referring repeatedly to analogous statements in [4].

2. Function spaces

Let $1 < p < \infty$ and let $W_{\text{loc}}^{m,p}(\overline{\mathbb{R}_+^n} \setminus \mathcal{O})$ denote the space of functions u defined on $\mathbb{R}_+^n$ and such that $\eta u \in W^{m,p}(\mathbb{R}_+^n)$ for all smooth η with compact support in $\overline{\mathbb{R}^n} \setminus \mathcal{O}$. Also let $\mathring{W}_{\text{loc}}^{m,p}(\overline{\mathbb{R}_+^n} \setminus \mathcal{O})$ be the subspace of $W_{\text{loc}}^{m,p}(\overline{\mathbb{R}_+^n} \setminus \mathcal{O})$, which contains functions subject to

$$\partial_{x_n}^k u = 0 \quad \text{on } \partial\mathbb{R}_+^n \setminus \mathcal{O} \text{ for } k = 0, \ldots, m-1. \tag{6}$$

We introduce a family of seminorms in $\mathring{W}^{m,p}_{\text{loc}}(\overline{\mathbb{R}^n_+} \setminus \mathcal{O})$ by

$$\mathfrak{M}^m_p(u; K_{ar,br}) = \Big(\sum_{k=0}^{m} \int_{K_{ar,br}} |\nabla_k u(x)|^p |x|^{pk-n} dx\Big)^{1/p}, \quad r > 0, \tag{7}$$

where $K_{\rho,r} = \{x \in \mathbb{R}^n_+ \,:\, \rho < |x| < r\}$, a and b are positive constants, $a < b$ and $\nabla_k u$ is the vector $\{\partial^\alpha_x u\}_{|\alpha|=k}$. One can easily see that (6) implies the equivalence of $\mathfrak{M}^m_p(u; K_{ar,br})$ and the seminorm

$$\Big(\int_{K_{ar,br}} |\nabla_m u(x)|^p |x|^{pm-n} dx\Big)^{1/p}.$$

With another choice of a and b we arrive at an equivalent family of seminorms. Clearly,

$$\mathfrak{M}^m_p(u; K_{a'r,b'r}) \le c_1(a,b,a',b') \int_{a'r/b}^{b'r/a} \mathfrak{M}^m_p(u; K_{a\rho,b\rho}) \frac{d\rho}{\rho}, \tag{8}$$

where c_1 is a continuous function of its arguments.

We say that a function v belongs to the space $\mathring{W}^{m,q}_{\text{comp}}(\overline{\mathbb{R}^n_+} \setminus \mathcal{O})$, $pq = p + q$, if $v \in \mathring{W}^{m,q}_{\text{loc}}(\overline{\mathbb{R}^n_+} \setminus \mathcal{O})$ and v has a compact support in $\overline{\mathbb{R}^n_+} \setminus \mathcal{O}$. By $W^{-m,p}_{\text{loc}}(\overline{\mathbb{R}^n_+} \setminus \mathcal{O})$ we denote the dual of $\mathring{W}^{m,q}_{\text{comp}}(\overline{\mathbb{R}^n_+} \setminus \mathcal{O})$ with respect to the inner product in $L^2(\mathbb{R}^n_+)$. We supply $W^{-m,p}_{\text{loc}}(\overline{\mathbb{R}^n_+} \setminus \mathcal{O})$ with the seminorms

$$\mathfrak{M}^{-m}_p(f; K_{ar,br}) = \sup \Big| \int_{\mathbb{R}^n_+} f\, \overline{v}\, |x|^{-n} dx \Big|, \tag{9}$$

where the supremum is taken over all functions $v \in \mathring{W}^{m,q}_{\text{comp}}(\overline{\mathbb{R}^n_+} \setminus \mathcal{O})$ supported by $ar \le |x| \le br$ and such that $\mathfrak{M}^m_p(v; K_{ar,br}) \le 1$. By a standard argument it follows from (8) that

$$\mathfrak{M}^{-m}_p(f; K_{a'r,b'r}) \le c_2(a,b,a'b') \int_{a'r/b}^{b'r/a} \mathfrak{M}^{-m}_p(f; K_{a\rho,b\rho}) \frac{d\rho}{\rho}, \tag{10}$$

where c_2 depends continuously on its arguments.

3. Statement of the Dirichlet problem in $\mathbb{R}^n_+$

We consider the Dirichlet problem

$$\mathcal{L}(x, \partial_x) u = f(x) \quad \text{in } \mathbb{R}^n_+, \tag{11}$$

$$\partial^k_{x_n} u\big|_{x_n=0} = 0 \quad \text{for } k = 0, 1, \ldots, m-1 \quad \text{on } \mathbb{R}^{n-1} \setminus \mathcal{O} \tag{12}$$

for the differential operator

$$\mathcal{L}(x, \partial_x) u = \sum_{|\alpha|,|\beta| \le m} (-\partial_x)^\alpha \big(\mathcal{L}_{\alpha\beta}(x) \partial^\beta_x u\big) \tag{13}$$

with measurable complex-valued coefficients $\mathcal{L}_{\alpha\beta}$ in $\mathbb{R}^n_+$.

We also need a differential operator with constant coefficients

$$L(\partial_x) = (-1)^m \sum_{|\alpha|=|\beta|=m} L_{\alpha\beta}\partial_x^{\alpha+\beta}, \tag{14}$$

where $\Re L(\xi) > 0$ for $\xi \in \mathbb{R}^n \setminus \mathcal{O}$. It will be convenient to require that the coefficient of $L(\partial_x)$ in $\partial_{x_n}^{2m}$ is equal to $(-1)^m$.

We treat $\mathcal{L}(x, \partial_x)$ as a perturbation of $L(\partial_x)$ and characterize this perturbation by the function

$$\Omega(r) = \sup_{x \in K_{r/e,r}} \Big(\sum_{|\alpha|=|\beta|=m} |\mathcal{L}_{\alpha\beta}(x) - L_{\alpha\beta}| + \sum_{|\alpha+\beta|<2m} x_n^{2m-|\alpha+\beta|} |\mathcal{L}_{\alpha\beta}(x)| \Big), \tag{15}$$

which is assumed to be smaller than a certain positive constant depending on n, m, p and the coefficients $L_{\alpha\beta}$. It is straightforward that

$$\Omega(r) \le \int_{r/e}^{re} \Omega(t) \frac{dt}{t}. \tag{16}$$

By the classical Hardy inequality

$$\mathfrak{M}_p^{-m}((\mathcal{L} - L)(u); K_{r/e,r}) \le c\, \Omega(r) \mathfrak{M}_p^m(u; K_{r/e,r}), \tag{17}$$

where c depends only on n, m and p. Therefore, the boundedness of $\Omega(r)$ implies that the operator $\mathcal{L}(x, \partial_x)$ maps $\mathring{W}_{\text{loc}}^{m,p}(\overline{\mathbb{R}_+^n} \setminus \mathcal{O})$ into $W_{\text{loc}}^{-m,p}(\overline{\mathbb{R}_+^n} \setminus \mathcal{O})$.

In what follows we always require that the right-hand side f in (11) belongs to $W_{\text{loc}}^{-m,p}(\overline{\mathbb{R}_+^n} \setminus \mathcal{O})$ and consider a solution u of (11) in the space $\mathring{W}_{\text{loc}}^{m,p}(\overline{\mathbb{R}_+^n} \setminus \mathcal{O})$. This solution satisfies

$$\int_{\mathbb{R}_+^n} \sum_{|\alpha|,|\beta| \le m} \mathcal{L}_{\alpha\beta}(x) \partial_x^\beta u(x) \partial_x^\alpha \overline{v}(x) dx = \int_{\mathbb{R}_+^n} f \overline{v}(x) dx \tag{18}$$

for all $v \in \mathring{W}_{\text{comp}}^{m,q}(\overline{\mathbb{R}_+^n} \setminus \mathcal{O})$, $pq = p + q$. The integral on the right is understood in the distribution sense.

4. Formulation of the main result

In the next statement we make use of the notation introduced in Section 3. We also need the Poisson kernel E of the equation

$$\sum_{|\alpha|=|\beta|=m} L_{\alpha\beta}\partial_x^{\alpha+\beta} E(x) = 0 \quad \text{in } \mathbb{R}_+^n, \tag{19}$$

which is positive homogeneous of degree $m - n$ and subject to the Dirichlet conditions on the hyperplane $x_n = 0$:

$$\partial_{x_n}^j E = 0 \quad \text{for} \quad 0 \le j \le m-2, \quad \text{and} \quad \partial_{x_n}^{m-1} E = \delta(x'), \tag{20}$$

where δ is the Dirac function. In principle, the function E can be calculated using the Fourier transform in x' (see [1], Chapter 1, Section 2).

In the case of the polyharmonic operator $(-\Delta)^m$ one verifies directly that $E(x) = \mathrm{const}\, x_n^m |x|^{-n}$. The constant factor can be evaluated by the identity

$$(\partial_{x_n}^m E)(x', 0) = m\Gamma(n/2)\pi^{-n/2}|x'|^{-n}$$

which is found in Example 11.6.4[5]. Eventually,

$$E(x) = \frac{\Gamma(n/2)}{(m-1)!\pi^{n/2}} \frac{x_n^m}{|x|^n}. \tag{21}$$

In what follows, by c and C (sometimes enumerated) we denote different positive constants which depend only on m, n, p and the coefficients $L_{\alpha,\beta}$.

Theorem 1. *Assume that $\Omega(r)$ does not exceed a sufficiently small positive constant depending on m, n, p and $L_{\alpha\beta}$. There exist positive constants C and c depending on the same parameters such that the following assertions are valid.*

(i) *There exists $\mathcal{Z} \in \mathring{W}^{m,p}_{\mathrm{loc}}(\overline{\mathbb{R}^n_+} \setminus \mathcal{O})$ subject to $\mathcal{L}(x, \partial_x)\mathcal{Z} = 0$ on B_e^+ and satisfying*

$$(r\partial_r)^k \mathcal{Z}(x) = \exp\Big(\int_r^1 (\mathcal{T}(\rho) + \Upsilon(\rho))\frac{d\rho}{\rho}\Big)\Big((m-n)^k E(x) + r^{m-n} v_k(x)\Big), \tag{22}$$

where $k = 0, 1, \dots, m$, $r = |x| < 1$ and Υ is a measurable function on $(0,1)$ satisfying

$$|\Upsilon(r)| \tag{23}$$
$$\leq C\,\Omega(r)\Big(r^{-n}\int_0^r e^{C\int_\rho^r \Omega(s)\frac{ds}{s}}\Omega(\rho)\rho^{n-1}d\rho + r\int_r^e e^{C\int_r^\rho \Omega(s)\frac{ds}{s}}\Omega(\rho)\rho^{-2}d\rho\Big),$$

and

$$\begin{aligned}\mathcal{T}(\rho) &= \rho^n \int_{S_+^{n-1}} \sum_{|\beta|=m} (\mathcal{L}_{(0',m),\beta}(\xi) - L_{(0',m),\beta})E^{(\beta)}(\xi)d\theta_\xi \\ &+ \rho^n \int_{S_+^{n-1}} \sum_{|\beta|+k<2m} \mathcal{L}_{(0',k),\beta}(\xi)\frac{\xi_n^{m-k}}{(m-k)!}E^{(\beta)}(\xi)d\theta_\xi\end{aligned}$$

with $\rho = |\xi|$, $\theta = \xi/|\xi|$. The functions v_k belong to $L^p_{\mathrm{loc}}((0,\infty); \mathring{W}^{m-k,p}(S_+^{n-1}))$ and satisfy

$$\Big(\int_{r/e}^r \big(||v_k(\rho,\cdot)||^p_{W^{m-k,p}(S_+^{n-1})} + ||\rho\partial_\rho v_k(\rho,\cdot)||^p_{W^{m-k-1,p}(S_+^{n-1})}\big)\frac{d\rho}{\rho}\Big)^{1/p}$$
$$\leq c\Big(r^{-1}\int_0^r e^{C\int_\rho^r \Omega(s)\frac{ds}{s}}\Omega(\rho)d\rho + r^n\int_r^e e^{C\int_r^\rho \Omega(s)\frac{ds}{s}}\Omega(\rho)\rho^{-n-1}d\rho\Big), \tag{24}$$

where $k = 0, \dots, m-1$, S_+^{n-1} is the upper hemisphere and $\mathring{W}^{m-k,p}(S_+^{n-1})$ is the completion of $C_0^\infty(S_+^{n-1})$ in the norm of the Sobolev space $W^{m,p}(S_+^{n-1})$. In the case $k = m$ estimate (24) holds without the second norm in the left-hand side.

(ii) *Let*

$$I_f := \int_0^e \rho^{m+n}\exp\Big(C\int_\rho^1 \Omega(s)\frac{ds}{s}\Big)\mathfrak{M}_p^{-m}(f; K_{\rho/e,\rho})\frac{d\rho}{\rho} < \infty \tag{25}$$

and let $u \in \mathring{W}^{m,p}_{\mathrm{loc}}(\overline{\mathbb{R}^n_+} \setminus \mathcal{O})$ *be a solution of* $\mathcal{L}(x, \partial_x)u = f$ *on* B^+_e *subject to*

$$\Big(\int_{K_{r/e,r}} |u(x)|^p |x|^{-n} dx \Big)^{1/p} = o\Big(r^{m-n-1} \exp\Big(-\mathcal{C} \int_r^1 \Omega(\rho) \frac{d\rho}{\rho} \Big) \Big) \tag{26}$$

as $r \to 0$. *Then for* $x \in B^+_1$

$$u(x) = C\mathcal{Z}(x) + w(x)\,, \tag{27}$$

where the constant C *satisfies*

$$|C| \le c(I_f + ||u||_{L^p(K_{1,e})}) \tag{28}$$

and the function $w \in \mathring{W}^{m,p}_{\mathrm{loc}}(\overline{\mathbb{R}^n_+} \setminus \mathcal{O})$ *is subject to*

$$\mathfrak{M}^m_p(w; K_{r/e,r}) \le c\, r^m \Big(\int_0^r r^{-n} \rho^{m+n} e^{\mathcal{C} \int_\rho^r \Omega(s) \frac{ds}{s}} \mathfrak{M}^{-m}_p(f; K_{\rho/e,\rho}) \frac{d\rho}{\rho}$$

$$+ \int_r^e \rho^m e^{\mathcal{C} \int_r^\rho \Omega(s) \frac{ds}{s}} \mathfrak{M}^{-m}_p(f; K_{\rho/e,\rho}) \frac{d\rho}{\rho} + e^{\mathcal{C} \int_r^1 \Omega(s) \frac{ds}{s}} ||u||_{L^p(K_{1,e})} \Big) \tag{29}$$

for $r < 1$.

The proof of this theorem will be given in Sections 5–18.

5. Reduction of problem (11), (12) to the Dirichlet problem in a cylinder

We write problem (11), (12) in the variables

$$t = -\log|x| \quad \text{and} \quad \theta = x/|x|. \tag{30}$$

The mapping $x \to (t, \theta)$ transforms $\mathbb{R}^n_+$ onto the cylinder $\Pi = S^{n-1}_+ \times \mathbb{R}$.

We shall need the spaces $\mathring{W}^{m,p}_{\mathrm{loc}}(\Pi)$ and $W^{-m,p}_{\mathrm{loc}}(\Pi)$ which are the images of $\mathring{W}^{m,p}_{\mathrm{loc}}(\overline{\mathbb{R}^n_+} \setminus \mathcal{O})$ and $W^{-m,p}_{\mathrm{loc}}(\overline{\mathbb{R}^n_+} \setminus \mathcal{O})$ under mapping (30). They can be defined independently as follows.

The space $\mathring{W}^{m,p}_{\mathrm{loc}}(\Pi)$ consists of functions whose derivatives up to order m belong to $L^p(D)$ for every compact subset D of $\overline{\Pi}$ and whose derivatives up to order $m-1$ vanish on $\partial\Pi$. The seminorm $\mathfrak{M}^m_p(u; K_{e^{-1-t},e^{-t}})$ in $\mathring{W}^{m,p}_{\mathrm{loc}}(\overline{\mathbb{R}^n_+} \setminus \mathcal{O})$ is equivalent to the seminorm $||u||_{W^{m,p}(\Pi_t)}$, $t \in \mathbb{R}$, where $\Pi_t = \{(\theta, \tau) \in \Pi \;:\; \tau \in (t, t+1)\}$. The space $W^{-m,p}_{\mathrm{loc}}(\Pi)$ consists of the distributions f on Π such that the seminorm

$$||f||_{W^{-m,p}(\Pi_t)} = \sup \Big| \int_{\Pi_t} f\, \overline{v} d\tau d\theta \Big| \tag{31}$$

is finite for every $t \in \mathbb{R}$. The supremum in (31) is taken over all $v \in \mathring{W}^{m,q}_{\mathrm{loc}}(\Pi)$, $pq = p+q$, supported by $\overline{\Pi_t}$ and subject to $||v||_{W^{m,q}(\Pi_t)} \le 1$. The seminorm (31) is equivalent to $\mathfrak{M}^{-m}_p(f;; K_{e^{-1-t},e^{-t}})$.

In the variables (t, θ) the operator L takes the form

$$L(\partial_x) = e^{2mt} \mathbf{A}(\theta, \partial_\theta, -\partial_t)\,, \tag{32}$$

where $\mathbf{A}$ is an elliptic partial differential operator of order $2m$ on Π with smooth coefficients. We introduce the operator $\mathbb{N}$ by

$$L(\partial_x) - \mathcal{L}(x, \partial_x) = e^{2mt}\mathbb{N}(\theta, t, \partial_\theta, -\partial_t). \tag{33}$$

Now problem (11), (12) can be written as

$$\begin{cases} \mathbf{A}(\theta, \partial_\theta, -\partial_t)u = \mathbb{N}(\theta, t, \partial_\theta, -\partial_t)u + e^{-2mt}f & \text{on } \Pi \\ u \in \mathring{W}^{p,m}_{\text{loc}}(\Pi)\,, \end{cases} \tag{34}$$

where $f \in W^{-m,p}_{\text{loc}}(\Pi)$. We do not mark the dependence on the new variables t, θ in u and f.

Let $W^{-m,p}(S^{n-1}_+)$ denote the dual of $\mathring{W}^{m,q}(S^{n-1}_+)$ with respect to the inner product in $L^2(S^{n-1}_+)$. We introduce the operator pencil

$$\mathcal{A}(\lambda) : \mathring{W}^{m,p}(S^{n-1}_+) \to W^{-m,p}(S^{n-1}_+) \tag{35}$$

by

$$\mathcal{A}(\lambda)U(\theta) = r^{-\lambda+2m}L(\partial_x)r^\lambda U(\theta) = \mathbf{A}(\theta, \partial_\theta, \lambda)U(\theta). \tag{36}$$

The following properties of $\mathcal{A}$ and its adjoint are standard and their proofs can be found, for example in [5], Section 10.3. The operator (35) is Fredholm for all $\lambda \in \mathbb{C}$ and its spectrum consists of eigenvalues with finite geometric multiplicities. These eigenvalues are

$$m, m+1, m+2, \dots \quad \text{and} \quad m-n, m-n-1, m-n-2, \dots\,, \tag{37}$$

and there are no generalized eigenvectors. The only eigenvector (up to a constant factor) corresponding to the eigenvalue $m-n$ is $|x|^{n-m}E(x) = E(\theta)$, where E is the Poisson kernel defined in Section 4.

We introduce the operator pencil $\overline{\mathcal{A}}(\lambda)$ defined on $\mathring{W}^{m,p}(S^{n-1}_+)$ by the formula $\overline{\mathcal{A}}(\lambda)U(\theta) = r^{-\lambda+2m}\overline{L}(\partial_x)r^\lambda U(\theta)$. This pencil has the same eigenvalues as the pencil $\mathcal{A}(\lambda)$. The only eigenvector (up to a constant factor) of $\overline{\mathcal{A}}$ corresponding to the eigenvalue m is $|x|^{-m}x_n^m = \theta_n^m$.

Using the definitions of the above pencils and Green's formula for L and $\overline{L}$ one can show that $(\mathcal{A}(\lambda))^* = \overline{\mathcal{A}}(2m - n - \overline{\lambda})$, where $*$ denotes passage to the adjoint operator in $L^2(S^{n-1}_+)$.

6. Properties of the unperturbed Dirichlet problems in $\mathbb{R}^n_+$ and Π

Let us consider the Dirichlet problem

$$\begin{cases} L(\partial_x)u = f & \text{in } \mathbb{R}^n_+, \\ u \in \mathring{W}^{p,m}_{\text{loc}}(\overline{\mathbb{R}^n_+} \setminus \mathcal{O})\,. \end{cases} \tag{38}$$

Proposition 1. (i) *Let $f \in W^{-m,p}_{\text{loc}}(\overline{\mathbb{R}^n_+} \setminus \mathcal{O})$ be subject to*

$$\int_0^1 \rho^{m+n}\mathfrak{M}_p^{-m}(f; K_{\rho/e,\rho})\frac{d\rho}{\rho} + \int_1^\infty \rho^m \mathfrak{M}_p^{-m}(f : K_{\rho/e,\rho})\frac{d\rho}{\rho} < \infty\,. \tag{39}$$

Then problem (38) *has a solution* $u \in \mathring{W}^{m,p}_{\mathrm{loc}}(\overline{\mathbb{R}^n_+} \setminus \mathcal{O})$ *satisfying*

$$\begin{aligned} &\mathfrak{M}^m_p(u; K_{r/e,r}) \\ &\le c\Big(\int_0^r r^{m-n}\rho^{m+n}\mathfrak{M}^{-m}_p(f; K_{\rho/e,\rho})\frac{d\rho}{\rho} + \int_r^\infty r^m\rho^m\mathfrak{M}^{-m}_p(f; K_{\rho/e,\rho})\frac{d\rho}{\rho}\Big). \end{aligned} \tag{40}$$

Estimate (40) *implies*

$$\mathfrak{M}^m_p(u; K_{r/e,r}) = \begin{cases} o(r^{m-n}) & \text{if } r \to 0 \\ o(r^m) & \text{if } r \to \infty. \end{cases} \tag{41}$$

Solution $u \in \mathring{W}^{m,p}_{\mathrm{loc}}(\overline{\mathbb{R}^n_+} \setminus \mathcal{O})$ *of problem* (38) *subject to* (41) *is unique.*

(ii) *Let* $f \in W^{-m,p}_{\mathrm{loc}}(\overline{\mathbb{R}^n_+} \setminus \mathcal{O})$ *be subject to*

$$\int_0^1 \rho^{m+n+1}\mathfrak{M}^{-m}_p(f; K_{\rho/e,\rho})\frac{d\rho}{\rho} + \int_1^\infty \rho^{m+n}\mathfrak{M}^{-m}_p(f; K_{\rho/e,\rho})\frac{d\rho}{\rho} < \infty\,. \tag{42}$$

Then problem (38) *has a solution* $u \in \mathring{W}^{m,p}_{\mathrm{loc}}(\overline{\mathbb{R}^n_+} \setminus \mathcal{O})$ *satisfying*

$$\begin{aligned} \mathfrak{M}^m_p(u; K_{r/e,r}) \ \le\ & c\Big(\int_0^r r^{m-n-1}\rho^{m+n+1}\mathfrak{M}^{-m}_p(f; K_{\rho/e,\rho})\frac{d\rho}{\rho} \\ &+ \int_r^\infty r^{m-n}\rho^{m+n}\mathfrak{M}^{-m}_p(f; K_{\rho/e,\rho})\frac{d\rho}{\rho}\Big). \end{aligned} \tag{43}$$

Estimate (43) *implies*

$$\mathfrak{M}^m_p(u; K_{r/e,r}) = \begin{cases} o(r^{m-n-1}) & \text{if } r \to 0 \\ o(r^{m-n}) & \text{if } r \to \infty. \end{cases} \tag{44}$$

Solution $u \in \mathring{W}^{m,p}_{\mathrm{loc}}(\overline{\mathbb{R}^n_+} \setminus \mathcal{O})$ *of problem* (38) *subject to* (44) *is unique.*

Proof. (i) Let us assume that f is supported by $\{x \in \overline{\mathbb{R}^n_+} : 1/2 \le |x| \le 4\}$. We set

$$u(x) = \int_{\mathbb{R}^n_+} \mathcal{G}(x,y)f(y)dy\,, \tag{45}$$

where $\mathcal{G}$ is Green's function of problem (38). Using standard estimates of $\mathcal{G}$ and its derivatives, one arrives at

$$\mathfrak{M}^m_p(u; K_{r/e,r}) \le c(r^m + r^{m-n})\mathfrak{M}^{-m}_p(f; K_{1/4,8})\,.$$

By (10) this inequality can be written in the form (40). We check by dilation that the same holds for f supported by $\rho/2 \le |x| \le 4\rho$ where ρ is an arbitrary positive number.

Now, we remove the restriction on the support of f. By a partition of unity we represent f as the series $f = \sum f_k$, where $f_k \in W^{-m,p}(\mathbb{R}^n_+)$ is supported by $2^{k-1} \le |x| \le 2^{k+2}$, $k = 0, \pm 1, \ldots$, and

$$\sum_{k=-\infty}^{\infty} \mathfrak{M}^{-m}_p(f_k; K_{\rho/e,\rho}) \le c\mathfrak{M}^{-m}_p(f; K_{\rho/e,\rho})\,. \tag{46}$$

Denote by u_k the solution of problem (38) given by (45) with f replaced by f_k. It follows from (46) that the series $u = \sum u_k$ satisfies (40). Hence, u is a required solution.

The uniqueness of u follows from Theorem 3.9.1 in [3], where $k_+ = m$ and $k_- = m - n$.

(ii) The proof of existence is the same as in (i) with the only difference that representation (45) is replaced by

$$u(x) = \int_{\mathbb{R}^n_+} \mathcal{G}(x,y)f(y)dy + \frac{E(x)}{m!}\int_{\mathbb{R}^n_+} y_n^m f(y)dy\,.$$

Uniqueness is a consequence of Theorem 3.9.1 in [3] where $k_+ = m - n$ and $k_- = m - n - 1$. The proof is complete.

Let us turn to the Dirichlet problem

$$\begin{cases} \mathbf{A}(\theta, \partial_\theta, -\partial_t)u = e^{-2mt}f & \text{on } \Pi \\ u \in \mathring{W}^{p,m}_{\text{loc}}(\Pi)\,. \end{cases} \tag{47}$$

The next statement follows directly from Proposition 1 by the change of variables (30).

Proposition 2. (i) *Let $f \in W^{-m,p}_{\text{loc}}(\Pi)$ be subject to*

$$\int_0^\infty e^{-(m+n)\tau}||f||_{W^{-m,p}(\Pi_\tau)}d\tau + \int_{-\infty}^0 e^{-m\tau}||f||_{W^{-m,p}(\Pi_\tau)}d\tau < \infty\,. \tag{48}$$

Then problem (47) has a solution $u \in \mathring{W}^{m,p}_{\text{loc}}(\Pi)$ satisfying the estimate

$$\begin{aligned} ||u||_{W^{m,p}(\Pi_t)} \le c\Big(\int_t^\infty & e^{(n-m)t-(m+n)\tau}||f||_{W^{-m,p}(\Pi_\tau)}d\tau \\ & + \int_{-\infty}^t e^{-m(t+\tau)}||f||_{W^{-m,p}(\Pi_\tau)}d\tau\Big). \end{aligned} \tag{49}$$

Estimate (49) implies

$$||u||_{W^{m,p}(\Pi_t)} = \begin{cases} o(e^{(n-m)t}) & \textit{if } t \to +\infty \\ o(e^{-mt}) & \textit{if } t \to -\infty. \end{cases} \tag{50}$$

The solution $u \in \mathring{W}^{m,p}_{\text{loc}}(\Pi)$ of problem (47) subject to (50) is unique.

(ii) *Let $f \in W^{-m,p}_{\text{loc}}(\Pi)$ be subject to*

$$\int_0^\infty e^{-(m+n+1)\tau}||f||_{W^{-m,p}(\Pi_\tau)}d\tau + \int_{-\infty}^0 e^{-(m+n)\tau}||f||_{W^{-m,p}(\Pi_\tau)}d\tau < \infty\,. \tag{51}$$

Then problem (47) has a solution $u \in \mathring{W}^{m,p}_{\text{loc}}(\Pi)$ satisfying the estimate

$$\begin{aligned} ||u||_{W^{m,p}(\Pi_t)} \quad \le \quad & c\Big(\int_t^\infty e^{(n+1-m)t-(m+n+1)\tau}||f||_{W^{-m,p}(\Pi_\tau)}d\tau \\ & + \int_{-\infty}^t e^{(n-m)t-(m+n)\tau}||f||_{W^{-m,p}(\Pi_\tau)}d\tau\Big). \end{aligned} \tag{52}$$

Estimate (52) *implies*

$$\|u\|_{W^{m,p}(\Pi_t)} = \begin{cases} o(e^{(n-m+1)t}) & \text{if } t \to +\infty \\ o(e^{(n-m)t}) & \text{if } t \to -\infty. \end{cases} \tag{53}$$

The solution $u \in \mathring{W}^{m,p}_{\mathrm{loc}}(\Pi)$ *of problem* (47) *subject to* (53) *is unique.*

The following assertion can be interpreted as a description of the asymptotic behavior of solutions to problem (47) at $\pm\infty$.

Proposition 3. *Let* $f \in W^{-m,p}_{\mathrm{loc}}(\Pi)$ *be subject to*

$$\int_{\mathbb{R}} e^{-(m+n)\tau} \|f\|_{W^{-m,p}(\Pi_\tau)} d\tau < \infty . \tag{54}$$

Also let u_1 *and* u_2 *be solutions from* Proposition 2 (i) *and* (ii) *respectively. Then*

$$u_2 - u_1 = Ce^{(n-m)t} E(\theta) , \tag{55}$$

where C *is a constant.*

Proof. By Proposition 2 (i) and (ii)

$$\|u_2 - u_1\|_{W^{m,p}(\Pi_t)} = o(e^{(n-m+1)t}) \quad \text{as } t \to +\infty$$

and

$$\|u_2 - u_1\|_{W^{m,p}(\Pi_t)} = o(e^{-mt}) \quad \text{as } t \to -\infty.$$

By the local regularity result (see [1], Section 15) the same relations remain valid for $\|u_2 - u_1\|_{W^{2m,2}(\Pi_t)}$. Now (55) follows from Proposition 3.8.1 in [KM].

Returning to the variables x we derive from Proposition 3 the following description of the asymptotic behavior of solutions to problem (38) both at infinity and near the origin.

Proposition 4. *Let* $f \in W^{-m,p}_{\mathrm{loc}}(\overline{\mathbb{R}^n_+} \setminus \mathcal{O})$ *be subject to*

$$\int_0^\infty \rho^{m+n} \mathfrak{M}_p^{-m}(f; K_{\rho/e,\rho}) \frac{d\rho}{\rho} < \infty . \tag{56}$$

Also let u_1 *and* u_2 *be solutions from* (i) *and* (ii) *in Proposition* 1 *respectively. Then*

$$u_2(x) - u_1(x) = C\, E(x) , \tag{57}$$

where C *is a constant.*

We show that the constant C in (57) can be found explicitly.

Proposition 5. *The constant* C *in* (57) *is given by*

$$C = \frac{-1}{m!} \int_{\mathbb{R}^n_+} f(x) x_n^m dx . \tag{58}$$

Proof. Integrating by parts we check the identity

$$\int_{\mathbb{R}^n_+} L(\partial_x)(\zeta E(x))\, x_n^m\, dx = -m!\,, \tag{59}$$

where ζ is a smooth function equal to 1 in a neighborhood of the origin and zero for large $|x|$.

By (57)

$$\int_{\mathbb{R}^n_+} L(\zeta(u_2 - u_1))x_n^m dx = C \int_{\mathbb{R}^n_+} L(\zeta E(x))x_n^m dx. \tag{60}$$

It follows from (59) that the right-hand side is equal to $-C\,m!$. Using (44), we see that

$$\int_{\mathbb{R}^n_+} L(\zeta u_1)x_n^m dx = 0.$$

Similarly, by (41)

$$\int_{\mathbb{R}^n_+} L((\zeta - 1)u_2)x_n^m dx = 0,$$

which together with (60) lead to (58).

Proposition 6. *The constant C in* (55) *is given by*

$$C = \frac{-1}{m!} \int_\Pi e^{-(m+n)t} f(t,\theta)\, \theta_n^m\, dt d\theta\,.$$

The *Proof* results from Proposition 5.

The following uniqueness result is a consequence of Proposition 3.

Corollary 1. *Let $u \in \mathring{W}^{m,p}_{\text{loc}}(\Pi)$ be a solution of* (47) *with $f = 0$. Suppose that u is subject to*

$$||u||_{W^{m,p}(\Pi_t)} = \begin{cases} o(e^{(n-m+1)t}) & \text{if } t \to +\infty \\ o(e^{-mt}) & \text{if } t \to -\infty. \end{cases} \tag{61}$$

Then $u = \text{const}\, e^{(n-m)t} E(\theta)$.

Proof. Let $\zeta = \zeta(t)$ be a smooth function on $\mathbb{R}$ equal to 1 for $t > 1$ and 0 for $t < 0$. Then $u = u_2 - u_1$, where $u_2 = \zeta u$ and $u_1 = (\zeta - 1)u$. The functions u_1 and u_2 satisfy (47) with $f = \mathbf{A}(\zeta u) - \zeta \mathbf{A} u$. Now the result follows from Proposition 3.

7. Properties of the perturbed Dirichlet problems in Π and $\mathbb{R}^n_+$

Now we turn to the Dirichlet problem (34). By (17), the perturbation $\mathbb{N}$ of the operator $\mathbf{A}$ satisfies

$$||\mathbb{N}||_{\mathring{W}^{m,p}(\Pi_t) \to W^{-m,p}(\Pi_t)} \leq c\omega(t)\,, \tag{62}$$

where we use the notation $\omega(t) = \Omega(e^{-t})$. As before, we assume that Ω does not exceed a sufficiently small constant depending on n, m, p and $L_{\alpha\beta}$.

The next statement generalizes Proposition 2.

Proposition 7. *There exist positive constants c and C such that the following two assertions hold:*
(i) *Let* $f \in W_{\text{loc}}^{-m,p}(\Pi)$ *be subject to*

$$\int_0^\infty e^{-(m+n)\tau + C\int_0^\tau \omega(s)ds}||f||_{W^{-m,p}(\Pi_\tau)}d\tau + \int_{-\infty}^0 e^{-m\tau + C\int_\tau^0 \omega(s)ds}||f||_{W^{-m,p}(\Pi_\tau)}d\tau < \infty\,. \tag{63}$$

Then problem (34) *has a solution* $u \in \mathring{W}_{\text{loc}}^{m,p}(\Pi)$ *satisfying the estimate*

$$||u||_{W^{m,p}(\Pi_t)} \le c\Big(\int_t^\infty e^{(n-m)t-(m+n)\tau+C\int_t^\tau \omega(s)ds}||f||_{W^{-m,p}(\Pi_\tau)}d\tau + \int_{-\infty}^t e^{-m(t+\tau)+C\int_\tau^t \omega(s)ds}||f||_{W^{-m,p}(\Pi_\tau)}d\tau\Big). \tag{64}$$

Estimate (64) *implies*

$$||u||_{W^{m,p}(\Pi_t)} = \begin{cases} o(e^{(n-m)t-C\int_0^t \omega(s)ds}) & \text{if } t \to +\infty \\ o(e^{-mt-C\int_t^0 \omega(s)ds}) & \text{if } t \to -\infty. \end{cases} \tag{65}$$

Solution $u \in \mathring{W}_{\text{loc}}^{m,p}(\Pi)$ *of problem* (34) *subject to* (65) *is unique.*
(ii) *Let* $f \in W_{\text{loc}}^{-m,p}(\Pi)$ *be subject to*

$$\int_0^\infty e^{-(m+n+1)\tau + C\int_0^\tau \omega(s)ds}||f||_{W^{-m,p}(\Pi_\tau)}d\tau + \int_{-\infty}^0 e^{-(m+n)\tau + C\int_\tau^0 \omega(s)ds}||f||_{W^{-m,p}(\Pi_\tau)}d\tau < \infty\,. \tag{66}$$

Then problem (34) *has a solution* $u \in \mathring{W}_{\text{loc}}^{m,p}(\Pi)$ *satisfying the estimate*

$$||u||_{W^{m,p}(\Pi_t)} \le c\Big(\int_t^\infty e^{(n+1-m)t-(m+n+1)\tau+C\int_t^\tau \omega(s)ds}||f||_{W^{-m,p}(\Pi_\tau)}d\tau + \int_{-\infty}^t e^{(n-m)t-(m+n)\tau+C\int_\tau^t \omega(s)ds}||f||_{W^{-m,p}(\Pi_\tau)}d\tau\Big). \tag{67}$$

Estimate (67) *implies*

$$||u||_{W^{m,p}(\Pi_t)} = \begin{cases} o(e^{(n-m+1)t-C\int_0^t \omega(s)ds}) & \text{if } t \to +\infty \\ o(e^{(n-m)t-C\int_t^0 \omega(s)ds}) & \text{if } t \to -\infty. \end{cases} \tag{68}$$

Solution $u \in \mathring{W}_{\text{loc}}^{m,p}(\Pi)$ *of problem* (34) *subject to* (68) *is unique.*

Proof. Let $k_+ = m$, $k_- = m - n$ in the case (i) and $k_+ = m - n$, $k_- = m - n - 1$ in the case (ii). Repeating the proof of Theorem 5.3.2 in [3] with Proposition 2 playing the role of Theorem 3.5.5 in [3], we construct a solution u satisfying

$$||u||_{W^{m,p}(\Pi_t)} \le c\int_{\mathbb{R}} g_\omega(t,\tau)||f||_{W^{-m,p}(\Pi_\tau)}d\tau\,,$$

where g_ω is a certain positive Green's function of the ordinary differential operator

$$-(\partial_t + k_+)(\partial_t + k_-) - c\omega(t)\,.$$

According to Proposition 6.3.1 in [3] this Green's function satisfies

$$g_\omega(t,\tau) \le c\, e^{k_\pm(\tau-t)\pm C\int_\tau^t \omega(s)ds} \quad \text{for } t \gtrless \tau,$$

which completes the proof of existence.

We turn to the proof of uniqueness. Let $u \in \mathring{W}^{m,p}_{\mathrm{loc}}(\Pi)$ be a solution of problem (34) with $f = 0$ subject either to estimate (65) or (68). Clearly, these estimates are valid for $p = 2$. The result follows from Theorem 10.8.13 in [3], where $\ell = 2m$ and $q = m$.

By the change of variables (30) on can formulate Proposition 7 as follows

Proposition 8. *There exists a positive constant C such that the following two assertions hold:*

(i) *Let $f \in W^{-m,p}_{\mathrm{loc}}(\overline{\mathbb{R}^n_+} \setminus \mathcal{O})$ be subject to*

$$\int_0^1 \rho^{m+n} e^{C\int_\rho^1 \Omega(s)\frac{ds}{s}} \mathfrak{M}_p^{-m}(f; K_{\rho/e,\rho})\frac{d\rho}{\rho} + \int_1^\infty \rho^m e^{C\int_1^\rho \Omega(s)\frac{ds}{s}} \mathfrak{M}_p^{-m}(f : K_{\rho/e,\rho})\frac{d\rho}{\rho} < \infty\,. \tag{69}$$

Then problem (11), (12) *has a solution $u \in \mathring{W}^{m,p}_{\mathrm{loc}}(\overline{\mathbb{R}^n_+} \setminus \mathcal{O})$ satisfying*

$$\mathfrak{M}_p^m(u; K_{r/e,r}) \le c\Big(\int_0^r r^{m-n}\rho^{m+n} e^{C\int_\rho^r \Omega(s)\frac{ds}{s}} \mathfrak{M}_p^{-m}(f; K_{\rho/e,\rho})\frac{d\rho}{\rho} + \int_r^\infty r^m \rho^m e^{C\int_r^\rho \Omega(s)\frac{ds}{s}} \mathfrak{M}_p^{-m}(f; K_{\rho/e,\rho})\frac{d\rho}{\rho}\Big). \tag{70}$$

Estimate (70) *implies*

$$\mathfrak{M}_p^m(u; K_{r/e,r}) = \begin{cases} o(r^{m-n} e^{-C\int_r^1 \Omega(s)\frac{ds}{s}}) & \text{if } r \to 0 \\ o(r^m e^{-C\int_1^r \Omega(s)\frac{ds}{s}}) & \text{if } r \to \infty. \end{cases} \tag{71}$$

Solution $u \in \mathring{W}^{m,p}_{\mathrm{loc}}(\overline{\mathbb{R}^n_+} \setminus \mathcal{O})$ of problem (11), (12) *subject to* (71) *is unique.*

(ii) *Let $f \in W^{-m,p}_{\mathrm{loc}}(\overline{\mathbb{R}^n_+} \setminus \mathcal{O})$ be subject to*

$$\int_0^1 \rho^{m+n+1} e^{C\int_\rho^1 \Omega(s)\frac{ds}{s}} \mathfrak{M}_p^{-m}(f; K_{\rho/e,\rho})\frac{d\rho}{\rho} + \int_1^\infty \rho^{m+n} e^{C\int_1^\rho \Omega(s)\frac{ds}{s}} \mathfrak{M}_p^{-m}(f; K_{\rho/e,\rho})\frac{d\rho}{\rho} < \infty\,. \tag{72}$$

Then problem (11), (12) *has a solution* $u \in \mathring{W}^{m,p}_{\text{loc}}(\overline{\mathbb{R}^n_+} \setminus \mathcal{O})$ *satisfying*

$$\begin{aligned}\mathfrak{M}^m_p(u; K_{r/e,r}) \;\le\; & c\Big(\int_0^r r^{m-n-1}\rho^{m+n+1} e^{C\int_\rho^r \Omega(s)\frac{ds}{s}}\,\mathfrak{M}^{-m}_p(f; K_{\rho/e,\rho})\frac{d\rho}{\rho} \\ & + \int_r^\infty r^{m-n}\rho^{m+n} e^{C\int_r^\rho \Omega(s)\frac{ds}{s}}\,\mathfrak{M}^{-m}_p(f; K_{\rho/e,\rho})\frac{d\rho}{\rho}\Big). \qquad (73)\end{aligned}$$

Estimate (73) *implies*

$$\mathfrak{M}^m_p(u; K_{r/e,r}) = \begin{cases} o(r^{m-n-1}e^{-C\int_r^1 \Omega(s)\frac{ds}{s}}) & \textit{if } r \to 0 \\ o(r^{m-n}e^{-C\int_1^r \Omega(s)\frac{ds}{s}}) & \textit{if } r \to \infty. \end{cases} \qquad (74)$$

Solution $u \in \mathring{W}^{m,p}_{\text{loc}}(\overline{\mathbb{R}^n_+} \setminus \mathcal{O})$ *of problem* (11), (12) *subject to* (74) *is unique.*

8. Reduction of problem (34) to a first order system in t

Let

$$\int_{\mathbb{R}} e^{-(m+n)\tau + C|\int_0^\tau \omega(s)ds|}\,||f||_{W^{-m,p}(\Pi_\tau)}d\tau < \infty\,. \qquad (75)$$

This condition implies both (66) and (63) hence there exist the solutions u_1 and u_2 from Proposition 7 (i) and (ii) respectively. Clearly, the difference $u_1 - u_2$ satisfies the homogeneous problem (34) and the relation

$$||u_1 - u_2||_{W^{m,p}(\Pi_t)} = \begin{cases} o(e^{(n-m+1)t - C\int_0^t \omega(s)ds}) & \text{if } t \to +\infty \\ o(e^{-mt - C\int_t^0 \omega(s)ds}) & \text{if } t \to -\infty. \end{cases} \qquad (76)$$

Here and in Sections 9–18 we show that there exists a solution $\mathcal{Z}$ of the homogeneous problem (34), unique up to a constant factor, such that $u_1 - u_2 = C_f\mathcal{Z}$, where C_f is a constant depending on f. We also give an asymptotic representation of $\mathcal{Z}$ at infinity. We start with reducing problem (34) to a first order system in t. To this end we write (34) in a slightly different form. First we obtain a representation of the right-hand side f by using the following standard assertion

Lemma 1. *One can represent* $f \in W^{-m,p}_{\text{loc}}(\Pi)$ *as*

$$f = e^{2mt}\sum_{j=0}^{m}(-\partial_t)^{m-j} f_j\,, \qquad (77)$$

where $f_j \in L^p_{\text{loc}}(\mathbb{R}; W^{-j,p}(S^{n-1}_+))$. *This representation can be chosen to satisfy*

$$c_1\mathfrak{M}^{-m}_p(f; K_{e^{-1-t},e^{-t}}) \le e^{2mt}\sum_{j=0}^{m}||f_j||_{W^{-j,p}(\Pi_t)} \le c_2\mathfrak{M}^{-m}_p(f; K_{e^{-2-t},e^{1-t}})\,,$$

where c_1 *and* c_2 *are constants depending only on* n, m *and* p.

One verifies directly that

$$r^{|\alpha|}\partial_x^\alpha u = \sum_{l=0}^{|\alpha|} Q_{\alpha l}(\theta, \partial_\theta)(r\partial_r)^l u$$

and

$$r^{2m}\partial_x^\alpha (r^{-2m+|\alpha|}u) = \sum_{l=0}^{|\alpha|} P_{\alpha l}(\theta, \partial_\theta)(r\partial_r)^l u$$

where $Q_{\alpha l}(\theta, \partial_\theta)$ and $P_{\alpha l}(\theta, \partial_\theta)$ are differential operators of order $|\alpha| - l$ with smooth coefficients. Furthermore, integrating by parts in

$$\int_{\mathbb{R}^n_+} \partial_x^\alpha (r^{-2m+|\alpha|}u) r^{2m-n}\overline{v}dx$$

we obtain

$$(-1)^{|\alpha|}\sum_{l=0}^{|\alpha|} Q_{\alpha l}(r\partial_r + 2m - n)^l = \sum_{l=0}^{|\alpha|} P^*_{\alpha l}(-r\partial_r)^l \,. \tag{78}$$

Now we write $\mathbf{A}$ in the form

$$\mathbf{A}(\theta, \partial_\theta, -\partial_t) = \sum_{j=0}^{m} (-\partial_t)^{m-j} \mathcal{A}_j(-\partial_t)\,,$$

where

$$\mathcal{A}_j(-\partial_t) = \sum_{k=0}^{m} A_{jk}(-\partial_t)^{m-k}$$

with

$$A_{jk} = (-1)^m \sum_{|\alpha|=|\beta|=m} P_{\alpha,m-j}(\theta, \partial_\theta) L_{\alpha\beta} Q_{\beta,m-k}(\theta, \partial_\theta)\,.$$

It is clear that

$$A_{jk} : \mathring{W}^{k,p}(S^{n-1}_+) \to W^{-j,p}(S^{n-1}_+) \tag{79}$$

are differential operators of order $\le j + k$ on S^{n-1}_+ with smooth coefficients. Since $Q_{\alpha,|\alpha|} = P_{\alpha,|\alpha|} = \theta^\alpha$, we have

$$A_{00} = L(\theta). \tag{80}$$

We also write

$$\mathbb{N}(\theta, t, \partial_\theta, -\partial_t)u = \sum_{j=0}^{m} (-\partial_t)^{m-j} \left(\mathcal{N}_j(t, -\partial_t)u\right), \tag{81}$$

where

$$\mathcal{N}_j(t, -\partial_t) = \sum_{k=0}^{m} \mathcal{N}_{jk}(t)(-\partial_t)^{m-k} \tag{82}$$

with

$$\mathcal{N}_{jk} = \sum_{m-j\le|\alpha|\le m} \; \sum_{m-k\le|\beta|\le m} (-1)^{|\alpha|} P_{\alpha,m-j} N_{\alpha\beta} Q_{\beta,m-k}\,. \tag{83}$$

We use the notation

$$N_{\alpha\beta}(e^{-t}\theta) = L_{\alpha\beta} - \mathcal{L}_{\alpha\beta}(e^{-t}\theta)$$

if $|\alpha| = |\beta| = m$ and

$$N_{\alpha\beta}(e^{-t}\theta) = -e^{(|\alpha+\beta|-2m)t}\mathcal{L}_{\alpha\beta}(e^{-t}\theta)$$

if $|\alpha+\beta| < 2m$. By (83) the operators

$$\mathcal{N}_{jk}(t) \,:\, \mathring{W}^{k,p}(S_+^{n-1}) \to W^{-j,p}(S_+^{n-1})$$

are continuous. By (83) and (78), for almost all $r > 0$

$$\begin{aligned}
&\int_{S_+^{n-1}} \sum_{j,k\le m} \mathcal{N}_{jk}(-\partial_t)^{m-k} u \partial_t^{m-j}(e^{(2m-n)t}\overline{v})d\theta \\
&= \int_{S_+^{n-1}} \sum_{j,k\le m}\ \sum_{m-j\le|\alpha|\le m}\ \sum_{m-k\le|\beta|\le m} (-1)^{|\alpha|} N_{\alpha\beta} Q_{\beta,m-k}(-\partial_t)^{m-k}u \\
&\qquad\qquad \times P^*_{\alpha,m-j}\partial_t^{m-j}(e^{(2m-n)t}\overline{v})d\theta \\
&= r^n \int_{S_+^{n-1}} \sum_{|\alpha|,|\beta|\le m} (L_{\alpha\beta} - \mathcal{L}_{\alpha\beta}(x))\partial_x^\beta u \partial_x^\alpha \overline{v} d\theta \\
&\qquad - r^n \int_{S_+^{n-1}} \sum_{|\alpha+\beta|<2m} \mathcal{L}_{\alpha\beta}(x)\partial_x^\beta u\partial_x^\alpha\overline{v}d\theta\,, \qquad (84)
\end{aligned}$$

where u and v are in $\mathring{W}^{m,p}_{\text{loc}}(\overline{\mathbb{R}^n_+})$.

Using the operators $\mathcal{A}_j(-\partial_t)$ and $\mathcal{N}_j(t,-\partial_t)$, and (77) we write problem (34) in the form

$$\sum_{j=0}^{m}(-\partial_t)^{m-j}\mathcal{A}_j(-\partial_t)u(t) = \sum_{j=0}^{m}(-\partial_t)^{m-j}\big(\mathcal{N}_j(t,-\partial_t)u + f_j(t)\big) \quad \text{on } \mathbb{R}, \qquad (85)$$

where we consider u and f_j as functions on $\mathbb{R}$ taking values in function spaces on S_+^{n-1}. By (15) and (78)

$$||\mathcal{N}_{jk}(t)||_{\mathring{W}^{k,p}(S_+^{n-1})\to W^{-j,p}(S_+^{n-1})} \le c\omega(t)\,. \qquad (86)$$

Clearly, $\mathcal{N}_j$ acts from $W^{m,p}_{\text{loc}}(\Pi)$ to $L^p_{\text{loc}}(\mathbb{R}; W^{-j,p}(S_+^{n-1}))$.

Let $\mathcal{U} = \text{col}(\mathcal{U}_1,\dots,\mathcal{U}_{2m})$, where

$$\mathcal{U}_k = (-\partial_t)^{k-1}u, \quad k=1,\dots,m, \qquad (87)$$

$$\mathcal{U}_{m+1} = \mathcal{A}_0(-\partial_t)u - \mathcal{N}_0(t,-\partial_t)u - f_0 \qquad (88)$$

and

$$\mathcal{U}_{m+j} = -\partial_t\mathcal{U}_{m+j-1} + \mathcal{A}_{j-1}(-\partial_t)u - \mathcal{N}_{j-1}(t,-\partial_t)u - f_{j-1} \qquad (89)$$

for $j = 2,\dots,m$. With this notation (85) takes the form

$$-\partial_t\mathcal{U}_{2m} + \mathcal{A}_m(-\partial_t)u - \mathcal{N}_m(t,-\partial_t)u - f_m = 0. \qquad (90)$$

Using (87) we write (88) as

$$(A_{00} - \mathcal{N}_{00}(t))(-\partial_t)^m u = \mathcal{U}_{m+1} - \sum_{k=0}^{m-1} (A_{0,m-k} - \mathcal{N}_{0,m-k}(t))\mathcal{U}_{k+1} + f_0 \,. \tag{91}$$

Since the function

$$\mathcal{N}_{00}(t) = \sum_{|\alpha|=|\beta|=m} N_{\alpha\beta}(e^{-t}\theta)\theta^{\alpha+\beta}$$

is bounded by $c\omega(t)$, equation (91) is uniquely solvable with respect to $(-\partial_t)^m u$ and

$$(-\partial_t)^m u = \mathcal{S}(t)\mathcal{U} + (A_{00} - \mathcal{N}_{00}(t))^{-1} f_0 \,, \tag{92}$$

where

$$\mathcal{S}(t)\mathcal{U} = (A_{00} - \mathcal{N}_{00}(t))^{-1}\Big(\mathcal{U}_{m+1} - \sum_{k=0}^{m-1} (A_{0,m-k} - \mathcal{N}_{0,m-k}(t))\mathcal{U}_{k+1}\Big). \tag{93}$$

From (87) it follows that

$$-\partial_t \mathcal{U}_k = \mathcal{U}_{k+1} \quad \text{for } k = 1, \dots, m-1. \tag{94}$$

By (92) we have

$$-\partial_t \mathcal{U}_m = \mathcal{S}(t)\mathcal{U} + (A_{00} - \mathcal{N}_{00}(t))^{-1} f_0. \tag{95}$$

Using (92), we write (89) as

$$\begin{aligned} -\partial_t \mathcal{U}_{m+j} &= \mathcal{U}_{m+j+1} - \sum_{k=0}^{m-1} (A_{j,m-k} - \mathcal{N}_{j,m-k}(t))\mathcal{U}_{k+1} \\ &-(A_{j0} - \mathcal{N}_{j0}(t))(\mathcal{S}(t)\mathcal{U} + (A_{00} - \mathcal{N}_{00}(t))^{-1} f_0) + f_j \end{aligned} \tag{96}$$

for $j = 1, \dots, m-1$ and (90) takes the form

$$\begin{aligned} &-\partial_t \mathcal{U}_{2m} + \sum_{k=0}^{m-1} (A_{m,m-k} - \mathcal{N}_{m,m-k}(t))\mathcal{U}_{k+1} \\ &+(A_{m0} - \mathcal{N}_{m0}(t))(\mathcal{S}(t)\mathcal{U} + (A_{00} - \mathcal{N}_{00}(t))^{-1} f_0) - f_m = 0. \end{aligned} \tag{97}$$

The relations (94), (95)–(97) can be written as the first order evolution system

$$(-\mathcal{I}\partial_t - \mathfrak{A})\mathcal{U}(t) - \mathfrak{N}(t)\mathcal{U}(t) = \mathcal{F}(t) \quad \text{on } \mathbb{R}, \tag{98}$$

where

$$\mathcal{F}(t) = \mathrm{col}(0, \dots, 0, \mathcal{F}_m(t), \mathcal{F}_{m+1}(t), \dots, \mathcal{F}_{2m}(t)) \tag{99}$$

with

$$\mathcal{F}_m(t) = (A_{00} - \mathcal{N}_{00}(t))^{-1} f_0(t), \tag{100}$$

$$\mathcal{F}_{m+j}(t) = f_j(t) - (A_{j0} - \mathcal{N}_{j0}(t))(A_{00} - \mathcal{N}_{00}(t))^{-1} f_0(t), \quad j = 1, \dots, m\,. \tag{101}$$

The operator $\mathfrak{N}$ is given by

$$\mathfrak{N}(t)\mathcal{U} = \mathrm{col}(0, \dots, 0, \mathfrak{N}_m(t)\mathcal{U}, \mathfrak{N}_{m+1}(t)\mathcal{U}, \dots, \mathfrak{N}_{2m}(t)\mathcal{U}), \tag{102}$$

where

$$\mathfrak{N}_m(t)\mathcal{U} = (A_{00} - \mathcal{N}_{00}(t))^{-1}\Big(\sum_{k=0}^{m-1} \mathcal{N}_{0,m-k}(t))\mathcal{U}_{k+1} + \mathcal{N}_{00}(t)\mathcal{S}(t)\mathcal{U}\Big) \tag{103}$$

and

$$\begin{aligned}\mathfrak{N}_{m+j}(t)\mathcal{U} = &\sum_{k=0}^{m-1} \mathcal{N}_{j,m-k}(t)\mathcal{U}_{k+1} + \mathcal{N}_{j0}(t)\mathcal{S}(t)\mathcal{U}\\ &- A_{j0}(A_{00} - \mathcal{N}_{00}(t))^{-1}\Big(\sum_{k=0}^{m-1} \mathcal{N}_{0,m-k}(t)\mathcal{U}_{k+1} + \mathcal{N}_{00}(t)\mathcal{S}(t)\mathcal{U}\Big)\end{aligned} \tag{104}$$

for $j = 1, \ldots, m$.

In (98), by $\mathcal{I}$, we denote the identity operator. We also use the operator matrix

$$\mathfrak{A} = \mathcal{J} - \mathfrak{L} \tag{105}$$

with $\mathcal{J} = \{(\mathcal{J})_{jk}\}_{j,k=1}^{2m}$ given by

$$\begin{array}{cc} & \begin{matrix} & & & & (m+1) & & \end{matrix} \\ \begin{matrix} \\ \\ \\ (m) \\ \\ \\ \\ \end{matrix} & \begin{pmatrix} 0 & I & 0 & \cdots & \cdots & \cdots & 0 \\ 0 & 0 & I & \cdots & \vdots & \cdots & 0 \\ \cdots & \cdots & \cdots & \ddots & \vdots & \cdots & \cdots \\ 0 & \cdots & \cdots & \cdots & A_{00}^{-1} & \cdots & 0 \\ \cdots & \cdots & \cdots & \cdots & \cdots & \ddots & \cdots \\ 0 & 0 & 0 & \cdots & \cdots & \cdots & I \\ 0 & 0 & 0 & \cdots & \cdots & \cdots & 0 \end{pmatrix} \end{array}$$

and with $\mathfrak{L} = \{\mathfrak{L}_{jk}\}_{j,k=1}^{2m}$ equal to

$$\begin{pmatrix} 0 & \cdots & 0 & 0 & \cdots & 0 \\ \vdots & \ddots & \vdots & \cdots & \vdots & \vdots \\ 0 & \cdots & 0 & 0 & \cdots & 0 \\ A_{00}^{-1}A_{0,m} & \cdots & A_{00}^{-1}A_{0,1} & 0 & \cdots & 0 \\ A_{1,m} - A_{1,0}A_{00}^{-1}A_{0,m} & \cdots & A_{1,1} - A_{1,0}A_{00}^{-1}A_{0,1} & 0 & \cdots & 0 \\ \vdots & \cdots & \vdots & \cdots & \ddots & \vdots \\ A_{m,m} - A_{m,0}A_{00}^{-1}A_{0,m} & \cdots & A_{m,1} - A_{m,0}A_{00}^{-1}A_{0,1} & 0 & \cdots & 0 \end{pmatrix}$$

We put

$$\mathcal{D} = \mathring{W}^{m,p}(S_+^{n-1}) \times \cdots \times \mathring{W}^{1,p}(S_+^{n-1}) \times L^p(S_+^{n-1}) \times (W^{-m,p}(S_+^{n-1}))^{m-1}$$

and

$$\mathcal{R} = \mathring{W}^{m-1,p}(S_+^{n-1}) \times \cdots \times \mathring{W}^{1,p}(S_+^{n-1}) \times L^p(S_+^{n-1}) \times (W^{-m,p}(S_+^{n-1}))^{m}\,.$$

By (79) the operator $\mathfrak{A} : \mathcal{D} \to \mathcal{R}$ is continuous.

9. Linearization of the pencil $\mathcal{A}(\lambda)$

Here we find a correspondence between $\mathcal{A}(\lambda)$ and the linear pencil $\lambda\mathcal{I} - \mathfrak{A}$.

Lemma 2. *Let the row vector*

$$e(\lambda) = (e_1(\lambda), \dots, e_{2m}(\lambda))$$

be given by

$$e_{2m-j}(\lambda) = \lambda^j, \quad j = 0, \dots, m-1, \tag{106}$$

$$e_m(\lambda) = \sum_{j=0}^{m} \lambda^{m-j} A_{j0}\,, \tag{107}$$

$$e_{m-k}(\lambda) = \sum_{s=0}^{k} \sum_{j=0}^{m} \lambda^{k+m-s-j} A_{js}, \quad k = 1, \dots, m-1. \tag{108}$$

Then for all $\lambda \in \mathbb{C}$ *the equality*

$$e(\lambda)(\lambda\mathcal{I} - \mathfrak{A}) = (\mathcal{A}(\lambda), 0, \dots, 0) \tag{109}$$

is valid.

The *Proof* follows by direct substitution of (106)–(108) in (109).

We introduce the operator matrix $\mathcal{E}(\lambda) = \{\mathcal{E}_{pq}(\lambda)\}_{p,q=1}^{2m}$ as

$$\begin{array}{c} \\ \\ \\ \\ \\ (m+1) \\ \\ \\ \end{array}\begin{pmatrix} e_1(\lambda) & e_2(\lambda) & \cdots & \overset{(m)}{e_m(\lambda)} & \cdots & e_{2m-1}(\lambda) & e_{2m}(\lambda) \\ -I & 0 & \cdots & 0 & \cdots & 0 & 0 \\ 0 & -I & \cdots & 0 & \cdots & 0 & 0 \\ \vdots & \vdots & \cdots & \vdots & \cdots & \vdots & \vdots \\ 0 & 0 & \cdots & -A_{00} & \cdots & 0 & 0 \\ \vdots & \vdots & \cdots & \vdots & \cdots & \vdots & \vdots \\ 0 & 0 & \cdots & 0 & \cdots & -I & 0 \end{pmatrix} \tag{110}$$

One can check directly that $\mathcal{E}^{-1}(\lambda)$ is given by

$$\begin{array}{c} \\ \\ \\ (m) \\ \\ \\ \\ \end{array}\begin{pmatrix} 0 & -I & \cdots & \overset{(m+1)}{0} & \cdots & 0 & 0 \\ 0 & 0 & \cdots & 0 & \cdots & 0 & 0 \\ \vdots & \vdots & \cdots & \vdots & \cdots & \vdots & \vdots \\ 0 & 0 & \cdots & -A_{00}^{-1} & \cdots & 0 & 0 \\ \vdots & \vdots & \cdots & \vdots & \cdots & \vdots & \vdots \\ 0 & 0 & \cdots & 0 & \cdots & 0 & -I \\ I & e_1(\lambda) & \cdots & e_m(\lambda) & \cdots & e_{2m-2}(\lambda) & e_{2m-1}(\lambda) \end{pmatrix}$$

Lemma 3. *For all* $\lambda \in \mathbb{C}$

$$\mathcal{E}(\lambda)(\lambda\mathcal{I} - \mathfrak{A}) = \operatorname{diag}(\mathcal{A}(\lambda), I, \dots, I)\begin{pmatrix} J(\lambda) & 0 \\ -\mathcal{B}(\lambda) & J(\lambda) - \mathcal{M} \end{pmatrix} \tag{111}$$

where the $m \times m$ *matrices* $J(\lambda)$, $\mathcal{M}$ *and* $\mathcal{B}(\lambda)$ *are defined by*

$$J(\lambda) = \begin{pmatrix} I & 0 & \dots & 0 & 0 \\ -\lambda & I & \dots & 0 & 0 \\ \vdots & \vdots & & \vdots & \vdots \\ 0 & 0 & \dots & I & 0 \\ 0 & 0 & \dots & -\lambda & I \end{pmatrix},$$

$$\mathcal{M} = \begin{pmatrix} 0 & 0 & \dots & 0 & 0 \\ A_{10}A_{00}^{-1} & 0 & \dots & 0 & 0 \\ A_{20}A_{00}^{-1} & 0 & \dots & 0 & 0 \\ \vdots & \vdots & & \vdots & \vdots \\ A_{m-1,0}A_{00}^{-1} & 0 & \dots & 0 & 0 \end{pmatrix}$$

and

$$\mathcal{B}(\lambda) = \begin{pmatrix} A_{0m} & A_{0,m-1} & \dots & A_{02} & A_{01} \\ A_{1,m} & A_{1,m-1} & \dots & A_{12} & A_{11} \\ \vdots & \vdots & \dots & \vdots & \vdots \\ A_{m-1,m} & A_{m-1,m-1} & \dots & A_{m-1,2} & A_{m-1,1} \end{pmatrix}$$

$$- \begin{pmatrix} 0 & 0 & \dots & 0 & -\lambda A_{00} \\ A_{10}A_{00}^{-1}A_{0,m} & A_{10}A_{00}^{-1}A_{0,m-1} & \dots & A_{10}A_{00}^{-1}A_{02} & A_{10}A_{00}^{-1}A_{01} \\ \vdots & \vdots & \dots & \vdots & \vdots \\ A_{m-1,0}A_{00}^{-1}A_{0,m} & A_{m-1,0}A_{00}^{-1}A_{0,m-1} & \dots & A_{m-1,0}A_{00}^{-1}A_{02} & A_{m-1,0}A_{00}^{-1}A_{01} \end{pmatrix}$$

Proof. By Lemma 2 the left-hand side of (111) is a triangular matrix with the diagonal $\mathcal{A}(\lambda), I, \dots, I$. One can directly verify that it is equal to the right-hand side in (111).

Clearly, the matrix $J(\lambda)$ has the inverse

$$J(\lambda)^{-1} = \begin{pmatrix} I & 0 & 0 & \dots & 0 & 0 \\ \lambda & I & 0 & \dots & \vdots & \vdots \\ \lambda^2 & \lambda & I & \dots & \vdots & \vdots \\ \vdots & \vdots & \vdots & \dots & I & 0 \\ \lambda^{m-1} & \lambda^{m-2} & \lambda^{m-3} & \dots & \lambda & I \end{pmatrix}. \tag{112}$$

In the next lemma we evaluate the inverse of the last matrix in (111). We show, in particular, that this inverse is a polynomial operator matrix.

Lemma 4. *The following formula is valid:*

$$\begin{pmatrix} J(\lambda) & 0 \\ -\mathcal{B}(\lambda) & J(\lambda)-\mathcal{M} \end{pmatrix}^{-1} = \begin{pmatrix} J^{-1}(\lambda) & 0 \\ Q(\lambda) & J^{-1}(\lambda)(I+\mathcal{M}) \end{pmatrix}, \tag{113}$$

where the elements of the matrix $Q(\lambda) = \{Q_{jk}(\lambda)\}_{j,k=1}^{m}$ *are given by*

$$Q_{jk}(\lambda) = \sum_{l=0}^{j-1} \sum_{q=k-1}^{m} \lambda^{q+l+1-k} A_{j-l-1,m-q}. \tag{114}$$

Proof. Let us look for $(J(\lambda)-\mathcal{M})^{-1}$ in the form $J^{-1}(\lambda)+S(\lambda)$, where $S(\lambda)$ has non-zero elements only in the first column and $S_{11}(\lambda)=0$. We have

$$(J(\lambda)-\mathcal{M})(J^{-1}(\lambda)+S(\lambda)) = I + J(\lambda)S(\lambda) - \mathcal{M}J^{-1}(\lambda).$$

Hence

$$S(\lambda) = J^{-1}(\lambda)\mathcal{M}.$$

Therefore we arrive at (113) with

$$Q(\lambda) = (J^{-1}(\lambda)+S(\lambda))\mathcal{B}(\lambda)J^{-1}(\lambda).$$

One can check that the last equality gives (114).

Lemma 5.

(i) *The operator*

$$\lambda\mathcal{I} - \mathfrak{A} : \mathcal{D} \to \mathcal{R} \tag{115}$$

is Fredholm for all $\lambda \in \mathbb{C}$.

(ii) *The spectra of the operator* $\mathfrak{A}$ *and the pencil* $\mathcal{A}(\lambda)$ *coincide and consist of eigenvalues of the same multiplicity.*

Proof. Let

$$\begin{aligned} \mathfrak{B} &= W^{-m,p}(S_+^{n-1}) \times \mathring{W}^{m-1,p}(S_+^{n-1}) \\ &\quad \times \cdots \times \mathring{W}^{1,p}(S_+^{n-1}) \times L^p(S_+^{n-1}) \times (W^{-m,p}(S_+^{n-1}))^{m-1}. \end{aligned}$$

The operator

$$\mathcal{E}(\lambda) : \mathcal{R} \to \mathfrak{B}$$

is an isomorphism for all $\lambda \in \mathbb{C}$. Analogously, one verifies that the operator

$$\left\{ \begin{matrix} J(\lambda) & 0 \\ -\mathcal{B}(\lambda) & J(\lambda)-\mathcal{M} \end{matrix} \right\} : \mathcal{D} \to \mathcal{D}$$

is isomorphic for all $\lambda \in \mathbb{C}$. Hence and by (111) the polynomial operator functions

$$\lambda\mathcal{I} - \mathfrak{A} : \mathcal{D} \to \mathcal{R}$$

and

$$\operatorname{diag}(\mathcal{A}(\lambda), I, \dots, I) : \mathcal{D} \to \mathfrak{B}$$

are equivalent and therefore these functions have the same spectrum, and the geometric, partial and algebraic multiplicities of their eigenvalues coincide (see, for example, [3], Appendix).

10. Spectral properties of $\mathfrak{A}$

We put

$$\phi(\theta) = E(\theta) \quad \text{and} \quad \psi(\theta) = -(m!)^{-1}\theta_n^m .$$

By (59) and (36)

$$\int_\Pi \mathcal{A}(-\partial_t)(\eta(t)e^{(n-m)t}\phi(\theta))\, e^{(m-n)t}\psi(\theta)d\theta dt = 1 , \tag{116}$$

where η is a smooth function equal to 1 for large positive t and 0 for large negative t. The equality (116) can be written as

$$\int_{S_+^{n-1}} \mathcal{A}'(m-n)\phi(\theta)\, \psi(\theta)d\theta = -1 . \tag{117}$$

We introduce the vector

$$\Phi = \mathrm{col}(\Phi_k)_{k=1}^{2m} = \begin{pmatrix} J^{-1}(m-n) & 0 \\ Q(m-n) & J^{-1}(m-n)(I+\mathcal{M}) \end{pmatrix} \mathrm{col}(\phi, 0, \dots, 0). \tag{118}$$

Owing to (111) and (113) we obtain

$$((m-n)\mathcal{I} - \mathfrak{A})\Phi = 0 . \tag{119}$$

Using (112) and the definitions of the matrices $\mathcal{M}$ and $\mathcal{B}$ we get

$$\Phi_k = (m-n)^{k-1}\phi, \quad k = 1, \dots, m, \tag{120}$$

$$\Phi_{m+k} = \sum_{p=0}^{k-1}\sum_{q=0}^{m} A_{k-p-1,m-q}(m-n)^{p+q}\phi \tag{121}$$

for $k = 1, \dots, m$.

We introduce the vector $\Psi = \mathrm{col}(\Psi_k)_{k=1}^{2m}$, by

$$\Psi = \mathcal{E}^*(m-n)\mathrm{col}(\psi, 0, \dots, 0) \tag{122}$$

where $\mathcal{E}^*(\lambda)$ is the adjoint of $\mathcal{E}(\overline{\lambda})$. Since ψ is the eigenfunction of the pencil $(\mathcal{A}(\lambda))^*$ corresponding to the eigenvalue $\lambda = m$, it follows from (111) that

$$((m-n)\mathcal{I} - \mathfrak{A}^*)\Psi = 0. \tag{123}$$

By (110)

$$\Psi_k = \sum_{p=0}^{m-k}\sum_{q=0}^{m} A^*_{qp}(m-n)^{2m-k-q-p}\psi$$

for $k = 1, \dots, m-1$,

$$\Psi_m = \sum_{q=0}^{m} A^*_{q0}(m-n)^{m-q}\psi$$

and $\Psi_{m+k} = (m-n)^{m-k}\psi$ for $k = 1, \dots, m$.

Clearly, $\Phi \in \mathcal{D}$, $\Psi \in \mathcal{R}^*$, where

$$\mathcal{R}^* = W^{1-m,q}(S_+^{n-1}) \times \cdots \times W^{-1,q}(S_+^{n-1}) \times L^q(S_+^{n-1}) \times (\mathring{W}^{m,q}(S_+^{n-1}))^m .$$

Proposition 9. *The biorthogonality condition*

$$(\Phi, \Psi)_{L^2(S_+^{n-1})} = -1 \tag{124}$$

is valid.

Proof. By (111) and (113)

$$\big((\lambda\mathcal{I} - \mathfrak{A})\Phi_\lambda, \Psi_\lambda\big) = (\mathcal{A}(\lambda)\phi, \psi)_{(L^2(S_+^{n-1}))}, \tag{125}$$

where

$$\Phi_\lambda = \begin{pmatrix} J^{-1}(\lambda) & 0 \\ Q(\lambda) & J^{-1}(\lambda)(I + \mathcal{M}) \end{pmatrix} \operatorname{col}(\phi, 0, \dots, 0)$$

and $\Psi_\lambda = \mathcal{E}^*(\lambda)\operatorname{col}(\psi_k, 0, \dots, 0)$. Taking the first derivative of (125) with respect to λ, setting $\lambda = m - n$ and using (119) and (123) together with (117) we arrive at (124).

We introduce the spectral projector $\mathcal{P}$ corresponding to the eigenvalue $\lambda = m - n$:

$$\mathcal{P}\mathcal{F} = -(\mathcal{F}, \Psi)_{L^2(S_+^{n-1})}\Phi\,. \tag{126}$$

This operator maps $\mathcal{R}$ into $\mathcal{D}$.

11. Equivalence of equation (85) and system (98)

We introduce some vector function spaces to be used in the subsequent study of system (98).

Let $\mathbb{S}(a,b)$ be the space of vector functions $\mathcal{U}$ on the interval (a,b) with values in $\mathcal{D}$ such that

$$||\mathcal{U}||_{\mathbb{S}(a,b)} = \Big(\int_a^b (||\mathcal{U}(\tau)||^p_{\mathcal{D}} + ||\partial_\tau \mathcal{U}(\tau)||^p_{\mathcal{R}})d\tau\Big)^{1/p} < \infty\,.$$

More explicitly:

$$\begin{aligned} ||\mathcal{U}||_{\mathbb{S}(a,b)} =& \Big(\int_a^b \Big(\sum_{j=1}^{m+1} ||\mathcal{U}_j(\tau)||^p_{\mathring{W}^{m+1-j,p}(S_+^{n-1})} + \sum_{j=m+2}^{2m} ||\mathcal{U}_j(\tau)||^p_{W^{-m,p}(S_+^{n-1})} \\ &+ \sum_{j=1}^{m} ||\partial_\tau \mathcal{U}_j(\tau)||^p_{\mathring{W}^{m-j,p}(S_+^{n-1})} + \sum_{j=m+1}^{2m} ||\partial_\tau \mathcal{U}_j(\tau)||^p_{W^{-m,p}(S_+^{n-1})}\Big)d\tau\Big)^{1/p}. \end{aligned}$$

By $\mathbb{S}_{\text{loc}}(\mathbb{R})$ we denote the space of functions defined on $\mathbb{R}$ with finite seminorms $||\mathcal{U}||_{\mathbb{S}(t,t+1)}$, $t \in \mathbb{R}$. Let $\mathcal{P}$ be the projector given by (126). Clearly,

$$||\mathcal{P}\mathcal{U}||_{\mathbb{S}(a,b)} \le c||\mathcal{U}||_{\mathbb{S}(a,b)}\,. \tag{127}$$

By $L^p(a,b;B)$ and $L^p_{\text{loc}}(\mathbb{R};B)$ we denote the L^p and L^p_{loc} spaces of vector functions on (a,b) and $\mathbb{R}$ which take values in a Banach space B.

Let $W_0^{m,p}((a,b)\times S_+^{n-1})$ be the subspace of the Sobolev space $W^{m,p}((a,b)\times S_+^{n-1})$ containing functions vanishing on $(a,b)\times\partial S_+^{n-1}$ together with their derivatives up to order $m-1$. The space of vector functions

$$\mathcal{U}(t)=\mathrm{col}(u(t),\dots,\partial_t^{m-1}u(t),u_{m+1},\dots,u_{2m}(t)) \tag{128}$$

with $u\in W_0^{m,p}((a,b)\times S_+^{n-1})$,

$$u_{m+1}\in L^p(a,b;L^p(S_+^{n-1})),\quad \partial_t u_{m+1}\in L^p(a,b;W^{-m,p}(S_+^{n-1}))$$

and

$$u_{m+j},\ \partial_t u_{m+j}\in L^p(a,b;W^{-m,p}(S_+^{n-1})),\quad j=2,\dots,m,$$

will be denoted by $\mathbf{S}(a,b)$. The norm in $\mathbf{S}(a,b)$ is defined by

$$\begin{aligned}||\mathcal{U}||_{\mathbf{S}(a,b)}&=||u||_{W^{m,p}((a,b)\times S_+^{n-1})}+||u_{m+1}||_{L^p(a,b;L^p(S_+^{n-1}))}\\&+\sum_{j=2}^m||u_{m+j}||_{L^p(a,b;W^{-m,p}(S_+^{n-1}))}+\sum_{j=1}^m||\partial_t u_{m+j}||_{L^p(a,b;W^{-m,p}(S_+^{n-1}))}.\end{aligned}$$

The space $\mathbf{S}(a,b)$ is embedded into $\mathbb{S}(a,b)$ and for $\mathcal{U}\in\mathbf{S}$

$$c_1||\mathcal{U}||_{\mathbf{S}(a,b)}\le||\mathcal{U}||_{\mathbb{S}(a,b)}\le c_2||\mathcal{U}||_{\mathbf{S}(a,b)}.$$

The space $\mathbf{S}_{\mathrm{loc}}(\mathbb{R})$ is defined as the set of vector functions $\mathcal{U}$ such that their restrictions to every finite interval (a,b) belong to $\mathbf{S}(a,b)$. The seminorms in this space are $||\mathcal{U}||_{\mathbf{S}(t,t+1)}$, $t\in\mathbb{R}$.

By $\mathbb{X}(a,b)$ we denote the space

$$\mathbb{X}(a,b)=\{\mathcal{V}\,:\,\mathcal{V}=(\mathcal{I}-\mathcal{P})\mathcal{U},\ \mathcal{U}\in\mathbf{S}(a,b)\} \tag{129}$$

endowed with the norm

$$||\mathcal{V}||_{\mathbb{X}(a,b)}=\inf||\mathcal{U}||_{\mathbf{S}(a,b)}\,,$$

where the infimum is taken over all $\mathcal{U}$ in (129).

We use the space $\mathbb{X}_{\mathrm{loc}}(\mathbb{R})=\{\mathcal{V}\,:\,\mathcal{V}=(\mathcal{I}-\mathcal{P})\mathcal{U},\quad \mathcal{U}\in\mathbf{S}_{\mathrm{loc}}(\mathbb{R}))\}$ and finally, we introduce the space $\mathbb{Y}_{\mathrm{loc}}(\mathbb{R})$ which consists of the vector functions $\mathcal{F}=\mathrm{col}(0,\dots,0,\mathcal{F}_m,\mathcal{F}_{m+1},\dots,\mathcal{F}_{2m})$ with finite seminorms

$$||\mathcal{F}||_{\mathbb{Y}(t,t+1)}=\Big(\sum_{j=0}^m\int_t^{t+1}||\mathcal{F}_{m+j}(\tau)||^p_{W^{-j,p}(S_+^{n-1})}d\tau\Big)^{1/p},\quad t\in\mathbb{R}\,.$$

We return to system (98). By (86) the operator $\mathfrak{N}(t)\,:\,\mathbb{S}_{\mathrm{loc}}(\mathbb{R})\to\mathbb{Y}_{\mathrm{loc}}(\mathbb{R})$ is continuous and

$$||\mathfrak{N}||_{\mathbb{S}(t,t+1)\to\mathbb{Y}(t,t+1)}\le c\,\omega(t)\,. \tag{130}$$

Furthermore,

$$c_1||\mathcal{F}||_{\mathbb{Y}(t,t+1)}\le\sum_{j=0}^m||f_j||_{W^{-j,p}(\Pi_t)}\le c_2||\mathcal{F}||_{\mathbb{Y}(t,t+1)}\,, \tag{131}$$

where c_1 and c_2 are positive constant.

We prove that equation (85) and system (98) is equivalent in a certain sense.

Lemma 6. *Let the functions* $f_j \in W^{-j,p}_{\mathrm{loc}}(\Pi)$ *and the vector function* $\mathcal{F} \in \mathbb{Y}_{\mathrm{loc}}(\mathbb{R})$ *be connected by* (99)–(101).

(i) *If* $u \in \mathring{W}^{m,p}_{\mathrm{loc}}(\Pi)$ *is a solution of* (85) *then the vector function* $\mathcal{U} \in \mathbb{S}_{\mathrm{loc}}(\mathbb{R})$ *given by* (87)–(89) *solves* (98).

(ii) *If* $\mathcal{U} \in \mathbb{S}_{\mathrm{loc}}(\mathbb{R})$ *is a solution of* (98) *then* $\mathcal{U} \in \mathbf{S}_{\mathrm{loc}}(\mathbb{R})$ *and the function* $u = \mathcal{U}_1$ *solves* (85).

Proof. (i) This assertion follows directly from the above reduction of (85) to the first order system (98).

(ii) By (94) and (95) we obtain $\mathcal{U}_k = (-\partial_t)^{k-1}\mathcal{U}_1$ for $k = 1, \dots, m$ and $\mathcal{S}(t)\mathcal{U} = (-\partial_t)^m \mathcal{U}_1$. Now (96) takes the form

$$-\partial_t \mathcal{U}_{m+j} = \mathcal{U}_{m+j+1} - \mathcal{A}_j(-\partial_t)\mathcal{U}_1 + \mathcal{N}_j(t, -\partial_t)\mathcal{U}_1 + \mathcal{F}_j$$

and (97) can be written as

$$-\partial_t \mathcal{U}_{2m} + \mathcal{A}_m(-\partial_t)\mathcal{U}_1 - \mathcal{N}_m(t, -\partial_t)\mathcal{U}_1 - \mathcal{F}_m = 0.$$

The last two equations imply (85) for $u = \mathcal{U}_1$.

12. Spectral splitting of the first order system (98)

Let $\mathcal{P}$ be the spectral projector (126). Applying $\mathcal{P}$ and $\mathcal{I} - \mathcal{P}$ to system (98) we arrive at

$$(\mathcal{I}\partial_t + \mathfrak{A})\mathbf{u} + \mathcal{P}\mathfrak{N}(t)(\mathbf{u} + \mathbf{v}) = -\mathcal{P}\mathcal{F} \quad \text{on } \mathbb{R} \tag{132}$$

and

$$(\mathcal{I}\partial_t + \mathfrak{A})\mathbf{v} + (\mathcal{I} - \mathcal{P})\mathfrak{N}(t)\mathbf{v} = (\mathcal{P} - \mathcal{I})(\mathcal{F} + \mathfrak{N}(t)\mathbf{u}) \quad \text{on } \mathbb{R}, \tag{133}$$

where

$$\mathbf{u}(t) = \mathcal{P}\mathcal{U}(t), \quad \mathbf{v}(t) = (\mathcal{I} - \mathcal{P})\mathcal{U}(t). \tag{134}$$

Clearly, $\mathbf{u}$ can be represented as $\mathbf{u}(t) = \kappa(t)\Phi$, where Φ is given by (118). Furthermore, $\mathbf{u} \in \mathbb{S}_{\mathrm{loc}}(\mathbb{R})$ if and only if $\kappa \in W^{1,p}_{\mathrm{loc}}(\mathbb{R})$. Thus we have split system (98) into the scalar equation (132) and the infinite-dimensional system (133). Equation (132) can be written as

$$\frac{d\kappa}{dt}(t) + (m - n)\kappa(t) - (\mathfrak{N}(t)(\mathbf{u} + \mathbf{v})(t), \Psi) = (\mathcal{F}(t), \Psi)$$

where Ψ is defined in Section 10.

In the next lemma we establish the equivalence of equation (85) and the split system (132), (133).

Lemma 7.

(i) *Let* $f_j \in L^p_{\mathrm{loc}}(\mathbb{R}; W^{-j,p}(S^{n-1}_+))$, $j = 0, \dots, m$, *and let* $u \in \mathring{W}^{m,p}_{\mathrm{loc}}(\Pi)$ *be a solution of* (85). *Then the vector function* $\mathcal{U}$ *given by* (87)–(89) *belongs to* $\mathbf{S}_{\mathrm{loc}}(\mathbb{R})$ *and the vector functions* (134) *satisfy* (132) *and* (133) *with* $\mathcal{F}$ *given by* (99)–(101).

(ii) *Let* $\mathcal{F} \in \mathbb{Y}_{\mathrm{loc}}(\mathbb{R})$. *Assume that*

$$\mathbf{u}(t) = (\mathbf{u}_1(t), \dots, \mathbf{u}_{2m}(t)) = \kappa(t)\Phi\,,$$

$\kappa \in W^{1,2}_{\rm loc}(\mathbb{R})$, *and* $\mathbf{v} = (\mathbf{v}_1, \ldots, \mathbf{v}_{2m}) \in \mathbb{S}_{\rm loc}(\mathbb{R})$, *such that* $\mathcal{P}\mathbf{v}(t) = 0$ *for all* $t \in \mathbb{R}$, *satisfy* (132) *and* (133). *Then* $\mathbf{u} + \mathbf{v} \in \mathbf{S}_{\rm loc}(\mathbb{R})$ *and*

$$u = \mathbf{u}_1 + \mathbf{v}_1 \in \mathring{W}^{m,p}_{\rm loc}(\Pi)$$

solves (85) *with* $f_0 = (A_{00} - \mathcal{N}_{00})\mathcal{F}_m$ *and*

$$f_j = \mathcal{F}_{m+j} + (A_{j0} - \mathcal{N}_{j0})\mathcal{F}_m, \quad j = 1, \ldots, m\,.$$

Moreover, $(-\partial_t)^j u = \mathbf{u}_{j+1} + \mathbf{v}_{j+1}$ *for* $j = 1, \ldots, m-1$.

Proof. (i) It suffices to use Lemma 6(i) and to apply the projectors $\mathcal{P}$ and $\mathcal{I} - \mathcal{P}$ to system (98).

(ii) We put $\mathcal{U} = \mathbf{u} + \mathbf{v}$. Clearly, $\mathcal{U} \in \mathbb{S}_{\rm loc}(\mathbb{R})$ and equalities (98) and (134) hold. Now the result follows from Lemma 6(ii).

13. Solvability of the unperturbed infinite-dimensional part of the split system

We consider the case $\mathfrak{N} = 0$. In other words, we deal with the system

$$(-\mathcal{I}\partial_t - \mathfrak{A})\mathbf{v} = (\mathcal{I} - \mathcal{P})\mathcal{F} \quad \text{on } \mathbb{R}. \tag{135}$$

Lemma 8. (i) (Existence) *Let* $\mathcal{F} \in \mathbb{Y}_{\rm loc}(\mathbb{R})$. *Suppose that*

$$\int_0^\infty e^{(m-n-1)\tau} ||\mathcal{F}||_{\mathbb{Y}(\tau,\tau+1)} d\tau + \int_{-\infty}^0 e^{m\tau} ||\mathcal{F}||_{\mathbb{Y}(\tau,\tau+1)} d\tau < \infty\,. \tag{136}$$

Then equation (135) *has a solution* $\mathbf{v} \in \mathbb{X}_{\rm loc}(\mathbb{R})$ *satisfying*

$$\begin{aligned} ||\mathbf{v}||_{\mathbb{X}(t,t+1)} \quad \le \quad & c\Big(\int_t^\infty e^{(n-m+1)(t-\tau)} ||\mathcal{F}||_{\mathbb{Y}(\tau,\tau+1)} d\tau \\ & + \int_{-\infty}^t e^{-m(t-\tau)} ||\mathcal{F}||_{\mathbb{Y}(\tau,\tau+1)} d\tau \Big), \end{aligned} \tag{137}$$

where c *is a constant independent of* $\mathcal{F}$.

(ii) (Uniqueness) *Let* $\mathbf{v} \in \mathbb{X}_{\rm loc}(\mathbb{R})$ *satisfy* (135) *with* $\mathcal{F} = 0$. *Also let*

$$||\mathbf{v}||_{\mathbb{S}(t,t+1)} = \begin{cases} o(e^{(n-m+1)t}) & \text{if } t \to +\infty \\ o(e^{-mt}) & \text{if } t \to -\infty. \end{cases} \tag{138}$$

be valid. Then $\mathbf{v} = 0$.

Proof. (i) Let $f_0 = A_{00}\mathcal{F}_m$ and

$$f_j = \mathcal{F}_{m+j} + A_{j0}\mathcal{F}_m\,, \quad j = 1, \ldots, m.$$

Clearly, $f_j \in L^p_{\rm loc}(\mathbb{R}; W^{-j,p}(S^{n-1}_+))$ and

$$\sum_{j=0}^m ||f_j||_{L^p_{\rm loc}(t,t+1;W^{-j,p}(S^{n-1}_+))} \le c||\mathcal{F}||_{L^p(t,t+1;\mathbb{Y})}\,.$$

Let ζ be a smooth function on $\mathbb{R}$, equal to 1 for $t > 1$ and 0 for $t < 0$. For a fixed $a \in \mathbb{R}$ we represent f_j as $f_{ja}^{(-)} + f_{ja}^{(+)}$, where

$$f_{ja}^{(-)}(t) = \zeta(t-a) f_j(t), \quad f_{ja}^{(+)}(t) = (1 - \zeta(t-a)) f_j(t)\,.$$

Then the functions

$$f_a^{(\pm)}(t) = e^{2mt} \sum_{j=0}^{m} (-\partial_t)^{m-j} f_{ja}^{(\pm)}(t)$$

satisfy (51) and (48) respectively because of (136). By Proposition 2 there exist solutions $u_a^{(\pm)} \in W_{\text{loc}}^{m,p}(\mathbb{R})$ subject to (52) and (49) with f replaced by $f_a^{(\pm)}$. We put $u_a = u_a^{(-)} + u_a^{(+)}$. Then u_a satisfies (34) and

$$\begin{aligned} ||u_a||_{W^{m,p}(a,a+1)} \leq\; & c\Big(\int_a^{\infty} e^{(m-n-1)(a-\tau)} ||\mathcal{F}||_{\mathbb{Y}(\tau,\tau+1)} d\tau \\ & + \int_{-\infty}^{a} e^{-m(a-\tau)} ||\mathcal{F}||_{\mathbb{Y}(\tau,\tau+1)} d\tau \Big), \end{aligned} \tag{139}$$

We introduce the vector function $\mathcal{U}_a = \text{col}(\mathcal{U}_1, \ldots, \mathcal{U}_{2m})$ by (87)–(89), where $\mathcal{N} = 0$ and u is replaced by u_a, and put $\mathbf{v}_a = (\mathcal{I} - \mathcal{P})\mathcal{U}_a$. Clearly, $\mathbf{v}_a$ belongs to $\mathbb{X}_{\text{loc}}(\mathbb{R})$ and satisfies (138). Let us show that $\mathbf{v}_a$ does not depend on a. In fact, let a and b be different real numbers. Then the function $u_a - u_b$ satisfies the homogeneous problem (34) and relations (61). Hence and by Corollary 1 we have $u_a - u_b = Ce^{(n-m)t} E(\theta)$. The last equality implies $\mathbf{v}_a = \mathbf{v}_b$ because of the definition of $\mathcal{P}$. Thus, we can use the notation $\mathbf{v}$ for the vector function $\mathbf{v}_a$. Since $||\mathbf{v}||_{\mathbb{S}(a,a+1)} \leq c||u_a||_{W^{m,p}(a,a+1)}$, estimate (137) follows from (139).

(ii) We put $u = \mathbf{v}_1$. Since $(\mathcal{I}\partial_t + \mathfrak{A})\mathbf{v} = 0$ it follows by (94) and (95) with $\mathcal{N} = 0$ that $\mathbf{v}_k = (-\partial_t)^{k-1} u$ for $k = 1, \ldots, m$ and

$$(-\partial_t)^m u = A_{00}^{-1} \Big(\mathbf{v}_{m+1} - \sum_{k=0}^{m-1} A_{0,m-k} (-\partial_t)^k u \Big)\,.$$

Hence $\mathbf{v}_{m+1} = \mathcal{A}_0(-\partial_t)u$. (Note that $u \in W_{\text{loc}}^{m,p}(\Pi)$ because $\mathbf{v} \in \mathbb{S}_{\text{loc}}(\mathbb{R})$). Now relation (96) with $\mathcal{N} = 0$ takes the form

$$-\partial_t \mathbf{v}_{m+j} = \mathbf{v}_{m+j+1} - \mathcal{A}_j(-\partial_t)u\,, \tag{140}$$

where $j = 1, \ldots, m-1$, and (97) becomes

$$-\partial_t \mathbf{v}_{2m} + \mathcal{A}_m(-\partial_t)u = 0. \tag{141}$$

Using (140) and (141) we obtain $\mathcal{A}(-\partial_t)u = 0$. Furthermore, by (138) the function u satisfies (61). By Corollary 1 we arrive at $u(t) = Ce^{(n-)mt} E(\theta)$ and using the definition (126) of $\mathcal{P}$ we get $\mathbf{v} = 0$. The proof is complete.

14. Solvability of the infinite-dimensional part of the perturbed split system

Here we study the system

$$(\mathcal{I}\partial_t + \mathfrak{A})\mathbf{v} + (\mathcal{I} - \mathcal{P})\mathfrak{N}(t)\mathbf{v} = (\mathcal{I} - \mathcal{P})F \quad \text{on } \mathbb{R}. \tag{142}$$

We introduce the operator $\mathfrak{L}$ which assigns the solution $\mathbf{v} \in \mathbb{S}_{\text{loc}}(\mathbb{R})$ subject to (138) to the right-hand side in (135) satisfying the conditions of Lemma 8. Estimate (137) can be written as

$$||\mathfrak{L}(\mathcal{I} - \mathcal{P})\mathcal{F}||_{\mathbb{X}(t,t+1)} \leq c \int_{-\infty}^{\infty} g(t-\tau)||\mathcal{F}||_{\mathbb{Y}(\tau,\tau+1)} d\tau, \tag{143}$$

where

$$g(t) = \begin{cases} e^{-mt} & \text{for } t \geq 0 \\ e^{-(m-n-1)t} & \text{for } t < 0. \end{cases} \tag{144}$$

Lemma 9. *Let c, $c > 0$, and*

$$\delta := \sup_{\tau \in \mathbb{R}} \omega(\tau) \tag{145}$$

satisfy the inequality $(1 + c)\delta \leq (n + 1)/8$. Then the series

$$g_\omega(t,\tau) = g(t-\tau) + \sum_{k=1}^{\infty} c^k \int_{\mathbb{R}^k} g(t-\tau_1)\omega(\tau_1)g(\tau_1 - \tau_2)\omega(\tau_2) \cdots \omega(\tau_k)g(\tau_k - \tau)d\tau_1\, d\tau_2 \ldots d\tau_k \tag{146}$$

is convergent and admits the estimate

$$g_\omega(t,\tau) \leq \begin{cases} c_1 e^{-m(t-\tau)+c_1\int_\tau^t \omega(s)ds} & \textit{for } t \geq \tau \\ c_1 e^{(n-m+1)(t-\tau)+c_1\int_t^\tau \omega(s)ds} & \textit{for } t < \tau, \end{cases} \tag{147}$$

where $c_1 = 2(1 + c)$.

Proof. We denote the right-hand side in (147) by $g_*(t,\tau)$ and justify the inequality

$$g_*(t,\tau) \geq g(t-\tau) + c\int_{\mathbb{R}} g(t-s)\omega(s)g_*(s,\tau)ds\,. \tag{148}$$

Consider the case $t \geq \tau$. We have

$$\int_\tau^t g(t-s)\omega(s)g_*(s,\tau)ds = e^{-m(t-\tau)}\Big(e^{c_1\int_\tau^t \omega(s)ds} - 1\Big).$$

Furthermore,

$$\int_t^\infty g(t-s)\omega(s)g_*(s,\tau)ds \leq \frac{c_1\delta}{n+1-c_1\delta} e^{-m(t-\tau)+c_1\int_\tau^t \omega(s)ds}$$

and

$$\int_{-\infty}^\tau g(t-s)\omega(s)g_*(s,\tau)ds \leq \frac{c_1\delta}{n+1-c_1\delta} e^{-m(t-\tau)}\,.$$

From the last three relations we derive (148), taking into account that

$$c_1\delta < n+1 \quad \text{and} \quad c_1 \geq 1 + c + 2\frac{c_1 c\delta}{n+1-c_1\delta} \tag{149}$$

by the assumptions of Lemma. The case $\tau > t$ is considered analogously.

Now, iterating (148) we arrive at (147).

The following assertion which concerns the variable coefficient case is similar to Lemma 8.

Lemma 10. *There exist positive constants δ_0 and c_0 depending only on n, m, p and L such that for all $\delta \leq \delta_0$, where δ is given by* (145)*, the following assertions hold:*

(i) *Let F belong to $\mathbb{Y}_{\rm loc}(\mathbb{R})$ and be subject to*

$$\int_0^\infty e^{(m-n-1)\tau + c_0\int_0^\tau \omega(s)ds} ||F||_{\mathbb{Y}(\tau,\tau+1)} d\tau + \int_{-\infty}^0 e^{m\tau + c_0\int_\tau^0 \omega(s)ds} ||F||_{\mathbb{Y}(\tau,\tau+1)} d\tau < \infty . \tag{150}$$

Then system (142) *has a solution $\mathbf{v} \in \mathbb{X}_{\rm loc}(\mathbb{R})$ satisfying*

$$||\mathbf{v}||_{\mathbb{S}(t,t+1)} \leq c \int_t^\infty e^{(n-m+1)(t-\tau)+c_0\int_t^\tau \omega(s)ds} ||F||_{\mathbb{Y}(\tau,\tau+1)} d\tau \\ \int_{-\infty}^t e^{-m(t-\tau)+c_0\int_\tau^t \omega(s)ds} ||F||_{\mathbb{Y}(\tau,\tau+1)} d\tau . \tag{151}$$

(ii) *The solution $\mathbf{v} \in \mathbb{X}_{\rm loc}(\mathbb{R})$ to* (142) *subject to*

$$||\mathbf{v}||_{\mathbb{S}(t,t+1)} = \begin{cases} o\Big(e^{(n-m+1)t - c_0\int_0^t \omega(\tau)d\tau}\Big) & \text{as } t \to +\infty \\ o\Big(e^{-mt - c_0\int_t^0 \omega(\tau)d\tau}\Big) & \text{as } t \to -\infty \end{cases} \tag{152}$$

is unique. (*We note that* (150) *together with* (151) *imply* (152).)

Proof. Let c be the constant in (143). Then one can take

$$\delta_0 = \frac{n+1}{8(1+c)} \quad \text{and} \quad c_0 = 4(1+c).$$

(i) Formally, the solution $\mathcal{U}$ of (142) can be written as the series

$$\sum_{k=0}^\infty (\mathfrak{L}(\mathcal{I}-\mathcal{P})\mathfrak{N})^k \mathfrak{L}(\mathcal{I}-\mathcal{P})F, \tag{153}$$

where $\mathfrak{L}$ is the operator defined at the end of Section 13. We introduce the sequence

$$F^{(k)} = \mathfrak{N}\,\mathfrak{L}\big((\mathcal{I}-\mathcal{P})\mathfrak{N}\,\mathfrak{L}\big)^k(\mathcal{I}-\mathcal{P})F\,, \quad k = 0, 1, \ldots \tag{154}$$

Clearly, $F^{(k)} = \text{col}(0,\dots,0,F_m^{(k)},\dots,F_{2m}^{(k)})$ and by (143) and (130) $F^{(k)} \in \mathbb{Y}_{\text{loc}}(\mathbb{R})$. Now, (153) can be written as

$$\mathfrak{L}(\mathcal{I}-\mathcal{P})F + \mathfrak{L}(\mathcal{I}-\mathcal{P})\sum_{k=0}^{\infty} F^{(k)}. \tag{155}$$

We show that the series

$$\sum_{k=0}^{\infty} F^{(k)} \tag{156}$$

converges in $\mathbb{Y}_{\text{loc}}(\mathbb{R})$. We have $F^{(0)} = \mathfrak{N}\,\mathfrak{L}\,(\mathcal{I}-\mathcal{P})F$ and

$$F^{(k)} = \mathfrak{N}\,\mathfrak{L}(\mathcal{I}-\mathcal{P})F^{(k-1)}, \;\; k=1,\dots$$

By (143) and (130)

$$\| F^{(k)} \|_{\mathbb{Y}(t,t+1)} \le c\omega(t)\int_{\mathbb{R}} g(t-\tau)\, \| F^{(k-1)} \|_{\mathbb{Y}(\tau,\tau+1)}\, d\tau.$$

Therefore,

$$\| F^{(k)} \|_{\mathbb{Y}(t,t+1)} \le c^{k+1}\omega(t)\int_{\mathbb{R}^{k+1}} g(t-\tau_1)\omega(\tau_1)g(\tau_1-\tau_2)\omega(\tau_2)\dots\omega(\tau_k)g(\tau_k-\tau)$$
$$\times \| F \|_{\mathbb{Y}(\tau,\tau+1)}\, d\tau_1\dots d\tau_k d\tau, \;\; k=0,1,\dots \tag{157}$$

This implies

$$\sum_{k=0}^{\infty} ||F^{(k)}||_{\mathbb{Y}(t,t+1)} \le c\,\omega(t)\int_{\mathbb{R}} g_\omega(t,\tau)\, \| F \|_{\mathbb{Y}(\tau,\tau+1)}\, d\tau\,, \tag{158}$$

where g_ω is given by (146). Hence, series (156) converges in $\mathbb{Y}_{\text{loc}}(\mathbb{R})$ to a function F_*. Since

$$g_\omega(t,\tau) = g(t-\tau) + c\int_{\mathbb{R}} g(t-s)\omega(s)g_\omega(s,\tau)ds\,, \tag{159}$$

it follows from (158), (147) and (150) that

$$\int_{\mathbb{R}} g(-\tau)\, \| F_* \|_{\mathbb{Y}(\tau,\tau+1)}\, d\tau < \infty.$$

Therefore, $(\mathcal{I}-\mathcal{P})F_*$ belongs to the domain of $\mathfrak{L}$. Thus, series (155) is well defined, and we denote it by $\mathbf{v}$. Estimates (143), (158) together with (159) imply

$$\| \mathbf{v} \|_{\mathbb{X}(t,t+1)} \le c\int_{\mathbb{R}} g_\omega(t,\tau)\, \| F \|_{\mathbb{Y}(\tau,\tau+1)}\, d\tau\,.$$

Owing to (147) we arrive at (151). Clearly, $\mathbf{v}$ is a solution of (142).

(ii) Let $\mathbf{v} \in \mathbb{X}_{\text{loc}}(\mathbb{R})$ solve the equation

$$(\mathcal{I}\partial_t + \mathfrak{A})\mathbf{v} = (\mathcal{P}-\mathcal{I})\mathfrak{N}(t)\mathbf{v} \quad \text{on } \mathbb{R}. \tag{160}$$

Using (152) one checks directly that the right-hand side in (160) satisfies the conditions of Lemma 8(i). Therefore, by the same lemma and by (130) we arrive at

$$||\mathbf{v}||_{\mathbb{S}(t,t+1)} \le c \int_{\mathbb{R}} g(t-\tau)\omega(\tau)||\mathbf{v}||_{\mathbb{S}(\tau,\tau+1)} d\tau \,, \tag{161}$$

where g is given by (144).

By (152) there exists the least constants A_+ and A_- in

$$||\mathbf{v}||_{\mathbb{S}(t,t+1)} \le \begin{cases} A_+ e^{(n-m+1)t - c_0 \int_0^t \omega(\tau)d\tau} & \text{as } t \ge 0 \\ A_- e^{-mt - c_0 \int_t^0 \omega(\tau)d\tau} & \text{as } t < 0 \end{cases} \tag{162}$$

Without loss of generality we assume that $A_+ \le A_-$. Suppose that $A_+ > 0$ and let $t \ge 0$. Using (162) we estimate the right-hand side in (161) by

$$\begin{aligned} & c\,A_+\Big(e^{(n-m+1)t}\int_t^\infty \omega(\tau)e^{-c_0\int_0^\tau \omega(s)ds}d\tau + e^{-mt}\int_{-\infty}^0 \omega(\tau)e^{-c_0\int_\tau^0 \omega(s)ds}d\tau \\ & \quad + e^{-mt}\int_0^t \omega(\tau)e^{(n+1)\tau - c_0\int_0^\tau \omega(s)ds}d\tau\Big) \\ \le\ & c\,A_+ e^{(n-m+1)t - c_0\int_0^t \omega(s)ds}\Big(\frac{1}{c_0} + \frac{1}{c_0}e^{-(n+1)t + c_0\int_0^t \omega(s)ds} + \frac{\delta}{n+1-c_0\delta}\Big) \end{aligned}$$

provided $c_0\delta < n+1$. By the above assumptions

$$c(2/c_0 + \delta/(n+1-c_0\delta)) < 1\,.$$

Therefore the constant A_+ in (162) can be diminished. Thus, $A_+ = 0$ and therefore, $\mathbf{v} = 0$.

15. Scalar integro-differential equation

Lemma 10 enables one to introduce the operator $\mathfrak{M}$ whose domain consists of the vector functions $(\mathcal{I}-\mathcal{P})F$ with $F \in \mathbb{Y}_{\mathrm{loc}}(\mathbb{R})$ subject to (150). The vector function $\mathfrak{M}(\mathcal{I}-\mathcal{P})F$ is equal to the solution $\mathbf{v}$ from the same lemma. Using this operator one can write (132) as

$$\begin{aligned} (\mathcal{I}\partial_t + \mathfrak{A})\mathbf{u} + \mathcal{P}\mathfrak{N}(t)\mathbf{u} + \mathcal{P}\mathfrak{N}(t)\mathfrak{M}\mathfrak{N}(t)\mathbf{u} \\ = -\mathcal{P}\big(\mathcal{F} + \mathfrak{N}(t)\mathfrak{M}(\mathcal{P}-\mathcal{I})(\mathcal{F}\big) \quad \text{on } \mathbb{R}. \end{aligned} \tag{163}$$

Representing $\mathbf{u}$ as

$$\mathbf{u}(t) = \exp\big((n-m)t + \int_0^t \lambda(\tau)d\tau\big)\ h(t)\Phi\,,$$

where

$$\lambda(t) = (\mathfrak{N}(t)\Phi, \Psi)\,,$$

we derive from (163) the following integro-differential equation for h:

$$\dot{h}(t) - \mathcal{K}(h)(t) = \mathfrak{f}(t)\,, \tag{164}$$

where

$$\mathcal{K}(h)(t) = (\mathfrak{N}(t)\mathfrak{M}_{\tau\to t}(e^{(m-n)(t-\tau)-\int_\tau^t \lambda(s)ds}\mathfrak{N}(\tau)h(\tau)\Phi)(t), \Psi)$$

and

$$\mathfrak{f}(t) = e^{(m-n)t-\int_0^t \lambda(\tau)d\tau}(\mathcal{F}(t) + \mathfrak{N}(t)\mathfrak{M}(\mathcal{P}-\mathcal{I})\mathcal{F}(t), \Psi)\,. \tag{165}$$

Using (151) together with (130) we obtain the estimates

$$||\mathcal{K}(h)||_{W^{1,p}(t,t+1)} \le c\omega(t)\int_{\mathbb{R}} \sigma(t,\tau)\omega(\tau)||h||_{L^\infty(\tau,\tau+1)}d\tau \tag{166}$$

and

$$||\mathfrak{f}||_{L^p(t,t+1)} \le c\left(||\mathcal{F}||_{\mathbb{Y}(t,t+1)} + \omega(t)\int_{\mathbb{R}} \sigma(t,\tau)\omega(\tau)||\mathcal{F}||_{\mathbb{Y}(\tau,\tau+1)}d\tau\right), \tag{167}$$

where

$$\sigma(t,\tau) = \begin{cases} e^{-n(t-\tau)+c_2\int_\tau^t \omega(s)ds} & \text{for } t \ge \tau \\ e^{t-\tau+c_2\int_t^\tau \omega(s)ds} & \text{for } t < \tau \end{cases} \tag{168}$$

Here c_2 is a positive constant, which depends on n, m, p and the coefficients of the operator L.

Lemma 11. *The function* $\lambda(t) = (\mathfrak{N}(t)\Phi, \Psi)$ *admits the representation*

$$\lambda(t) = \sum_{j=0}^m\sum_{k=0}^m (\mathcal{N}_{m-j,m-k}(t)(m-n)^k\phi, (m-n)^j\psi) + O(\omega(t)^2), \tag{169}$$

where ϕ *and* ψ *are the same functions as in* Section 10.

Proof. By (93) and by (120), (121)

$$\begin{aligned} \mathcal{S}(t)\Phi &= (A_{00} - \mathcal{N}_{00}(t))^{-1}\Big(A_{00}(m-n)^m\phi + \sum_{k=0}^{m-1} \mathcal{N}_{0,m-k}(t)(m-n)^k\phi\Big) \\ &= (m-n)^m\phi + (A_{00} - \mathcal{N}_{00}(t))^{-1}\sum_{k=0}^{m} \mathcal{N}_{0,m-k}(t)(m-n)^k\phi\,. \end{aligned}$$

Now, using (102)–(104) we obtain

$$\mathfrak{N}_m(t)\Phi = A_{00}^{-1}\sum_{k=0}^m \mathcal{N}_{0,m-k}(t)(m-n)^k\phi + O(\omega^2(t))$$

and

$$\begin{aligned} \mathfrak{N}_{m+j}(t)\Phi &= \sum_{k=0}^m \mathcal{N}_{j,m-k}(t)(m-n)^k\phi \\ &- A_{j0}A_{00}^{-1}\sum_{k=0}^m \mathcal{N}_{0,m-k}(t)(m-n)^k\phi + O(\omega^2(t))\,. \end{aligned}$$

Therefore,

$$(\mathfrak{N}(t)\Phi, \Psi) = \sum_{j=0}^{m}\sum_{k=0}^{m} (A_{00}^{-1}\mathcal{N}_{0,m-k}(t)(m-n)^k\phi, A_{j0}^*(m-n)^{m-j}\psi)$$

$$+\sum_{j=1}^{m}\sum_{k=0}^{m} ((\mathcal{N}_{j,m-k}(t) - A_{j0}A_{00}^{-1}\mathcal{N}_{0,m-k})(m-n)^k\phi, (m-n)^{m-j}\psi) + O(\omega^2(t))$$

$$= \sum_{j=0}^{m}\sum_{k=0}^{m} (\mathcal{N}_{j,m-k}(t)(m-n)^k\phi, (m-n)^{m-j}\psi) + O(\omega^2(t)) .$$

Clearly, the right-hand sides in the last equality and (169) coincide.

16. Homogeneous equation (164)

We start with a uniqueness result for the equation

$$\dot{z}(t) + (\mathcal{K}z)(t) = 0 \quad t \in \mathbb{R}. \tag{170}$$

Lemma 12. *There exist positive constants δ_0 and c_3 depending only on n, m, p and L such that: if $\delta \le \delta_0$ and $z \in W^{1,p}_{\mathrm{loc}}(\mathbb{R})$ is a solution of* (170) *subject to*

$$z(t) = \begin{cases} o\Big(e^{t-c_3\int_0^t \omega(s)ds}\Big) & \text{as } t \to +\infty \\ o\Big(e^{-nt-c_3\int_t^0 \omega(s)ds}\Big) & \text{as } t \to -\infty \end{cases} \tag{171}$$

and $z(t_0) = 0$ for some t_0 then $z(t) = 0$ for all $t \in \mathbb{R}$.

Proof. Without loss of generality we set $t_0 = 0$. Integrating (170) and using (166) we obtain

$$\nu(t) \le c\Big|\int_0^t \omega(\tau)\int_{\mathbb{R}} \sigma(\tau, s)\omega(s)\nu(s)dsd\tau\Big| , \tag{172}$$

where $\nu(t) = ||z||_{L^\infty(t,t+1)}$. We set

$$A = \sup_{t\ge 0} e^{-t+c_3\int_0^t \omega(s)ds}\nu(t) + \sup_{t<0} e^{nt+c_3\int_t^0 \omega(s)ds}\nu(t) .$$

Let c_2 be the same constant as in (168). We may suppose that $c_3 > c_2$.

For $t \ge 0$ we estimate the right-hand side of (172) by

$$cA\int_0^t \omega(\tau)\Big\{\int_\tau^\infty \omega(s)e^{\tau+c_2\int_\tau^s \omega(x)dx - c_3\int_0^s \omega(x)dx}ds$$

$$+\int_0^\tau \omega(s)e^{n(s-\tau)+s+c_2\int_s^\tau \omega(x)dx - c_3\int_0^s \omega(x)dx}ds$$

$$+\int_{-\infty}^0 \omega(s)e^{-n\tau+c_2\int_s^\tau \omega(x)dx - c_3\int_s^0 \omega(x)dx}ds\Big\}d\tau$$

Direct calculations give that the right hand-side is majorized by

$$cA\int_0^t \omega(\tau)\Big\{\frac{1}{c_3-c_2}e^{\tau-c_3\int_0^\tau \omega(x)dx}$$
$$+\frac{1}{c_3+c_2}e^{-n\tau+c_2\int_0^\tau \omega(x)dx}+\frac{\delta}{n+1-(c_2+c_3)\delta}e^{\tau-c_3\int_0^\tau \omega(x)dx}\Big\}d\tau\,.$$

Supposing that $(c_2+c_3)\delta < n+1$ we conclude that the right-hand side is less than

$$cA\Big\{\frac{1}{c_3-c_2}+\frac{1}{c_2+c_3}+\frac{\delta}{n+1-(c_2+c_3)\delta}\Big\}\int_0^t \omega(\tau)e^{\tau-c_3\int_0^\tau \omega(x)dx}d\tau$$
$$\le cA\Big\{\frac{1}{c_3-c_2}+\frac{1}{c_2+c_3}+\frac{\delta}{n+1-(c_2+c_3)\delta}\Big\}\frac{\delta}{n-c_3\delta}e^{t-c_3\int_0^t \omega(x)dx}\,.$$

Therefore, assuming that δ is sufficiently small, one can choose $c_2 \ge 4(1+c)$ and c_3 satisfying the above restrictions and such that

$$\sigma_+ = c\Big\{\frac{1}{c_3-c_2}+\frac{1}{c_2+c_3}+\frac{\delta}{n+1-(c_2+c_3)\delta}\Big\}\frac{\delta}{n-c_3\delta} < 1.$$

This implies

$$\sup_{t\ge 0} e^{-t+c_3\int_0^t \omega(s)ds}\nu(t) \le \sigma_+ A\,.$$

Analogously, one verifies that

$$\sup_{t<0} e^{nt+c_3\int_t^0 \omega(s)ds}\nu(t) \le \sigma_- A$$

with some $\sigma_- < 1$. Therefore, $A=0$.

Lemma 13. *Equation* (170) *has a solution* $z \in W^{1,\infty}_{\rm loc}(\mathbb{R})$ *given by*

$$z(t) = \exp\Big(\int_{t_0}^t \Lambda(\tau)d\tau\Big)\,, \tag{173}$$

where Λ *is a locally summable function satisfying*

$$|\Lambda(t)| \le c\,\chi(t)\,, \tag{174}$$

where

$$\chi(\tau) = \omega(t)\Big(\int_{-\infty}^t e^{n(\tau-t)+C\int_\tau^t \omega(s)ds}\omega(\tau)d\tau + \int_t^\infty e^{t-\tau+C\int_t^\tau \omega(s)ds}\omega(\tau)d\tau\Big).$$

Proof. Let ε be a sufficiently small number depending on n, m and L and let $\mathcal{B}_\varepsilon = \{\Lambda \in L^\infty(\mathbb{R}) : |\Lambda(t)| \le \varepsilon\omega(t)\}$. Inserting (173) into (170) we arrive at the equation for Λ:

$$\Lambda(t) + G(\Lambda)(t) = 0\,,\quad t \in \mathbb{R}\,, \tag{175}$$

where

$$G(\Lambda)(t) = \mathcal{K}_{\tau\to t}\Big(\exp\Big(\int_t^\tau \Lambda(s)ds\Big)\Big)\,.$$

Using (166) with $p = 2$ and assuming that δ is sufficiently small we obtain for $\Lambda \in \mathcal{B}_\varepsilon$:

$$|G(\Lambda)(t)| \leq c\omega(t) \int_{\mathbb{R}} \sigma(t,\tau)\omega(\tau) e^{\varepsilon |\int_t^\tau \omega(s)ds|} d\tau \leq c_1 \delta \omega(t), \tag{176}$$

where c_1 is a constant depending only on n, m and L. We suppose that $c_1\delta \leq \varepsilon$. This guarantees, in particular, that G maps $\mathcal{B}_\varepsilon$ into itself.

Now let Λ_1 and Λ_2 be functions from $\mathcal{B}_\varepsilon$. By (166) we have

$$\begin{aligned} &|G(\Lambda_2)(t) - G(\Lambda_1)(t)| \\ &\leq c\omega(t) \int_{\mathbb{R}} \sigma(t,\tau)\omega(\tau) \sup_{\tau \in (t,t+1)} \Big| \exp\Big(\int_t^\tau \Lambda_2(s)ds\Big) - \exp\Big(\int_t^\tau \Lambda_1(s)ds\Big) \Big| d\tau \,. \end{aligned}$$

Since

$$\begin{aligned} \Big| \exp\Big(\int_t^\tau \Lambda_2(s)ds\Big) \quad &- \quad \exp\Big(\int_t^\tau \Lambda_1(s)ds\Big) \Big| \\ &\leq \quad e^{\varepsilon\delta|t-\tau|} \Big| \int_t^\tau \omega(s)ds\Big| \sup_{s\in\mathbb{R}} \frac{|\Lambda_2(s) - \Lambda_1(s)|}{\omega(s)} \,, \end{aligned}$$

we obtain

$$|G(\Lambda_2)(t) - G(\Lambda_1)(t)| \leq c_2 \delta \omega(t) \sup_{s\in\mathbb{R}} \frac{|\Lambda_2(s) - \Lambda_1(s)|}{\omega(s)}$$

with some constant c_2 depending on n, m and L. Assuming that $c_2\delta < 1$ we get the existence of $\Lambda \in \mathcal{B}_\varepsilon$ satisfying (175) by the Banach fixed point theorem.

Estimate (174) results from (176) and (175).

The next statement directly follows from Lemma 13.

Corollary 2. *Suppose that*

$$\Big| \int_{\mathbb{R}} \chi(\tau) d\tau \Big| < \infty \,.$$

Then the solution z from Lemma 13 *admits the asymptotic representation*

$$z(t) = 1 + O\Big(\int_t^\infty \chi(\tau)d\tau\Big) \quad \text{as } t \to +\infty \,.$$

We denote by $z(t,\tau)$ the solution of (170) subject to (171) and such that $z(\tau,\tau) = 1$. By Lemma 12 this solution is unique and by Lemma 13 such a solution exists and satisfies

$$e^{-c|\int_\tau^t \chi(s)ds|} \leq |z(t,\tau)| \leq e^{c|\int_\tau^t \chi(s)ds|} \tag{177}$$

with c depending only on n, m and L.

17. Representation of solutions of the homogeneous problem (34)

Lemma 14. *There exists a nontrivial solution* $\mathfrak{Z} \in \mathring{W}^{m,p}_{\mathrm{loc}}(\Pi)$ *to the homogeneous problem* (34) *subject to*

$$||\mathfrak{Z}||_{W^{m,p}(\Pi_t)} = \begin{cases} o(e^{(n-m+1)t - C\int_0^t \omega(s)ds}) & \text{if } t \to +\infty \\ o(e^{-mt - C\int_t^0 \omega(s)ds}) & \text{if } t \to -\infty. \end{cases} \tag{178}$$

This solution is unique up to a constant factor and

$$(-\partial_t)^k \mathfrak{Z}(t,\theta) = C(m-n)^k \exp\Big((n-m)t + \int_0^t \lambda(\tau)d\tau\Big) z(t)(\phi(\theta) + v_k(t,\theta)), \tag{179}$$

where $C = \mathrm{const}$, $k = 0, \dots, m$, *and* z *is the function from* Lemma 13. *For* $k < m$ *the remainder* v_k *satisfies*

$$\begin{aligned} &||v_k||_{W^{m-k,p}(\Pi_t)} + ||\partial_t v_k||_{W^{m-k-1,p}(\Pi_t)} \\ &\le c\Big(\int_{-\infty}^t e^{n(\tau - t + C\int_\tau^t \omega(s)ds}\omega(\tau)d\tau + \int_t^\infty e^{t-\tau+C\int_t^\tau \omega(s)ds}\omega(\tau)d\tau\Big). \end{aligned} \tag{180}$$

If $k = m$, *then the second term on the right in the last inequality should be omitted.*

Proof. We introduce the vector function $\mathcal{U} = (\mathcal{U}_1, \dots, \mathcal{U}_{2m})$ with $\mathcal{U}_k$ given by (87)–(89) where $f_0 = f_1 = \dots = f_{m-1} = 0$. By Lemma 6 the function $u \in \mathring{W}^{m,p}_{\mathrm{loc}}(\Pi)$ solves the homogeneous equation (85) (or equivalently (34)) if and only if $\mathcal{U} \in \mathbb{S}_{\mathrm{loc}}(\mathbb{R})$ is a solution of (98) with $\mathcal{F} = 0$.

(i) *Existence.* Let

$$\mathbf{u}(t) = \exp\big((n-m)t + \int_0^t \lambda(\tau)d\tau\big) z(t)\mathbf{\Phi}, \tag{181}$$

where $z(t)$ is the solution of (170) from Lemma 13. We are looking for a solution $\mathcal{U}$ of the homogeneous system (98) in the form $\mathcal{U}(t) = \mathbf{u}(t) + \mathbf{v}(t)$, where $\mathcal{P}\mathbf{v}(t) = 0$. Then $\mathbf{v}$ satisfies (133) with $\mathcal{F} = 0$. By (130) and Lemma 13 the $\mathbb{Y}(t, t+1)$-seminorm of $\mathcal{N}\mathbf{u}$ is majorized by

$$c\omega(t)\exp\big((n-m)t + \Re\int_o^t \lambda(\tau)d\tau + c\int_0^t \chi(\tau)d\tau\big).$$

By Lemma 10 system (133) has a solution $\mathbf{v}$ satisfying

$$\begin{aligned} &||\mathbf{v}||_{\mathbb{S}(t,t+1)} \le ce^{(n-m)t + \Re\int_0^t \lambda(\tau)d\tau} \times \\ &\Big(\int_{-\infty}^t e^{n(\tau-t)+C\int_\tau^t \omega(s)ds}\omega(\tau)d\tau + \int_t^\infty e^{t-\tau+C\int_t^\tau \omega(s)ds}\omega(\tau)d\tau\Big). \end{aligned}$$

Hence, $\mathfrak{Z} = \mathcal{U}_1$ is the required solution of equation (34). The solution of the homogeneous system (98) constructed above will be denoted by $\mathcal{U}_* = \mathbf{u}_* + \mathbf{v}_*$ where $\mathcal{P}\mathbf{v}_* = 0$.

(ii) *Uniqueness.* Suppose that the $W^{m,p}(\Pi_t)$-seminorm of a solution $u = \mathfrak{Z}$ of the homogeneous equation (34) is subject to (178). Consider the vector function $\mathcal{U} - c\mathcal{U}_*$, where c is a arbitrary constant. We represent $\mathbf{u}$ in the form (181) with a certain z. Similarly, let $\mathbf{u}_*$ be given by (181) with z_* instead of z. Clearly, $z - cz_*$ satisfies (170) and (171). Choosing c to satisfy $z(0) - cz_*(0) = 0$ and using Lemma 12 one obtains $z(t) - cz_*(t) = 0$ for all t. Now, applying Lemma 10(ii) to the vector function $\mathbf{v} - c\mathbf{v}_*$, which solves the homogeneous system (142), we conclude that $\mathbf{v} - c\mathbf{v}_* = 0$. The proof is complete.

Corollary 3. *Let $f \in W^{-m,p}_{\text{loc}}(\Pi)$ be subject to*

$$\mathcal{J}_f := \int_{\mathbb{R}} e^{-(n+m)\tau + c|\int_0^\tau \omega(s)ds|} ||f||_{W^{-m,p}(\Pi_\tau)} d\tau < \infty\,, \tag{182}$$

Also let u_1 and u_2 be solutions of problem (34) *from* Proposition 7 (i) *and* (ii) *respectively. Then*

$$u_2 - u_1 = CZ(t), \tag{183}$$

where C is a constant satisfying

$$|C| \le c\, \mathcal{J}_f\,. \tag{184}$$

Proof. It follows from Proposition 7 that $u_2 - u_1$ is the solution of the homogeneous equation (34) satisfying (178). Now, (183) holds by Lemma 14. In order to prove (184) we write

$$|C|\,||\mathfrak{Z}||_{L^p(\Pi_0)} \le ||u_2||_{L^p(\Pi_0)} + ||u_1||_{L^p(\Pi_0)}\,.$$

Using estimates (67) and (64) we see that the right-hand side is majorized by $c\mathcal{J}_f$. By (179) and (177)

$$||\mathfrak{Z}||_{L^p(\Pi_0)} \ge c||z||_{L^p(0,1)} \ge c_1\,.$$

The proof is complete.

Lemma 15. *There exists a nontrivial solution $\mathcal{Z} \in \mathring{W}^{m,p}(\overline{\mathbb{R}^n_+} \setminus \mathcal{O})$ to the homogeneous problem* (11), (12) *subject to*

$$\mathfrak{M}^m_p(\mathcal{Z}; K_{r/e,r}) = \begin{cases} o(r^{m-n-1} e^{-C\int_r^1 \Omega(\rho)\frac{d\rho}{\rho}}) & \textit{if } r \to 0 \\ o(r^m e^{-C\int_1^r \Omega(\rho)\frac{d\rho}{\rho}}) & \textit{if } r \to \infty. \end{cases} \tag{185}$$

This solution is unique up to a constant factor and admits the representation

$$\begin{aligned}&(r\partial_r)^k \mathcal{Z}(x) \\ &= C\, \exp\Big(\int_r^1 (\mathcal{T}(\rho) + \Upsilon(\rho))\frac{d\rho}{\rho}\Big)\Big((m-n)^k E(x) + r^{m-n} v_k(x)\Big)\end{aligned} \tag{186}$$

with the same notation as in the statement of Theorem 1 *and with v_k subject to* (24).

Proof. By (169)

$$\lambda(t) = \sum_{j=0}^m \sum_{k=0}^m (\mathcal{N}_{j,k}(t)(-\partial_t)^{m-k})(e^{(n-m)t}\phi), \partial_t^{m-j}(e^{(m-n)t}\psi)) + O(\omega(t)^2), \tag{187}$$

where ϕ and ψ are the same functions as in Section 10. Setting $u = e^{(n-m)t}\phi$ and $v = e^{-mt}\psi$ in (84) we arrive at

$$\lambda(-\log r) = \frac{r^n}{m!}\int_{S_+^{n-1}} \sum_{|\alpha|,|\beta|\le m} (\mathcal{L}_{\alpha\beta} - L_{\alpha\beta}(x))\partial_x^\beta E(x)\partial_x^\alpha x_n^m \, d\theta$$
$$+ \frac{r^n}{m!}\int_{S_+^{n-1}} \sum_{|\alpha+\beta|<2m} \mathcal{L}_{\alpha\beta}(x)\partial_x^\beta E(x)\partial_x^\alpha x_n^m \, d\theta + O(\Omega(r)^2)\,.$$

This can be written as $\lambda(-\log r) = \mathcal{T}(r) + O(\Omega(r)^2)$. The result follows from Lemma 14 by the change of variables $(t,\theta) \to x$.

Corollary 4. *Let $f \in W_{\rm loc}^{-m,p}(\overline{\mathbb{R}_+^n} \setminus \mathcal{O})$ be subject to*

$$\mathcal{I}_f := \int_0^\infty \rho^{m+n} e^{-C|\int_\rho^1 \Omega(s)\frac{ds}{s}|} \mathfrak{M}_p^{-m}(f; K_{\rho/e,\rho}) \frac{d\rho}{\rho} < \infty\,. \tag{188}$$

Also let u_1 and u_2 be solutions of problem (11), (12) *from* (i) *and* (ii) *in Proposition* 8 *respectively. Then*

$$u_2(x) - u_1(x) = C\mathcal{Z}(x)\,, \tag{189}$$

where $\mathcal{Z}$ is defined in Lemma 15 *and C is a constant subject to*

$$|C| \le c\mathcal{I}_f\,.$$

The *Proof* follows directly from Corollary 3.

18. End of proof of Theorem 1

Assertion (i) follows from Lemma 15. In order to obtain (ii) we introduce the cut-off function $\eta \in C_0^\infty(B_2)$, $\eta(x) = 1$ for $|x| \le 3/2$. The function ηu satisfies the zero Dirichlet conditions on $\mathbb{R}^{n-1} \setminus \mathcal{O}$ and the equation $\mathcal{L}(x,\partial_x)(\eta u) = f_1$ on $\mathbb{R}_+^n$ with $f_1 = \eta f + [\mathcal{L},\eta]u$. Clearly,

$$\mathfrak{M}_p^{-m}(f_1; K_{r/e,r}) = \mathfrak{M}_p^{-m}(f; K_{r/e,r}) \tag{190}$$

if $r < 3/2$ and $r > 2e$. By the standard local estimate for solutions of the Dirichlet problem

$$\mathfrak{M}_p^m(u; K_{3r/2,2r}) \le c\,(r^{2m}\mathfrak{M}_p^{-m}(f; K_{r,er}) + r^{-n/p}||u||_{L^p(K_{r,er})}) \tag{191}$$

we have

$$\mathfrak{M}_p^{-m}([\mathcal{L},\eta]u; K_{3/2,2}) \le c\,(\mathfrak{M}_p^{-m}(f; K_{1,e}) + ||u||_{L^p(K_{1,e})}).$$

Hence, for $r \in (3/2, 2e)$

$$\mathfrak{M}_p^{-m}(f_1; K_{r/e,r}) \le c(\mathfrak{M}_p^{-m}(f; K_{1/2,e}) + ||u||_{L^p(K_{1,e})}). \tag{192}$$

Therefore,

$$\mathcal{I}_{f_1} \le c(I_f + ||u||_{L^p(K_{1,e})}).$$

By (10) and finiteness of I_f,

$$\mathfrak{M}_p^{-m}(f; K_{\rho/e^2,e\rho}) \le c \int_{r/e^2}^{e^2 r} \mathfrak{M}_p^{-m}(f; K_{\rho/e,\rho}) \frac{d\rho}{\rho}$$
$$= o\Big(r^{-m-n} \exp\Big(-C\int_r^1 \Omega(\rho)\frac{d\rho}{\rho}\Big)\Big) \quad \text{as } r \to 0.$$

This along with (191) and (26) implies (71) with u replaced by ηu. Therefore ηu is the solution of problem (11), (12) (with f_1 instead of f) from Proposition 8(ii). The result follows from Corollary 4.

19. Corollaries of the main result

Clearly, Theorem 1 remains valid if Ω is replaced by its nondecreasing majorant

$$\Omega^\diamond(r) = \sup_{x\in B_r^+} \Big(\sum_{|\alpha|=|\beta|=m} |\mathcal{L}_{\alpha\beta}(x) - L_{\alpha\beta}| + \sum_{|\alpha+\beta|<2m} x_n^{2m-|\alpha+\beta|} |\mathcal{L}_{\alpha\beta}(x)|\Big). \quad (193)$$

Corollary 5. *Let $\mathcal{Z}$ be the same solution as in* Theorem 1. *Then*

$$\partial_x^\alpha \mathcal{Z}(x) = \exp\Big(\int_r^1 \mathcal{T}(\rho)\frac{d\rho}{\rho} + \Psi^\diamond(r)\Big)\Big(\partial_x^\alpha E(x) + r^{m-n-|\alpha|} v_\alpha(x)\Big), \quad (194)$$

where $|x|<1$ and the function $\Psi^\diamond$ satisfies

$$\Psi^\diamond(r) \le C\int_r^e \Omega^\diamond(\rho)^2 \frac{d\rho}{\rho}$$

and

$$|\partial_r \Psi^\diamond(r)| \le C\Omega^\diamond(r)\int_r^e e^{C\int_r^\rho \Omega^\diamond(s)\frac{ds}{s}} \Omega^\diamond(\rho)\frac{d\rho}{\rho^2}.$$

For $|\alpha| \le m-1$ the function v_α belongs to $\mathring{W}^{1,p}_{\rm loc}(\overline{\mathbb{R}^n_+}\setminus \mathcal{O})$ and satisfies

$$\Big(r^{-n}\int_{K_{r/e,r}} (r|\nabla v_\alpha(x)| + |v_\alpha(x)|)^p dx\Big)^{1/p} \le c r^{1-\varepsilon}\int_r^e \Omega^\diamond(\rho)\frac{d\rho}{\rho^{2-\varepsilon}} \quad (195)$$

for $r<1$. If $|\alpha| = m$, the term $r|\nabla v_\alpha(x)|$ should be removed. By ε, we denote a sufficiently small number depending on n, m, p and $L_{\alpha\beta}$.

The *Proof* is the same as that of Corollary 5 in [4].

Corollary 6. *If $\Omega(r)\to 0$ as $r\to 0$ then the right-hand side in* (24) *tends to* 0 *as $r \to 0$ and*

$$\mathfrak{M}_p^m(w; K_{r/e,r}) = o(r^{m-n} e^{-C\int_r^1 \Omega(s)\frac{ds}{s}}).$$

In the case $p>n$ the solution u in Theorem 1 (ii) *satisfies*

$$\partial_x^\alpha u(x) = \exp\Big(\int_{|x|}^1 (\mathcal{T}(\rho)+\Upsilon(\rho))\frac{d\rho}{\rho}\Big) \times \Big(C\partial_x^\alpha E(x) + o(|x|^{m-n-|\alpha|})\Big) \quad as\ |x|\to 0 \quad (196)$$

uniformly with respect to $x/|x|$. *Here* α *is an arbitrary multi-index of order* $\le m-1$. *The function* $\Psi^\diamond(r)$ *is the same as in Corollary* 5. *Moreover,* (196) *remains valid also for* $|\alpha| = m$ *but then* $\Phi = o(1)$ *should be understood as*

$$r^{-n/p}||\Phi||_{L^p(K_{r/e,r})} \to 0 \quad as\ r \to 0.$$

The *Proof* is the same as that of Corollary 7 in [4].

Corollary 7. *Let* $p > n$ *and*

$$\int_0^1 \Omega(\rho)^2 \frac{d\rho}{\rho} < \infty. \tag{197}$$

Then the solution u *from* Theorem 1 (ii) *satisfies*

$$\partial_x^\alpha u(x) = \exp\Big(\int_{|x|}^1 \mathcal{T}(\rho)\frac{d\rho}{\rho}\Big)\Big(C\partial_x^\alpha E(x) + o(|x|^{m-n-|\alpha|})\Big) \tag{198}$$

for $|\alpha| \le m-1$ *uniformly with respect to* $x/|x|$. *The same is true for* $|\alpha| = m$ *if the symbol* $o(1)$ *is understood as in* Theorem 1 (ii).

The *Proof* is the same as that of Corollary 8 in [4].

20. Second order elliptic equations

Example 1. Consider the equation with complex-valued measurable coefficients

$$-\sum_{i,j=1}^n \partial_{x_i}(a_{ij}(x)\partial_{x_j}u) = 0 \quad \text{in } B_3^+$$

complemented by the boundary condition

$$u(x',0) = 0 \quad \text{for } |x'| < 3. \tag{199}$$

We assume that there exists a constant symmetric matrix $\{a_{ij}\}_{i,j=1}^n$ with positive definite real part such that the function

$$\Omega^\diamond(r) = \sup_{B_r^+} \sum_{i,j=1}^n |a_{ij}(x) - a_{ij}|$$

is sufficiently small in B_3^+. In view of Theorem 1(ii) and Corollary 5

$$\begin{aligned} u(x) &= \exp\Big\{\int_{r<|y|<1}\sum_{i=1}^n (a_{ni}(y) - a_{ni})\partial_{y_i}E(y)dy + O\Big(\int_{|x|}^1 \Omega^\diamond(\rho)^2\frac{d\rho}{\rho}\Big)\Big\} \\ &\times\Big(C\Big(E(x) + O\big(|x|^{2-n-\varepsilon}\int_{|x|}^3 \Omega^\diamond(\rho)\rho^{\varepsilon-2}d\rho\big)\Big) + O(|x|^{1-\varepsilon})\Big), \end{aligned} \tag{200}$$

where ε is a small positive number depending on n and the coefficients a_{ij}. Here $E(x)$ stands for the Poisson kernel of the equation

$$\sum_{i,j=1}^n a_{ij}\partial_{x_i}\partial_{x_j}v = 0 \quad \text{in } \mathbb{R}_+^n,$$

i.e.

$$E(x) = (\det\{a_{ij}\})^{-1/2}|S^{n-1}|^{-1}x_n(\sum_{k,l=1}^{n} b_{kl}x_k x_l)^{-n/2}$$

where $\{b_{lj}\}$ is the inverse of $\{a_{ij}\}$ (see [2], Section 6.2). Setting this expression of $E(x)$ into (200), we arrive at (4) where $\delta = 1$, $G = \mathbb{R}^n_+$ and $\mathcal{Q}$ is given by (3). The case of a domain with smooth boundary mentioned in the introduction can be easily reduced to the present one by changing variables.

References

[1] S. Agmon, A. Douglis and L. Nirenberg, Estimates near the boundary for solutions of elliptic partial differential equations satisfying general boundary conditions I, Comm. Pure Appl. Math. **12**(1959), 623–727.

[2] L. Hörmander, *The Analysis of Linear Partial Differential Operators*, Vol. 1, Springer, 1983.

[3] V. Kozlov and V. Maz'ya, *Differential Equations with Operator Coefficients*, Monographs in Mathematics, Springer-Verlag, 1999.

[4] V. Kozlov and V. Maz'ya, Asymptotic formula for solutions to the Dirichlet problem for elliptic equations with discontinuous coefficients near the boundary. To appear.

[5] V. Kozlov, V. Maz'ya and J. Rossmann, *Conical singularities of solutions to elliptic equations*, Mathematical Surveys and Monographs, **85**, AMS, Providence, RI, 2001.

Vladimir Kozlov
Department of Mathematics
Linköping University
SE-581 83 Linköping, Sweden
E-mail address: `vlkoz@mai.liu.se`

Vladimir Maz'ya
Department of Mathematics
Linköping University
SE-581 83 Linköping, Sweden
E-mail address: `vlmaz@mai.liu.se`

D. Haroske, T. Runst, H.-J. Schmeisser (eds.): Function Spaces, Differential Operators and Nonlinear Analysis. The Hans Triebel Anniversary Volume.

Weighted Hardy Spaces on a Domain and its Application

Akihiko Miyachi

Dedicated to Prof. Hans Triebel on the occasion of his 65th birthday

Abstract. We give basic properties of the Hardy spaces on an open subset $\Omega \subset \mathbb{R}^n$ with respect to a doubling measure on Ω, and apply it to give the transplantation theorem for Jacobi series in Hardy spaces on a finite interval of $\mathbb{R}$.

Notation

The following notations will be used throughout the article. The *ball* $B(y,t)$ with $y \in \mathbb{R}^n$ and $\infty > t > 0$ is defined by $B(y,t) = \{z \in \mathbb{R}^n \mid |z-y| < t\}$. For a ball $Q = B(y,t)$, we write its center and radius as $x(Q) = y$ and $r(Q) = t$. If $Q = B(y,t)$ and $\infty > a > 0$, then $aQ = B(y, at)$. For a function f on $\mathbb{R}^n$ and for $\infty > t > 0$, the function $(f)_t$ on $\mathbb{R}^n$ is defined by $(f)_t(x) = t^{-n} f(t^{-1}x)$. If $p(x, y, \cdots)$ is a proposition containing the variables x, y, $\cdots$, then the symbol $\mathbf{1}\{p(x,y,\cdots)\}$ is defined to be 1 if $p(x,y,\cdots)$ holds and 0 if $p(x,y,\cdots)$ does not hold. The letter $\mathbb{N}$ denotes the set of positive integers. We write $\mathcal{P}_k$, $k \in \mathbb{N} \cup \{0\}$, to denote the set of polynomial functions on $\mathbb{R}^n$ of degree not exceeding k. If f is a bounded measurable function on $\mathbb{R}^n$ with compact support and if $\int f(x) x^\alpha dx = 0$ for $|\alpha| \leqq k$, $k \in \mathbb{N} \cup \{0\}$, then we write $f \perp \mathcal{P}_k$; we use the same notation for compactly supported distributions f with the obvious modification of definition. We use the letter c to denote various positive constants. The symbol $c(\alpha, \beta, \cdots)$ denotes a positive constant which depends only on the parameters α, β, $\cdots$.

1. Weighted Hardy spaces on a domain

Throughout this section, Ω denotes an arbitrary open subset of $\mathbb{R}^n$.

In this section, we give the definition and basic theorems for weighted Hardy spaces on Ω. Weighted Hardy spaces on the whole space $\mathbb{R}^n$ have been studied by García-Cuerva [5] and by Strömberg-Torchinsky [13]. Here we give a purely real variable method which works on general Ω. These results are already announced in [10]. We shall give the proofs of the theorems in Sections 3 and 4.

We begin with the definition of the measures to be considered.

Definition. We define $\mathcal{M}(\Omega)$ to be the set of all those measures λ on Borel subsets of Ω such that $0 < \lambda(B) < \infty$ for all balls B with its closure included in Ω. For $0 < \sigma < \infty$, we define Double (Ω, σ) as the set of all those $\lambda \in \mathcal{M}(\Omega)$ for which there exists $1 < T < \infty$ and $1 \leqq A < \infty$ such that the inequality

$$\lambda(B) \leqq At^{\sigma}\lambda(t^{-1}B) \tag{1.1}$$

holds for all balls B with $TB \subset \Omega$ and for all $t \geqq 1$. We define Double (Ω) as the set of the union of all Double (Ω, σ) for $0 < \sigma < \infty$.

It is easy to see that, for $\Omega \neq \emptyset$, we have Double $(\Omega, \sigma) \neq \emptyset$ only if $\sigma \geqq n$. It is also easy to see that if $\lambda \in$ Double (Ω, σ) then for every $1 < T < \infty$ there exists $A = A_T \in [1, \infty)$ such that (1.1) holds for all B with $TB \subset \Omega$ and for all $t \geqq 1$.

We shall associate two maximal functions to a distribution on Ω.

First, let ϕ be a function satisfying

$$\phi \in C_0^{\infty}(B(0,1)) \qquad \text{and} \qquad \int \phi(x)dx = 1. \tag{1.2}$$

For $f \in \mathcal{D}'(\Omega)$ and for $0 < \epsilon \leqq 1$, we define $f^{+(\phi),\Omega,\epsilon}(x)$, $x \in \Omega$, by

$$f^{+(\phi),\Omega,\epsilon}(x) = \sup\{|\langle f, (\phi)_t(x - \cdot)\rangle| \mid 0 < t < \epsilon \operatorname{dis}(x, \Omega^c)\}.$$

To define the second maximal function, we recall the definition of the Lipschitz space $\Lambda(s)$. For $0 < s < \infty$ and for $f \in L^1_{\mathrm{loc}}(\mathbb{R}^n)$, we define

$$\|f\|_{\Lambda(s)} = \sup_B \left[\inf\left\{|B|^{-1-s/n} \int_B |f(x) - P(x)|dx \mid P \in \mathcal{P}_{[s]}\right\}\right],$$

where the sup is taken over all balls B in $\mathbb{R}^n$ and $[s]$ denotes the integer satisfying $[s] \leqq s < [s] + 1$. We define $\Lambda(s)$, $0 < s < \infty$, as the set of all those $f \in L^1_{\mathrm{loc}}(\mathbb{R}^n)$ such that $\|f\|_{\Lambda(s)} < \infty$. For the space $\Lambda(s)$, see [17, §5.3, pp. 246–252] and [16, §2.5.7, pp. 89–91], where essentially the same space is denoted by $\mathcal{C}^s = B^s_{\infty,\infty}$.

The second maximal function is defined as follows. For a ball B, we set

$$\mathcal{T}_s(B) = \left\{\psi \in C_0^{\infty}(B) \mid \|\psi\|_{\Lambda(s)} \leqq r(B)^{-n-s}\right\}.$$

For $f \in \mathcal{D}'(\Omega)$, $0 < s < \infty$, and $1 \leqq b < \infty$, we define $f_s^{*\Omega,b}(x)$, $x \in \Omega$, by

$$f_s^{*\Omega,b}(x) = \sup\left\{|\langle f, \psi\rangle| \mid \psi \in \bigcup_B \{\mathcal{T}_s(B) \mid bB \subset \Omega,\ B \ni x\}\right\},$$

where B denotes a ball. If $\Omega = \mathbb{R}^n$, then $f_s^{*\Omega,b}$ does not depend on b and is simply written as f_s^*.

The following is the first fundamental theorem.

Theorem 1.1. *Let* $\lambda \in \mathrm{Double}(\Omega, \sigma)$, $n \leqq \sigma < \infty$, *and let* $0 < p \leqq \infty$. *Let* ϕ *be a function satisfying* (1.2) *and let* $0 < \epsilon < 1/3$. *Let* $\infty > b > 3$ *and* $0 < s < \infty$, *and suppose* $n + s \geqq \sigma$ *and* $n + s > \sigma/p$. *Then*

$$\left\| f^{+(\phi),\Omega,\epsilon} \right\|_{L^p(\Omega,\lambda)} \approx \| f_s^{*\Omega,b} \|_{L^p(\Omega,\lambda)}$$

for all $f \in \mathcal{D}'(\Omega)$.

With this theorem in mind, we define the weighted Hardy space as follows.

Definition. Let $\lambda \in \mathrm{Double}(\Omega)$ and $0 < p < \infty$. Let ϕ be a function satisfying (1.2) and $0 < \epsilon < 1/3$. For $f \in \mathcal{D}'(\Omega)$, we define

$$\|f\|_{H^p(\Omega,\lambda)} = \left\| f^{+(\phi),\Omega,\epsilon} \right\|_{L^p(\Omega,\lambda)}.$$

We define $H^p(\Omega, \lambda)$ as the set of all those $f \in \mathcal{D}'(\Omega)$ such that $\|f\|_{H^p(\Omega,\lambda)} < \infty$.

By Theorem 1.1, the quasinorm $\|f\|_{H^p(\Omega,\lambda)}$ does not depend, up to equivalence, on ϕ and ϵ, and $H^p(\Omega, \lambda)$ is defined independent of the choices of ϕ and ϵ.

Theorem 1.1 also shows that if b and s satisfy the conditions of the theorem then $\|f_s^{*\Omega,b}\|_{L^p(\Omega,\lambda)}$ is a quasinorm equivalent to $\|f\|_{H^p(\Omega,\lambda)}$. This last fact implies, in particular, that the canonical embedding $H^p(\Omega, \lambda) \hookrightarrow \mathcal{D}'(\Omega)$ is continuous.

The atomic decomposition for $H^p(\Omega, \lambda)$ is given in the next two theorems. In these theorems $L^\infty(\mathbb{R}^n)$ and $\|\cdot\|_{L^\infty}$ denote the L^∞ space and the L^∞ norm with respect to the Lebesgue measure.

Theorem 1.2. *Let* $\lambda \in \mathrm{Double}(\Omega, \sigma)$, $n \leqq \sigma < \infty$, *and* $0 < p < \infty$. *Let* m *be a positive integer satisfying* $n + m > \max\{\sigma, \sigma/p\}$. *Let* $\infty > b > 3$ *and* $0 < s < \infty$, *and suppose* $n + s \geqq \sigma$ *and* $n + s > \sigma/p$. *Finally let* $\infty > \bar{\delta} > \max\{b, 18\}$ *and* $\infty > \delta > 18$. *Then every* $f \in H^p(\Omega, \lambda)$ *can be written as follows:*

(i) $f = \sum_i g_i + \sum_j h_j$ *where both series converge unconditionally in* $H^p(\Omega, \lambda)$;

(ii) $g_i \in L^\infty(\mathbb{R}^n)$, $\mathrm{supp}\, g_i \subset B_i$, *with* B_i *a ball satisfying* $\mathrm{dis}\,(x(B_i), \Omega^c) = \bar{\delta} r(B_i)$, *and, for each* $0 < q < \infty$ *and for all* $x \in \Omega$,

$$\left(\sum_i \|g_i\|_{L^\infty}^q \chi_{B_i}(x) \right)^{1/q} \leqq c_q f_s^{*\Omega,b}(x) \tag{1.3}$$

(*If* $\Omega = \mathbb{R}^n$, *then* $\sum_i g_i = 0$.);

(iii) $h_j \in L^\infty(\mathbb{R}^n)$, $\mathrm{supp}\, h_j \subset Q_j$, *with* Q_j *a ball satisfying* $\delta Q_j \subset \Omega$, $h_j \perp \mathcal{P}_{m-1}$, *and, for each* $0 < q < \infty$ *and for all* $x \in \Omega$,

$$\left(\sum_j \|h_j\|_{L^\infty}^q \chi_{Q_j}(x) \right)^{1/q} \leqq c_q f_s^{*\Omega,b}(x). \tag{1.4}$$

The two constants c_q *can be taken depending only on* n, b, s, m, δ, $\bar{\delta}$, *and* q.

Theorem 1.3. *Let λ, σ, p, and m be the same as in Theorem 1.2, and let $\infty > \bar{\delta}, \delta > 18$. Suppose $\{g_i\}$ and $\{h_j\}$ satisfy the conditions* (ii) *and* (iii) *of Theorem 1.2 except that instead of* (1.3) *and* (1.4) *they satisfy*

$$F = \sum_i \|g_i\|_{L^\infty} \chi_{B_i} + \sum_j \|h_j\|_{L^\infty} \chi_{Q_j} \in L^p(\Omega, \lambda).$$

(If $\Omega = \mathbb{R}^n$, we assume $\sum_i g_i = 0$.) Then the two series $\sum_i g_i$ and $\sum_j h_j$ converge unconditionally in $H^p(\Omega, \lambda)$ and

$$\left\| \sum_i g_i + \sum_j h_j \right\|_{H^p(\Omega,\lambda)} \leqq c \|F\|_{L^p(\Omega,\lambda)}.$$

2. Preliminaries

In this section, we give several lemmas which will be used to prove the theorems of Section 1.

Throughout this section, we assume Ω is an open subset of $\mathbb{R}^n$ and $\lambda \in$ Double(Ω, σ) and $n \leqq \sigma < \infty$.

We begin with the definition of the Hardy-Littlewood maximal function. For a nonnegative Borel function f on Ω and for $\infty > b > 1$, we define $M_\lambda^{\Omega,b}(f)(x)$, $x \in \Omega$, by

$$M_\lambda^{\Omega,b}(f)(x) = \sup \left\{ \lambda(B)^{-1} \int_B f(y) d\lambda(y) \mid bB \subset \Omega,\ B \ni x \right\},$$

where B denotes a ball.

Lemma 2.1. *If $b > 3$, then it holds for all nonnegative Borel functions f on Ω:*

(1) $\lambda(\{x \in \Omega \mid M_\lambda^{\Omega,b}(f)(x) > s\}) \leqq cs^{-1} \|f\|_{L^1(\Omega,\lambda)}$ for all $s > 0$;

(2) $\|M_\lambda^{\Omega,b}(f)\|_{L^p(\Omega,\lambda)} \leqq c_p \|f\|_{L^p(\Omega,\lambda)}$ for $1 < p \leqq \infty$.

Proof. This lemma is well known in the case $\Omega = \mathbb{R}^n$. In the general case, the number 3 in the restriction $b > 3$ comes from the following covering lemma: If $\{B_1, \cdots, B_N\}$ is a finite family of balls, then there exists a disjoint subfamily $\{B_{i(1)}, \cdots, B_{i(M)}\}$ such that for each B_j there exists a $B_{i(k)}$ satisfying $B_j \cap B_{i(k)} \neq \emptyset$ and $r(B_{i(k)}) \geqq r(B_j)$, and hence $\bigcup_{j=1}^N B_j \subset \bigcup_{k=1}^M 3B_{i(k)}$. Details are left to the reader.

Lemma 2.2. *If $\infty > b > 9$ and if $1 < p, q < \infty$, then the inequality*

$$\left\| \left(\sum_j (M_\lambda^{\Omega,b}(f_j))^q \right)^{1/q} \right\|_{L^p(\Omega,\lambda)} \leqq c \left\| \left(\sum_j f_j^q \right)^{1/q} \right\|_{L^p(\Omega,\lambda)}$$

holds for all sequences $\{f_j\}_j$ of nonnegative Borel functions on Ω.

Proof. If $\Omega = \mathbb{R}^n$, this lemma is due to Fefferman and Stein [4]. Here we shall briefly see that $b > 9$ works in the general case. Using the covering lemma as mentioned in the proof of Lemma 2.1, we can prove that the inequality

$$\int_\Omega \mathbf{1}\{M_\lambda^{\Omega,b}(f)(x) > s\}w(x)d\lambda(x) \leqq cs^{-1}\int_\Omega f(x)M_\lambda^{\Omega,b/3}(w)(x)d\lambda(x)$$

holds for all nonnegative Borel functions f and w and for all $s > 0$. If $b > 9$, then using the above inequality and the $L^r(\Omega,\lambda)$ boundedness, $r > 1$, of $M_\lambda^{\Omega,b/3}$ (Lemma 2.1) and following the duality argument of [4] we can prove the inequality of Lemma 2.2 for $q < p < \infty$. The proof for $q = p$ immediately reduces to Lemma 2.1 and the proof for $p < q$ can be done in the same way as in [4]. Details are left to the reader.

Lemma 2.3. *Let $\infty > b > 9$ and let $\{B_j = B(x_j, r_j)\}$ be a sequence of balls satisfying $bB_j \subset \Omega$. Let $0 < p,\, q < \infty$ and $\infty > \alpha > \max\{\sigma/p, \sigma/q\}$. Then the following inequality holds for all $\{a_j\} \subset [0,\infty)$:*

$$\left\|\left(\sum_j a_j{}^q(1+r_j^{-1}|x-x_j|)^{-\alpha q}\mathbf{1}\{|x-x_j| < b^{-1}\mathrm{dis}\,(x_j,\Omega^c)\}\right)^{1/q}\right\|_{L^p(\Omega,\lambda)}$$
$$\leqq c\left\|\left(\sum_j a_j^q\chi_{B_j}\right)^{1/q}\right\|_{L^p(\Omega,\lambda)}.$$

Proof. We first prove

$$M_\lambda^{\Omega,b}(\chi_{B_j})(x) \geqq c(1+r_j^{-1}|x-x_j|)^{-\sigma}\mathbf{1}\{|x-x_j| < b^{-1}\mathrm{dis}\,(x_j,\Omega^c)\}. \tag{2.1}$$

To prove this, suppose first $x \notin B_j$ and $b|x-x_j| < \mathrm{dis}\,(x_j,\Omega^c)$. Then for sufficiently small $\epsilon > 0$ the ball $Q = B(x_j, |x-x_j|+\epsilon)$ satisfies $Q \ni x$, $bQ \subset \Omega$, $Q \supset B_j$, and, by the doubling condition, $\lambda(Q) \leqq c(r_j^{-1}|x-x_j|)^\sigma\lambda(B_j)$. Thus

$$M_\lambda^{\Omega,b}(\chi_{B_j})(x) \geqq \lambda(Q)^{-1}\lambda(B_j) \geqq c(r_j^{-1}|x-x_j|)^{-\sigma}.$$

If $x \in B_j$, then obviously $M_\lambda^{\Omega,b}(\chi_{B_j})(x) = 1$. Combining these estimates, we obtain (2.1).

Now applying Lemma 2.2 to $f_j = a_j\chi_{B_j}$ and using (2.1), we obtain the inequality of Lemma 2.3 for $\alpha = \sigma$. The general case reduces to the case $\alpha = \sigma$ if we replace p and q by $\alpha p/\sigma$ and $\alpha q/\sigma$ respectively. Lemma 2.3 is proved.

Before we give the next lemma, we introduce some notations. We write

$$T(\Omega) = \{(x,t) \in \Omega \times (0,\infty) \mid \mathrm{dis}\,(x,\Omega^c) > t\}.$$

(If $\Omega = \mathbb{R}^n$, then $T(\Omega) = T(\mathbb{R}^n) = \mathbb{R}^n \times (0,\infty)$.) If B is a ball, we write

$$Q(B) = B \times (0, r(B)) \subset \mathbb{R}^n \times (0,\infty).$$

For $(x,t) \in \mathbb{R}^n \times (0,\infty)$, we write $\delta_{(x,t)}$ to denote the Dirac measure concentrated at the point (x,t).

Lemma 2.4. *Let ν be a measure on the Borel subsets of $T(\Omega)$ and let $0 \leqq A,\ \delta < \infty$. Suppose the inequality*

$$\nu(T(\Omega) \cap Q(3B)) \leqq A\lambda(B)^{1+\delta}$$

holds for all balls B with $B \subset \Omega$. Then for each $1 < p < \infty$ and for all nonnegative Borel functions f on Ω we have

$$\int_{T(\Omega)} \left(\lambda(B(x,t))^{-1} \int_{B(x,t)} f\, d\lambda\right)^{p(1+\delta)} d\nu(x,t) \leqq c(\delta,p)A\left(\int_\Omega f^p\, d\lambda\right)^{1+\delta}.$$

Lemma 2.4 can be proved in the same way as in [3] or [18, pp. 582–583], where the case corresponding to $\Omega = \mathbb{R}^n$ is treated.

Lemma 2.5. *Let $0 \leqq A,\ \delta < \infty$, let $\{m_j\}$ be a sequence of nonnegative real numbers, and let $\{(x_j, r_j)\}$ be a sequence in $T(\Omega)$. Suppose that the inequality*

$$\sum_j m_j \delta_{(x_j,r_j)}(Q(3B)) \leqq A\lambda(B)^{1+\delta}$$

holds for all balls B with $B \subset \Omega$. Then for all nonnegative Borel functions g on Ω we have

$$\sum_j m_j \inf\{g(y) \mid y \in B(x_j, r_j)\} \leqq c(\delta)A\left(\int_\Omega g^{1/(1+\delta)}\, d\lambda\right)^{1+\delta}.$$

Proof. Since

$$\inf\{g(y) \mid y \in B(x_j,r_j)\} \leqq \left(\lambda(B(x_j,r_j))^{-1} \int_{B(x_j,r_j)} g^{1/2(1+\delta)}\, d\lambda\right)^{2(1+\delta)},$$

we obtain the desired inequality by applying the inequality of Lemma 2.4 to $f = g^{1/2(1+\delta)}$, $\nu = \sum_j m_j \delta_{(x_j,r_j)}$, and $p = 2$. Lemma 2.5 is proved.

Remark 1. Lemmas 2.4 and 2.5 hold for arbitrary $\lambda \in \mathcal{M}(\Omega)$; doubling condition is not necessary.

Lemma 2.6.

(1) *Let $0 < t,\ \delta < \infty$ and let η be a function such that $\eta \in C_0^\infty(B(0,1))$, $\int \eta(x)dx = 1$, and $\int \eta(x)x^\alpha dx = 0$ for $1 \leqq |\alpha| \leqq [t]$. For $f \in \Lambda(t)$ and for integers j, set*

$$f_j = f * (\eta)_{\delta 2^{-j}} - f * (\eta)_{\delta 2^{-j+1}}.$$

Then, for every multi-index α, there exists $c_\alpha = c(n,t,\eta,\alpha)$ such that

$$|\partial_x^\alpha f_j(\delta 2^{-j} x)| \leqq c_\alpha \|f\|_{\Lambda(t)} (\delta 2^{-j})^t.$$

(2) *Let $0 < t,\ \delta < \infty$, $0 \leqq M_0,\ M_1 < \infty$, and let m be an integer satisfying $m > t$. Let $\{g_j\}_{j\in\mathbb{Z}}$ be a sequence of C^m functions on $\mathbb{R}^n$. Suppose*

$$\|g_j\|_{L^\infty} \leqq M_0(\delta 2^{-j})^t$$

and

$$\sum_{|\alpha|=m} \|\partial_x^\alpha g_j(\delta 2^{-j}x)\|_{L^\infty} \leqq M_1(\delta 2^{-j})^t.$$

Then for all N_1, $N_2 \in \mathbb{N}$ *we have*

$$\left\| \sum_{j=-N_1}^{N_2} g_j \right\|_{\Lambda(t)} \leqq c(n,t,m) M_0^{1-t/m} M_1^{t/m}.$$

This lemma asserts the well-known fact that $\Lambda(t)$ is obtained by the real method of interpolation between L^∞ and C^m. The proof is left to the reader. Cf. [15, §2.7, pp. 200–202] or [12, Chapt. VI, §5.3, pp. 253–257]

The next lemma is also well known. Cf., for example, [12, Chapt. I, §3.2, pp. 14–16]

Lemma 2.7. *If* U *is a proper open subset of* $\mathbb{R}^n$ *and if* $\infty > u \geqq 3$, *then there exists a sequence of balls* $\{B_\nu\}$ *and a sequence of functions* $\{\varphi_\nu\} \subset C_0^\infty(\mathbb{R}^n)$ *which have the following properties:*

- (i) $r(B_\nu) = u^{-1}\mathrm{dis}\,(x(B_\nu), U^c)$;
- (ii) $\bigcup_\nu B_\nu = U$;
- (iii) *For each* v *satisfying* $0 < v < u$, *the overlap of the balls* $\{vB_\nu\}$ *is bounded by* $c(n,u,v)$, *i.e., we have* $\sum_\nu \chi_{B_\nu}(x) \leqq c(n,u,v)$ *for all* x;
- (iv) $\mathrm{supp}\,\varphi_\nu \subset 2B_\nu$, $0 \leqq \varphi_\nu(x) \leqq 1$ *for all* $x \in \mathbb{R}^n$, *and* $\sum_\nu \varphi_\nu(x) = 1$ *for all* $x \in U$;
- (v) $\|\partial^\alpha \varphi_\nu\|_{L^\infty} \leqq c(n,\alpha) r(B_\nu)^{-|\alpha|}$ *for every multi-index* α.

With a slight abuse of notation, we shall write $\mathcal{W}_u(U)$ to denote the sequence of balls $\{B_\nu\}$ which satisfies (i), (ii), and (iii) of the above lemma.

3. Proof of Theorem 1.1

The basic idea of the argument of this section goes back to Uchiyama [18] and [19].

Throughout this section, ϕ denotes a function satisfying (1.2).

Lemma 3.1. *Let* B *be a ball, let* $0 < s,\ t < \infty$, *and let* $0 < \alpha < 1$. *Then there exist sequences* $\{(x_j, r_j)\}_{j\in\mathbb{N}} \subset Q(2B)$ *and* $\{m_j\}_{j\in\mathbb{N}} \subset [0,\infty)$ *and there exists* $\epsilon \in (0,1)$ *which have the following properties:*

- (i) *There exists* $c = c(n,s,t,\phi,\alpha)$ *such that*

$$\sum_{j=1}^{\infty} m_j \delta_{(x_j,\epsilon r_j)}(Q(R)) \leqq c|R|^{(n+s)/n}$$

for all balls R;

(ii) *For each* $\psi \in C_0^\infty(B)$ *and for each* $\{\xi_j\}_{j\in\mathbb{N}}$ *with* $\xi_j \in B(x_j, \epsilon r_j)$*, there exists a finite sequence* $\{a_j\}_{j=1,\cdots,M} \subset \mathbb{C}$ *such that* $|a_j| \leqq m_j\|\psi\|_{\Lambda(s)}$ *for* $j = 1, \cdots, M$ *and*

$$\left\| \psi - \sum_{j=1}^{M} a_j(\phi)_{r_j}(\xi_j - \cdot) \right\|_{\Lambda(u)} \leqq \alpha\|\psi\|_{\Lambda(u)} \tag{3.1}$$

for both $u = s$ *and* $u = t$*. The number* ϵ *can be taken depending only on* n*,* s*,* t*,* ϕ*, and* α*, independent of* B*.*

Remark 2. Observe that the function $\sum_{j=1}^{M} a_j(\phi)_{r_j}(\xi_j - \cdot)$ in (ii) belongs to $C_0^\infty(6B)$.

Proof of Lemma 3.1. By dilation and translation, we may assume $B = B(0,1)$.

Take a function $\eta \in C_0^\infty(B(0,1))$ such that $\int \eta(x)dx = 1$ and $\int \eta(x)x^\beta dx = 0$ for $1 \leqq |\beta| \leqq \max\{[s],[t]\}$. Let $0 < \epsilon, \delta < 1$. For each $k \in \mathbb{N} \cup \{0\}$, take a finite number of disjoint nonempty Borel sets $E_{k,i} \subset B(0,2)$ such that $\operatorname{diam} E_{k,i} < \epsilon\delta 2^{-k}$ and $\bigcup_i E_{k,i} = B(0,2)$. For each $E_{k,i}$, take a point $x_{k,i} \in E_{k,i}$. The numbers ϵ and δ shall be fixed afterwards.

Let $\psi \in C_0^\infty(B(0,1))$ and $\xi_{k,i} \in B(x_{k,i}, \epsilon\delta 2^{-k})$ be given.

We define θ_k, u_k, v_k, and ψ_N ($k \in \mathbb{N} \cup \{0\}$, $N \in \mathbb{N}$) as follows:

$$\theta_0 = \psi * \eta,$$
$$\theta_k = \psi * \left((\eta)_{2^{-k}} - (\eta)_{2^{-k+1}}\right) \qquad (k \geqq 1),$$
$$u_k(y) = \int \theta_k(x)(\phi)_{\delta 2^{-k}}(x-y)dx,$$
$$v_k(y) = \sum_i |E_{k,i}|\theta_k(\xi_{k,i})(\phi)_{\delta 2^{-k}}(\xi_{k,i} - y),$$
$$\psi_N = \sum_{k=0}^{N} v_k.$$

We shall prove the following:

$$\operatorname{supp}\theta_k \subset B(0,2), \tag{3.2}$$
$$|E_{k,i}||\theta_k(\xi_{k,i})| \leqq c_1|E_{k,i}|2^{-ks}\|\psi\|_{\Lambda(s)}, \tag{3.3}$$
$$\|\psi * (\eta)_{2^{-N}} - \psi_N\|_{\Lambda(s)} \leqq c_2(\delta^{1/4} + \delta^{-s}\epsilon^{1/4})\|\psi\|_{\Lambda(s)}, \tag{3.4}$$
$$\|\psi * (\eta)_{2^{-N}} - \psi_N\|_{\Lambda(t)} \leqq c_3(\delta^{1/4} + \delta^{-t}\epsilon^{1/4})\|\psi\|_{\Lambda(t)}, \tag{3.5}$$

where $c_1 = c(n,s,\eta)$, $c_2 = c(n,s,\eta,\phi)$, and $c_3 = c(n,t,\eta,\phi)$.

Before we proceed to the proofs of (3.2)–(3.5), we shall see that the claim of the lemma follows from them. Notice first that $(x_{k,i}, \delta 2^{-k}) \in Q(B(0,2))$ and that ψ_N is a function of the form appearing in the left-hand side of (3.1). If R is a ball and $(x_{k,i}, \epsilon\delta 2^{-k}) \in Q(R)$, then $E_{k,i} \subset B(x_{k,i}, \epsilon\delta 2^{-k}) \subset 2R$; thus for all balls R we have

$$\sum_{k,\,i} c_1|E_{k,i}|2^{-ks}\delta_{(x_{k,i},\epsilon\delta 2^{-k})}(Q(R))$$
$$\leqq \sum_{\epsilon\delta 2^{-k}<r(R)} \sum_{E_{k,i}\subset 2R} c_1|E_{k,i}|2^{-ks} \leqq \sum_{\epsilon\delta 2^{-k}<r(R)} c_1|2R|2^{-ks} \leqq c_{\epsilon,\delta}|R|^{(n+s)/n}.$$

We fix δ and ϵ such that $c_2(\delta^{1/4}+\delta^{-s}\epsilon^{1/4}) \leqq \alpha/2$ and $c_3(\delta^{1/4}+\delta^{-t}\epsilon^{1/4}) \leqq \alpha/2$. If we take N sufficiently large such that $\|\psi-\psi*(\eta)_{2^{-N}}\|_{\Lambda(u)} \leqq (\alpha/2)\|\psi\|_{\Lambda(u)}$ for both $u=s$ and $u=t$, then by (3.4) and (3.5) we have $\|\psi-\psi_N\|_{\Lambda(u)} \leqq \alpha\|\psi\|_{\Lambda(u)}$ for $u=s,t$. Thus writing $\{(x_j,r_j)\}_j = \{(x_{k,i},\delta 2^{-k})\}_{k,i}$, $\{m_j\}_j = \{c_1|E_{k,i}|2^{-ks}\}_{k,i}$, and $\{a_j\}_{j\leqq M} = \{|E_{k,i}|\theta_k(\xi_{k,i})\}_{k\leqq N,i}$, we obtain the sequences which have the properties of the lemma.

We now prove (3.2)–(3.5). The claim (3.2) is easy to see. The inequality (3.3) follows from Lemma 2.6 (1). The main part is the proof of (3.4) and (3.5). Since (3.4) and (3.5) can be proved just in the same way, we shall give only the proof of (3.5). To simplify the notation, we assume $\|\psi\|_{\Lambda(t)}=1$.

By Lemma 2.6 (1), we have

$$|\partial_x^\beta \theta_k(2^{-k}x)| \leqq c_\beta 2^{-kt} \tag{3.6}$$

for every β. From this, by using Lemma 2.6 (2), we have

$$\|\theta_k\|_{\Lambda(1/2)} \leqq c2^{k(1/2-t)}. \tag{3.7}$$

We shall prove the following estimates:

$$\|\theta_k - u_k\|_{L^\infty} \leqq c\delta^{1/2}2^{-kt}, \tag{3.8}$$
$$\|u_k - v_k\|_{L^\infty} \leqq c\epsilon^{1/2}2^{-kt}, \tag{3.9}$$
$$|\partial_y^\beta u_k(2^{-k}y)| \leqq c_\beta 2^{-kt}, \tag{3.10}$$
$$|\partial_y^\beta v_k(\delta 2^{-k}y)| \leqq c_\beta 2^{-kt}, \tag{3.11}$$

where the last two estimates hold for every multi-index β.

First, using (3.7), we have

$$\begin{aligned}|\theta_k(y)-u_k(y)| &= \left|\int(\theta_k(y)-\theta_k(x))(\phi)_{\delta 2^{-k}}(x-y)dx\right| \\ &\leqq \int|\theta_k(y)-\theta_k(x)||(\phi)_{\delta 2^{-k}}(x-y)|dx \\ &\leqq c2^{k(1/2-t)}(\delta 2^{-k})^{1/2},\end{aligned}$$

which proves (3.8).

To prove (3.9), observe that we can write, by virtue of (3.2),

$$u_k(y)-v_k(y) = \sum_i \int_{E_{k,i}} w_{k,i}(x,y)dx$$

with

$$w_{k,i}(x,y) = \theta_k(x)(\phi)_{\delta 2^{-k}}(x-y) - \theta_k(\xi_{k,i})(\phi)_{\delta 2^{-k}}(\xi_{k,i}-y).$$

For all $x \in E_{k,i}$, we have

$$\begin{aligned}|w_{k,i}(x,y)| &\leqq |\theta_k(x) - \theta_k(\xi_{k,i})||(\phi)_{\delta 2^{-k}}(x-y)| \\ &\quad + |\theta_k(\xi_{k,i})||(\phi)_{\delta 2^{-k}}(x-y) - (\phi)_{\delta 2^{-k}}(\xi_{k,i}-y)| \\ &\leqq c2^{k(1/2-t)}(\epsilon\delta 2^{-k})^{1/2}(\delta 2^{-k})^{-n} + c2^{-kt}(\delta 2^{-k})^{-n-1}\epsilon\delta 2^{-k} \\ &\leqq c(\delta 2^{-k})^{-n}2^{-kt}\epsilon^{1/2},\end{aligned}$$

where we used (3.7) and (3.6) with $\beta = 0$. Integrating this, we obtain

$$\left|\int_{E_{k,i}} w_{k,i}(x,y)dx\right| \leqq c|E_{k,i}|(\delta 2^{-k})^{-n}2^{-kt}\epsilon^{1/2}.$$

On the other hand, if $\int_{E_{k,i}} w_{k,i}(x,y)dx \neq 0$, then $\operatorname{dis}(E_{k,i},y) < \delta 2^{-k} + \epsilon\delta 2^{-k}$ and hence $E_{k,i} \subset B(y, \delta 2^{-k} + 2\epsilon\delta 2^{-k}) \subset B(y, 3\delta 2^{-k})$. Hence

$$\begin{aligned}|u_k(y) - v_k(y)| &\leqq \sum_i \left|\int_{E_{k,i}} w_{k,i}(x,y)dx\right| \\ &\leqq c \sum_{E_{k,i}\subset B(y,3\delta 2^{-k})} |E_{k,i}|(\delta 2^{-k})^{-n}2^{-kt}\epsilon^{1/2} \\ &\leqq c2^{-kt}\epsilon^{1/2},\end{aligned}$$

which proves (3.9).

To prove (3.10), observe that we can write

$$\partial_y^\beta u_k(2^{-k}y) = \int \partial_y^\beta \theta_k\big(2^{-k}(y+z)\big)(\phi)_\delta(z)dz.$$

From this and (3.6), we obtain (3.10).

Finally to prove (3.11), we write

$$\partial_y^\beta v_k(\delta 2^{-k}y) = \sum_i |E_{k,i}|\theta_k(\xi_{k,i})(\delta 2^{-k})^{-n}\partial_y^\beta\phi(\delta^{-1}2^k\xi_{k,i} - y).$$

Observe that, if $\partial_y^\beta\phi(\delta^{-1}2^k\xi_{k,i} - y) \neq 0$, then $|\xi_{k,i} - \delta 2^{-k}y| < \delta 2^{-k}$ and hence $E_{k,i} \subset B(\delta 2^{-k}y, \delta 2^{-k} + 2\epsilon\delta 2^{-k}) \subset B(\delta 2^{-k}y, 3\delta 2^{-k})$. Hence, using (3.6) with $\beta = 0$, we obtain

$$\begin{aligned}|\partial_y^\beta v_k(\delta 2^{-k}y)| &\leqq \sum_i |E_{k,i}||\theta_k(\xi_{k,i})|(\delta 2^{-k})^{-n}|\partial_y^\beta\phi(\delta^{-1}2^k\xi_{k,i} - y)| \\ &\leqq c_\beta \sum_{E_{k,i}\subset B(\delta 2^{-k}y,3\delta 2^{-k})} |E_{k,i}|2^{-kt}(\delta 2^{-k})^{-n} \leqq c_\beta 2^{-kt},\end{aligned}$$

which proves (3.11). Thus (3.8)–(3.11) are proved.

Now we prove (3.5). We have

$$\psi * (\eta)_{2^{-N}} - \psi_N = \sum_{k=0}^{N} (\theta_k - u_k) + \sum_{k=0}^{N} (u_k - v_k).$$

From (3.6) and (3.10), we have

$$|\partial_x^\beta (\theta_k - u_k)(2^{-k}x)| \leqq c_\beta 2^{-kt}.$$

From this and (3.8), using Lemma 2.6 (2), we obtain

$$\left\| \sum_{k=0}^{N} (\theta_k - u_k) \right\|_{\Lambda(t)} \leqq c_m \delta^{(1-t/m)/2} \tag{3.12}$$

for each $m \in \mathbb{N}$ satisfying $m > t$. From (3.10) and (3.11), we have

$$|\partial_y^\beta (u_k - v_k)(\delta 2^{-k}y)| \leqq c_\beta 2^{-kt}.$$

From this and (3.9), using Lemma 2.6 (2) again, we obtain

$$\left\| \sum_{k=0}^{N} (u_k - v_k) \right\|_{\Lambda(t)} \leqq c_m \epsilon^{(1-t/m)/2} \delta^{-t} \tag{3.13}$$

for each $m \in \mathbb{N}$ satisfying $m > t$. If we take m so large that $m > 2t$, then the estimates (3.12) and (3.13) imply (3.5). Lemma 3.1 is proved.

Lemma 3.2. *Let B be a ball and let $0 < s, t < \infty$. Then there exist sequences $\{(x_i, r_i)\}_{i\in\mathbb{N}} \subset T(B)$ and $\{m_i\}_{i\in\mathbb{N}} \subset [0,\infty)$ and a number $\epsilon \in (0,1)$ which have the following properties:*

(i) *There exists $c = c(n, s, t, \phi)$ such that*

$$\sum_{i=1}^{\infty} m_i \delta_{(x_i, \epsilon r_i)}(Q(R)) \leqq c|R|^{(n+s)/n}$$

for all balls R;

(ii) *For each $\psi \in C_0^\infty(B)$ and for each $\{\xi_i\}_{i\in\mathbb{N}}$ with $\xi_i \in B(x_i, \epsilon r_i)$, there exists $\{a_{k,i}\} \subset \mathbb{C}$ ($i = 1, \cdots, M(k)$; $k = 0, 1, \cdots$) such that $|a_{k,i}| \leqq 2^{-k} m_i \|\psi\|_{\Lambda(s)}$ and*

$$\lim_{N\to\infty} \left\| \psi - \sum_{k=0}^{N} \sum_{i=1}^{M(k)} a_{k,i} (\phi)_{r_i}(\xi_i - \cdot) \right\|_{\Lambda(t)} = 0.$$

Proof. We take the balls $\{B_\nu\} = \mathcal{W}_{100}(B)$ and the partition of unity $\{\varphi_\nu\}$, $\varphi_\nu \in C_0^\infty(2B_\nu)$, as given in Lemma 2.7.

We use the following facts. First, if $\psi \in C_0^\infty(B)$, then $\psi = \sum_\nu \psi\varphi_\nu$, $\psi\varphi_\nu \in C_0^\infty(2B_\nu)$, $\psi\varphi_\nu \neq 0$ for only finitely many ν's, and $\sup \|\psi\varphi_\nu\|_{\Lambda(a)} \leqq c_a \|\psi\|_{\Lambda(a)}$ for each $a > 0$. Secondly, if $f_\nu \in C_0^\infty(12B_\nu)$ and $f_\nu \neq 0$ for only finitely many ν's, then $\|\sum_\nu f_\nu\|_{\Lambda(a)} \leqq c_a \sup \|f_\nu\|_{\Lambda(a)}$ for each $a > 0$. For these facts, see, e.g., [8, Section III]. Cf. also [17, §2.4.7, pp. 124–129] and [*ibid.*, §7.2.2, pp. 285–287].

Using the above facts and applying Lemma 3.1 to each ball $2B_\nu$, we obtain $(x_{\nu,j}, r_{\nu,j}) \in Q(4B_\nu) \subset T(B)$, $m_{\nu,j} \in [0,\infty)$, and $\epsilon \in (0,1)$ which have the following properties:

(i)′ There exists a constant $c = c(n,s,t,\phi)$ such that

$$\sum_j m_{\nu,j}\delta_{(x_{\nu,j},\epsilon r_{\nu,j})}(Q(R)) \leqq c|R|^{(n+s)/n}$$

for all ν and for all balls R;

(ii)′ For each $\psi \in C_0^\infty(B)$ and for each $\{\xi_{\nu,j}\}_{\nu,j}$ with $\xi_{\nu,j} \in B(x_{\nu,j}, \epsilon r_{\nu,j})$, there exists $\{a_{\nu,j}\}_{\nu,j} \subset \mathbb{C}$ such that $a_{\nu,j} \neq 0$ for only finitely many (ν,j)'s, $|a_{\nu,j}| \leqq m_{\nu,j}\|\psi\|_{\Lambda(s)}$, and

$$\|\psi - \sum_\nu \sum_j a_{\nu,j}(\phi)_{r_{\nu,j}}(\xi_{\nu,j} - \cdot)\|_{\Lambda(u)} \leqq 2^{-1}\|\psi\|_{\Lambda(u)}$$

for both $u = s$ and $u = t$.

We show that the inequality

$$\sum_\nu \sum_j m_{\nu,j}\delta_{(x_{\nu,j},\epsilon r_{\nu,j})}(Q(R)) \leqq c|R|^{(n+s)/n} \tag{3.14}$$

holds for all balls R. To see this, write the left-hand side of (3.14) as

$$\sum_\nu \sum_j = \sum_\nu{}' \sum_j + \sum_\nu{}'' \sum_j,$$

where $\sum'$ is the sum over those ν's satisfying $4B_\nu \cap R \neq \emptyset$ and $r(B_\nu) \leqq r(R)$, and $\sum''$ is the sum over those ν's satisfying $4B_\nu \cap R \neq \emptyset$ and $r(B_\nu) > r(R)$. For ν's in $\sum'$, we have $4B_\nu \subset 9R$. Hence, using the fact $(x_{\nu,j}, r_{\nu,j}) \in Q(4B_\nu)$ and (i)′, we obtain

$$\begin{aligned}\sum_\nu{}' \sum_j &\leqq \sum_\nu{}' \sum_j m_{\nu,j}\delta_{(x_{\nu,j},\epsilon r_{\nu,j})}(Q(4B_\nu)) \leqq \sum_\nu{}' c|4B_\nu|^{(n+s)/n} \\ &\leqq c\Big(\sum_\nu{}' |4B_\nu|\Big)^{(n+s)/n} \leqq c|R|^{(n+s)/n}.\end{aligned}$$

The number of ν's in $\sum''$ is bounded by c (by Lemma 2.7 (iii)). Hence, by (i)′,

$$\sum_\nu{}'' \sum_j \leqq \sum_\nu{}'' c|R|^{(n+s)/n} \leqq c|R|^{(n+s)/n}.$$

Thus we proved (3.14).

Notice that the function $\sum_\nu \sum_j a_{\nu,j}(\phi)_{r_{\nu,j}}(\xi_{\nu,j} - \cdot)$ appearing in (ii)$'$ belongs again to $C_0^\infty(B)$. Hence repeated application of (ii)$'$ gives $a_{k,\nu,j} \in \mathbb{C}$ for $k \in \mathbb{N} \cup \{0\}$ such that $|a_{k,\nu,j}| \leqq m_{\nu,j} 2^{-k} \|\psi\|_{\Lambda(s)}$, the number of (ν, j)'s satisfying $a_{k,\nu,j} \neq 0$ is finite for each k, and

$$\left\| \psi - \sum_{k=0}^{N} \sum_{\nu} \sum_{j} a_{k,\nu,j}(\phi)_{r_{\nu,j}}(\xi_{\nu,j} - \cdot) \right\|_{\Lambda(u)} \leqq 2^{-N-1} \|\psi\|_{\Lambda(u)}$$

for $u = s$, t and for all $N \in \mathbb{N} \cup \{0\}$.

Thus, writing $\{(x_i, r_i)\}_i = \{(x_{\nu,j}, r_{\nu,j})\}_{\nu,j}$, $\{m_i\}_i = \{m_{\nu,j}\}_{\nu,j}$, and $\{a_{k,i}\}_i = \{a_{k,\nu,j}\}_{\nu,j}$, we obtain the desired sequences. Lemma 3.2 is proved.

Lemma 3.3. *Let Ω be an open subset of $\mathbb{R}^n$ and let $\lambda \in$ Double (Ω, σ), $n \leqq \sigma < \infty$. Let $0 < s < \infty$, $\infty > b > 3$, and $0 < a \leqq 1$. Assume $n + s \geqq \sigma$. Then the inequality*

$$f_s^{*\Omega,b}(x) \leqq c \left[M_\lambda^{\Omega,b} \left((f^{+(\phi),\Omega,a})^{\sigma/(n+s)} \right)(x) \right]^{(n+s)/\sigma}$$

holds for all $f \in \mathcal{D}'(\Omega)$ and for all $x \in \Omega$.

Remark 3. For the measure λ of this lemma, there exists $A \in [1, \infty)$ such that the inequality $\lambda(B) \leqq At^\sigma \lambda(t^{-1}B)$ holds for all balls B satisfying $(b/3)B \subset \Omega$ and for all $t \geqq 1$. The constant c of the lemma can be taken depending only on n, s, ϕ, a, σ, and A.

Proof of Lemma 3.3. Replacing ϕ by $(\phi)_a$, we may assume $a = 1$. We write $f^+ = f^{+(\phi),\Omega,1}$. We shall prove that there exists a constant $c = c(n, s, \phi, \sigma, A)$ (here A is the constant mentioned in the above remark) such that the inequality

$$|\langle f, \psi \rangle| \leqq c \|\psi\|_{\Lambda(s)} r(B)^{n+s} \left(\lambda(B)^{-1} \int_B (f^+)^{\sigma/(n+s)} d\lambda \right)^{(n+s)/\sigma} \tag{3.15}$$

holds for all balls B satisfying $bB \subset \Omega$ and for all $\psi \in C_0^\infty(B)$. The inequality of the lemma is an immediate consequence of (3.15).

We first prove (3.15) under the additional assumption that f is of order s on B, by which we mean that there exists a constant K such that $|\langle f, \psi \rangle| \leqq K \|\psi\|_{\Lambda(s)}$ for all $\psi \in C_0^\infty(B)$.

Let $B = B(x, r)$ be a ball satisfying $bB \subset \Omega$ and let $\psi \in C_0^\infty(B)$. Let $\{(x_i, r_i)\}$, $\{m_i\}$, ϵ, $\{\xi_i\}$, and $\{a_{k,i}\}$ be as mentioned in Lemma 3.2 with $s = t$. Then, since f is of order s on B, we have

$$\langle f, \psi \rangle = \sum_{k=0}^{\infty} \sum_{i=1}^{M(k)} a_{k,i} \langle f, (\phi)_{r_i}(\xi_i - \cdot) \rangle.$$

Notice that

$$\mathrm{dis}\,(\xi_i, \Omega^c) \geqq \mathrm{dis}\,(x, \Omega^c) - |x - x_i| - |x_i - \xi_i| > br - r - \epsilon r_i > r_i.$$

Hence $|\langle f, (\phi)_{r_i}(\xi_i - \cdot)\rangle| \leqq f^+(\xi_i)$ and

$$\begin{aligned}|\langle f, \psi\rangle| &\leqq \sum_{k=0}^{\infty}\sum_{i=1}^{M(k)} |a_{k,i}||\langle f, (\phi)_{r_i}(\xi_i - \cdot)\rangle| \leqq \sum_{k=0}^{\infty}\sum_{i=1}^{\infty} 2^{-k} m_i \|\psi\|_{\Lambda(s)} f^+(\xi_i)\\ &= 2\sum_{i=1}^{\infty} m_i\|\psi\|_{\Lambda(s)} f^+(\xi_i).\end{aligned}$$

Since ξ_i can be taken arbitrary in $B(x_i, \epsilon r_i)$, we have

$$|\langle f, \psi\rangle| \leqq 2\sum_{i=1}^{\infty} m_i\|\psi\|_{\Lambda(s)} \inf\{f^+(y) \mid y \in B(x_i, \epsilon r_i)\}. \tag{3.16}$$

Let $R = B(y,t)$ be a ball satisfying $R \subset B$. Then $B = B(x,r) \subset B(y,2r)$ and $(b/3)B(y,2r) \subset B(x, r + 2br/3) \subset B(x, br) \subset \Omega$ (since $b > 3$). Hence by the doubling condition on λ we have $\lambda(B) \leqq \lambda(B(y,2r)) \leqq A(2r/t)^\sigma \lambda(R)$, and hence $t^\sigma \leqq (2r)^\sigma A\lambda(R)\lambda(B)^{-1}$. Thus

$$\sum_{i=1}^{\infty} m_i \delta_{(x_i,\epsilon r_i)}(Q(3R)) \leqq ct^{n+s} \leqq c(2r)^{n+s}\left(A\lambda(R)\lambda(B)^{-1}\right)^{(n+s)/\sigma}. \tag{3.17}$$

Now (3.15) follows from (3.16) and (3.17) by the use of Lemma 2.5.

Next, we show that (3.15) holds for general $f \in \mathcal{D}'(\Omega)$. In this general case, there exist t, depending on f and B, such that f is of order t on B. Using Lemma 3.2 with this t and following the same argument as above, we again obtain (3.15) but now with c depending on t and hence on f and B. But (3.15) thus obtained implies at least that f is of order s on B as far as $\int_B (f^+)^{\sigma/(n+s)} d\lambda < \infty$. Hence the proof reduces to the first case. Lemma 3.3 is proved.

Proof of Theorem 1.1. By Lemma 3.3 and Lemma 2.1 (2), we obtain

$$\|f_s^{*\Omega,b}\|_{L^p(\Omega,\lambda)} \leqq c\|f^{+(\phi),\Omega,\epsilon}\|_{L^p(\Omega,\lambda)}.$$

Conversely, using the obvious pointwise inequalities

$$f^{+(\phi),\Omega,\epsilon} \leqq cf_s^{*\Omega,1/\epsilon} \qquad \text{and} \qquad f^{+(\phi),\Omega,1/b} \leqq cf_s^{*\Omega,b},$$

and using Lemma 3.3 again, we obtain

$$\begin{aligned}\|f^{+(\phi),\Omega,\epsilon}\|_{L^p(\Omega,\lambda)} &\leqq c\|f_s^{*\Omega,1/\epsilon}\|_{L^p(\Omega,\lambda)}\\ &\leqq c\|f^{+(\phi),\Omega,1/b}\|_{L^p(\Omega,\lambda)}\\ &\leqq c\|f_s^{*\Omega,b}\|_{L^p(\Omega,\lambda)}.\end{aligned}$$

Theorem 1.1 is proved.

4. Proofs of Theorems 1.2 and 1.3

We first prove Theorem 1.3.

Let $\{g_i\}$, $\{h_j\}$, and F be as mentioned in Theorem 1.3. Take b and s sufficiently large. We shall prove the following estimates:

$$(g_i)_s^{*\Omega,b}(x) \leqq c\|g_i\|_{L^\infty}\left(1+\frac{|x-x(B_i)|}{r(B_i)}\right)^{-n}\mathbf{1}\left\{|x-x(B_i)|<\frac{2}{\delta}\mathrm{dis}\,(x(B_i),\Omega^c)\right\}; \quad (4.1)$$

for each $0<\epsilon<1$

$$(h_j)_s^{*\Omega,b}(x) \leqq c_\epsilon\|h_j\|_{L^\infty}\left(1+\frac{|x-x(Q_j)|}{r(Q_j)}\right)^{-n-m+\epsilon}\mathbf{1}\left\{|x-x(Q_j)|<\frac{2}{\delta}\mathrm{dis}\,(x(Q_j),\Omega^c)\right\}. \quad (4.2)$$

First, the estimate $(g_i)_s^{*\Omega,b}(x) \leqq c\|g_i\|_{L^\infty}(1+r(B_i)^{-1}|x-x(B_i)|)^{-n}$ is easy to see. If we take s such that $s \geqq m$, then it is also easy to see that the estimate $(h_j)_s^{*\Omega,b}(x) \leqq c_\epsilon\|h_j\|_{L^\infty}(1+r(Q_j)^{-1}|x-x(Q_j)|)^{-n-m+\epsilon}$ holds for every $\epsilon\in(0,1)$. These prove the estimates (4.1) and (4.2) except for the factors $\mathbf{1}\{\cdots\}$.

We shall show that (4.1) and (4.2) in fact hold with the factors $\mathbf{1}\{\cdots\}$. If $x\in\Omega$ and if B is a ball satisfying $B\ni x$, $bB\subset\Omega$, and $B\cap B_i\neq\emptyset$, then we have $1/4<\mathrm{dis}\,(x(B_i),\Omega^c)/\mathrm{dis}\,(x(B),\Omega^c)<4$ and hence

$$\begin{aligned}|x-x(B_i)| &< 2r(B)+r(B_i)\\ &\leqq 2b^{-1}\mathrm{dis}\,(x(B),\Omega^c)+\bar{\delta}^{-1}\mathrm{dis}\,(x(B_i),\Omega^c)\\ &< 2\bar{\delta}^{-1}\mathrm{dis}\,(x(B_i),\Omega^c).\end{aligned}$$

(Recall that we are assuming b is sufficiently large.) Hence, $(g_i)_s^{*\Omega,b}(x)\neq 0$ only if $|x-x(B_i)|<2\bar{\delta}^{-1}\mathrm{dis}\,(x(B_i),\Omega^c)$. This shows that we have the $\mathbf{1}\{\cdots\}$ factor in (4.1). The validity of the $\mathbf{1}\{\cdots\}$ factor in (4.2) can be seen in the same way. Thus (4.1) and (4.2) are proved.

By (4.1) and (4.2) with ϵ taken so small that $n+m-\epsilon>\max\{\sigma/p,\sigma\}$ and by Lemma 2.3, we see that

$$\left\|\sum_i (g_i)_s^{*\Omega,b}+\sum_j (h_j)_s^{*\Omega,b}\right\|_{L^p(\Omega,\lambda)} \leqq c\|F\|_{L^p(\Omega,\lambda)}<\infty.$$

From this all the conclusions of the theorem easily follow. Theorem 1.3 is proved.

We proceed to the proof of Theorem 1.2. We shall prove this theorem in four steps.

First Step. We recall the scheme of atomic decomposition given in [9, Section 3]. Suppose $f\in\mathcal{D}'(\mathbb{R}^n)$, $\infty>s>0$, and $m\in\mathbb{N}$.

For integers k, set

$$U(k)=\{x\in\mathbb{R}^n \mid f_s^*(x)>2^k\}.$$

This is an open subset of $\mathbb{R}^n$. For k satisfying $U(k) \neq \mathbb{R}^n$, let $\mathcal{W}_u(U(k))$ be the family of balls given in Lemma 2.7; we take u sufficiently large. We write $\{\varphi_I^{U(k)} \mid I \in \mathcal{W}_u(U(k))\}$, $\varphi_I^{U(k)} \in C_0^\infty(2I)$, to denote the partition of unity on $U(k)$ as given in Lemma 2.7 (iv), (v). For k with $U(k) \neq \mathbb{R}^n$ and for $J \in \mathcal{W}_u(U(k+1))$ and $I \in \mathcal{W}_u(U(k))$, we define $P_{J,I}^k \in \mathcal{P}_{m-1}$ as follows: If $2J \cap 2I \neq \emptyset$, then $P_{J,I}^k$ is the unique polynomial in $\mathcal{P}_{m-1}$ such that

$$f\varphi_J^{U(k+1)}\varphi_I^{U(k)} - P_{J,I}^k \chi_{5J}\chi_{5I} \perp \mathcal{P}_{m-1};$$

if $2J \cap 2I = \emptyset$, then we set $P_{J,I}^k = 0$. We define g^k and h_J^k $(J \in \mathcal{W}_u(U(k+1)))$ by

$$g^k = f - \sum_{J,\,I} f\varphi_J^{U(k+1)}\varphi_I^{U(k)} + \sum_{J,\,I} P_{J,I}^k \chi_{5J}\chi_{5I},$$

$$h_J^k = -\sum_K f\varphi_K^{U(k+2)}\varphi_J^{U(k+1)} + \sum_K P_{K,J}^{k+1}\chi_{5K}\chi_{5J} + f\varphi_J^{U(k+1)} - \sum_I P_{J,I}^k \chi_{5J}\chi_{5I},$$

where I, J, and K run on $\mathcal{W}_u(U(k))$, $\mathcal{W}_u(U(k+1))$, and $\mathcal{W}_u(U(k+2))$, respectively.

Then, by the argument given in [9, Section 3], we see that the following holds for each k satisfying $U(k) \neq \mathbb{R}^n$ and for all $J \in \mathcal{W}_u(U(k+1))$ and all $I \in \mathcal{W}_u(U(k))$:

(1) $\|P_{J,I}^k \chi_{5J}\chi_{5I}\|_{L^\infty} \leqq c2^k$;

(2) All the sums in the definitions of g^k and h_J^k converge unconditionally in $\mathcal{D}'(\mathbb{R}^n)$;

(3) g^k and h_J^k belong to $L^\infty(\mathbb{R}^n)$, and $\|g^k\|_{L^\infty} \leqq c2^k$ and $\|h_J^k\|_{L^\infty} \leqq c2^k$;

(4) $\operatorname{supp} h_J^k \subset 6J$ and $h_J^k \perp \mathcal{P}_{m-1}$;

(5) $g^{k+1} - g^k = \sum_{J \in \mathcal{W}_u(U(k+1))} h_J^k$; this series converges unconditionally in $\mathcal{D}'(\mathbb{R}^n)$.

Second Step. We shall prove Theorem 1.2 in the case $\Omega = \mathbb{R}^n$. Suppose the assumptions of Theorem 1.2 are satisfied with $\Omega = \mathbb{R}^n$ and suppose $f \in H^p(\mathbb{R}^n, \lambda)$.

Let $U(k)$, g^k, and h_J^k be as given in First Step. In the present case, $U(k) \neq \mathbb{R}^n$ for all $k \in \mathbb{Z}$. We shall prove that $\lim_{k\to\infty} g^k = f$ in $\mathcal{D}'(\mathbb{R}^n)$ and that the series $\sum_{k\in\mathbb{Z}} \sum_{J \in \mathcal{W}_u(U(k+1))} h_J^k$ converges unconditionally in $\mathcal{D}'(\mathbb{R}^n)$. Once these are proved, we obtain

$$f = \lim_{N\to\infty} \sum_{k=-N}^{N} (g^{k+1} - g^k) = \sum_{k=-\infty}^{\infty} \sum_{J \in \mathcal{W}_u(U(k+1))} h_J^k$$

(since $\lim_{k\to-\infty} g^k = 0$ by (3) of First Step), and it is easy to see that the last series $\sum_k \sum_J h_J^k$ has all the properties of the series $\sum_j h_j$ of Theorem 1.2. (The unconditional convergence of the series in $H^p(\mathbb{R}^n, \lambda)$ is seen by the use of Theorem 1.3.)

To prove $\lim_{k\to\infty} g^k = f$ in $\mathcal{D}'(\mathbb{R}^n)$, we write $f - g^k = \sum_{J,I} f_{J,I}^k$ with

$$f_{J,I}^k = f\varphi_J^{U(k+1)}\varphi_I^{U(k)} - P_{J,I}^k \chi_{5J}\chi_{5I}.$$

Fix a $\psi \in C_0^\infty(\mathbb{R}^n)$. We have

$$\langle f - g^k, \psi \rangle = \sum_{J,I} \langle f_{J,I}^k, \psi \rangle,$$

where the sum is taken over all those (J, I) such that

$$J \in \mathcal{W}_u(U(k+1)), \quad I \in \mathcal{W}_u(U(k)), \quad 2J \cap 2I \neq \emptyset, \quad 5J \cap \operatorname{supp} \psi \neq \emptyset. \tag{4.3}$$

Suppose $J \in \mathcal{W}_u(U(k+1))$ and $5J \cap \operatorname{supp} \psi \neq \emptyset$. If $r(J) > 1$, then for $x \in 5J \cap \operatorname{supp} \psi$ we have $B(x, 1) \subset 6J \subset U(k+1)$ and hence

$$2^{(k+1)p} \inf\{\lambda(B(x,1)) \mid x \in \operatorname{supp} \psi\} \leqq \int_{U(k+1)} (f_s^*)^p d\lambda < \infty. \tag{4.4}$$

The above inf is positive since $\operatorname{supp} \psi$ is compact and since the function $x \mapsto \lambda(B(x, 1))$ is positive and lower semicontinuous. Thus (4.4) cannot hold if k is sufficiently large. Thus, if k is sufficiently large, we must have $r(J) \leqq 1$.

If $J \in \mathcal{W}_u(U(k+1))$ and $5J \cap \operatorname{supp} \psi \neq \emptyset$ and if $r(J) \leqq 1$, then

$$J \subset \{x \mid \operatorname{dis}(x, \operatorname{supp} \psi) \leqq 6\} = E, \quad \text{say}, \tag{4.5}$$

and, by virtue of the doubling condition on λ,

$$\begin{aligned} \lambda(J) &\geqq A^{-1} r(J)^\sigma \lambda(B(x(J), 1)) \\ &\geqq A^{-1} r(J)^\sigma \inf_{x \in E} \lambda(B(x, 1)) \\ &= v r(J)^\sigma, \end{aligned} \tag{4.6}$$

where $v = A^{-1} \inf_{x \in E} \lambda(B(x, 1)) > 0$ by the same reason as above.

Now let (J, I) satisfy (4.3) and suppose k is sufficiently large so that (4.5) and (4.6) also hold. Using the fact $f_{J,I}^k \perp \mathcal{P}_{m-1}$, we can prove that there exists a constant c_ψ depending on ψ such that

$$|\langle f_{J,I}^k, \psi \rangle| \leqq c_\psi 2^k r(J)^{n+m}$$

(see [9, pp. 220–221]). Combining this estimate with (4.6), we obtain

$$|\langle f_{J,I}^k, \psi \rangle| \leqq c_\psi 2^k \left(v^{-1} \lambda(J)\right)^{(n+m)/\sigma}. \tag{4.7}$$

Using (4.7) and using the fact that for each fixed $J \in \mathcal{W}_u(U(k+1))$ the number of $I \in \mathcal{W}_u(U(k))$ satisfying $2J \cap 2I \neq \emptyset$ is bounded by c, we obtain, for large k,

$$
\begin{aligned}
|\langle f-g^k,\psi\rangle| &\leqq \sum_{(J,I):(4.3)} c_{\psi,v}2^k\lambda(J)^{(n+m)/\sigma}\\
&\leqq c_{\psi,v}2^k \sum_{\substack{J\in\mathcal{W}_u(U(k+1))\\ J\subset E}} \lambda(J)^{(n+m)/\sigma}\\
&\leqq c_{\psi,v}2^k\Bigg(\sum_{\substack{J\in\mathcal{W}_u(U(k+1))\\ J\subset E}} \lambda(J)\Bigg)^{(n+m)/\sigma}\\
&\leqq c_{\psi,v}2^k\lambda(E\cap U(k+1))^{(n+m)/\sigma}\\
&\leqq c_{\psi,v}\left(\int_{E\cap U(k+1)}(f_s^*)^{\sigma/(n+m)}d\lambda\right)^{(n+m)/\sigma}.
\end{aligned}
$$

Since $\sigma/(n+m)<p$ and $f_s^*\in L^p(\mathbb{R}^n,\lambda)$, we now see that $\langle f-g^k,\psi\rangle\to 0$ as $k\to\infty$.

To see the unconditional convergence of $\sum_k\sum_J h_J^k$, fix a $\psi\in C_0^\infty(\mathbb{R}^n)$. By (3) and (4) of First Step, we have

$$|\langle h_J^k,\psi\rangle|\leqq c2^k\int_{6J}|\psi(x)|dx.$$

On the other hand, using the moment condition $h_J^k\perp\mathcal{P}_{m-1}$ and arguing in the same way as for (4.7), we can deduce the estimate

$$|\langle h_J^k,\psi\rangle|\leqq c_\psi 2^k\left(v^{-1}\lambda(J)\right)^{(n+m)/\sigma}$$

for some $v>0$ and for all large k. These two estimates imply that

$$\sum_k\sum_J|\langle h_J^k,\psi\rangle|<\infty.$$

Thus we proved the unconditional convergence of $\sum_k\sum_J h_J^k$ in $\mathcal{D}'(\mathbb{R}^n)$. Theorem 1.2 for $\Omega=\mathbb{R}^n$ is proved.

Third Step. We consider the decomposition of $f\in\mathcal{D}'(\mathbb{R}^n)$ with $\operatorname{supp} f$ included in a ball. Let $f\in\mathcal{D}'(\mathbb{R}^n)$, $f\neq 0$, $\infty>s>0$, $m\in\mathbb{N}$, and $B_0=B(x_0,t_0)$ be a ball. Suppose $\operatorname{supp} f\subset B_0$. Also suppose there exists a measure $\lambda\in\operatorname{Double}(10B_0,\sigma)$ with $n\leqq\sigma\leqq n+m$ such that $\int_{2B_0}(f_s^*)^{\sigma/(n+m)}d\lambda<\infty$.

We define $U(k)$, g^k, and h_J^k as in First Step.

It is easy to see the following:

(6) If $x\notin(3/2)B_0$, then $f_s^*(x)\leqq c\inf\{f_s^*(y)\mid y\in 2B_0\}(t_0^{-1}|x-x_0|)^{-n}$;

(7) For each integer k, the set $U(k)$ is bounded, $\operatorname{supp} g^k\subset(\operatorname{supp} f)\cup(U(k+1))^a$ (where E^a denotes the closure of the set E), and $f-g^k\perp\mathcal{P}_{m-1}$.

We take the integer l such that

$$2^{l-1}<\sup\{f_s^*(x)\mid x\notin(3/2)B_0\}\leqq 2^l.$$

Then we have the following decomposition of f:

(8) $f = g^l + \sum_{k=l}^{\infty} \sum_{J \in \mathcal{W}_u(U(k+1))} h_J^k$ with the double series $\sum_k \sum_J h_J^k$ converging unconditionally in $\mathcal{D}'(\mathbb{R}^n)$;

(9) $g^l \in L^\infty(\mathbb{R}^n)$, $\operatorname{supp} g^l \subset 2B_0$, and $\|g^l\|_{L^\infty} \leqq c \inf\{f_s^*(y) \mid y \in 2B_0\}$;

(10) $f - g^l \perp \mathcal{P}_{m-1}$.

In fact, (9) and (10) easily follow from the results in First Step and from (6) and (7) above. The claim (8) can be proved in the same way as in Second Step.

Fourth Step. We shall prove Theorem 1.2 for $\Omega \neq \mathbb{R}^n$. Suppose the assumptions of Theorem 1.2 are satisfied with $\Omega \neq \mathbb{R}^n$ and suppose $f \in H^p(\Omega, \lambda)$.

Let $\{B_\nu\} = \mathcal{W}_v(\Omega)$, $B_\nu = B(x_\nu, r_\nu)$, and $\{\varphi_\nu\}$, $\varphi_\nu \in C_0^\infty(2B_\nu)$, be as given in Lemma 2.7; we take the number v sufficiently large. Take a function $\eta \in C_0^\infty(\mathbb{R}^n)$ such that $0 \leqq \eta(x) \leqq 1$ for all $x \in \mathbb{R}^n$, $\eta(x) = 1$ for $x \in B(0,1)$, and $\operatorname{supp} \eta \subset B(0,2)$. For each ν, set $\eta_\nu(x) = \eta\big(r_\nu^{-1}(x - x_\nu)\big)$, take the unique polynomial $P_\nu \in \mathcal{P}_{m-1}$ such that $f\varphi_\nu - P_\nu \eta_\nu \perp \mathcal{P}_{m-1}$, and set $f_\nu = f\varphi_\nu - P_\nu \eta_\nu$.

It is easy to see that

$$\|P_\nu \chi_{2B_\nu}\|_{L^\infty} \leqq c \inf\{f_s^{*\Omega,b}(y) \mid y \in 4B_\nu\}$$

and

$$(f_\nu)_s^*(x) \leqq c f_s^{*\Omega,b}(x) \quad \text{for all} \quad x \in 4B_\nu. \tag{4.8}$$

We apply the decomposition of Third Step to each f_ν and $2B_\nu$ (notice that $\operatorname{supp} f_\nu \subset 2B_\nu$) to obtain, say,

$$f_\nu = g_\nu^{l(\nu)} + \sum_{k=l(\nu)}^{\infty} \sum_{J \in \mathcal{W}_u(U_\nu(k+1))} h_{\nu,J}^k;$$

we take the number u sufficiently large. We have $\operatorname{supp} g_\nu^{l(\nu)} \subset 4B_\nu$ and

$$\|g_\nu^{l(\nu)}\|_{L^\infty} \leqq c \inf\{f_s^{*\Omega,b}(y) \mid y \in 4B_\nu\}$$

by (9) of Third Step and by (4.8). We also have $g_\nu^{l(\nu)} \perp \mathcal{P}_{m-1}$ since $f_\nu \perp \mathcal{P}_{m-1}$ and since $f_\nu - g_\nu^{l(\nu)} \perp \mathcal{P}_{m-1}$ ((10) of Third Step).

We have

$$f = \sum_\nu P_\nu \eta_\nu + \sum_\nu f_\nu = \sum_\nu P_\nu \eta_\nu + \sum_\nu g_\nu^{l(\nu)} + \sum_\nu \sum_{k=l(\nu)}^{\infty} \sum_{J \in \mathcal{W}_u(U_\nu(k+1))} h_{\nu,J}^k.$$

It is easy to see that the last decomposition has all the properties as mentioned in Theorem 1.2 if we write

$$\sum_i g_i = \sum_\nu P_\nu \eta_\nu$$

and

$$\sum_j h_j = \sum_\nu g_\nu^{l(\nu)} + \sum_\nu \sum_{k=l(\nu)}^{\infty} \sum_{J \in \mathcal{W}_u(U_\nu(k+1))} h_{\nu,J}^k.$$

Theorem 1.2 is proved.

5. Transplantation theorem for Jacobi series

In this section, we give an application of the results of Section 1 to the transplantation theorem for Jacobi series.

For an interval $I \subset \mathbb{R}$, let $L^p(I)$ denote the L^p space on I with respect to the Lebesgue measure.

We first recall which kind of assertion the transplantation theorem is. Let $\{u_n\}$ and $\{v_n\}$ be two orthonormal series in $L^2(I)$. Let X be a function space on I equipped with the quasinorm $\|\cdot\|_X$. (We shall not be concerned with the precise definition of the function space on I; a typical example we have in mind is $L^p(I)$ or the L^p space on I with respect to a weighted measure $w(x)dx$.) The transplantation theorem is the assertion that

$$\Big\|\sum_n \xi_n u_n\Big\|_X \approx \Big\|\sum_n \xi_n v_n\Big\|_X \quad \text{for all} \quad \{\xi_n\} \in l^2,$$

where we set $\|f\|_X = \infty$ if $f \in L^2(I) \setminus X$. Notice that both series $\sum_n \xi_n u_n$ and $\sum_n \xi_n v_n$ converge in $L^2(I)$ as long as $\{\xi_n\} \in l^2$.

Several transplantation theorems are already known. Guy [6, Lemma 8C, p. 185] gave the transplantation theorem related to Hankel transforms. Askey and Wainger [2] treated the orthonormal series defined through the ultraspherical polynomials. Askey [1] and Muckenhoupt [11] treated the orthonormal series defined through Jacobi polynomials. Kanjin [7] treated the series defined through Laguerre polynomials.

As far as we know, in the transplantation theorems known so far, the space X is the L^p space with $1 < p < \infty$ (or sometimes with $p = 1$). The purpose of this section is to give a transplantation theorem in the Hardy space H^p, $0 < p \leqq 1$. We shall consider the orthonormal series defined through Jacobi polynomials. In our theorem, the use of certain weighted Hardy spaces of Section 1 will be essential. We shall begin with the definition of Jacobi polynomials.

For $\infty > \alpha, \beta > -1$, the Jacobi polynomial $P_n^{(\alpha,\beta)}(x)$ $(n = 0, 1, 2, \cdots)$ is defined by

$$(1-x)^\alpha (1+x)^\beta P_n^{(\alpha,\beta)}(x) = \frac{(-1)^n}{2^n\, n!} \frac{d^n}{dx^n}\Big[(1-x)^{n+\alpha}(1+x)^{n+\beta}\Big].$$

Then $P_n^{(\alpha,\beta)}(x)$ is indeed a polynomial in x of degree n with real coefficients. They satisfy the following orthogonal relation

$$\int_{-1}^{1} P_n^{(\alpha,\beta)}(x) P_k^{(\alpha,\beta)}(x)(1-x)^\alpha(1+x)^\beta dx = 0 \quad \text{if} \quad n \neq k.$$

For details about Jacobi polynomials, see [14, Chapter IV].

Changing variables $x = \cos\theta$, we define $\varphi_n^{(\alpha,\beta)}(\theta)$, $0 < \theta < \pi$, by

$$\varphi_n^{(\alpha,\beta)}(\theta) = t_n^{(\alpha,\beta)}(1-\cos\theta)^{\alpha/2+1/4}(1+\cos\theta)^{\beta/2+1/4} P_n^{(\alpha,\beta)}(\cos\theta),$$

where $t_n^{(\alpha,\beta)} > 0$ is taken so that $\|\varphi_n^{(\alpha,\beta)}\|_{L^2(0,\pi)} = 1$. Then $\{\varphi_n^{(\alpha,\beta)}\}$ is an orthonormal series in $L^2(0,\pi)$.

We define h_n by

$$h_n(\theta) = \begin{cases} \pi^{-1/2} & (n = 0) \\ (\pi/2)^{-1/2} \cos n\theta & (n = 1, 2, \cdots). \end{cases}$$

Then $\{h_n\}$ is an orthonormal series in $L^2(0,\pi)$. It is known that $h_n = \varphi_n^{(-1/2,-1/2)}$ (see [14, p. 59]).

For a, $b \in \mathbb{R}$, we write $H^p_{a,b}(0,\pi)$ to denote the space $H^p(\Omega,\lambda)$ of Section 1 with $\Omega = (0,\pi)$ and with the measure λ defined by

$$\lambda(E) = \lambda_{a,b}(E) = \int_E \theta^a(\pi-\theta)^b d\theta.$$

Observe that $\lambda_{a,b} \in \text{Double}\,((0,\pi),1)$.

Now our transplantation theorem reads as follows.

Theorem 5.1. *If* $0 < p \leqq 1$, $\infty > \alpha$, $\beta > -1/2+2([1/p]-1)$, *and* $-1 < a$, $b < p-1$, *then for all* $\{\xi_n\} \in l^2$

$$\left\| \sum_{n=0}^{\infty} \xi_n \varphi_n^{(\alpha,\beta)} \right\|_{H^p_{a,b}(0,\pi)} \approx \left\| \sum_{n=0}^{\infty} \xi_n h_n \right\|_{H^p_{a,b}(0,\pi)}.$$

The details of the proof of this theorem will be given in a different occasion.

Notice that the assumption of Theorem 5.1 excludes the unweighted case $a = b = 0$. We shall observe the key point where the weight $\theta^a(\pi-\theta)^b$ plays an essential role.

If we write

$$f = \sum \xi_n \varphi_n^{(\alpha,\beta)} \qquad \text{and} \qquad g = \sum \xi_n h_n,$$

then

$$f = \sum (g, h_n)\varphi_n^{(\alpha,\beta)} \qquad \text{and} \qquad g = \sum (f, \varphi_n^{(\alpha,\beta)}) h_n,$$

where $(\cdot,\cdot)$ denotes the inner product in $L^2(0,\pi)$. In order to prove Theorem 5.1, it is sufficient to show that the transplantation operator

$$S : g \mapsto \sum_{n=0}^{\infty} (g, h_n)\varphi_n^{(\alpha,\beta)}$$

and its adjoint are bounded with respect to the quasinorm of $H^p_{a,b}(0,\pi)$. At least formally, the operator S can be written as

$$(Sg)(\theta) = \int_0^\pi K(\theta,t)g(t)dt$$

with

$$K(\theta,t) = \lim_{r\uparrow 1} \sum_{n=0}^{\infty} r^n \varphi_n^{(\alpha,\beta)}(\theta) h_n(t).$$

It is known that the kernel $K(\theta,t)$ is singular along $\theta = t$ and the singularity is essentially a constant times $1/(\theta - t)$ (see [1]). Thus the claim of Theorem 5.1 is that a certain singular integral defines a bounded operator in the $H^p_{a,b}(0,\pi)$.

To see a typical feature of the $H^p_{a,b}(0,\pi)$ estimate for singular integral operators, we consider, instead of S or its adjoint, the Hilbert transform

$$(Tg)(\theta) = \text{p.v.} \int_0^\pi \frac{g(t)}{\theta - t} dt$$

and consider the case $p = 1$.

By virtue of Theorem 1.2, the estimate $\|Tg\|_{H^1_{a,b}(0,\pi)} \leqq c\|g\|_{H^1_{a,b}(0,\pi)}$ for all $g \in L^2(0,\pi)$ is equivalent to the estimate $\|Tg\|_{H^1_{a,b}(0,\pi)} \leqq c$ for all atoms g, where the atom of $H^1_{a,b}(0,\pi)$ is defined as follows.

There are two kinds of atoms. The atom of the first kind is a function $g \in L^\infty(\mathbb{R})$ such that there exists an interval $I = I_g = (t-\epsilon, t+\epsilon)$ with $\epsilon = 19^{-1}\min\{t, \pi - t\}$ satisfying $\operatorname{supp} g \subset I$ and $\|g\|_{L^\infty} \leqq \lambda_{a,b}(I)^{-1}$. The atom of the second kind is a function $h \in L^\infty(\mathbb{R})$ such that there exists an interval $I = I_h = (t-\epsilon, t+\epsilon)$ with $\epsilon \leqq 19^{-1}\min\{t, \pi - t\}$ satisfying $\operatorname{supp} h \subset I$ and $\|h\|_{L^\infty} \leqq \lambda_{a,b}(I)^{-1}$ and that $\int h(x)dx = 0$.

We shall consider a typical example of the first kind atom. Let $I = (t - t/20, t + t/20)$ with $0 < t < \pi/4$ and let $g = \lambda_{a,b}(I)^{-1}\chi_I \approx t^{-a-1}\chi_I$. We shall estimate $(Tg)^+(x) = (Tg)^{+(\phi),(0,\pi),1/10}(x)$ for $3t/2 < x < \pi/2$.

We immediately see that $(Tg)(x) \approx t^{-a}x^{-1}$ for $3t/2 - t/6 < x < \pi$, from which we also see $(Tg)^+(x) \approx t^{-a}x^{-1}$ for $3t/2 < x < \pi/2$. (A slightly weaker estimate such as $(Tg)^+(x) \leqq ct^{-a}x^{-1}\log(x/t)$ or $(Tg)^+(x) \leqq ct^{-a}x^{-1}(x/t)^\epsilon$ will also give the estimate (5.1) below.) Thus

$$\begin{aligned}\int_{3t/2}^{\pi/2} (Tg)(x)d\lambda_{a,b}(x) &\approx \int_{3t/2}^{\pi/2} (Tg)^+(x)d\lambda_{a,b}(x)\\ &\approx \int_{3t/2}^{\pi/2} t^{-a}x^{-1+a}dx.\end{aligned}$$

If $a < 0$, then we have

$$\int_{3t/2}^{\pi/2} (Tg)^+(x)d\lambda_{a,b}(x) \leqq c. \tag{5.1}$$

Observe that if $a \geqq 0$ then

$$\int_{3t/2}^{\pi/2} (Tg)^+ d\lambda_{a,b}(x) \approx \begin{cases} \log(1/t) & \text{if } \ a = 0 \\ t^{-a} & \text{if } \ a > 0 \end{cases}$$

and hence (5.1) does not hold.

We can in fact prove that the Hilbert transform T and also the general multiplier transformation

$$\sum_{n=0}^{\infty} \xi_n \cos n\theta \mapsto \sum_{n=0}^{\infty} \sigma_n \xi_n \cos n\theta$$

with $\{\sigma_n\}$ satisfying

$$|\triangle^k \sigma_n| = O(n^{-k}) \quad (k = 0, 1, \cdots)$$

define bounded operators in $H^p_{a,b}(0, \pi)$ if $0 < p \leqq 1$ and $-1 < a, b < p - 1$. Details will be discussed elsewhere.

References

[1] Askey, R., A transplantation theorem for Jacobi series, Illinois J. of Math. **13**(1969), 583–590.

[2] Askey, R., and Wainger, S., A transplantation theorem between ultraspherical series, Illinois J. Math. **10**(1966), 322–344.

[3] Duren, P. L., Extension of a theorem of Carleson, Bull. Amer. Math. Soc. **75**(1969), 143–146.

[4] Fefferman, C., and Stein, E. M., Some maximal inequalities, Amer. J. Math. **93**(1971), 107–115.

[5] García-Cuerva, J., Weighted H^p spaces, Dissert. Mathematicae 162, Warszawa, 1979.

[6] Guy, D. L., Hankel multiplier transformations and weighted p-norms, Trans. Amer. Math. Soc. **95**(1960), 137–189.

[7] Kanjin, Y., A transplantation theorem for Laguerre series, Tôhoku Math. J. **43**(1991), 537–555.

[8] Miyachi, A., Maximal functions for distributions on open sets, Hitotsubashi J. Arts Sci. **28**(1987), 45–58.

[9] Miyachi, A., H^p spaces over open subsets of $\mathbb{R}^n$, Studia Math. **95**(1990), 205–228.

[10] Miyachi, A., Weighted Hardy spaces on a domain, Proceedings of the Second ISAAC Congress, Vol. 1, pp. 59–64, H. G. W. Begehr et al. eds., Kluwer Academic Publ., 2000.

[11] Muckenhoupt, B., Transplantation theorems and multiplier theorems for Jacobi series, Mem. Amer. Math. Soc., Vol. 64, No. 356, 1986.

[12] Stein, E. M., *Harmonic Analysis: Real-Variable Methods, Orthogonality, and Oscillatory Integrals,* Princeton Univ. Press, 1993.

[13] Strömberg, J.-O., and Torchinsky, A., *Weighted Hardy Spaces,* Lecture Notes in Math. 1381, Springer-Verlag, 1989.

[14] Szegö, G., *Orthogonal Polynomials*, Amer. Math. Soc. Colloquium Publ., Vol. XXIII, 1939.

[15] Triebel, H., *Interpolation Theory, Function Spaces, Differential Operators,* North-Holland Math. Library Vol. 18, North-Holland Publ., Amsterdam, 1978.

[16] Triebel, H., *Theory of Function Spaces,* Monographs in Math. Vol. 78, Birkhäuser Verlag, Basel, 1983.

[17] Triebel, H., *Theory of Function Spaces II,* Monographs in Math. Vol. 84, Birkhäuser Verlag, Basel, 1992.

[18] Uchiyama, A., A maximal function characterization of H^p on the space of homogeneous type, Trans. Amer. Math. Soc. **262**(1980), 579–592.

[19] Uchiyama, A., On the radial maximal function of distributions, Pacific J. Math. **121**(1986), 467–483.

Akihiko Miyachi
Department of Mathematics
Tokyo Woman's Christian University
Zempukuji 2-6-7
Suginami-ku, Tokyo 167-8585, Japan
E-mail address: `miyachi@twcu.ac.jp`

D. Haroske, T. Runst, H.-J. Schmeisser (eds.): Function Spaces, Differential Operators and Nonlinear Analysis. The Hans Triebel Anniversary Volume.

The General Blow-up for Nonlinear PDE's

Stanislav Pohozaev[1]

Dedicated to Prof. Hans Triebel on the occasion of his 65th birthday

Abstract. Using the nonlinear capacity approach, introduced by the author, we prove some nonexistence results for nonlinear partial differential inequalities of elliptic, parabolic and hyperbolic types.

Introduction

In this paper we give some applications of nonlinear capacity technique, introduced in [22], to nonexistence problems for nonlinear partial differential inequalities.

Our main goal is to show that, under reasonable assumptions on the pair (A, f), sharp nonexistence theorems hold for the problem

$$\begin{cases} A(u) \geq f(u), & x \in Q \subseteq \mathbb{R}^N, \\ u \in S_{\text{loc}}(Q). \end{cases}$$

Let us speak now about functional spaces of solutions. For differential inequalities, in contrast to the case of equations, there is no theory of regular solutions. This is a principal difference, and that is why the definition of the class of solutions plays an essential role. It is easy to construct examples of differential inequalities where solutions exist in one functional class and do not in another.
We shall consider some examples of nonlinear elliptic and parabolic types.
The results were obtained jointly with E. Mitidieri, A. Tesei.

1. Quasilinear elliptic problems

The results below were obtained jointly with E. Mitidieri [12, 14].

The aim of this section is to illustrate the applications of our method by means of a simple example. Let us begin with specifying the functional class of solutions that we shall consider throughout.

Let $p > 1$ and $q \geq 0$. We define S as

$$S := W^{1,p}_{\text{loc}}(\mathbb{R}^N) = \left\{ u \colon \mathbb{R}^N \to \mathbb{R}_+, \quad u^q, |Du|^p \in L^1_{\text{loc}}(\mathbb{R}^N) \right\},$$

where $D_i u$, $i = 1, \dots, N$, are understood in the sense of distributions.

[1]This work was partially supported by INTAS 00-0136 and INTAS 971-30551

Our first result is the following.

Theorem 1.1. *If either*

$$0 < p-1 < q \le \frac{N(p-1)}{N-p}, \qquad N > p, \tag{1.1}$$

or

$$0 \le q \le p-1, \quad p > 1, \qquad N \ge 1, \tag{1.2}$$

then the problem

$$\begin{cases} -\operatorname{div}(|Du|^{p-2}Du) \ge u^q, & x \in \mathbb{R}^N, \\ u \ge 0, \ u \not\equiv 0, & x \in \mathbb{R}^N, \end{cases} \tag{1.3}$$

has no solutions of class $W^{1,p}_{loc}(\mathbb{R}^N)$.

Proof. We shall organize the proof in case when (1.1) holds in two steps. The first one deals with the case q such that $p-1 < q < \frac{N(p-1)}{N-p}$, while in the second one we consider the critical case $q = \frac{N(p-1)}{N-p}$.

First step. Let $\varphi \in C_0^1(\mathbb{R}^N)$ be a standard cut-off function that we shall specify later. Let $\alpha < 0$ be a parameter (α will also be chosen later). Without loss of generality we may suppose that $u > 0$ on $\mathbb{R}^N$ (otherwise, we consider $u_\varepsilon := u + \varepsilon$ and let $\varepsilon \downarrow 0$ in the course). Multiplying (1.3) by $u^\alpha \varphi$ and integrating by parts, we obtain

$$\begin{aligned} \int_{\mathbb{R}^N} u^{q+\alpha}\varphi\,dx &\le \alpha \int_{\mathbb{R}^N} |Du|^p u^{\alpha-1}\varphi\,dx + \int_{\mathbb{R}^N} |Du|^{p-2}(Du, D\varphi)u^\alpha\,dx \\ &\le \alpha \int_{\mathbb{R}^N} |Du|^p u^{\alpha-1}\varphi\,dx + \int_{\mathbb{R}^N} |Du|^{p-1}|D\varphi|u^\alpha\,dx \end{aligned}$$

and, by the Young inequality with parameter $\varepsilon > 0$,

$$\begin{aligned} &\int_{\mathbb{R}^N} u^{q+\alpha}\varphi\,dx + |\alpha| \int_{\mathbb{R}^N} |Du|^p u^{\alpha-1}\varphi\,dx \\ &\le \frac{\varepsilon^p(p-1)}{p} \int_{\mathbb{R}^N} |Du|^p u^{\alpha-1}\varphi\,dx + \frac{1}{p\varepsilon^p} \int_{\mathbb{R}^N} u^{\alpha+p-1}\frac{|D\varphi|^p}{\varphi^{p-1}}\,dx\,. \end{aligned} \tag{1.4}$$

By putting $\theta_\varepsilon = |\alpha| - \frac{\varepsilon^p(p-1)}{p} > 0$ and $\theta'_\varepsilon = \frac{1}{\varepsilon^p p}$, we get

$$\int_{\mathbb{R}^N} u^{q+\alpha}\varphi\,dx + \theta_\varepsilon \int_{\mathbb{R}^N} |Du|^p u^{\alpha-1}\varphi\,dx \le \theta'_\varepsilon \int_{\mathbb{R}^N} u^{\alpha+p-1}\frac{|D\varphi|^p}{\varphi^{p-1}}\,dx\,. \tag{1.5}$$

Now, by choosing $\frac{1}{\varkappa} + \frac{1}{\varkappa'} = 1$ and using the Young inequality again, we obtain:

$$\begin{aligned} &\int_{\mathbb{R}^N} u^{q+\alpha}\varphi\,dx + \theta_\varepsilon \int_{\mathbb{R}^N} |Du|^p u^{\alpha-1}\varphi\,dx \\ &\le \theta''_\varepsilon \int_{\mathbb{R}^N} u^{(\alpha+p-1)\varkappa}\varphi\,dx + \widetilde{\theta}_\varepsilon \int_{\mathbb{R}^N} \frac{|D\varphi|^{p\varkappa'}}{\varphi^{p\varkappa'-1}}\,dx\,. \end{aligned} \tag{1.6}$$

Choosing $\varkappa = \frac{\alpha+q}{\alpha+p-1}$ (note that it is possible by (1.1) and small $\alpha < 0$) we get

$$c_\varepsilon \int_{\mathbb{R}^N} u^{q+\alpha}\varphi\, dx + \theta_\varepsilon \int_{\mathbb{R}^N} |Du|^p u^{\alpha-1}\varphi\, dx \le \widetilde{\theta}_\varepsilon \int_{\mathbb{R}^N} \frac{|D\varphi|^{p\varkappa'}}{\varphi^{p\varkappa'-1}}\, dx\,. \tag{1.7}$$

Next we estimate the term $\int_{\mathbb{R}^N} u^q\varphi\, dx$. By multiplying (1.3) by φ and integrating by parts we find

$$\begin{aligned}\int_{\mathbb{R}^N} u^q\varphi\, dx &\le \int_{\mathbb{R}^N} |Du|^{p-2}(Du, D\varphi)\, dx \le \int_{\mathbb{R}^N} |Du|^{p-1}|D\varphi|\, dx\\ &\le \int_{\mathbb{R}^N} |Du|^{p-1} u^{\frac{(\alpha-1)(p-1)}{p}} \varphi^{\frac{1}{p}} u^{\frac{(1-\alpha)(p-1)}{p}} \varphi^{-\frac{1}{p}} |D\varphi|\, dx \qquad (1.8)\\ &\le \left(\int_{\mathbb{R}^N} |Du|^p u^{\alpha-1}\varphi\, dx\right)^{\frac{p-1}{p}} \left(\int_{\mathbb{R}^N} u^{(1-\alpha)(p-1)} \frac{|D\varphi|^p}{\varphi^{p-1}}\, dx\right)^{\frac{1}{p}}.\end{aligned}$$

By Hölder inequality with parameter $(\frac{1}{a} + \frac{1}{a'} = 1)$ we get

$$\begin{aligned}\int_{\mathbb{R}^N} u^q\varphi\, dx &\le \left(\int_{\mathbb{R}^N} |Du|^p u^{\alpha-1}\varphi\, dx\right)^{\frac{p-1}{p}} \qquad (1.9)\\ &\quad \times \left(\int_{\mathbb{R}^N} u^{a(1-\alpha)(p-1)}\varphi\, dx\right)^{\frac{1}{ap}} \left(\int_{\mathbb{R}^N} \frac{|D\varphi|^{pa'}}{\varphi^{pa'-1}}\, dx\right)^{\frac{1}{a'p}}.\end{aligned}$$

By choosing a such that $a(1-\alpha)(p-1) = q+\alpha$ (note that this choice is also possible by (1.1) for $q > p-1$) and combining (1.9) with (1.7), we obtain

$$\begin{aligned}\int_{\mathbb{R}^N} u^q\varphi\, dx &\le \left(\frac{\widetilde{\theta}_\varepsilon}{\theta_\varepsilon}\right)^{\frac{p-1}{p}} \left(\int_{\mathbb{R}^N} \frac{|D\varphi|^{p\varkappa'}}{\varphi^{p\varkappa'-1}}\, dx\right)^{\frac{p-1}{p}}\\ &\quad \times \left(\frac{\widetilde{\theta}_\varepsilon}{c_\varepsilon}\right)^{\frac{1}{ap}} \left(\int_{\mathbb{R}^N} \frac{|D\varphi|^{p\varkappa'}}{\varphi^{p\varkappa'-1}}\, dx\right)^{\frac{1}{ap}} \left(\int_{\mathbb{R}^N} \frac{|D\varphi|^{pa'}}{\varphi^{pa'-1}}\, dx\right)^{\frac{1}{a'p}}\\ &= \overline{C}_\varepsilon \left(\int_{\mathbb{R}^N} \frac{|D\varphi|^{p\varkappa'}}{\varphi^{p\varkappa'-1}}\, dx\right)^{\frac{p-1}{p}+\frac{1}{ap}} \left(\int_{\mathbb{R}^N} \frac{|D\varphi|^{pa'}}{\varphi^{pa'-1}}\, dx\right)^{\frac{1}{a'p}}, \qquad (1.10)\end{aligned}$$

where $\overline{C}_\varepsilon = \left(\frac{\widetilde{\theta}_\varepsilon}{c_\varepsilon}\right)^{\frac{1}{ap}} \left(\frac{\widetilde{\theta}_\varepsilon}{\theta_\varepsilon}\right)^{\frac{p-1}{p}}$. Next we choose the function φ. Let $\xi \in C_0^\infty(\mathbb{R}^N)$ and

$$\xi(x) = \xi_0\left(\frac{|x|}{R}\right), \qquad R > 0\,, \quad 0 \le \xi_0 \le 1 \tag{1.11}$$

with $\xi_0 \in C^\infty(\mathbb{R}_+)$ and

$$\xi_0(t) = \begin{cases} 1, & 0 \le t \le 1,\\ 0, & 2 \le t. \end{cases}$$

We define φ by $\varphi := \xi^\lambda$ with λ chosen sufficiently large. With this choice of φ, by a standard charge of variables, it follows from (1.10) that

$$\int_{B_R} u^q(x)\,dx \le \widetilde{C} R^\sigma\,, \tag{1.12}$$

where $\sigma = (N - p\varkappa')(\frac{p-1}{p} + \frac{1}{ap}) + (\frac{N-pa'}{pa'})$. Taking into account our choice of $\varkappa'$ and a, we find

$$\sigma = \frac{q(N-p) - N(p-1)}{q-p+1}\,. \tag{1.13}$$

Since $\sigma < 0$, inequality (1.12) implies

$$\int_{\mathbb{R}^N} u^q(x)\,dx = 0\,.$$

This contradicts our assumption that $u > 0$ thereby concluding the proof in the case $\sigma < 0$.

Second step. We pass to the case $q = \frac{N(p-1)}{N-p}$, i.e. $\sigma = 0$. By our choice of φ we have

$$\begin{aligned}
\int_{B_R} \varphi u^q(x)\,dx &\le \int_{S(D\varphi)} |Du|^{p-1}|D\varphi|\,dx \\
&\le \left(\int_{S(D\varphi)} |Du|^p u^{\alpha-1}\varphi\,dx\right)^{\frac{p-1}{p}} \left(\int_{S(D\varphi)} u^{(1-\alpha)(p-1)}\frac{|D\varphi|^p}{\varphi^{p-1}}\,dx\right)^{\frac{1}{p}} \\
&\le \left(\int_{S(D\varphi)} |Du|^p u^{\alpha-1}\varphi\,dx\right)^{\frac{p-1}{p}} \left(\int_{S(D\varphi)} u^{m(1-\alpha)(p-1)}\varphi\,dx\right)^{\frac{1}{pm}} \\
&\quad \times \left(\int_{S(D\varphi)} \frac{|D\varphi|^{pm'}}{\varphi^{pm'-1}}\,dx\right)^{\frac{1}{pm'}},
\end{aligned}$$

where $\frac{1}{m} + \frac{1}{m'} = 1$ and $S(D\varphi) = \operatorname{supp}(D\varphi)$. By choosing $m = \frac{q}{(1-\alpha)(p-1)} \ge 1$ (note that this is possible), using (1.7) and the definition of φ, we obtain finally

$$\begin{aligned}
&\int_{B_R} u^q\,dx \\
&\le C\left(\int_{S(D\varphi)} \frac{|D\varphi|^{p\varkappa'}}{\varphi^{p\varkappa'-1}}\,dx\right)^{\frac{p-1}{p}} \left(\int_{S(D\varphi)} \frac{|D\varphi|^{pm'}}{\varphi^{pm'-1}}\,dx\right)^{\frac{1}{pm'}} \left(\int_{R\le|x|<2R} u^q\,dx\right)^{\frac{1}{pm}}.
\end{aligned} \tag{1.14}$$

Thus by the usual change of variable we get

$$\int_{B_R} u^q\,dx \le C_0 R^\tau \left(\int_{R<|x|<2R} u^q\,dx\right)^{\frac{1}{pm}}, \tag{1.15}$$

where, by taking into account that $q = \frac{N(p-1)}{N-p}$, we have

$$\tau = (N - p\varkappa')\frac{p-1}{p} + \frac{N - pm'}{pm'} = 0\,.$$

By (1.12) we know that

$$\int_{\mathbb{R}^N} u^q\, dx < \infty\,.$$

Inequality (1.15) implies that there exists a sequence $\{R_k\}$ such that $R_k \to \infty$ and

$$\lim_{k\to\infty} \int_{B_{R_k}} u^q\, dx = 0\,.$$

This completes the proof of the theorem in the case when (1.1) holds.

Next we consider case (1.2). Let us first suppose that $0 \le q < p-1$. From (1.5) we know that

$$\int_{\mathbb{R}^N} u^{q+\alpha}\varphi\, dx \le C\left(\int_{\mathbb{R}^N} u^{(\alpha+p-1)\varkappa}\varphi\, dx\right)^{\frac{1}{\varkappa}} \left(\int_{\mathbb{R}^N} \frac{|D\varphi|^{p\varkappa'}}{\varphi^{p\varkappa'-1}}\, dx\right)^{\frac{1}{\varkappa'}}, \tag{1.16}$$

where $\frac{1}{\varkappa} + \frac{1}{\varkappa'} = 1$. Choosing $\alpha = q(N-p) - N(p-1) < 0$ (note that in this case $\alpha < 1-p < -q$ due to the choice of α) and

$$\varkappa = \frac{q+\alpha}{\alpha+p-1} > 1,$$

we obtain

$$\int_{\mathbb{R}^N} u^{q+\alpha}\varphi\, dx \le C\int_{\mathbb{R}^N} \frac{|D\varphi|^{p\varkappa'}}{\varphi^{p\varkappa'-1}}\, dx\,. \tag{1.17}$$

Now, proceeding as in the proof of the first step of case (1.1), we arrive at

$$\int_{B_R} u^{q+\alpha}\, dx \le CR^{N-p\varkappa'},$$

with

$$N - p\varkappa' = \frac{N(q-p+1) - pq - p\alpha}{q-p+1} < 0\,.$$

This completes the proof of case (1.2) if the strict inequality holds.

Suppose now that $q = p-1$. By choosing $\alpha = 1-p$ in (1.5) we get

$$\int_{\mathbb{R}^N} \varphi\, dx \le \widetilde{C}\int_{\mathbb{R}^N} \frac{|D\varphi|^p}{\varphi^{p-1}}\, dx\,.$$

Hence,

$$R^N \le CR^{N-p},$$

which leads easily to a contradiction. □

The result of Theorem 1.1 is sharp. Indeed, suppose that the second inequality in (1.1) of Theorem 1.1 does not hold, i.e.,

$$q > \frac{N(p-1)}{N-p}, \qquad N > p\,.$$

It is very easy to verify that the function

$$u(x) := \frac{\varepsilon}{(1+|x|^{\frac{p}{p-1}})^{\frac{p-1}{q-p+1}}}, \qquad x \in \mathbb{R}^N,$$

is a positive solution of (1.3) provided that we choose ε such that

$$0 < \varepsilon < \left(\frac{p}{q-p+1}\right)^{\frac{p-1}{q-p+1}} \left(\frac{q(N-p) - N(p-1)}{q-p+1}\right)^{\frac{1}{q-p+1}}.$$

Let us now consider some generalizations of Theorem 1.1.

Let $A\colon \mathbb{R}_+ \to \mathbb{R}_+$ be a given function and suppose that there exist $c_1, c_2 > 0$ and $p > 1$ such that for any $t \geq 0$ we have

$$c_1 t^{p-1} \leq A(t)t \leq c_2 t^{p-1}. \tag{1.18}$$

The following result can be obtained by slightly adapting the proof of Theorem 1.1.

Theorem 1.2. *Suppose that $p > 1$ and A satisfies (1.18). If either*

$$p - 1 < q \leq \frac{N(p-1)}{N-p}, \qquad N > p, \tag{1.19}$$

or

$$0 < q \leq p - 1, \qquad N \geq 1, \tag{1.20}$$

then the problem

$$\begin{cases} -\operatorname{div}(A(|Du|)Du) \geq u^q, & x \in \mathbb{R}^N, \\ u \geq 0, \ u \not\equiv 0, & x \in \mathbb{R}^N, \end{cases} \tag{1.21}$$

has no solution in $W^{1,p}_{loc}(\mathbb{R}^N)$.

Remark 1.3. *It is not difficult to realize that, if the function A is such that for any $(x, s, t) \in \mathbb{R}^N \times \mathbb{R}_+ \times \mathbb{R}^N$ we have*

$$c_1 |t|^{p-2} \leq A(x, s, t) \leq c_2 |t|^{p-2}, \tag{1.22}$$

where $c_1, c_2 > 0$ and $p > 1$, the corresponding problem

$$-\operatorname{div}(A(x, u, Du)Du) \geq u^q, \qquad x \in \mathbb{R}^N, \tag{1.23}$$

has no solutions $u \geq 0$, $u \not\equiv 0$ on $\mathbb{R}^N$ which belong to $W^{1,p}_{loc}(\mathbb{R}^N)$ provided that (1.19) or (1.20) of Theorem 1.2 are satisfied.

Another important class of operators that can be studied are the so-called "mean curvature type" operators.

Definition 1.4. *Let $A\colon \mathbb{R}_+ \to \mathbb{R}_+$ be a continuous function. Suppose that there exists $C > 0$ such that for any $t \geq 0$ we have*

$$0 < A(t) \leq C. \tag{1.24}$$

Then the operator T defined by

$$T(u) = \operatorname{div}(A(|Du|)Du), \qquad u \in W^{1,2}_{loc}(\mathbb{R}^N), \tag{1.25}$$

is called a mean curvature type operator, associated with the function A.

Some important examples are [19, 20]:

$$A(t) = \frac{1}{\sqrt{1+t^2}} \qquad (\textit{mean curvature operator}), \tag{1.26}$$

$$A(t) = \frac{1}{(1+|t|^k)^s}, \quad k, s > 0 \qquad (\textit{generalized mean curvature operator}). \tag{1.27}$$

A slight modification of the proof of Theorem 1.1 gives the following result.

Theorem 1.5. *If either*

$$1 < q \leq \frac{N}{N-2}, \qquad N > 2,$$

or

$$0 < q \leq 1, \qquad N \geq 1,$$

then the problem

$$\begin{cases} -\mathrm{div}\,(A(|Du|)Du) \geq u^q, & x \in \mathbb{R}^N, \\ u \geq 0, \ u \not\equiv 0, & x \in \mathbb{R}^N, \end{cases} \tag{1.28}$$

has no solutions in $W^{1,2}_{loc}(\mathbb{R}^N)$.

Moreover, if $A(t) = \frac{1}{\sqrt{1+t^2}}$, $N > 2$ *and* $q > \frac{N}{N-2}$, *then the function*

$$u(x) := \frac{\varepsilon}{(1+|x|^2)^{\frac{1}{q-1}}}$$

is a solution of class $C^2(\mathbb{R}^N)$ *to* (1.28) *provided that* $\varepsilon > 0$ *is chosen sufficiently small.*

There are some other possible ways to generalize Theorem 1.3. A typical situation is when we replace the right-hand side of the inequality appearing in (1.23) by

$$f(x,u) = a(x)u^q, \qquad x \in \mathbb{R}^N,$$

where $a \geq 0$ is a function such that

$$a(x) \geq C_0|x|^\gamma \qquad \forall x: \ |x| \geq R_0 \geq 0, \quad \gamma \in \mathbb{R}. \tag{1.29}$$

The following result can be proved by repeating the proof of Theorem 1.1.

Theorem 1.6. *Let* $N > p > 1$. *Suppose that* A *and* a *satisfy, respectively,* (1.22) *and* (1.29). *If*

$$p - 1 < q \leq \frac{(N+\gamma)(p-1)}{N-p}, \tag{1.30}$$

then the problem

$$\begin{cases} -\mathrm{div}\,(A(x,u,Du)Du) \geq a(x)u^q, & x \in \mathbb{R}^N, \\ u \geq 0, \ u \not\equiv 0, & x \in \mathbb{R}^N, \end{cases} \tag{1.31}$$

has no solutions in $W^{1,p}_{loc}(\mathbb{R}^N)$.

Remark 1.7. *The interested reader may formulate results similar to those of Theorem 1.6 when A generates an operator of mean curvature type (see* [18, 12]*). Other possible generalizations can be obtained by considering the generating function A to be strongly dependent on its arguments. A simple example in this direction is given by*

$$A(x,u,Du) := |x|^{\alpha_1}|u|^{q_1}|Du|^{p-2}. \tag{1.32}$$

In this case the nonexistence theorem for the inequality

$$\begin{cases} -\operatorname{div}(A(x,u,Du)Du) \geq a(x)u^q, & x \in \mathbb{R}^N, \\ u \geq 0, \ u \not\equiv 0, & x \in \mathbb{R}^N, \end{cases} \tag{1.33}$$

will be formulated in terms of an algebraic condition involving N, α_1, q_1, p, γ *and* q.

2. Nonlinear parabolic problems

These results were obtained jointly with A. Tesei.

We investigate blow-up of nonnegative solutions to systems of parabolic inequalities of the following type:

$$\begin{cases} \partial_t u \geq \sum_{i=1}^n \partial_{x_i}\Big[A(x,t,u,v,\nabla u,\nabla v)\ \partial_{x_i}u\Big] + b(x,t,u,v,\nabla u,\nabla v)\, v^r \\ \\ \partial_t v \geq \sum_{i=1}^n \partial_{x_i}\Big[A(x,t,u,v,\nabla u,\nabla v)\ \partial_{x_i}v\Big] + c(x,t,u,v,\nabla u,\nabla v)\, u^s \end{cases} \tag{2.34}$$

in $\mathbb{R}^n \times (0,\infty)$; here $r > 0$, $s > 0$ and A, b, c are locally bounded positive functions.

Our concern is to investigate the existence of *critical exponents* for blow-up (see [10, 1], [25]). To this purpose we make use of the method of nonlinear capacity (see [22]), already used in [23] to deal with the case of a single parabolic inequality. In the particular case $A = b = c = 1$ our results reduce to those in [2] (see Theorem 2.5 and Corollary 2.6 below); however, the present method applies to a wider class of problems (*e.g.*, see Theorem 2.7).

2.1. Mathematical framework and results

Let S_T denote the strip $\mathbb{R}^n \times (0,T]$ ($T \in (0,\infty]$); set $S \equiv S_\infty$. By a *classical* solution to system (2.34) in S_T we mean any couple of positive functions $u, v \in C^2(S_T) \cap C(\bar{S}_T)$ such that system (2.34) is satisfied pointwise in S_T (here the function A is supposed to be smooth). We shall make use below of the following more general definitions.

Definition 2.1. *By a* strong solution *to system (2.34) in* S_T *we mean any couple of nonnegative functions* $u, v \in C(S_T)$ *whose distributional derivatives of first order in time and of second order in the space variables are defined almost everywhere in* S_T*, such that system (2.34) is satisfied almost everywhere in* S_T.

Definition 2.2. *Let $\alpha, \beta \in (-1, 0)$. By a* solution of class $P_{\alpha,\beta}$ *to system (2.34) in S_T we mean any strong solution in S_T such that for any test function $\psi \geq 0$ with support in $\bar{S}_T$ there holds:*

(i)

$$\begin{cases} \int\int_{\operatorname{supp} u \cap \operatorname{supp} v} A\, u^{\alpha}\, v^{\beta} \mid \nabla u \mid\mid \nabla v \mid \psi < \infty \\ \int\int_{\operatorname{supp} u \cap \operatorname{supp} v} A\, u^{\alpha}\, v^{\beta+1} \mid \nabla u \mid\mid \nabla \psi \mid < \infty \\ \int\int_{\operatorname{supp} u \cap \operatorname{supp} v} A\, u^{\alpha+1}\, v^{\beta} \mid \nabla v \mid\mid \nabla \psi \mid < \infty\, ; \end{cases} \tag{2.35}$$

(ii)

$$\begin{aligned} &\mid \alpha \mid (\alpha+1) \int\int_{\operatorname{supp} u \cap \operatorname{supp} v} A\, u^{\alpha-1}\, v^{\beta+1} \mid \nabla u \mid^2 \psi \\ &+ \mid \beta \mid (\beta+1) \int\int_{\operatorname{supp} u \cap \operatorname{supp} v} A\, u^{\alpha+1}\, v^{\beta-1} \mid \nabla v \mid^2 \psi \\ &+ (\alpha+1) \int\int_{\operatorname{supp} u \cap \operatorname{supp} v} b\, u^{\alpha}\, v^{\beta+1+r}\, \psi \\ &+ (\beta+1) \int\int_{\operatorname{supp} u \cap \operatorname{supp} v} c\, u^{\alpha+1+s}\, v^{\beta}\, \psi \\ &\leq 2(\alpha+1)(\beta+1) \int\int_{\operatorname{supp} u \cap \operatorname{supp} v} A\, u^{\alpha}\, v^{\beta} \mid \nabla u \mid\mid \nabla v \mid \psi \\ &+ (\alpha+1) \int\int_{\operatorname{supp} u \cap \operatorname{supp} v} A\, u^{\alpha}\, v^{\beta+1} \mid \nabla u \mid\mid \nabla \psi \mid \\ &+ (\beta+1) \int\int_{\operatorname{supp} u \cap \operatorname{supp} v} A\, u^{\alpha+1}\, v^{\beta} \mid \nabla v \mid\mid \nabla \psi \mid \\ &- \int\int_{\operatorname{supp} u \cap \operatorname{supp} v} u^{\alpha+1}\, v^{\beta+1} \partial_t \psi. \end{aligned} \tag{2.36}$$

Due to condition (2.35) and to the assumptions $\alpha > -1$, $\beta > -1$, every integral in the right-hand side of inequality (2.36) is finite, thus Definition 2.2 is well posed. Moreover, every integral in the left-hand side is also finite.

Any solution to system (2.34) is said to be *global*, if it is a solution in S_T for any $T > 0$.

Clearly, any classical solution is strong. Concerning the relationship between Definitions 2.1 and 2.2, the following result can be proved.

Proposition 2.3. *Let (u, v) be a strong solution to system (2.34) in S_T, such that $\partial_t u \in L^1_{loc}(S_T)$, $\partial_t v \in L^1_{loc}(S_T)$ and the pointwise limit $(u(\cdot,0), v(\cdot,0)) := \lim_{t\to 0^+}(u(\cdot,t), v(\cdot,t))$ is defined and continuous in $\mathbb{R}^n$. Let condition (2.35) be satisfied. Then (u, v) is a solution of class $P_{\alpha,\beta}$ $(\alpha, \beta \in (-1, 0))$.*

Let us define:

$$D = D(x,t,u,v,\nabla u,\nabla v) := \left(\frac{a^{\rho\sigma}}{b^{\rho-1}c}\right)^{\frac{1}{\rho(\sigma-1)}}, \tag{2.37}$$

where $\rho > 1$, $\sigma > 1$ and

$$a = a(x,t,u,v,\nabla u,\nabla v) := \max\{A\,,1\}\,. \tag{2.38}$$

Our main nonexistence result can be stated as follows.

Theorem 2.4. *Let $\alpha, \beta \in (-1,0)$ satisfy the following condition:*

$$|\,\alpha\,| + |\,\beta\,| > 1\,. \tag{2.39}$$

Let there exist $\rho > 1$, $\sigma > 1$ such that

$$\begin{cases} [\alpha\,(\sigma-1)+\sigma\,]\,\rho = 1+s \\ [\beta\,(\sigma-1)+\sigma\,]\,\rho = (1+r)\,(\rho-1)\,. \end{cases} \tag{2.40}$$

Moreover, suppose that

$$R^{n-\frac{2}{\sigma-1}} \int\!\!\int_{\{1\le\eta\le 2\}} \left[\sup_{u,v\ge 0;\,p,q\in\mathbb{R}^n} D(R\xi,\,R^2\tau,u,v,p,q)\right] d\xi\,d\tau \longrightarrow 0 \tag{2.41}$$

as $R \to \infty$, where

$$\eta := |\xi|^2 + \tau \qquad (\xi \in \mathbb{R}^n, \tau > 0)\,. \tag{2.42}$$

Then the only global solution of class $P_{\alpha,\beta}$ to system (2.34) is trivial.

Let us apply the above theorem to the following system:

$$\begin{cases} \partial_t u \ge \Delta u + v^r \\ \partial_t v \ge \Delta v + u^s \end{cases} \tag{2.43}$$

in $\mathbb{R}^n \times (0,\infty)$ $(r > 0, s > 0)$. The following result can be proved.

Theorem 2.5. *Let*

$$r\,s > 1\,, \tag{2.44}$$

$$\frac{\gamma+1}{r\,s-1} > \frac{n}{2}\,, \tag{2.45}$$

where $\gamma := \max\{r,s\}$. Then there exist $\alpha, \beta \in (-1,0)$ such that the only global solution of class $P_{\alpha,\beta}$ to system (2.43) is trivial.

The following corollary – which applies in particular to the Cauchy problem for systems of parabolic *equations* – is an immediate consequence of Theorem 2.5 and Proposition 2.3.

Corollary 2.6. *Let the assumptions of Theorem 2.5 be satisfied. Then the only global classical solution to system (2.43) is trivial.*

It is easily seen that the proof of Theorem 2.5 still holds true for any system of the form (2.34) such that

$$\sup_{u,v\geq 0;\, p,q\in\mathbb{R}^n} D(R\xi,\, R^2\tau, u, v, p, q) \leq C \qquad \text{for any } \xi \in \mathbb{R}^n, \tau > 0$$

for some constant $C > 0$. For instance, this is the case if A is bounded and b, c are constant, as in the following theorem.

Theorem 2.7. *Let the assumptions of Theorem 2.5 be satisfied. Then there exist $\alpha, \beta \in (-1, 0)$ such that the only global solution of class $P_{\alpha,\beta}$ to the system:*

$$\begin{cases} \partial_t u \geq \sum_{i=1}^n \partial_{x_i}\left[\frac{\partial_{x_i} u}{(1+|\nabla u|^2+|\nabla v|^2)^\theta}\right] + v^r \\ \\ \partial_t v \geq \sum_{i=1}^n \partial_{x_i}\left[\frac{\partial_{x_i} v}{(1+|\nabla u|^2+|\nabla v|^2)^\theta}\right] + u^s \end{cases} \tag{2.46}$$

in $\mathbb{R}^n \times (0, \infty)$ is trivial ($r > 0$, $s > 0$, $\theta > 0$).

The proof is the same as that of Theorem 2.5, thus it will be omitted.

2.2. Proofs

Let us first prove Theorem 2.4. For this purpose we need the following lemma (in the following we always set $\int\int_{\text{supp}\, u \cap \text{supp}\, v} \equiv \int\int$ for shortness).

Lemma 2.8. *Let (u, v) be a solution of class $P_{\alpha,\beta}$ to (2.34) in S_T, with $\alpha, \beta \in (-1, 0)$ satisfying condition (2.39). Then there exist $k_1 > 0$, $k_2 > 0$ (depending on α, β) with the following property: For any $\rho > 1$, $\sigma > 1$ satisfying condition (2.40) and any test function $\psi \geq 0$ with support in $\bar{S}_T$ there holds:*

$$\begin{aligned} &k_1 \int\int A\, u^{\alpha-1}\, v^{\beta+1} \mid \nabla u \mid^2 \psi + k_2 \int\int A\, u^{\alpha+1}\, v^{\beta-1} \mid \nabla v \mid^2 \psi \\ &+ \int\int b\, u^\alpha\, v^{\beta+1+r}\, \psi + \int\int c\, u^{\alpha+1+s}\, v^\beta\, \psi \\ &\leq \left(\frac{1}{\delta}\right)^{\frac{\sigma}{\sigma-1}} \int\int_{\text{supp}\,\psi} D\left(\frac{\chi^\sigma}{\psi}\right)^{\frac{1}{\sigma-1}}, \end{aligned} \tag{2.47}$$

where

$$\begin{aligned} \delta &:= \frac{1}{2} \min\{\alpha+1\,,\, \beta+1\}\,, \\ \chi &:= \frac{1}{2}\Big(\frac{\alpha+1}{k_1} + \frac{\beta+1}{k_2}\Big) \frac{\mid \nabla\psi \mid^2}{\psi} + \mid \partial_t \psi \mid \end{aligned} \tag{2.48}$$

and D is the function (2.37).

Proof. (*i*) Concerning the integrand of the first term in the right-hand side of inequality (2.36), by Young inequality we have:

$$\begin{aligned} &A\, u^\alpha\, v^\beta \mid \nabla u \mid\mid \nabla v \mid \psi \\ &\quad\leq \frac{1}{2}\Big(k\, A\, u^{\alpha-1}\, v^{\beta+1} \mid \nabla u \mid^2 \psi + \frac{A}{k}\, u^{\alpha+1}\, v^{\beta-1} \mid \nabla v \mid^2 \psi\Big) \end{aligned}$$

for any $k > 0$. Integrating on $\operatorname{supp} u \cap \operatorname{supp} v$ both members of the above inequality and inserting the resulting inequality in (2.36) we obtain:

$$\begin{aligned}
&\Big(|\alpha| - k(\beta+1)\Big)(\alpha+1)\iint A\,u^{\alpha-1}\,v^{\beta+1}\,|\nabla u|^2\,\psi \\
&\quad + \Big(|\beta| - \frac{1}{k}(\alpha+1)\Big)(\beta+1)\iint A\,u^{\alpha+1}\,v^{\beta-1}\,|\nabla v|^2\,\psi \\
&\quad + (\alpha+1)\iint b\,u^{\alpha}\,v^{\beta+1+r}\,\psi + (\beta+1)\iint c\,u^{\alpha+1+s}\,v^{\beta}\,\psi \\
&\leq (\alpha+1)\iint A\,u^{\alpha}\,v^{\beta+1}\,|\nabla u|\,|\nabla\psi| \\
&\quad + (\beta+1)\iint A\,u^{\alpha+1}\,v^{\beta}\,|\nabla v|\,|\nabla\psi| - \iint u^{\alpha+1}\,v^{\beta+1}\partial_t\psi.
\end{aligned} \tag{2.49}$$

Due to condition (2.39), the interval $(\frac{|\alpha|}{\beta+1}, \frac{\alpha+1}{|\beta|})$ is nonempty. Fix k in this interval; set

$$k_1 := |\alpha| - k(\beta+1),$$

$$k_2 := |\beta| - \frac{1}{k}(\alpha+1).$$

Then $k_1 > 0$, $k_2 > 0$ and inequality (2.49) reads:

$$\begin{aligned}
&(\alpha+1)k_1\iint A\,u^{\alpha-1}\,v^{\beta+1}\,|\nabla u|^2\,\psi + (\beta+1)k_2\iint A\,u^{\alpha+1}\,v^{\beta-1}\,|\nabla v|^2\,\psi \\
&\quad + (\alpha+1)\iint b\,u^{\alpha}\,v^{\beta+1+r}\,\psi + (\beta+1)\iint c\,u^{\alpha+1+s}\,v^{\beta}\,\psi \\
&\leq (\alpha+1)\iint A\,u^{\alpha}\,v^{\beta+1}\,|\nabla u|\,|\nabla\psi| + (\beta+1)\iint A\,u^{\alpha+1}\,v^{\beta}\,|\nabla v|\,|\nabla\psi| \\
&\quad - \iint u^{\alpha+1}\,v^{\beta+1}\partial_t\psi\,.
\end{aligned} \tag{2.50}$$

(*ii*) Now consider the first term in the right-hand side of inequality (2.50); by Young inequality we have on $\operatorname{supp} u \cap \operatorname{supp}\psi$:

$$\begin{aligned}
&A\,u^{\alpha}\,v^{\beta+1}\,|\nabla u|\,|\nabla\psi| \\
&\leq \frac{1}{2}\Big(k_1\,A\,u^{\alpha-1}\,v^{\beta+1}\,|\nabla u|^2\,\psi + \frac{A}{k_1}\,u^{\alpha+1}\,v^{\beta+1}\,\frac{|\nabla\psi|^2}{\psi}\Big).
\end{aligned}$$

Similarly, concerning the second term in the right-hand side of the same inequality there holds:

$$\begin{aligned}
&A\,u^{\alpha+1}\,v^{\beta}\,|\nabla v|\,|\nabla\psi| \\
&\leq \frac{1}{2}\Big(k_2\,A\,u^{\alpha+1}\,v^{\beta-1}\,|\nabla v|^2\,\psi + \frac{A}{k_2}\,u^{\alpha+1}\,v^{\beta+1}\,\frac{|\nabla\psi|^2}{\psi}\Big)
\end{aligned}$$

on $\operatorname{supp} v \cap \operatorname{supp} \psi$. Integrating on $\operatorname{supp} u \cap \operatorname{supp} v \cap \operatorname{supp} \psi$ the above inequalities and inserting the resulting inequalities in (2.50) we obtain:

$$\begin{aligned}\frac{\alpha+1}{2} k_1 \iint A\, u^{\alpha-1} v^{\beta+1} \mid \nabla u \mid^2 \psi + \frac{\beta+1}{2} k_2 \iint A\, u^{\alpha+1} v^{\beta-1} \mid \nabla v \mid^2 \psi \\ + (\alpha+1) \iint b\, u^{\alpha} v^{\beta+1+r} \psi + (\beta+1) \iint c\, u^{\alpha+1+s} v^{\beta} \psi \qquad (2.51)\\ \leq \iint_{\operatorname{supp} \psi} a\, u^{\alpha+1} v^{\beta+1} \chi \,,\end{aligned}$$

a and χ being the functions defined in (2.38), respectively (2.48).

(*iii*) Finally, let us estimate the right-hand side of inequality (2.51). For any $\sigma > 1$, $\delta > 0$ there holds:

$$a\, u^{\alpha+1} v^{\beta+1} \chi \leq \frac{\delta}{\sigma} E^{-(\sigma-1)} u^{(\alpha+1)\sigma} v^{(\beta+1)\sigma} \psi + \frac{\sigma-1}{\sigma} a^{\frac{\sigma}{\sigma-1}} E \left(\frac{\chi^{\sigma}}{\delta \psi} \right)^{\frac{1}{\sigma-1}},$$

where

$$E = E(x,t,u,v) := \left(\frac{1}{b^{\rho-1} c} \right)^{\frac{1}{\rho(\sigma-1)}}.$$

Moreover, for any $\rho > 1$ by Young inequality we have:

$$\begin{aligned} u^{(\alpha+1)\sigma} v^{(\beta+1)\sigma} &\leq u^{\alpha} v^{\beta} \left\{ \frac{1}{\rho} \left(\frac{c}{b} \right)^{\frac{\rho-1}{\rho}} u^{[\alpha(\sigma-1)+\sigma]\rho} + \frac{\rho-1}{\rho} \left(\frac{b}{c} \right)^{\frac{1}{\rho}} v^{\frac{[\beta(\sigma-1)+\sigma]\rho}{\rho-1}} \right\} \\ &= \frac{1}{\rho} \left(\frac{c}{b} \right)^{\frac{\rho-1}{\rho}} u^{\alpha+1+s} v^{\beta} + \frac{\rho-1}{\rho} \left(\frac{b}{c} \right)^{\frac{1}{\rho}} u^{\alpha} v^{\beta+1+r}, \end{aligned}$$

due to condition (2.40). Combining the two above inequalities we obtain:

$$a\, u^{\alpha+1} v^{\beta+1} \chi \leq \delta \left\{ b\, u^{\alpha} v^{\beta+1+r} \psi + c\, u^{\alpha+1+s} v^{\beta} \psi \right\} + D \left(\frac{\chi^{\sigma}}{\delta \psi} \right)^{\frac{1}{\sigma-1}},$$

where $D = a^{\frac{\sigma}{\sigma-1}} E$ is the function (2.37). Integrating the above inequality and inserting the resulting inequality in (2.51) we obtain:

$$\begin{aligned}\frac{\alpha+1}{2} k_1 \iint A\, u^{\alpha-1} v^{\beta+1} \mid \nabla u \mid^2 \psi + \frac{\beta+1}{2} k_2 \iint A\, u^{\alpha+1} v^{\beta-1} \mid \nabla v \mid^2 \psi \\ + (\alpha+1-\delta) \iint b\, u^{\alpha} v^{\beta+1+r} \psi + (\beta+1-\delta) \iint c\, u^{\alpha+1+s} v^{\beta} \psi \\ \leq \left(\frac{1}{\delta} \right)^{\frac{1}{\sigma-1}} \iint_{\operatorname{supp} \psi} D \left(\frac{\chi^{\sigma}}{\psi} \right)^{\frac{1}{\sigma-1}} \qquad (2.52)\end{aligned}$$

for any $\delta > 0$; choosing δ as in (2.46) the conclusion follows. □

Now we can prove Theorem 2.4.

Proof of Theorem 2.4. Let (u,v) be a global solution of class $P_{\alpha,\beta}$ to (2.34) with $\alpha, \beta \in (-1,0)$ satisfying condition (2.39). We shall prove the following claim: There exists $k_3 > 0$ (depending on α, β) such that for any $R > 0$

$$\begin{aligned} k_1 \int\!\!\int_{B_R} A\, u^{\alpha-1}\, v^{\beta+1} \mid \nabla u \mid^2 + k_2 \int\!\!\int_{B_R} A\, u^{\alpha+1}\, v^{\beta-1} \mid \nabla v \mid^2 \\ + \int\!\!\int_{B_R} b\, u^{\alpha}\, v^{\beta+1+r} + \int\!\!\int_{B_R} c\, u^{\alpha+1+s}\, v^{\beta} \\ \leq k_3\, R^{n-\frac{2}{\sigma-1}} \int\!\!\int_{\{1\leq\eta\leq 2\}} \Big[\sup_{u,v\geq 0;\, p,q\in\mathbb{R}^n} D(R\xi,\, R^2\tau, u, v, p, q) \Big]\, d\xi\, d\tau\, ; \end{aligned} \tag{2.53}$$

here η is defined by (2.42) and

$$B_R := \operatorname{supp} u \bigcap \operatorname{supp} v \bigcap \{(x,t) \in S \mid \mid x \mid^2 + t \leq R^2\} \qquad (R > 0).$$

From inequality (2.53) the conclusion follows easily (see [23] for details in the scalar case).

The proof of inequality (2.53) makes use of inequality (2.47) with a proper choice of the test function ψ. Consider any smooth function $\psi_0 : [0, \infty) \to [0, 1]$ with the following properties:

(i) $\psi_0 \equiv 1$ in $[0, 1]$, $\psi_0 \equiv 0$ in $[2, \infty)$, ψ_0 nonincreasing;
(ii) there holds:

$$\sup_{\eta\in[1,2]} \frac{\chi_0^{\sigma}(\eta)}{\psi_0(\eta)} < \infty\ ,$$

where

$$\chi_0(\eta) := 2\Big(\frac{\alpha+1}{k_1} + \frac{\beta+1}{k_2}\Big) \frac{\eta\, |\psi_0'(\eta)|^2}{\psi_0(\eta)} + |\psi_0'(\eta)|\, . \tag{2.54}$$

Observe that $\chi_0(\eta) \equiv 0$ if $\eta \notin [1, 2]$, due to property (i).

Introducing the scaled variables

$$\xi := \frac{x}{R}\,, \qquad \tau := \frac{t}{R^2}$$

there holds: $\mid x \mid^2 + t = R^2\, \eta$, thus

$$\{(x,t) \in S \mid \mid x \mid^2 + t \leq R^2\} = \{(\xi, \tau) \in S \mid \eta \leq 1\}\, .$$

Define

$$\psi_R(x,t) := \psi_0\,\Big(\frac{\mid x \mid^2 + t}{R^2}\Big) = \psi_0\,(\eta)\, ;$$

then $\psi_R \geq 0$, $\psi_R \equiv 1$ in B_R, $\operatorname{supp} \psi_R \subseteq \{\eta \leq 2\}$. Denote by χ_R the function (2.48) with $\psi = \psi_R$. It is easily seen that

$$\chi_R(x,t) = 2\Big(\frac{\alpha+1}{k_1} + \frac{\beta+1}{k_2}\Big) \frac{|\psi_0'(\eta)|^2}{\psi_0(\eta)}\, \frac{|x|^2}{R^4} + \frac{|\psi_0'(\eta)|}{R^2} \leq \frac{\chi_0(\eta)}{R^2}\,,$$

where χ_0 is the function (2.54). Then setting $\psi = \psi_R$ in inequality (2.47) we obtain:

$$k_1 \int\int_{B_R} A\, u^{\alpha-1}\, v^{\beta+1} \mid \nabla u \mid^2 + k_2 \int\int_{B_R} A\, u^{\alpha+1}\, v^{\beta-1} \mid \nabla v \mid^2$$

$$+ \int\int_{B_R} b\, u^{\alpha}\, v^{\beta+1+r} + \int\int_{B_R} c\, u^{\alpha+1+s}\, v^{\beta}$$

$$\leq \left(\frac{1}{\delta}\right)^{\frac{\sigma}{\sigma-1}} R^{n-\frac{2}{\sigma-1}}$$

$$\times \int\int_{\{1\leq\eta\leq 2\}} \left[\sup_{u,v\geq 0;p,q\in\mathbb{R}^n} D(R\xi,\, R^2\tau, u, v, p, q)\right] \left[\frac{\chi_0^{\sigma}(\eta)}{\psi_0(\eta)}\right]^{\frac{1}{\sigma-1}} d\xi d\tau.$$

Hence defining

$$k_3 := \left(\frac{1}{\delta}\right)^{\frac{\sigma}{\sigma-1}} \sup_{\eta\in[1,2]} \left[\frac{\chi_0^{\sigma}(\eta)}{\psi_0(\eta)}\right]^{\frac{1}{\sigma-1}}$$

the conclusion follows. □

Let us now prove Theorem 2.5.

Proof of Theorem 2.5. In the present case $a = b = c = 1$, thus $D \equiv 1$ and condition (2.41) is satisfied if

$$1 < \sigma < 1 + \frac{2}{n}. \tag{2.55}$$

On the other hand, summing both equations of system (2.40) we obtain:

$$H = H(\alpha, \beta, \sigma) := \frac{\alpha\,(\sigma-1) + \sigma}{1+s} + \frac{\beta\,(\sigma-1) + \sigma}{1+r} = 1\,. \tag{2.56}$$

Set

$$\mathsf{T} := \{\alpha, \beta \in (-1, 0) \mid \mid \alpha \mid + \mid \beta \mid > 1\}\,.$$

We shall prove the following claim: There exist $(\bar{\alpha}, \bar{\beta}) \in \mathsf{T}$, $\bar{\sigma} \in (1, 1 + \frac{2}{n})$ such that

$$H(\bar{\alpha}, \bar{\beta}, \bar{\sigma}) = 1\,.$$

Then solving the second equation of system (2.40) with respect to ρ gives

$$\rho = \bar{\rho} := \frac{1+r}{1+r-\beta(\sigma-1)-\sigma} > 1\,.$$

Since the quadruple $(\bar{\alpha}, \bar{\beta}, \bar{\rho}, \bar{\sigma})$ satisfies all conditions of Theorem 2.4, the conclusion will follow.

In order to prove the above claim observe that for any $\alpha, \beta \in (-1, 0)$

$$H(\alpha, \beta, 1) = \frac{1}{1+s} + \frac{1}{1+r} < 1\,, \tag{2.57}$$

due to condition (2.44). On the other hand,

$$H(\alpha, \beta, 1 + \frac{2}{n}) = \frac{2}{n}\left(\frac{\alpha}{1+s} + \frac{\beta}{1+r}\right) + \left(1 + \frac{2}{n}\right)\left(\frac{1}{1+s} + \frac{1}{1+r}\right).$$

An elementary calculation shows that

$$\max_{(\alpha,\beta)\in\bar{\mathsf{T}}} H(\alpha,\beta,1+\frac{2}{n}) = \frac{2}{n}\,\frac{1}{1+\min\{r,s\}} + \frac{1}{1+s} + \frac{1}{1+r}\,.$$

Then by condition (2.45) there exists $(\bar{\alpha},\bar{\beta}) \in \mathsf{T}$ such that

$$H(\bar{\alpha},\bar{\beta},1+\frac{2}{n}) > 1\,. \tag{2.58}$$

On the other hand, since

$$\alpha\,(\sigma-1)+\sigma = (\alpha+1)\,\sigma-\alpha > 1\,,$$

$$\beta\,(\sigma-1)+\sigma = (\beta+1)\,\sigma-\beta > 1\,,$$

the function $H(\alpha,\beta,\cdot)$ is increasing in $(1,1+\frac{2}{n})$ for any $\alpha,\beta\in(-1,0)$; then by (2.57)–(2.58) the claim follows. The proof is complete. □

Proof of Corollary 2.6. Every classical solution to system (2.34) satisfies the assumptions of Proposition 2.3 for any $\alpha,\beta\in(-1,0)$. Then by Proposition 2.3 and Theorem 2.5 the conclusion follows. □

Finally, let us prove Proposition 2.3.

Proof of Proposition 2.3. The conclusion will follow, if we prove that inequality (2.36) is satisfied. To this purpose, let us multiply the second equation in (2.34) by $u^{\alpha+1}\,v^{\beta}\,\psi$ (where $\alpha,\beta\in(-1,0)$ and $\psi\geq 0$ is any test function with support in $\bar{S}_T$) and integrate by parts. We obtain easily:

$$\begin{aligned}&\int\!\!\int u^{\alpha+1}\,v^{\beta}\partial_t v\,\psi\\ &\geq |\,\beta\,|\int\!\!\int A\,u^{\alpha+1}\,v^{\beta-1}\,|\,\nabla v\,|^2\;\psi - (\alpha+1)\int\!\!\int A\,u^{\alpha}\,v^{\beta}\,|\,\nabla u\,|\,|\,\nabla v\,|\,\psi\\ &\quad-\int\!\!\int A\,u^{\alpha+1}\,v^{\beta}\;|\,\nabla v\,|\,|\,\nabla\psi\,| + \int\!\!\int c\,u^{\alpha+1+s}\,v^{\beta}\,\psi\,.\end{aligned} \tag{2.59}$$

Observe that the right-hand side of the above inequality is well defined, since the second and the third integral are finite by condition (2.35).

Next, multiplying the first equation in (2.34) by $u^{\alpha}\,v^{\beta+1}\,\psi$ and integrating by parts obtains:

$$\begin{aligned}&\frac{\beta+1}{\alpha+1}\int\!\!\int u^{\alpha+1}\,v^{\beta}\partial_t v\,\psi + \frac{1}{\alpha+1}\int\!\!\int u^{\alpha+1}\,v^{\beta+1}\partial_t\psi\\ &\quad+\frac{1}{\alpha+1}\int_{\mathbb{R}^n} u^{\alpha+1}(x,0)\,v^{\beta+1}(x,0)\,\psi(x,0)\\ &\leq \alpha\int\!\!\int A\,u^{\alpha-1}\,v^{\beta+1}\,|\,\nabla u\,|^2\;\psi + (\beta+1)\int\!\!\int A\,u^{\alpha}\,v^{\beta}\,|\,\nabla u\,|\,|\,\nabla v\,|\,\psi\\ &\quad+\int\!\!\int A\,u^{\alpha}\,v^{\beta+1}\;|\,\nabla u\,|\,|\,\nabla\psi\,| - \int\!\!\int b\,u^{\alpha}\,v^{\beta+1+r}\,\psi\,,\end{aligned}$$

whence

$$\begin{aligned}
&(\beta+1)\int\!\!\int u^{\alpha+1}\, v^{\beta}\partial_t v\, \psi \\
&\le \alpha(\alpha+1)\int\!\!\int A\, u^{\alpha-1}\, v^{\beta+1} \mid \nabla u \mid^2\, \psi \\
&\quad + (\alpha+1)(\beta+1)\int\!\!\int A\, u^{\alpha}\, v^{\beta} \mid \nabla u \mid\mid \nabla v \mid \psi \qquad (2.60)\\
&\quad + (\alpha+1)\int\!\!\int A\, u^{\alpha}\, v^{\beta+1} \mid \nabla u \mid\mid \nabla \psi \mid - (\alpha+1)\int\!\!\int b\, u^{\alpha}\, v^{\beta+1+r}\, \psi \\
&\quad - \int\!\!\int u^{\alpha+1}\, v^{\beta+1}\partial_t \psi\,.
\end{aligned}$$

From inequalities (2.59)–(2.60) we obtain:

$$\begin{aligned}
&\mid \beta \mid (\beta+1)\int\!\!\int A\, u^{\alpha+1}\, v^{\beta-1} \mid \nabla v \mid^2\, \psi \\
&\quad - (\alpha+1)(\beta+1)\int\!\!\int A\, u^{\alpha}\, v^{\beta} \mid \nabla u \mid\mid \nabla v \mid \psi \\
&\quad - (\beta+1)\int\!\!\int A\, u^{\alpha+1}\, v^{\beta} \mid \nabla v \mid\mid \nabla \psi \mid + (\beta+1)\int\!\!\int c\, u^{\alpha+1+s}\, v^{\beta}\, \psi \\
&\le \alpha(\alpha+1)\int\!\!\int A\, u^{\alpha-1}\, v^{\beta+1} \mid \nabla u \mid^2\, \psi \\
&\quad + (\alpha+1)(\beta+1)\int\!\!\int A\, u^{\alpha}\, v^{\beta} \mid \nabla u \mid\mid \nabla v \mid \psi \\
&\quad + (\alpha+1)\int\!\!\int A\, u^{\alpha}\, v^{\beta+1} \mid \nabla u \mid\mid \nabla \psi \mid - (\alpha+1)\int\!\!\int b\, u^{\alpha}\, v^{\beta+1+r}\, \psi \\
&\quad - \int\!\!\int u^{\alpha+1}\, v^{\beta+1}\partial_t \psi\,.
\end{aligned}$$

Due to condition (2.35) and to the assumptions $\alpha > -1$, $\beta > -1$, the left-hand side of the above inequality is either finite or positively diverging. Similarly, the right-hand side is either finite or negatively diverging. Hence both sides of (2.60) – thus every integral in (2.60) – are finite. Then the conclusion immediately follows. □

References

[1] K. Deng and H.A. Levine, *The role of critical exponents in blow-up theorems: the sequel.* J. Math. Anal. Appl. 243 (2000), 85–126.

[2] M. Escobedo and M.A. Herrero, *Boundedness and blow up for a semilinear reaction-diffusion system.* J. Diff. Equat. 89 (1991), 176–202.

[3] M. Escobedo and H.A. Levine, *Explosion et existence globale pour un système faiblement couplé d'Equations de Réaction Diffusion.* C. R. Acad. Sci. Paris. Sér. 1. 314 (1992), 735–739.

[4] M. Escobedo and H.A. Levine, *Critical blow up and global existence numbers for a weakly coupled system of reaction-diffusion equations.* Arch. Ration. Mech. and Anal. 129 (1995), 47–100.

[5] B. Gidas and J. Spruck, *Global and local behavior of positive solutions of nonlinear elliptic equations.* Commun. Pure and Appl. Math. 34 (1981), 525–598.

[6] G.G. Laptev, *The absence of global positive solutions of systems of semilinear elliptic inequalities in cones.* Russian Acad. Sci. Izv. Math. 64 (2000), 108–124.

[7] G.G. Laptev, *On the absence of solutions to a class of singular semilinear differential inequalities.* Proc. Steklov Inst. Math. 232 (2001), 223–235.

[8] G.G. Laptev, *Nonexistence of solutions to semilinear parabolic inequalities in cones.* Mat. Sb. 192(10) (2001), 51–70.

[9] G.G. Laptev, *Some nonexistence results for higher-order evolution inequalities in cone-like domains.* Electron. Res. Announc. Amer. Math. Soc. 7 (2001), 87–93.

[10] H.A. Levine, *The role of critical exponents in blow-up theorems.* SIAM Rev. 32 (1990), 262–288.

[11] H.A. Levine, *A Fujita type global existence — global nonexistence theorem for a weakly coupled system of reaction-diffusion equations.* Ztschr. angew. Math. und Phys. 42 (1991), 408–430.

[12] E. Mitidieri and S.I. Pohozaev, *Nonexistence of global positive solutions to quasilinear elliptic inequalities.* Dokl. Russ. Acad. Sci. 57 (1998), 250–253.

[13] E. Mitidieri and S.I. Pohozaev, *Nonexistence of positive solutions for a system of quasilinear elliptic equations and inequalities in* $\mathbb{R}^N$. Dokl. Russ. Acad. Sci. 59 (1999), 1351–1355.

[14] E. Mitidieri and S.I. Pohozaev, *Nonexistence of positive solutions for quasilinear elliptic problems on* $\mathbb{R}^N$. Proc. Steklov Inst. Math. 227 (1999), 192–222.

[15] E. Mitidieri and S.I. Pohozaev. *Nonexistence of weak solutions for some degenerate elliptic and parabolic problems on* $\mathbb{R}^n$. J. Evolution Equations 1 (2001), 189–220.

[16] E. Mitidieri and S.I. Pohozaev, *Nonexistence of weak solutions for some degenerate and singular hyperbolic problems on* $\mathbb{R}^N$. Proc. Steklov Inst. Math. 232 (2001), 240–259.

[17] E. Mitidieri and S.I. Pohozaev, *A priori Estimates and Nonexistence of Solutions to Nonlinear Partial Differential Equations and Inequalities.* Moscow, Nauka, 2001. (Proc. Steklov Inst. Math. 234).

[18] E. Mitidieri, G. Sweers and R. van der Vorst, *Nonexistence theorems for systems of quasilinear partial differential equations.* Diff. Integr. Equat. 8 (1995), 1331–1354.

[19] W.-M. Ni and J. Serrin, *Non-existence theorems for quasilinear partial differential equations.* Rend. Circ. Mat. Palermo. Ser. 2. Suppl. 8 (1985), 171–185.

[20] W.-M. Ni and J. Serrin, *Existence and nonexistence theorems for ground states of quasilinear partial differential equations. The anomalous case.* Atti Convegni Lincei 77 (1986), 231–257.

[21] S.I. Pohozaev, *On eigenfunctions to equation* $\Delta u + \lambda f(u) = 0$. Dokl. USSR Acad. Sci. 165 (1965), 36–39.

[22] S.I. Pohozaev, *Essential nonlinear capacities induced by differential operators.* Dokl. Russ. Acad. Sci. 357 (1997), 592–594.

[23] S.I. Pohozaev and A. Tesei, *Blow-up of nonnegative solutions to quasilinear parabolic inequalities.* Atti Accad. Naz. Lincei. Cl. Sci. Fis. Mat. Natur. Rend. Lincei. Ser. 9. 11 (2000), 99–109.

[24] S.I. Pohozaev and L. Veron, *Blow-up results for nonlinear hyperbolic inequalities.* Ann. Scuola Norm. Super. Pisa. Cl. Sci. Ser. 4. 29 (2001), 393–420.

[25] A.A. Samarskii, V.A. Galaktionov, S.P. Kurdyumov and A.P. Mikhailov, *Blow-up in Quasilinear Parabolic Equations.* Nauka, Moscow, 1987 (in Russian); English translation: Walter de Gruyter, Berlin/New York, 1995.

Stanislav Pohozaev
Steklov Mathematical Institute
Gubkina 8
117966 Moscow, Russia
E-mail address: pohozaev@mi.ras.ru

D. Haroske, T. Runst, H.-J. Schmeisser (eds.): Function Spaces, Differential Operators and Nonlinear Analysis. The Hans Triebel Anniversary Volume.

Laplace and Schrödinger Operators on Regular Metric Trees: The Discrete Spectrum Case

Michael Solomyak

Dedicated to Prof. Hans Triebel on the occasion of his 65th birthday

1. Introduction

Spectral theory of differential operators on metric trees is an interesting branch of such theory on general metric graphs. Among the trees, the so-called regular trees are of particular interest due to their very special geometry.

Let Γ be a tree rooted at some vertex o and having infinitely many edges. Below $|x|$ stands for the distance between a point $x \in \Gamma$ and the root o. For a vertex x, its *generation* is the number of vertices lying between o and x (including x but excluding o). We say that a tree Γ is *regular* if for any vertex x the quantity $|x|$ and the number of edges emanating from x depend only on the generation of x; see Definition 2.1 for the more detailed description.

The regular trees are highly symmetric. This allows one to construct an orthogonal decomposition of the space $\mathsf{L}^2(\Gamma)$ which reduces the Schrödinger operator $\mathbf{A}_V = -\boldsymbol{\Delta} + V$ with any symmetric (i.e. depending on $|x|$) potential V. We call this decomposition the *basic decomposition* of $\mathsf{L}^2(\Gamma)$. The parts of $\mathbf{A}_V$ in the components of the basic decomposition are denoted by $\mathcal{A}_{V,k}$. Each operator $\mathcal{A}_{V,k}$ appears with a multiplicity which rapidly grows as $k \to \infty$. The study of the spectrum $\sigma(\mathbf{A}_V)$ of the operator $\mathbf{A}_V$ is thus reduced to the study of the spectra of the parts $\mathcal{A}_{V,k}$. Each part can be identified with a differential operator acting in a weighted space $\mathsf{L}^2\big((t_k, R), g_k\big)$, $k = 0, 1, \dots$ where the intervals (t_k, R) and the weight functions g_k are determined by the geometry of Γ. In particular, the quantity $R = R(\Gamma) = \sup\{|x| : x \in \Gamma\} \le \infty$ is the *radius* of the tree.

According to general spectral theory,

$$\sigma(\mathbf{A}_V) = \overline{\cup_{k=0}^{\infty} \sigma(\mathcal{A}_{V,k})}.$$

However, the quantitative characteristics of $\sigma(\mathbf{A}_V)$ cannot be obtained automatically from the corresponding characteristics for the operators $\mathcal{A}_{V,k}$, due to the growing multiplicities of $\mathcal{A}_{V,k}$ as parts of $\mathbf{A}_V$.

The main purpose of this paper is to study the situations when the spectrum of the Schrödinger operator $\mathbf{A}_V$ is discrete. More exactly, we consider two typical cases: for the regular trees of finite radius we show that the classical Weyl asymptotic law holds for the eigenvalues of each operator $\mathcal{A}_{V,k}$ with any bounded potential V, including the basic case $V = 0$. If the tree Γ has finite total length, we show that the Weyl formula holds for the whole operator $\mathbf{A}_V$.

Another case is the operator $\mathbf{A}_V$ with growing potential on a tree of infinite radius. Under certain assumptions about the geometry of the tree and the behaviour of the potential, we find a version of the Weyl formula for such operators.

To make the general picture more complete, we also present, without proofs, some known results concerning the structure of $\sigma(\mathbf{A}_V)$ in the case when the spectrum is not discrete.

There are several papers devoted to differential operators on regular metric trees. In [4] the case of the *homogeneous trees* Γ_b was considered. A regular tree is called *homogeneous* if all its edges have equal length (say 1) and all vertices have the same number of edges (say b) emanating from them. In [4] the potential V was supposed to be periodic and even. It was shown that the spectrum $\sigma(\mathbf{A}_V)$ has the band-gap structure, with no more than one eigenvalue in each gap. Actually, this eigenvalue is of infinite multiplicity, but the multiplicities were not discussed in [4].

In [10] the weighted spectral problems of the form $-\mathbf{\Delta} f = \lambda V f$ on general (not necessarily regular) trees were investigated. The estimates and, under some additional assumptions, the asymptotic behaviour of the eigenvalues were found. For the regular trees, the basic decomposition of $\mathsf{L}^2(\Gamma)$ was discovered in [10], and much more advanced results for such trees were obtained with the help of this decomposition.

In the paper [5] the basic decomposition was re-discovered, and a detailed spectral analysis of the operators $\mathcal{A}_{V,k}$ for the regular trees Γ with finite radius was given. In particular, it was shown that if $R(\Gamma) < \infty$ but the total length of Γ is infinite, the operators $\mathcal{A}_{V,k}$ do not require the boundary condition at $t = R$. It was also shown that each operator $\mathcal{A}_{V,k}$ has compact resolvent, and the corresponding eigenvalue distribution function $N(\lambda; \mathcal{A}_{V,k})$ grows not faster than $O(\lambda^{1/2+\epsilon})$ for any $\epsilon > 0$. Our Theorem 5.3 considerably refines this result.

The main topic of [11] is the Hardy-type inequalities on regular trees. As a consequence, a necessary and sufficient condition of the positive definiteness of the Laplacian was established, see Theorem 5.6.

In [12] operators $\mathbf{A}_V$ with decaying symmetric potentials on the homogeneous trees Γ_b were investigated. For $V = 0$ (that is, for the free Laplacian) the spectrum was explicitly calculated. This result is presented here as Theorem 5.7. In accordance with the results of [4], the spectrum has infinitely many gaps. Perturbation of the operator $\mathbf{A}_0$ by a decaying potential may create eigenvalues in each gap, and in [12] their behaviour was investigated in detail both for positive and negative perturbations.

It is necessary to mention here also the papers [6] and [7], though formally they do not deal with the Laplacian on trees. In [6] the Hardy-type integral operators on trees were introduced in connection with the spectral analysis of the Neumann Laplacian in certain irregular domains. For these "ridged" domains, there exists a tree that serves as a "ridge", or a "skeleton", for the domain. There is a close relation between the approximation numbers of the Hardy-type integral operators and the eigenvalues of the problem $-\boldsymbol{\Delta} f = \lambda V f$ on the tree. It was the paper [6] which attracted the author's attention to operators on trees.

In [7] the behaviour of the approximation numbers of the Hardy-type integral operators on trees was studied in detail, not only in L^2-case but also in the general L^p-case. When applied to the Laplacian, the estimates obtained substantially refine some results of [10], Section 4. The asymptotic formulae found in [7] may provide an alternative approach to the proof of our Theorem 5.3, (ii).

The structure of the present paper is as follows. The short Sections 2–4 contain the necessary preliminary material: the definitions of regular and homogeneous trees and of the Laplace and Schrödinger operators on them, and the description of the basic orthogonal decomposition of $\mathsf{L}^2(\Gamma)$. Section 5 contains formulations of the main results, the proofs are given in Section 6. In the final Section 7 we discuss the extension of the presented results to the case of regular trees without boundary.

2. Regular rooted trees

2.1. Geometry of a tree

Let Γ be a rooted tree with the root o, the set of vertices $\mathcal{V} = \mathcal{V}(\Gamma)$ and the set of edges $\mathcal{E} = \mathcal{E}(\Gamma)$. We suppose that $\#\mathcal{V} = \#\mathcal{E} = \infty$. Unlike the combinatorial trees, whose edges are just pairs of vertices, each edge e of a metric tree is viewed as a non-degenerate line segment. The distance $\rho(x,y)$ between any two points $x, y \in \Gamma$ (and thus the metric topology on Γ) is introduced in a natural way. As it was already said in Introduction, $|x|$ stands for $\rho(x,o)$.

We write $y \preceq z$ if $|z| = |y| + \rho(y,z)$, and $y \prec z$ means that $y \preceq z$ and $z \neq y$. The relation $\prec$ defines on Γ a partial ordering. If $y \prec z$, we denote

$$\langle y, z\rangle := \{x \in \Gamma : \ y \preceq x \preceq z\}.$$

For any vertex y its *generation* $\operatorname{Gen}(y)$ is defined as

$$\operatorname{Gen}(y) = \#\{x \in \mathcal{V} : o \prec x \preceq y\}.$$

We assume that $\operatorname{Gen}(y) < \infty$ for any vertex y. For an edge e we define $\operatorname{Gen}(e)$ as the generation of its initial point.

The *branching number* $b(y)$ of a vertex y is defined as the number of edges emanating from y. We assume that $b(o) = 1$ and $b(y) > 1$ for $y \neq o$. We denote by e_y^- the only edge which terminates at a vertex $y \neq o$, and by $e_y^1, \dots, e_y^{b(y)}$ the edges emanating from any vertex $y \in \mathcal{V}$.

Definition 2.1. *We call a tree* Γ regular *if all the vertices of the same generation have equal branching numbers, and all the edges of the same generation are of the same length.*

In this paper we consider only the regular trees. Evidently, any such tree is fully determined by specifying two number sequences, $\{b_k\}$ and $\{t_k\}$, $k = 0, 1, \dots$ such that

$$b(y) = b_{\mathrm{Gen}(y)}, \ |y| = t_{\mathrm{Gen}(y)} \qquad \text{for each } y \in \mathcal{V}(\Gamma).$$

According to our assumptions, $b_0 = 1$ and $b_k \geq 2$ for any $k > 0$. It is clear that $t_0 = 0$ and the sequence $\{t_k\}$ is strictly increasing, and we denote

$$R = R(\Gamma) = \lim_{k\to\infty} t_k = \sup_{x\in\Gamma} |x|.$$

We call $R(\Gamma)$ the *radius* of the tree. Another important characteristic of a tree is its *total length* (in other terminology, *volume*)

$$|\Gamma| = \sum_{e\in\mathcal{E}(\Gamma)} |e|.$$

The natural measure dx on Γ is induced by the Lebesgue measure on the edges. Below $\mathsf{L}^2(\Gamma) = \mathsf{L}^2(\Gamma, dx)$.

2.2. Homogeneous trees

A rooted tree is called homogeneous if its edges are all of the same length (for definiteness, of the length one) and all the vertices $y \neq o$ have the same branching number b. A homogeneous tree is evidently regular. It is fully determined by specifying the parameter b and we use for it the notation Γ_b. For the tree Γ_b one has $t_k = k$, $k = 0, 1, \dots$, and $b_k = b$, $k = 1, 2, \dots$.

3. The Laplace and the Schrödinger operators on a regular tree

The notion of differential operator on any metric graph, in particular on a tree, is well known. Still, for the sake of completeness we present here the variational definitions of the Laplacian and of the Schrödinger operator on a tree.

We say that a scalar-valued function f on Γ belongs to the Sobolev space $\mathsf{H}^1 = \mathsf{H}^1(\Gamma)$ if f is continuous, $f \restriction e \in \mathsf{H}^1(e)$ for each edge e, and

$$\|f\|^2_{\mathsf{H}^1} := \int_\Gamma (|f'(x)|^2 + |f(x)|^2)dx < \infty. \tag{1}$$

The derivative of a function $f \restriction e$ at an interior point $x \in e$ is always taken in the direction compatible with the partial ordering on Γ. This agreement is indifferent for the definition (1) but we shall use it later.

The set $\mathsf{H}^1_{\mathrm{b}}$ of all boundedly supported functions $u \in \mathsf{H}^1$ is dense in H^1. Indeed, for any number $L > 0$ let $\varphi_L(t)$ be the continuous function on $\mathbb{R}_+$, which is 1 for $t \leq L$, is 0 for $t \geq L + 1$ and is linear on $[L, L + 1]$. Given a function $f \in \mathsf{H}^1(\Gamma)$, denote $f_L(x) = \varphi_L(|x|)f(x)$. Then $f_L \in \mathsf{H}^1_{\mathrm{b}}$ and an elementary calculation shows that $f_L \to f$ in $\mathsf{H}^1(\Gamma)$ as $L \to \infty$.

Along with $\mathsf{H}^1(\Gamma)$, let us introduce also its subspace of codimension one:

$$\mathsf{H}^{1,0} := \mathsf{H}^{1,0}(\Gamma) = \{f \in \mathsf{H}^1(\Gamma) : f(o) = 0\}.$$

We define the Dirichlet Laplacian $-\boldsymbol{\Delta}$ on Γ as the self-adjoint operator in $\mathsf{L}^2(\Gamma)$, associated with the quadratic form $\int_\Gamma |f'|^2 dx$ considered on the form domain $\mathrm{Quad}(-\boldsymbol{\Delta}) = \mathsf{H}^{1,0}(\Gamma)$. It is easy to describe the operator domain $\mathrm{Dom}(\boldsymbol{\Delta})$ and the action of $\boldsymbol{\Delta}$. Evidently $f \in \mathrm{Dom}(\boldsymbol{\Delta}) \Rightarrow f \restriction e \in \mathsf{H}^2(e)$ for each edge e and the Euler-Lagrange equation reduces on e to $\boldsymbol{\Delta} f = f''$. At the root we have the boundary condition $f(o) = 0$, since $\mathrm{Dom}(\boldsymbol{\Delta}) \subset \mathsf{H}^{1,0}(\Gamma)$. At each vertex $y \neq o$ the functions $f \in \mathrm{Dom}(\boldsymbol{\Delta})$ satisfy certain matching conditions. In order to describe them, denote by f_- the restriction $f \restriction e_y^-$ and by f_j, $j = 1, \dots, b(y)$ the restrictions $f \restriction e_y^j$. The matching conditions at $y \neq o$ are

$$f_-(y) = f_1(y) = \dots = f_b(y); \quad f_1'(y) + \dots + f_b'(y) = f_-'(y).$$

The first condition comes from the requirement $f \in \mathsf{H}^1(\Gamma)$ which includes continuity of f, and the second appears as the natural condition in the sense of Calculus of Variations. It is easy to check that the listed conditions are also sufficient for $f \in \mathrm{Dom}(\boldsymbol{\Delta})$.

Let V be a measurable, real-valued, bounded from below and *symmetric* (that is, depending only on $|x|$) function on Γ. Along with the Laplacian we shall be interested also in the Schrödinger operator with the potential $V(|x|)$:

$$\mathbf{A}_V f := -\boldsymbol{\Delta} f + V(|x|)f. \tag{2}$$

The operator $\mathbf{A}_V$ is defined via its quadratic form

$$\mathbf{a}_V[f] := \int_\Gamma \left(|f'(x)|^2 + V(|x|)|f(x)|^2\right) dx \tag{3}$$

considered on the natural domain $\mathrm{Dom}(\mathbf{a}_V) = \mathrm{Quad}(\mathcal{A}_V) = \mathsf{H}^{1,0}(\Gamma) \cap \mathsf{L}^2_V(\Gamma)$. On this domain the quadratic form (3) is bounded from below and closed in $\mathsf{L}^2(\Gamma)$, and the corresponding self-adjoint operator is taken as the realization of the operator (2). We do not need the precise description of the domain and the action of $\mathbf{A}_V$.

4. The basic decomposition of $\mathsf{L}^2(\Gamma)$

Consider two types of subtrees $T \subset \Gamma$. Namely, for any vertex y and for any edge $e = \langle z, w \rangle$ we set

$$T_y = \{x \in \Gamma : x \succeq y\}, \qquad T_e = e \cup T_w.$$

Evidently $T_o = \Gamma$.

For any subtree $T = T_y$ or $T = T_e$ its *branching function* $g_T(t)$ is defined as

$$g_T(t) = \#\{x \in T : |x| = t\}.$$

If $T = T_e$ and $\mathrm{Gen}(e) = k \geq 0$, then $g_T(t) = g_k(t)$ where

$$g_k(t) = \begin{cases} 0, & t < t_k, \\ 1, & t_k \leq t \leq t_{k+1}, \\ b_{k+1} \dots b_n, & t_n < t \leq t_{n+1},\ n > k. \end{cases}$$

In particular, $g_0(t) = g_\Gamma(t) = 1$ for $0 \leq t \leq t_1$ and

$$g_\Gamma(t) = b_1 \dots b_n, \qquad t_n < t \leq t_{n+1},\ n \geq 1.$$

So we see that

$$g_k(t) = (b_1 \dots b_k)^{-1} g_\Gamma(t), \qquad t > t_k,\ k > 0. \tag{4}$$

Note that

$$\int_\Gamma g_\Gamma(t)dt = |\Gamma|. \tag{5}$$

Given a subtree $T \subset \Gamma$, we say that a function $f \in \mathsf{L}^2(\Gamma)$ belongs to the set $\mathcal{F}_T$ if and only if $f = 0$ outside T and

$$f(x) = f(y) \qquad \text{if } x, y \in T \text{ and } |x| = |y|.$$

Evidently $\mathcal{F}_T$ is a closed subspace of $\mathsf{L}^2(\Gamma)$. Any function $f \in \mathcal{F}_{T_e}$, $\mathrm{Gen}(e) = k \geq 0$ can be naturally identified with the function $u := J_e f$ on (t_k, R), such that $f(x) = u(|x|)$ for each $x \in T_e$ and $f(x) = 0$ outside T_e. We have

$$\int_\Gamma |f(x)|^2 dx = \|u\|^2_{\mathsf{L}^2((t_k,R),g_k)} := \int_{t_k}^R |u(t)|^2 g_k(t)dt; \tag{6}$$
$$f \in \mathcal{F}_{T_e},\ u = J_e f.$$

This shows that the operator J_e defines an isometry of the subspace $\mathcal{F}_{T_e}$ onto the weighted space $\mathsf{L}^2\big((t_k, R); g_k\big)$. Along with (6), we have

$$\int_\Gamma |f'(x)|^2 dx = a_k[u] := \int_{t_k}^R |u'(t)|^2 g_k(t)dt; \tag{7}$$
$$f \in \mathcal{F}_{T_e} \cap \mathsf{H}^{1,0}(\Gamma),\ u = J_e f.$$

For the sake of brevity, below we use the notations $\mathcal{F}_y$ for $\mathcal{F}_{T_y}$ and $\mathcal{F}_y^j$ for $\mathcal{F}_{T_{e_y^j}}$, $j = 1, \dots, b = b(y)$. It is clear that the subspaces $\mathcal{F}_y^1, \dots, \mathcal{F}_y^b$ are mutually orthogonal and their orthogonal sum

$$\widetilde{\mathcal{F}_y} = \mathcal{F}_y^1 \oplus \dots \oplus \mathcal{F}_y^b$$

contains $\mathcal{F}_y$. Denote

$$\mathcal{F}'_y = \widetilde{\mathcal{F}_y} \ominus \mathcal{F}_y$$

Theorem 4.1. *Let Γ be a regular tree.*

(i) *The subspaces $\mathcal{F}'_y$, $y \in \mathcal{V}(\Gamma)$ are mutually orthogonal and orthogonal to $\mathcal{F}_\Gamma$. Moreover,*

$$\mathsf{L}^2(\Gamma) = \mathcal{F}_\Gamma \oplus \bigoplus_{y \in \mathcal{V}(\Gamma)} \mathcal{F}'_y. \tag{8}$$

(ii) *Let $V(t)$ be a real, measurable and bounded from below function on $\mathbb{R}_+$. Then the decomposition* (8) *reduces the Schrödinger operator* (2).

According to Theorem 4.1, the description of the spectrum $\sigma(\mathbf{A}_V)$ reduces to the similar problem for the parts of $\mathbf{A}_V$ in the components of the decomposition (8). These parts can be described in terms of auxiliary differential operators $\mathcal{A}_{V,k}$, $k = 0, 1, \dots$ acting in the spaces $\mathsf{L}^2((t_k, R), g_k)$.

For $f \in \mathrm{Dom}(\mathbf{a}) \cap \mathcal{F}_{T_e}$ the quadratic form $\mathbf{a}_V$, cf. (3), transforms as follows:

$$\mathbf{a}_V[f] = a_{V,k}[u] := \int_{t_k}^{R} \left(|u'(t)|^2 + V(t)|u(t)|^2\right) g_k(t) dt, \tag{9}$$
$$f \in \mathrm{Dom}(\mathbf{a}) \cap \mathcal{F}_{T_e}, \ \mathrm{Gen}(e) = k, \ u = J_e f.$$

For $V \equiv 0$ the quadratic form $a_{V,k}[u]$ turns into the quadratic form $a_k[u]$ defined in (7). We define $\mathcal{A}_{V,k}$ as the self-adjoint operator in $\mathsf{L}^2((t_k, T), g_k)$, associated with the quadratic form $a_{V,k}$. We drop the subindex V in these notation when dealing with the free Laplacian $-\mathbf{\Delta} = \mathbf{A}_0$.

The following result was actually proved in [4] and [11]; minor distinctions in the formulations are unessential.

Theorem 4.2. *Let Γ be a regular tree and V be a bounded from below, real-valued symmetric potential on Γ. Then the part of $\mathbf{A}_V$ in the subspace $\mathcal{F}_\Gamma$ is unitarily equivalent to the operator $\mathcal{A}_{V,0}$ and the part of $\mathbf{A}_V$ in each subspace $\mathcal{F}'_z$, Gen$(z) = k > 0$, is unitary equivalent to the orthogonal sum of $(b_k - 1)$ copies of $\mathcal{A}_{V,k}$.*

The next theorem is an immediate consequence of Theorems 4.1 and 4.2. Below $\mathcal{A}^{[r]}$ stands for the orthogonal sum of r copies of a self-adjoint operator $\mathcal{A}$. The symbol "$\sim$" means unitary equivalence.

Theorem 4.3. *Under the assumptions of Theorem 4.2 the operator $\mathbf{A}_{V,\Gamma}$ is unitary equivalent to the orthogonal sum of the operators $\mathcal{A}_{V,k}$, with growing multiplicities:*

$$\mathbf{A}_{V,\Gamma} \sim \mathcal{A}_{V,0} \oplus \bigoplus_{k=1}^{\infty} {\mathcal{A}_{V,k}}^{[b_1 \dots b_{k-1}(b_k - 1)]}. \tag{10}$$

5. Main results

5.1. The eigenvalue counting functions

For a self-adjoint, bounded from below operator $\mathcal{A}$ with discrete spectrum, we denote by $N(\lambda; \mathcal{A})$ the distribution function of its eigenvalues $\lambda_j(\mathcal{A})$ (counted according to their multiplicities),

$$N(\lambda; \mathcal{A}) = \#\{j : \lambda_j(\mathcal{A}) < \lambda\}, \qquad \lambda \in \mathbb{R}.$$

We start with the following simple but useful statement.

Theorem 5.1. *Let Γ be a regular tree and V be a symmetric measurable real-valued function, bounded from below. The spectrum $\sigma(\mathbf{A}_V)$ is discrete if and only if the spectrum of the operator $\mathcal{A}_{V,0}$ is discrete. If this is the case, then*

$$N(\lambda; \mathbf{A}_V) = N(\lambda; \mathcal{A}_{V,0}) + \sum_{k=1}^{\infty} b_1 \dots b_{k-1}(b_k - 1) N(\lambda; \mathcal{A}_{V,k}), \qquad \lambda \in \mathbb{R}. \quad (11)$$

Proof. Consider the Rayleigh quotient $a_{V,k}[u]/\|u\|^2_{L^2((t_k,R),g_k)}$, cf. (6) and (9). Due to the equality (4), this ratio does not change if we replace in its numerator and denominator the weight function $g_k(t)$ by $g_\Gamma(t)$. Now it follows from the variational principle that the spectrum of each operator $\mathcal{A}_{V,k}$, $k = 1, 2, \dots$ is discrete provided this is true for $k = 0$. Moreover, we see that

$$N(\lambda; \mathcal{A}_{V,k_2}) \le N(\lambda; \mathcal{A}_{V,k_1}), \qquad k_1 < k_2, \ \lambda > 0.$$

The discreteness of $\sigma(\mathcal{A}_{V,k})$ for all k implies the same property of $\sigma(\mathbf{A}_V)$. The converse is evident. The equality (11) is an immediate consequence of the relation (10).

The detailed study of the function $N(\lambda; \mathbf{A}_V)$ is hampered by the presence of the rapidly growing factors $b_1 \dots b_{k-1}(b_k - 1)$. These factors reflect the geometry of the tree and do not depend on the potential V. For this reason, sometimes we consider another counting function (introduced in [12]):

$$\widetilde{N}(\lambda; \mathbf{A}_V) := \sum_{k=0}^{\infty} N(\lambda; \mathcal{A}_{V,k}). \quad (12)$$

5.2. The spectrum of the Laplacian

The spectrum of the Laplacian on a regular tree depends on the behaviour of the sequences $\{t_k\}$ and $\{b_k\}$ and can be quite different. We present here several results in this direction. The proofs of those which are new are given in the next section. In other cases we give the relevant references.

Our first result is quite elementary and its proof is standard. The result applies to arbitrary metric graphs rather than to trees only.

Theorem 5.2. *Let Γ be a metric graph such that $\sup_{e \in \mathcal{E}(\Gamma)} |e| = \infty$. Then the spectrum of the Laplacian $-\boldsymbol{\Delta}$ on Γ coincides with $[0, \infty)$.*

Other results concern the regular trees. We start with the trees of finite radius, for which the information provided is rather complete.

Theorem 5.3. *Let Γ be a regular tree and $R(\Gamma) < \infty$.*

(i) *The spectrum of the Laplacian $-\boldsymbol{\Delta}$ on Γ is discrete. For each operator $\mathcal{A}_k$ its eigenvalues behave according to the Weyl law,*

$$\pi N(\lambda; \mathcal{A}_k) = \sqrt{\lambda}(R - t_k) + o(\sqrt{\lambda}), \qquad \lambda \to \infty. \quad (13)$$

(ii) *If* $|\Gamma| < \infty$*, then the Weyl asymptotic law holds for the operator* $-\boldsymbol{\Delta}$*:*

$$\pi N(\lambda; -\boldsymbol{\Delta}) = \sqrt{\lambda}|\Gamma| + o(\sqrt{\lambda}), \qquad \lambda \to \infty. \tag{14}$$

(iii) *If*

$$\widetilde{R}(\Gamma) := \sum_{k=0}^{\infty}(R - t_k) < \infty,$$

then

$$\pi \widetilde{N}(\lambda; -\boldsymbol{\Delta}) = \sqrt{\lambda}\widetilde{R}(\Gamma) + o(\sqrt{\lambda}), \qquad \lambda \to \infty. \tag{15}$$

Theorem 5.3 refines an earlier result of [5]. The case of general (i.e. not necessarily regular) trees was analyzed in [10], Theorem 4.1. For the trees with $|\Gamma| < \infty$, satisfying some additional assumptions, the Weyl asymptotics (14) follows from this theorem. However, the result of [10] does not cover the case of arbitrary regular trees of finite total length.

The next statement can be derived from Theorem 5.3 by means of the elementary variational arguments. We present it without proof.

Corollary 5.4. *Let* Γ *be a regular tree,* $R(\Gamma) < \infty$*, and the potential* $V(x)$ *(not necessarily symmetric) be bounded. Then the spectrum* $\sigma(-\boldsymbol{\Delta} + V)$ *is discrete. If, in addition,* $|\Gamma| < \infty$*, then the asymptotic formula* (14) *holds for its eigenvalues.*

If, in addition, the potential is symmetric, then the asymptotic formula (13) *holds for each operator* $\mathcal{A}_{V,k}$.

If for a regular tree Γ one has $R(\Gamma) < \infty$ but $|\Gamma| = \infty$, then the asymptotic behaviour of $N(\lambda; -\boldsymbol{\Delta})$ can be rather exotic. The following example can serve as an illustration.

Fix the numbers $q \in (0,1)$ and $b \in \mathbb{N}$. Consider the regular tree $\Gamma = \Gamma_{q,b}$ defined by the sequences $t_k = 1 - q^k$, $k = 0, 1, \dots$ and $b_k = b$, $k = 1, \dots$. Then $R(\Gamma) = 1$, so that the spectrum of the Laplacian on Γ is always discrete. Further, $g_0(t) = b^k$ for $t_k < t \le t_{k+1}$. The total length of Γ is

$$|\Gamma| = 1 - q + \sum_{k=1}^{\infty} b^k(q^k - q^{k+1}) = (1-q)\sum_{k=0}^{\infty}(bq)^k.$$

Hence, $|\Gamma| = \frac{1-q}{1-bq} < \infty$ if $bq < 1$ and $|\Gamma| = \infty$ otherwise. In the first case, Theorem 5.3 (ii) shows that the Weyl law (14) holds for the eigenvalues of $-\boldsymbol{\Delta}$. Besides, $\widetilde{R}(\Gamma_{q,b}) = (1-q)^{-1} < \infty$, and by Theorem 5.3 (iii) the asymptotic formula (15) holds for any $q < 1$ and any b.

Below we present the results for the function $N(\lambda, -\boldsymbol{\Delta})$, for $bq \ge 1$. The case $bq > 1$ was analyzed in [10], Example 8.2 (where one should take $\alpha = 0$). The result of [10] for $bq = 1$ was not complete.

Theorem 5.5. *Let* $\Gamma = \Gamma_{q,b}$.

(i) *If* $bq > 1$, *then there exists a bounded and bounded away from zero periodic function* ψ *with the period* $\ln(q^{-2})$ *such that*

$$\pi N(\lambda; -\mathbf{\Delta}) = \lambda^{\beta/2}\big(\psi(\ln\lambda) + o(1)\big), \qquad \lambda \to \infty \tag{16}$$

where $\beta = -\log_q b > 1$.

(ii) *If* $bq = 1$, *then*

$$\pi N(\lambda; -\mathbf{\Delta}) = \frac{1-q}{2\ln b}\sqrt{\lambda}\big(\ln\lambda + O(1)\big), \qquad \lambda \to \infty. \tag{17}$$

The proof given in the next section covers both cases. For $bq > 1$, it reproduces the argument from [10].

The results for the trees with $R(\Gamma) = \infty$ are much less exhaustive. We start with a criterion of positive definiteness of the Laplacian on a regular tree, proved in [11].

Theorem 5.6. *Let* Γ *be a regular tree and* $R(\Gamma) = \infty$. *Then the Laplacian on* Γ *is positive definite in* $\mathsf{L}^2(\Gamma)$ *if and only if*

$$\sup_{t>0}\left(\int_0^t g_\Gamma(s)ds \cdot \int_t^\infty \frac{ds}{g_\Gamma(s)}\right) < \infty. \tag{18}$$

The condition (18) is satisfied, in particular, for the homogeneous trees Γ_b. For them the spectrum can be described completely. The next result is proved in [12], Theorem 3.3. Introduce the number

$$\theta = \arccos\frac{2}{b^{1/2} + b^{-1/2}}.$$

Theorem 5.7. *The spectrum of the operator* $-\mathbf{\Delta}$ *on the tree* Γ_b *is of infinite multiplicity and consists of the bands* $\left[(\pi(l-1)+\theta)^2, (\pi l - \theta)^2\right]$, $l \in \mathbb{N}$ *and the eigenvalues* $\lambda_l = (\pi l)^2$.

So, in this case the spectrum has the *band-gap structure* which is typical for periodic problems. An analogue of Theorem 5.7 can be proved for regular trees for which the sequences $t_{k+1} - t_k$ and b_k are not necessarily constant, as for Γ_b, but periodic.

Suppose now that for a regular tree all the branching numbers are equal, $b_1 = b_2 = \cdots = b$, but the edge lengths $l_k = t_k - t_{k-1}$ are identically distributed random variables.

More precisely, let $[L_1, L_2]$ be a finite segment, $L_1 > 0$. Suppose that μ is a Borelian probability measure on $[L_1, L_2]$. Denote by μ^∞ the product of infinitely many copies of μ; this is a measure on the space of all sequences $\{l_k\}_{k\in\mathbb{N}}$ taking their values in $[L_1, L_2]$.

Theorem 5.8. *Let the measure μ be absolute continuous. Suppose that for each $k = 1, 2, \dots$ the lengths l_k are independent random variables with distribution μ. Then almost surely with respect to the measure μ^∞, the spectrum of the operator $-\boldsymbol{\Delta}$ on Γ contains no absolute continuous component.*

The proof, which we do not present in this paper, was obtained in cooperation with G. Berkolaiko, K. Naimark, and U. Smilansky. Its starting point is the equality (11). Then the spectrum of each operator $\mathcal{A}_k$ is analyzed with the help of Fürstenberg's Theorem on the product of random matrices.

Later the author had an opportunity to discuss this result with I. Goldsheid. Here is the information provided by him.

1. The result (for the components $\mathcal{A}_k$) was known to him and to S. Molchanov before.
2. Moreover, the spectrum $\sigma(\mathcal{A}_k)$ is almost surely pure point and the eigenfunctions exponentially decay as $t \to \infty$ (the property which is called Anderson localization).

5.3. Operators $-\boldsymbol{\Delta} + V$ with growing potential

Theorem 5.9. *Let Γ be a regular tree and $R(\Gamma) = \infty$. Denote by Ψ the counting function for the sequence $\{t_k\}$,*

$$\Psi(\lambda) = \#\{k : t_k < \lambda\}, \qquad \lambda > 0.$$

Let $V(t)$ be a non-negative, strictly monotonically increasing, unbounded continuous function on $\mathbb{R}_+$. Let Q stand for its inverse. Suppose that the functions Q and $\Psi \circ Q$ satisfy the Δ_2-condition

$$Q(2\lambda) \le CQ(\lambda), \qquad \lambda \ge \lambda_0; \tag{19}$$

$$\Psi(Q(2\lambda)) \le C\Psi(Q(\lambda)), \qquad \lambda \ge \lambda_0 \tag{20}$$

and that

$$\Psi(t) = o\big(t\sqrt{V(t)}\big), \qquad t \to \infty. \tag{21}$$

Then for the counting function $\widetilde{N}(\lambda; \mathbf{A}_V)$, cf. (12), *the asymptotic formula is valid:*

$$\pi \widetilde{N}(\lambda; \mathbf{A}_V) = (1 + o(1)) \sum_{k=0}^{\infty} \int_{t_k}^{\infty} (\lambda - V(t))_+^{1/2} dt, \qquad \lambda \to \infty. \tag{22}$$

The asymptotic formula (22) looks quite natural. The condition (19) is standard for this class of problems. The other two conditions are rather restrictive and we do not know whether they are sharp. Note that the condition (20) is automatically satisfied if we suppose that the function Ψ itself satisfies the Δ_2-condition. Note also that for $t_k = k^r$, $r > 0$ and $V(t) = t^\gamma$, $\gamma > 0$ the assumption (21) reduces to $r^{-1} < 1 + \gamma/2$.

6. Proofs

6.1. Proof of Theorem 5.2

It is enough to show that for any $r > 0$ the point $\lambda = r^2$ belongs to $\sigma(-\mathbf{\Delta})$. For this purpose we fix a non-negative function $\varphi \in C_0^\infty(-1,1)$ such that $\varphi(t) = 1$ on $(-1/2, 1/2)$. Further, choose an edge $e \in \mathcal{E}(\Gamma)$. In an appropriate coordinate system, e can be identified with the interval $(-l, l)$ where $l = |e|/2$. The function f on Γ,

$$f(t) = \varphi(t/l)\sin rt \text{ on } e, \qquad f(t) = 0 \text{ otherwise},$$

belongs to $\operatorname{Dom}(\mathbf{\Delta})$. An elementary calculation shows that

$$\|\mathbf{\Delta} f + r^2 f\| \leq \varepsilon(l)\|f\|, \qquad \varepsilon(l) \to 0 \text{ as } l \to \infty.$$

Choosing a sequence of edges e such that $|e| \to \infty$, we obtain a Weyl sequence for the operator $-\mathbf{\Delta}$ and the point $\lambda = r^2$. This implies that $\lambda \in \sigma(\mathbf{\Delta})$.

6.2. Auxiliary material

We shall use the variational techniques, in the spirit of the book [3]. We present the material we need in the form, convenient for the applications to the operators $\mathcal{A}_k$.

Let $w(t)$ be a monotonically growing function on a finite interval $[a, b)$. In our applications we shall take $[a, b) = [t_k, R)$ and $w(t) = g_k(t)$, which explains the nature of our assumptions about the function w. We suppose that $w(t) \geq 1$ and that the points t_k of discontinuity of w may accumulate at the point b only. Consider the Hilbert space $\mathcal{H}^{1,\bullet}\big((a,b), w\big)$ whose elements are the functions u on $[a, b)$, such that $u \in \mathsf{H}^1(a, b-\epsilon)$ for any $\epsilon > 0$, $u(a) = 0$, and

$$\|u\|^2_{\mathcal{H}^{1,\bullet}((a,b)),w)} := \int_a^b |u'(t)|^2 w(t)dt < \infty,$$

cf. (7). We write $\mathcal{H}^{1,\bullet}(a,b)$ instead of $\mathcal{H}^{1,\bullet}\big((a,b),1\big)$. The weighted Sobolev space with the weight w is defined as

$$\mathsf{H}^{1,\bullet}\big((a,b),w\big) = \mathcal{H}^{1,\bullet}\big((a,b),w\big) \cap \mathsf{L}^2\big((a,b),w\big).$$

Let us change the variables, taking

$$s = s(t) = \int_a^t \frac{d\tau}{w(\tau)}. \tag{23}$$

The variable s runs over the interval $[0, L)$ where

$$L = \int_a^b \frac{d\tau}{w(\tau)}.$$

Since $w(t) \geq 1$, we have $L \leq b - a < \infty$. Below $t(s)$ stands for the function on $[0, L)$, inverse to $s(t)$. The derivative $t'(s) = w(t(s))$ exists everywhere, except for the points $s_k = s(t_k)$.

Let $y(s) = u(t(s))$, then

$$\|u\|^2_{\mathcal{H}^{1,\bullet}((a,b),w)} = \int_0^L |y'(s)|^2 ds \tag{24}$$

and

$$\|u\|^2_{\mathsf{L}^2((a,b),w)} = \int_0^L W(s)|y(s)|^2 ds, \qquad W(s) = w^2(t(s)).$$

The function $W(s)$ is monotone, and

$$\int_0^L W(s)ds = \int_0^L w(t(s))t'(s)ds = \int_a^b w(t)dt; \tag{25}$$

$$\int_0^L \sqrt{W(s)}ds = \int_0^L t'(s)ds = b - a. \tag{26}$$

In the course of the proofs of Theorems 5.3 and 5.5 we make use of the following result. Its most important part (i) was obtained in [1], see Theorem 3.1 and, especially, Remark 3.1 there. See also an exposition in [2], Corollary 6.3. The part (ii) is new and we present it with proof.

Theorem 6.1.

(i) *Let $L \le \infty$ and let $W \in \mathsf{L}^{1/2}(0, L)$ be a monotone, non-negative function. Then the following inequality holds*

$$\int_0^L W(s)|y(s)|^2 ds \le C(W) \int_0^L |y'(s)|^2 ds, \qquad y \in \mathcal{H}^{1,\bullet}(0, L), \tag{27}$$

and therefore the quadratic form in the left-hand side generates in $\mathcal{H}^{1,\bullet}(0, L)$ a bounded self-adjoint operator, say T_W. Moreover, the operator T_W is compact and for its eigenvalues $\mu_j(T_W)$ the following estimate holds, with a constant factor which does not depend on L and on W:

$$\#\{j : \mu_j(T_W) > \lambda^{-1}\} \le C\sqrt{\lambda} \int_0^L \sqrt{W(s)}ds, \qquad \lambda > 0. \tag{28}$$

Also, the asymptotic formula is valid:

$$\#\{j : \mu_j(T_W) > \lambda^{-1}\} = \frac{\sqrt{\lambda}}{\pi} \int_0^L \sqrt{W(s)}ds + o(\sqrt{\lambda}), \qquad \lambda \to \infty. \tag{29}$$

(ii) *Suppose in addition that the function W satisfies the estimate*

$$W(s) \le C(L - s)^{-r} \tag{30}$$

with some $r \in (0, 2)$. Then the following uniform in λ remainder estimate in the asymptotic formula (29) is satisfied:

$$\left|\#\{j : \mu_j(T_W) > \lambda^{-1}\} - \frac{\sqrt{\lambda}}{\pi} \int_0^L \sqrt{W(s)}ds\right| \le C(L)(\lambda^{1/(4-r)} + 1), \qquad \lambda > 0. \tag{31}$$

Proof of (ii). For definiteness, we assume the function W to be increasing.

Suppose at first that W is bounded. Then we use the standard variational reasoning: divide $[0, L]$ into n equal parts, on each part (s_k, s_{k+1}) replace $W(s)$ by its inf and sup and solve the resulting eigenvalue problem under the Dirichlet or the Neumann boundary conditions. We obtain, denoting $h = L/n$:

$$\sum_{k=0}^{n-1}\left[\frac{h}{\pi}\sqrt{\lambda W(s_k+)}\right] \le \#\{j : \mu_j(T_W) > \lambda^{-1}\} \le n + \sum_{k=1}^{n}\left[\frac{h}{\pi}\sqrt{\lambda W(s_k-)}\right].$$

Roughening this inequality, we obtain:

$$-n + \frac{h\sqrt{\lambda}}{\pi}\sum_{k=0}^{n-1}\sqrt{W(s_k+)} \le \#\{j : \mu_j(T_W) > \lambda^{-1}\}$$

$$\le n + \frac{h\sqrt{\lambda}}{\pi}\sum_{k=1}^{n}\sqrt{W(s_k-)}.$$

We also have, due to the monotonicity of W:

$$h\sum_{k=0}^{n-1}\sqrt{W(s_k+)} \le \int_0^L \sqrt{W(s)}ds \le h\sum_{k=1}^{n}\sqrt{W(s_k-)}.$$

This yields

$$\left|\#\{j : \mu_j(T_W) > \lambda^{-1}\} - \frac{\sqrt{\lambda}}{\pi}\int_0^L \sqrt{W(t)}dt\right| \le C(\sqrt{\lambda}\frac{L\sqrt{W(L-)}}{n} + n). \quad (32)$$

Suppose now that $W(s)$ is unbounded. Then we choose a point $S < L$, insert the condition $y(S) = 0$, apply the inequality (32) on $(0, S)$ and use the estimate (28) on (S, L). We obtain

$$\left|\pi\#\{j : \mu_j(T_W) > \lambda^{-1}\} - \sqrt{\lambda}\int_0^L \sqrt{W(s)}ds\right|$$

$$\le C\Big(\sqrt{\lambda}\Big(\frac{L\sqrt{W(S)}}{n} + \int_S^L \sqrt{W(s)}ds\Big) + n + 1\Big).$$

Now we use the inequality (30) and then minimize the right-hand side over $S \in (0, L)$. This gives

$$\left|\pi\#\{j : \mu_j(T_W) > \lambda^{-1}\} - \sqrt{\lambda}\int_0^L \sqrt{W(s)}ds\right| \le C\Big(\sqrt{\lambda}(L/n)^{1-(r/2)} + n + 1\Big).$$

We arrive at (31), taking here $n = \left[\lambda^{1/(4-r)}\right] + 1$.

6.3. Proof of Theorem 5.3

(i) Fix $k = 0, 1, \dots$ and apply the construction in the beginning of Subsection 6.2 to the interval $[a, b) = [t_k, R)$ and the weight function $w(t) = g_k(t)$. Let $s_k(t)$ stands for the corresponding function (23) and $t_k(s)$ stands for its inverse. The assumptions of Theorem 6.1 are satisfied for the function $W_k(s) = g_k^2(t_k(s))$. According to (26), the relations (28) and (29) turn into

$$\#\{j : \mu_j(T_{W_k}) > \lambda^{-1}\} \le C\sqrt{\lambda}(R - t_k), \qquad \lambda > 0, \tag{33}$$

where the constant C does not depend on k, and

$$\#\{j : \mu_j(T_{W_k}) > \lambda^{-1}\} = \frac{\sqrt{\lambda}}{\pi}(R - t_k) + o(\sqrt{\lambda}), \qquad \lambda \to \infty. \tag{34}$$

Substituting $u(t) = y(s_k(t))$ in (27) (with $L = L_k = \int_{t_k}^{R}(g_k(\tau))^{-1}d\tau$ and $W = W_k$), we come to the inequality

$$\int_{t_k}^{R} |u(t)|^2 g_k(t)dt \le C a_k[u], \qquad u \in \mathcal{H}^{1,\bullet}((t_k, R), g_k)$$

where $a_k[u]$ is the quadratic form defined in (7). This shows that the operator $\mathcal{A}_k$ has bounded inverse and that the spectrum of $\mathcal{A}_k^{-1}$ coincides with the one of the operator T_{W_k}. The inequality (33) turns into

$$N(\lambda; \mathcal{A}_k) \le C\sqrt{\lambda}(R - t_k), \qquad \lambda > 0, k = 0, 1, \dots \tag{35}$$

and the asymptotic formula (34) turns into the formula (13).

(ii) By (11), we have

$$\lambda^{-1/2}N(\lambda; -\mathbf{\Delta}) = \Big(\lambda^{-1/2}N(\lambda; \mathcal{A}_0)\Big) + \sum_{k=1}^{\infty} b_1 \dots b_{k-1}(b_k - 1)\Big(\lambda^{-1/2}N(\lambda; \mathcal{A}_k)\Big).$$

As $\lambda \to \infty$, each term in big parentheses tends to $\pi^{-1}(R - t_k)$. Besides, by (35) the series is dominated by

$$C\big(R + \sum_{k=1}^{\infty} b_1 \dots b_{k-1}(b_k - 1)(R - t_k)\big) = C|\Gamma|.$$

Now (14) follows from the Lebesgue Theorem on the dominated convergence.

(iii) The proof of (15) is the same and we skip it.

Remark. It follows from (24) that for any $u \in \mathcal{H}^{1,\bullet}((t_k, R), g_k)$ its image y has a finite limit at $s = L_k$. Therefore, the same is true for the function u at the point $t = R$. The equalities (25) and (5) imply that necessarily $u(R-) = 0$ for any function $u \in \mathrm{Dom}(a_k)$, provided $|\Gamma| = \infty$. If $|\Gamma| < \infty$, then various boundary conditions at $t = R$ for functions are possible. This is consistent with the result of

[5], Theorem 5.2 where the boundary value problems for the differential equations on regular trees were studied from a different point of view.

6.4. Proof of Theorem 5.5

For the tree $\Gamma_{q,b}$ we have for $k = 0, 1, \dots$, cf. (6) and (7):

$$\|u\|^2_{L^2((t_k,R);g_k)} = \int_{1-q^k}^{1} |u(t)|^2 g_k(t)dt \qquad \text{and} \qquad a_k[u] = \int_{1-q^k}^{1} |u'(t)|^2 g_k(t)dt.$$

Substituting $t = 1 - q^k(1-s)$, $u(t) = v(s)$, and taking (4) into account, we obtain:

$$\|u\|^2_{L^2((t_k,R),g_k)} = q^{-k}b^k\|v\|^2_{L^2((0,R),g_\Gamma)}; \qquad a_k[u] = q^k b^k a_0[v].$$

This implies that for any k the operator $\mathcal{A}_k$ is unitarily equivalent to $q^{-2k}\mathcal{A}_0$ and therefore,

$$N(\lambda; \mathcal{A}_k) = N(\lambda q^{2k}; \mathcal{A}_0), \qquad \lambda > 0, \ k = 1, 2, \dots \tag{36}$$

This property of self-similarity is the key observation which allows us to handle the problem.

It follows from Theorem 4.3 and the formula (36) that

$$N(\lambda; -\boldsymbol{\Delta}) = N(\lambda; \mathcal{A}_0) + (1 - b^{-1}) \sum_{k=1}^{\infty} b^k N(\lambda q^{2k}; \mathcal{A}_0). \tag{37}$$

Now we are in a position to complete the proof of the statement (i). The function $N(\lambda; \mathcal{A}_0)$ satisfies the inequality

$$N(\lambda; \mathcal{A}_0) \le C\sqrt{\lambda}, \qquad \lambda > 0, \tag{38}$$

cf. (35). Denote $\mu = \ln \lambda$, $\eta = -2\ln q$, $\Phi(\mu) = \lambda^{-\beta/2}N(\lambda; -\boldsymbol{\Delta})$ and $\varphi(\mu) = \lambda^{-\beta/2}N(\lambda; \mathcal{A}_0)$, then the equality (37) turns into

$$\Phi(\mu) = \varphi(\mu) + (1 - b^{-1}) \sum_{k=1}^{\infty} \varphi(\mu - k\eta).$$

This yields

$$\Phi(\mu) - \Phi(\mu - \eta) = \varphi(\mu) - b^{-1}\varphi(\mu - \eta). \tag{39}$$

This is a particular case of the *Renewal Equation*, well known in probability. The function in the right-hand side of (39) is zero at $-\infty$ (since $N(\lambda; \mathcal{A}_0) = 0$ for small $\lambda > 0$) and exponentially decays at $+\infty$ (since $N(\lambda; \mathcal{A}_0)$ satisfies (38) and $\beta > 1$). Therefore, the Renewal Theorem applies, see e.g. [8], Chapter XI.1, or a modern exposition in [9]. The equation (39) involves the single shift (by η), hence this is the so-called lattice case. According to the Renewal Theorem, there exists an η-periodic function $\psi(\mu)$ which is bounded and bounded away from zero, such that

$$\Phi(\mu) = \psi(\mu) + o(1), \qquad \mu \to \infty.$$

This immediately leads to (16).

(ii) As in the proof of Theorem 5.3, we use the scheme presented in the beginning of Subsection 6.2. For the tree $\Gamma = \Gamma_{b^{-1},b}$ and $k = 0$ we have $[a, b) = [0, 1)$ and

$$w(t) = g_\Gamma(t) = b^j \text{ for } 1 - b^{-j} < t \leq 1 - b^{-j-1},\ j = 0, 1, \dots.$$

It follows that

$$w(t) \leq C(1-t)^{-1}. \tag{40}$$

Besides,

$$L = \int_0^1 \frac{dt}{w(t)} = \sum_{j=0}^{\infty} \frac{b^{-j} - b^{-j-1}}{b^j} = \frac{b}{b+1}.$$

For the function $s = s(t)$ defined by (23), we have

$$s(t_k) = \sum_{j=0}^{k-1} \frac{b^{-j} - b^{-j-1}}{b^j} = L - \frac{b^{-2k}}{b+1},$$

or

$$L - s(t_k) = \frac{(1-t_k)^2}{b+1}.$$

Since $s(t)$ is monotone, this implies

$$L - s(t) \geq c(1-t)^2,\ c > 0, \qquad t \in (0,1)$$

and therefore, $1 - t(s) \leq c^{-1/2}(L-s)^{1/2}$. It follows from here and from (40) that

$$W(s) = w^2(t(s)) \leq C(L-s)^{-1},$$

so that the inequality (30) is satisfied with $r = 1$. Correspondingly, the estimate (31) takes the form

$$\left|\pi N(\lambda; \mathcal{A}_0) - \lambda^{1/2}\right| \leq C(\lambda^{1/3} + 1), \qquad \lambda > 0. \tag{41}$$

Let us return to the equality (37) (where now $q = b^{-1}$). According to the estimate (35), with $R = 1$ and $k = 0$, we find that

$$N(\lambda; \mathcal{A}_0) = 0 \qquad \text{if } C^2\lambda < 1.$$

Therefore, summation in (37) is actually taken over such k that $b^{2k} \leq C^2\lambda$. Using for all such k the estimate (41), we obtain:

$$\left|\pi N(\lambda; \mathcal{A}_0) - \Big(1 + (1-q)\#\{k > 1 :\ b^{2k} \leq C^2\lambda\}\Big)\lambda^{1/2}\right| \\ \leq C \sum_{k: b^{2k} \leq C^2\lambda} (\lambda^{1/3} b^{k/3} + 1).$$

The sum in the right-hand side is of order $O(\lambda^{1/2})$, and the factor in front of $\lambda^{1/2}$ in the left-hand side differs from $\frac{(1-q)\ln\lambda}{2\ln b}$ by $O(1)$. This completes the proof of (17).

6.5. Proof of Theorem 5.9

We split the proof into several steps.

1. Denote

$$J(\lambda; V) = \int_0^\infty (\lambda - V(t))_+^{1/2} dt = \int_0^{Q(\lambda)} (\lambda - V(t))^{1/2} dt.$$

Under the assumption (19) one has

$$J(\lambda; q) \asymp Q(\lambda)\sqrt{\lambda}, \qquad \lambda \to \infty$$

where the symbol $\asymp$ means a two-sided estimate. Indeed, evidently $J(\lambda; V) \leq Q(\lambda)\sqrt{\lambda}$, and for $\lambda \geq 2\lambda_0$ one has

$$J(\lambda; V) = \frac{1}{2} \int_{V(0)}^{\lambda} \frac{Q(s)ds}{(\lambda - s)^{1/2}} \geq \frac{Q(\lambda/2)}{2} \int_{\lambda/2}^{\lambda} \frac{ds}{(\lambda - s)^{1/2}} \geq cQ(\lambda)\sqrt{\lambda}.$$

Later we shall need also the inequality

$$\int_r^\infty (\lambda - V(t))_+^{1/2} dt \geq cQ(\lambda)\sqrt{\lambda}, \qquad r \leq Q(\lambda/2),\ \lambda \geq 2\lambda_0. \tag{42}$$

Its proof, and also the value of c, are the same as in the preceding inequality.

2. Consider the Schrödinger operator $K_V y = -y'' + Vy$, $u(y) = 0$ in $\mathsf{L}^2(\mathbb{R}_+)$. Fix $\hat{\lambda} > V(0)$ and compare the values of $N(\hat{\lambda}; K_V)$ and $\#\{j : \mu_j(T_W) > 1\}$ for the operator T_W introduced in Theorem 6.1, with $L = Q(\hat{\lambda})$ and $W(s) = \hat{\lambda} - V(s)$. It follows from the decoupling principle and from the Birman-Schwinger principle that these two numbers differ no more than by 2. For estimating the number $\#\{j : \mu_j(T_W) > 1\}$, we use the inequality (32) (with $\lambda = 1$). The only difference with (32) is that now the function $W(s)$ is decreasing, and for this reason the term $W(L-)$ in the right-hand side must be replaced by $W(0)$. As a result, we obtain (replacing in the result $\hat{\lambda}$ by λ):

$$\begin{aligned} \Big|\pi N(\lambda; K_V) - J(\lambda; V)\Big| &\leq C\Big(\frac{Q(\lambda)\sqrt{\lambda - V(0)}}{n} + n + 1\Big) \\ &\leq C\Big(\frac{Q(\lambda)\sqrt{\lambda}}{n} + n + 1\Big). \end{aligned}$$

Let us stress that the factor C does not depend on λ and n. Minimizing the right-hand side of the last inequality over n, we come to the estimate

$$\Big|\pi N(\lambda; K_V) - J(\lambda; V)\Big| \leq C\Big(\big(Q(\lambda)\sqrt{\lambda}\big)^{1/2} + 1\Big) \tag{43}$$

where C is an absolute constant.

3. Consider the quadratic forms $a_{V,k}$ defined in (9). The corresponding operators $\mathcal{A}_{V,k}$ act in the weighted spaces $\mathsf{L}^2\big((t_k, \infty), g_k\big)$ (recall that in our case $R = \infty$). For us it is more convenient to deal with the operators acting in the "usual" L^2.

For this purpose we make a substitution which we describe for $k = 0$. The changes needed in the case $k > 0$, are evident.

Denote $y(t) = \sqrt{g_\Gamma(t)}u(t)$, then also $y'(t) = \sqrt{g_\Gamma(t)}u'(t)$, $t \neq t_k$, $k = 1, 2, \dots$, since g_Γ is a step function. Evidently,

$$\int_{\mathbb{R}_+} |u(t)|^2 g_\Gamma(t)dt = \int_{\mathbb{R}_+} |y(t)|^2 dt$$

and

$$a_{V,0}[u] = \widetilde{a}_{V,0}[y] := \sum_{k=0}^{\infty} \int_{t_k}^{t_{k+1}} \big(|y'(t)|^2 + V(t)|y(t)|^2\big)dt. \tag{44}$$

The domain $\mathrm{Dom}(a_{V,0})$ consists of all functions $y(t)$, such that $y \restriction (t_k, t_{k+1}) \in \mathsf{H}^1(t_k, t_{k+1})$ for each k, the sum in the last side of (44) is finite, $y(0) = 0$ and the matching conditions at the points t_k are fulfilled:

$$y(t_k+) = \sqrt{b_k}\, y(t_k-), \qquad k = 1, 2, \dots$$

It is not difficult to derive from (43) a similar estimate for the operator $\mathcal{A}_{V,0}$. Indeed, again the problem reduces to the interval $(0, Q(\lambda))$. The linear spaces $\mathrm{Dom}(a_{V,0})$ and $\mathrm{Quad}(K_V)$, restricted to the set of functions supported by this interval, differ by a subspace of dimension $2\Psi(Q(\lambda))$. Therefore, (43) implies

$$\Big|\pi N(\lambda; \mathcal{A}_{V,0}) - J(\lambda; V)\Big| \leq C\Big(\big(Q(\lambda)\sqrt{\lambda}\big)^{1/2} + \Psi(Q(\lambda))\Big).$$

The term 1 appearing in (43) can be dropped, since $\Psi(Q(\lambda)) \geq 1$ for any $\lambda > V(0)$.

Quite similarly,

$$\left|\pi N(\lambda; \mathcal{A}_{V,k}) - \int_{t_k}^{\infty} (\lambda - V(t))_+^{1/2}\right| \leq C\Big(\big(Q(\lambda)\sqrt{\lambda}\big)^{1/2} + \Psi(Q(\lambda))\Big).$$

4. Now we are in a position to complete the proof. We have

$$\widetilde{N}(\lambda; \mathbf{A}_V) = \sum_{k=0}^{\infty} N(\lambda; \mathcal{A}_{V,k}) = \sum_{k:0 \leq t_k < Q(\lambda)} N(\lambda; \mathcal{A}_{V,k}).$$

Therefore,

$$\left|\widetilde{N}(\lambda; \mathbf{A}_V) - \sum_{k=0}^{\infty} \int_{t_k}^{\infty} (\lambda - V(t))_+^{1/2} dt\right|$$
$$\leq C\Psi(Q(\lambda))\Big(\big(Q(\lambda)\sqrt{\lambda}\big)^{1/2} + \Psi(Q(\lambda))\Big). \tag{45}$$

Our next task is to estimate from below the sum appearing in the left-hand side of (45). We derive from (42):

$$\sum_{k=0}^{\infty}\int_{t_k}^{\infty}(\lambda - V(t))_+^{1/2}dt \geq \sum_{k:t_k\leq Q(\lambda/2)}\int_{t_k}^{\infty}(\lambda - V(t))_+^{1/2}dt$$
$$\geq c\Psi(Q(\lambda/2))Q(\lambda)\sqrt{\lambda} \geq c'\Psi(Q(\lambda))Q(\lambda)\sqrt{\lambda}.$$

The latter inequality is implied by the Δ_2-condition (20).

It follows from the assumption (21) that

$$\Psi(Q(\lambda))\Big(\big(Q(\lambda)\sqrt{\lambda}\big)^{1/2} + \Psi(Q(\lambda))\Big) = o\Big(\Psi(Q(\lambda))Q(\lambda)\sqrt{\lambda}\Big).$$

The desired asymptotic formula (22) immediately follows.

7. Regular trees without boundary

In conclusion, let us discuss the case when the tree Γ has no boundary.

Let Γ be a general metric tree. Choose a vertex $o \in \mathcal{E}(\Gamma)$ and suppose that there are d edges of Γ adjacent to o. Then Γ can be split into d rooted subtrees $\Gamma^1, \dots, \Gamma^d$ having the common root o. We say that the tree Γ is regular if and only if all the subtrees Γ^j are regular in the sense of Definition 2.1 and the corresponding sequences $\{t_k\}$ and $\{b_k\}$ are the same for all $j = 1, \dots, d$. Note that this definition is not invariant with respect to the choice of the vertex o.

Suppose now that all the subtrees $\{\Gamma^j\}$ are homogeneous, $\Gamma^1 = \ldots = \Gamma^d = \Gamma_b$ and $d = b + 1$. Then we say that the tree Γ is homogeneous. Unlike the case of arbitrary regular trees, this definition is invariant with respect to the choice of o.

The definitions of the Laplacian and of the Schrödinger operator extend to the trees without boundary in a natural way. The only difference is that now we have no boundary condition at o. Instead, the functions from the quadratic domain of the operator are required to be continuous at o.

Replacement of this continuity condition by the Dirichlet boundary condition $u(o) = 0$ means the passage to a subspace of codimension 1 of $\mathrm{Quad}(\mathbf{A}_V)$. Therefore, the character of the spectrum is not affected, and moreover, the eigenvalue distribution function $N(\lambda; \mathbf{A}_V)$ can change no more than by one. The new operator splits into the orthogonal sum of d copies of the operator studied in the main part of this paper. This allows one to reformulate immediately all the results for this new situation. It is unnecessary to present their precise formulations.

Note that the papers [4] and [5] deal with the operators on trees without boundary.

8. Acknowledgements

The work on this paper was supported by the Minerva center for non-linear physics and by the Israel Science Foundation.
I use this opportunity to express my deep thanks to G. Berkolaiko, K. Naimark and U. Smilansky for their participation in the proof of Theorem 10, and to I. Goldsheid for the useful discussion of this result. My special gratitude goes to U. Smilansky for his enthusiastic support of the study of differential operators on metric graphs.

References

[1] M.Sh. Birman and V.V. Borzov, *On the asymptotics of the discrete spectrum for certain singular differential operators*, Probl. Mat. Fiz. **5** (1971), 24–38; English transl. in Topics in Math. Phys. **5** (1972).

[2] M.Sh. Birman, A. Laptev, and M. Solomyak, *On the eigenvalue behaviour for a class of differential operators on semiaxis*, Math. Nachr. **195** (1998), 17–46.

[3] M.Sh. Birman and M. Solomyak, *Quantitative analysis in Sobolev imbedding theorems and applications to spectral theory*, Tenth Mathem. School, Izd. Inst. Mat. Akad. Nauk Ukrain 5–189, SSSR, Kiev, 1974 (Russian); English transl. in Amer. Math. Soc. Translations, (2), **114** (1980), 1–132.

[4] R. Carlson, *Hill's equation for a homogeneous tree*, Electron. J. Differential Equations 1997 (23), 30 pp. (electronic).

[5] R. Carlson, *Nonclassical Sturm-Liouville Problems and Schrödinger operators on radial trees*, Electron. J. Differential Equations 2000 (71), 24 pp. (electronic).

[6] W.D. Evans and D.J. Harris, *Fractals, trees and the Neumann Laplacian*, Math. Ann. **296** (1993), 493–527.

[7] W.D. Evans, D.J. Harris, and J. Lang, *The approximation numbers of Hardy-type operators on trees*, Proc. London Math. Soc. (3) **83** (2001), 390–418.

[8] W. Feller, *An introduction to the probability theory and its applications, Vol. II*, John Wiley and Sons, Inc., New York-London-Sidney-Toronto, 1971.

[9] M. Levitin and D. Vassiliev, *Spectral asymptotics, renewal theorem, and the Berry conjecture for a class of fractals*, Proc. London Math. Soc. (3) **72** (1996), 188–214.

[10] K. Naimark and M. Solomyak, *Eigenvalue estimates for the weighted Laplacian on metric trees*, Proc. London Math. Soc. (3) **80** (2000), 690–724.

[11] K. Naimark and M. Solomyak, *Geometry of the Sobolev spaces on the regular trees and Hardy's inequalities*, Russian Journal of Mathematical Physics, **8**, No 3 (2001).

[12] A.V. Sobolev and M. Solomyak, *Schrödinger operator on homogeneous metric trees: spectrum in gaps*, Preprint math.SP/0109016, 2001; Rev. Math. Phys. (submitted).

Michael Solomyak
Department of Mathematics
The Weizmann Institute of Science
Rehovot 76100, Israel
E-mail address: `solom@wisdom.weizmann.ac.il`

D. Haroske, T. Runst, H.-J. Schmeisser (eds.): Function Spaces, Differential Operators and Nonlinear Analysis. The Hans Triebel Anniversary Volume.

Inverse Boundary Problems in Two Dimensions

Gunther Uhlmann

Dedicated to Prof. Hans Triebel on the occasion of his 65th birthday

Abstract. In this paper we survey some of the recent progress on inverse boundary problems in two dimensions. The common theme is the use of inverse scattering for a $\overline{\partial}\partial$ type system in two dimensions.

1. Introduction

Let $\Omega \subseteq \mathbb{R}^n$ be a bounded domain with smooth boundary (many of the results we will describe are valid for domains with Lipschitz boundaries). The electrical conductivity of Ω is represented by a bounded and positive function $\gamma(x)$. In the absence of sinks or sources of current the equation for the potential is given by

$$\operatorname{div}(\gamma \nabla u) = 0 \text{ in } \Omega \tag{1.1}$$

since, by Ohm's law, $\gamma \nabla u$ represents the current flux.

Given a potential $f \in H^{\frac{1}{2}}(\partial\Omega)$ on the boundary the induced potential $u \in H^1(\Omega)$ solves the Dirichlet problem

$$\begin{aligned} \operatorname{div}(\gamma \nabla u) &= 0 \text{ in } \Omega, \\ u\big|_{\partial\Omega} &= f. \end{aligned} \tag{1.2}$$

The Dirichlet to Neumann (DN) map, or voltage to current map, is given by

$$\Lambda_\gamma(f) = \left(\gamma \frac{\partial u}{\partial \nu}\right)\Big|_{\partial\Omega} \tag{1.3}$$

where ν denotes the unit outer normal to $\partial\Omega$.

The inverse problem is to determine γ knowing Λ_γ. This problem is known also as *Electrical Impedance Tomography*. It arose originally in geophysical prospection [ZK94]. More recently it has been proposed as a valuable diagnostic tool in medicine [CIN99]. The mathematical formulation and the first mathematical results are due to Calderón [Cal80]. The survey paper [Uhl99] summarizes many of the developments in inverse boundary boundary problems up to 1997 which were pioneered by Calderón's contribution. In this paper we will discuss the progress in the last few years for the two-dimensional problem based on the $\overline{\partial}\partial$ method which was originally developed by Beals and Coifman [BC88] in order to solve, via inverse scattering, the nonlinear Davey-Stewartson system.

The main breakthrough in EIT in the two-dimensional case is due to A. Nachman who proved in [Nac96] that one can uniquely determine conductivities in $W^{2,p}(\Omega)$ for some $p > 1$. Moreover he proposed a method to reconstruct the conductivity from the DN map. Significant progress has been made in developing a numerical algorithm based on Nachman's reconstruction method [SMI00].

We describe an alternative approach to the identifiability result of Nachman developed in [BU97], approach that allows for less regular conductivities. Section 2 outlines this approach and an extension to complex conductivities. In Section 3 we make a further analysis of this method for the case of $C^{1+\epsilon}(\overline{\Omega})$ conductivities which is needed for Section 4. In Section 4 we discuss stability estimates proven in [BBR01] and a reconstruction method proposed in [KT01], both of which consider $C^{1+\epsilon}(\overline{\Omega})$ conductivities. In Section 5 we outline the use of the $\overline{\partial}\partial$ method in studying the DN map (or more generally the set of Cauchy data) for the Pauli Hamiltonian [KU02] and in Section 6 for any first order perturbation of the Laplacian with no zeroth order terms [CY98]. Finally in Section 7 we mention some open problems.

2. The $\overline{\partial}\partial$ system

In this section we describe an extension of Nachman's result to $W^{1,p}(\Omega)$ conductivities with $p > 2$, result due to Brown and the author [BU97]. We follow an earlier approach of Beals and Coifman [BC88] and L. Sung [Sung94a,b,c] who studied scattering for a first order system whose principal part is $\begin{pmatrix} \overline{\partial} & 0 \\ 0 & \partial \end{pmatrix}$.

The main result of [BU97] is:

Theorem 2.1. *Let $n = 2$. Let $\gamma \in W^{1,p}(\Omega)$, $p > 2$, γ strictly positive. Assume $\Lambda_{\gamma_1} = \Lambda_{\gamma_2}$. Then $\gamma_1 = \gamma_2$ in Ω.*

As mentioned earlier, the proof of Theorem 2.1 first reduces the conductivity equation to a first order system, which we will call the $\overline{\partial}\partial$ system. We define

$$q = -\frac{1}{2}\partial \log \gamma \tag{2.1}$$

and a matrix potential Q by

$$Q = \begin{pmatrix} 0 & q \\ \overline{q} & 0 \end{pmatrix}. \tag{2.2}$$

Let also D be the operator

$$D = \begin{pmatrix} \overline{\partial} & 0 \\ 0 & \partial \end{pmatrix}, \tag{2.3}$$

where $\overline{\partial} = \frac{1}{2}(\partial_{x_1} - i\partial_{x_2}), \partial = \frac{1}{2}(\partial_{x_1} + i\partial_{x_2})$.

An easy calculation shows that, if u satisfies the conductivity equation $\operatorname{div}(\gamma\nabla u) = 0$, then

$$\begin{pmatrix} v \\ w \end{pmatrix} = \gamma^{\frac{1}{2}} \begin{pmatrix} \partial u \\ \overline{\partial} u \end{pmatrix} \tag{2.4}$$

solves the system

$$D\begin{pmatrix} v \\ w \end{pmatrix} - Q\begin{pmatrix} v \\ w \end{pmatrix} = 0. \tag{2.5}$$

In [BU97] Brown and Uhlmann construct matrix solutions of (2.5) of the form

$$\psi(z,k) = m(z,k)\begin{pmatrix} e^{izk} & 0 \\ 0 & e^{-i\overline{z}k} \end{pmatrix}, \tag{2.6}$$

where $z = x_1 + ix_2$, $k \in \mathbb{C}$ with $m \to 1$ as $|z| \to \infty$ in a sense to be described below. A simple calculation shows that m from (2.6) satisfies in Ω the following equation

$$D_k m - Qm = 0, \tag{2.7}$$

where D_k is the operator

$$D_k = \begin{pmatrix} (\overline{\partial} - i\overline{k}) & 0 \\ 0 & (\partial + ik) \end{pmatrix}.$$

In order to explain the construction of m we need a few more definitions. Let

$$\Lambda_k(z) = \begin{pmatrix} e(z,k) & 0 \\ 0 & e(z,-\overline{k}) \end{pmatrix}, \quad e(z,k) = e^{i(zk+\overline{z}\overline{k})}$$

and for any matrix A, define the following operator

$$E_k A = E_k\begin{pmatrix} a_{11} & a_{12} \\ a_{21} & a_{22} \end{pmatrix} = \begin{pmatrix} a_{11} & e(z,-\overline{k})a_{12} \\ a_{21}e(z,k) & a_{22} \end{pmatrix}.$$

Notice that

$$D_k = E_k^{-1} D E_k. \tag{2.8}$$

Let D^{-1} be the operator

$$D^{-1} = \begin{pmatrix} \overline{\partial}^{-1} & 0 \\ 0 & \partial^{-1} \end{pmatrix},$$

where

$$\overline{\partial}^{-1} f(z) = \frac{1}{2\pi i}\int \frac{f(w)}{z-w} dw \wedge \overline{w}$$

and

$$\partial^{-1} f(z) = \frac{1}{2\pi i}\int \frac{f(w)}{\overline{z}-\overline{w}} dw \wedge \overline{w}.$$

We have from (2.8) that $D_k^{-1} = E_k^{-1} D^{-1} E_k$. We look for solutions of (2.7) among the solutions of the integral equation

$$(I - D_k^{-1}Q)m(z) = I, \tag{2.9}$$

where I is the 2×2 identity matrix. For a 2×2 matrix A, let A^d and A^{off} denote its diagonal respectively off-diagonal part. If

$$J = \frac{1}{2}\begin{pmatrix} -i & 0 \\ 0 & i \end{pmatrix}$$

we define the operator $\mathcal{J}$ by

$$\mathcal{J}A = [J, A] = 2JA^{off} = -2A^{off}J. \tag{2.10}$$

To end with the preliminary notation, we recall the definition of the weighted L^p space

$$L^p_\alpha(\mathbb{R}^2) = \{f \ : \ \int (1+|x|^2)^\alpha |f(x)|^p dx < \infty\}.$$

The next result gives the solvability of (2.7) in an appropriate space.

Theorem 2.2. *Let $Q \in L^p(\mathbb{R}^2)$, $p > 2$, and compactly supported. Assume that Q is a hermitian matrix. Choose r so that $\frac{1}{p} + \frac{1}{r} > \frac{1}{2}$ and then β so that $\beta r > 2$. Then the operator $(I - D_k^{-1}Q)$ is invertible in $L^r_{-\beta}$. Moreover the inverse is differentiable in k in the strong operator topology.*

Theorem 2.2 implies the existence of solutions of the form (2.6) with $m - 1 \in L^r_{-\beta}(\mathbb{R}^2)$.

We remark that the proof of 2.2 consists in showing that the integral equation (2.9) is of Fredholm type in $L^r_{-\beta}$. The fact that it has been a trivial kernel follows by showing that if $(I - D_k^{-1}Q)n(z,k) = 0$, then $n \in L^p$, for all $p > 2$, satisfies a pseudoanalytic equation in the z-variable. By the standard Liouville's theorem for pseudoanalytic equations with coefficients in L^p, $p > 2$, it follows that $n = 0$.

Next we compute $\frac{\partial}{\partial \overline{k}} m(z,k)$.

Theorem 2.3. *Let m be the solution of (2.7) with $m - 1 \in L^r_{-\beta}(\mathbb{R}^2)$. Then*

$$\frac{\partial}{\partial \overline{k}} m(z,k) - m(z,\overline{k})\Lambda_k(z)S_Q(k) = 0 \tag{2.11}$$

where the scattering data S_Q is given by ([BC88])

$$S_Q(k) = -\frac{1}{\pi}\mathcal{J}\int_{\mathbb{R}^2} E_k Q m d\mu, \tag{2.12}$$

where $d\mu$ denotes Lebesgue measure in $\mathbb{R}^2$.

A further calculation shows that

$$S_Q(k) = \frac{i}{\pi}\int_{\mathbb{R}^2}\begin{pmatrix} 0 & e(z,-\overline{k})q(z)m_{22}(z,k) \\ -e(z,k)\overline{q}(z)m_{11}(z,k) & 0 \end{pmatrix}. \tag{2.13}$$

The behavior of m in the k-variable is given by the following result:

Theorem 2.4. *Let $Q \in L^p(\mathbb{R}^2)$, $p > 2$, and compactly supported. Then there exists $R = R(Q)$ so that for all $q > \frac{2p}{p-2}$,*

$$\sup_z \|m(z,\cdot) - 1\|_{L^q\{k;|k|>R\}} \leq C\|Q\|^2_{L^p}$$

where the constants depend on p, q and the diameter of the support of Q.

Outline of proof of Theorem 2.1 We know [Ale88, KV84, Nac88, SyU88] that if $\gamma_i \in W^{1,p}(\Omega)$ and $\Lambda_{\gamma_1} = \Lambda_{\gamma_2}$, then $\partial^\alpha \gamma_1\Big|_{\partial\Omega} = \partial^\alpha \gamma_2\Big|_{\partial\Omega}$ $\forall\ |\alpha| \leq 1$. Therefore we can extend $\gamma_i \in W^{1,p}(\mathbb{R}^2)$, $\gamma_1 = \gamma_2$ in $\mathbb{R}^2 \setminus \Omega$ and $\gamma_i = 1$ outside a large ball. Thus $Q_i \in L^p(\mathbb{R}^2)$, $i = 1, 2$. The proof follows the following steps.

Step 1. $\Lambda_{\gamma_1} = \Lambda_{\gamma_2} \Rightarrow S_{Q_1} = S_{Q_2} := S$. With these extensions, we observe that for each j the scattering data $S_{Q_j}(k)$, $j = 1, 2$, has the representation

$$\begin{aligned}S_{Q_j}(k) &= -2J \int_{\mathbf{R}^2} \begin{pmatrix} 0 & \bar\partial \psi_j^{12} e^{-iz\bar k} \\ \partial \psi_j^{21} e^{i\bar z k} & 0 \end{pmatrix} d\mu(z) \\ &= -2J \left[\int_{\mathbf{R}^2\setminus\Omega} \begin{pmatrix} 0 & \bar\partial \psi_j^{12} e^{-iz\bar k} \\ \partial \psi_j^{21} e^{i\bar z k} & 0 \end{pmatrix} d\mu(z) + \int_{\partial\Omega} \begin{pmatrix} 0 & \bar\nu \psi_j^{12} e^{-iz\bar k} \\ \nu \psi_j^{21} e^{i\bar z k} & 0 \end{pmatrix} d\mu(z) \right].\end{aligned}$$

The formula for S_{Q_j} uses the complexified normal to the boundary

$$\nu(z) = \nu_1(z) + i\nu_2(z), \quad \overline{\nu}(z) = \nu_1(z) - i\nu_2(z) \tag{2.14}$$

with $(\nu_1(z), \nu_2(z))$ the unit outer normal at $z \in \partial\Omega$.

From this expression for S_{Q_j}, $j = 1, 2$, we see that if we can show

$$\psi_1(z,k) = \psi_2(z,k) \quad \text{in } \mathbf{R}^2\backslash\bar\Omega, \tag{2.15}$$

then $S_{Q_1} = S_{Q_2}$.

The last formula follows by using a very similar argument to Lemma 2.6 in [SyU86].

Step 2. Let $\widetilde{m} = m_1 - m_2$. Using the $\overline{\partial}$-equation (2.11) and Step 1 we conclude that

$$\frac{\partial}{\partial \overline{k}} \widetilde{m}(z,k) - \widetilde{m}(z,\overline{k}) \Lambda_k(z) S(k) = 0\ . \tag{2.16}$$

With the elements of $\widetilde{m}$ we form the following four functions

$$\begin{aligned} u_\pm(k,z)) &= \widetilde{m}_{11}(z,k) \pm \overline{\widetilde{m}_{12}(z,\overline{k})} \\ v_\pm(k,z)) &= \widetilde{m}_{21}(z,k) \pm \overline{\widetilde{m}_{22}(z,\overline{k})} \end{aligned}$$

each of which lies in $L^q(\mathbb{R}^2)$ in the k-variable and satisfies, for a fixed z, a pseudoanalytic equation in the k-variable,

$$\frac{\partial}{\partial \overline{k}} w(z,k) = r(z,k)\overline{w(z,k)} \tag{2.17}$$

where $r(z,k)$ is some component of S multiplied by a complex coefficient of norm 1.

Step 3. In [BU97] it was shown that, for $Q \in L^p_c$ for some $p > 2$ with $Q^* = Q$, we have that

$$\int trS_Q S_Q^* \leq \int trQQ^*.$$

This shows that $S_Q \in L^2$. Consequently, for each fixed z we have the map $k \to r(z,k)$ is in $L^2(\mathbb{R}^2)$.

Step 4. Prove that $u_\pm = v_\pm = 0$, hence $\widetilde{m} = 0$ or $m_1 = m_2$. Then it is easy to show $Q_1 = Q_2$ and therefore $\gamma_1 = \gamma_2$.

To do this we need the following generalization of the Liouville Theorem for pseudoanalytic functions proven in [BU97].

Lemma 2.5. *Let $f \in L^2(\mathbb{R}^2)$ and $w \in L^p(\mathbb{R}^2)$ for some finite p. Assume that $we^{\overline{\partial}^{-1}f}$ is analytic. Then $w = 0$.*

Let us define

$$\begin{aligned} \widetilde{u}_\pm &= u_\pm e^{\overline{\partial}^{-1}r} \\ \widetilde{v}_\pm &= v_\pm e^{\overline{\partial}^{-1}r}. \end{aligned} \tag{2.18}$$

It is easy to check that $\widetilde{u}_\pm$ and $\widetilde{v}_\pm$ are analytic. By the lemma above we conclude that $u_\pm = v_\pm = 0$ which in turn gives $m_1 = m_2$. It is easy to show $Q_1 = Q_2$ and therefore $\gamma_1 = \gamma_2$, concluding the proof of Theorem 2.1.

The idea of the proof of Lemma 2.5 is the observation that since $r \in L^2(\mathbb{R}^2)$, $f = \overline{\partial}^{-1} r$ is in $VMO(\mathbb{R}^2)$ (the space of functions with vanishing mean oscillation) and thus is $O(\log|z|)$ as $|z| \to \infty$. Hence $e^f w \in L^{\widetilde{p}}$ for $\widetilde{p} > p$. By Liouville's theorem it follows that $e^f w = 0$. The details can be found in [BU97].

Theorem 2.1 was extended to complex conductivities with small imaginary part in [Fra00], using the $\overline{\partial}\partial$ method. Complex conductivities with small imaginary part arise naturally when considering Maxwell's equations for time harmonic waves with small frequency.

3. $C^{1+\epsilon}$ conductivities

The difficulty in obtaining stability estimates or a reconstruction procedure for conductivities in $W^{1,p}$, $p > 2$, is to prove stability estimates for $w(z,k)$ as in (2.17), in terms of the scattering data $S_Q(k) \in L^2$ or to reconstruct $w(z,k)$ in terms of $S_Q(k)$. All of the other steps are constructive and stability estimates can be obtained. If one assumes that the conductivity is more regular then S_Q is in a better space than L^2. This allows to carry out a constructive proof and give stability estimates. We will assume that the potential Q is in $C^\epsilon(\mathbf{R}^2)$ for some $\epsilon > 0$ and derive the stronger decay properties of the scattering matrix than being in L^2. More precisely we have the following result proven in [BBR01]:

Lemma 3.1. *Let $Q \in C^\epsilon(\mathbb{R}^2)$ be compactly supported. Then $S_Q \in L^r(\mathbb{R}^2)$ for any $r > 4/(2+\epsilon)$.*

The proof of the Lemma is based on some unpublished results of Brown and the author [BU96] which we outline below. We construct more explicitly the solutions of the integral equation (2.9). In particular we show that if the potential Q is in $C^\epsilon(\mathbf{R}^2)$ and is compactly supported, then the special solutions are in $C^\epsilon(\mathbf{R}^2)$ with respect to z, uniformly in k. Next we observe that the diagonal part of m is $C^{1+\epsilon}(\mathbf{R}^2)$. This regularity is essential in proving the Lemma.

We will use the following norm for the space $C^\alpha(\mathbf{R}^2)$, $0 < \alpha < 1$, $\|f\|_{C^\alpha} = \|f\|_{L^\infty} + |f|_{C^\alpha}$ where we use $|f|_{C^\alpha}$ for the semi-norm

$$|f|_{C^\alpha} = \sup_{z \neq w} \frac{|u(z) - u(w)|}{|z - w|^\alpha}$$

and then $C^{1+\alpha}(\mathbf{R}^2)$ is the collection of functions for which the norm $\|u\|_{L^\infty(\mathbf{R}^2)} + \|\nabla u\|_{C^\alpha(\mathbf{R}^2)}$ is finite.

The strategy is to show that for large k, the special solutions are given by a convergent Neumann series. For small k, we appeal to the Fredholm theory as in the proof of Theorem 2.2. This method does not give quantitative estimates without a more thorough analysis. We remark that one feature of the operator D_k^{-1} is that the diagonal part of $D_k^{-1}F$ is smoother than the off-diagonal part. This is because the off-diagonal part is multiplied by an exponential. This elementary observation will be used several times.

We begin with some elementary but useful estimates for the action of $D_k^{-1}Q$ on the spaces $C^\epsilon(\mathbf{R}^2)$. There estimates rely on the well-known fact that when $\epsilon = 1 - 2/p$ (and then $p > 2$)

$$|f|_{C^\epsilon} \leq C\|f\|_{L^p}. \tag{3.1}$$

Proposition 3.2. *Let $F \in C^\epsilon(\mathbf{R}^2)$, and suppose F is supported in a ball of radius R. Then for $0 < \epsilon < 1/2$,*

$$\|D_k^{-1}F\|_{C^\epsilon(\mathbf{R}^2)} \leq C(R,\epsilon)\|F\|_{C^\epsilon}. \tag{3.2}$$

Also,

$$\|D_k^{-1}F^{off}\|_{L^\infty} \leq C(1+|k|)^{-\epsilon}\|F^{off}\|_{C^\epsilon} \tag{3.3}$$

and

$$\|D_k^{-1}F^d\|_{C^\epsilon} \leq C\|F\|_{L^\infty}. \tag{3.4}$$

All constants depend on R and ϵ.

Proof. The third estimate (3.4) is a consequence of the sharper estimate (3.1). Thanks to (3.4), we only need to prove (3.2) for $F = F^d$. The second estimate (3.3) will be established as a byproduct of our proof of (3.2).

When $|k| \leq 1$, the estimates (3.2) and (3.3) are easy. Thus we suppose $F = F^{off}$ and $|k| \geq 1$. We first note that

$$\|E_k F\|_{C^\epsilon} \leq C(1+|k|)^\epsilon\|F\|_{C^\epsilon}. \tag{3.5}$$

Next, we choose a smooth function φ supported in the unit ball, centered at 0, and with $\int \varphi = 1$. As usual, we set $\varphi_t(z) = t^{-2}\varphi(z/t)$. We let $0 < t \leq 1$ and split

$F = F_s + F_r$ where the smooth part is $F_s = \varphi_t * F$ and the rough part $F_r = F - F_s$. We observe that $\|F_r\|_{L^\infty} \le C|F|_{C^\epsilon} t^\epsilon$, and for $|\alpha| \ge 1$, $\|\frac{\partial^\alpha}{\partial x^\alpha} F_s\|_{L^\infty} \le C|F|_{C^\epsilon} t^{\epsilon - |\alpha|}$. Since $0 < t \le 1$, we have that F_s and F_r supported in a ball of radius $R+1$. We consider $D^{-1}E_k F_r$ and observe that since F_r is compactly supported, we may use (3.5) to conclude

$$\|D^{-1}E_k F_r\|_{C^\epsilon} \;\le\; C(R,\epsilon)\|E_k F_r\|_{L^\infty} \;\le\; Ct^\epsilon.$$

For $D^{-1}E_k F_r$, we observe that $D\Lambda_k^{-1}(z) = ik \begin{pmatrix} -1 & 0 \\ 0 & +1 \end{pmatrix} \Lambda_k^{-1}(z)$ and then integrating by parts gives

$$\begin{aligned} D^{-1}E_k F_s(z) &= \frac{1}{\pi}\int \begin{pmatrix} z-w & 0 \\ 0 & \bar z - \bar w \end{pmatrix}^{-1} \begin{pmatrix} -ik & 0 \\ 0 & ik \end{pmatrix}^{-1} D\Lambda_k^{-1}(w) F_s(w)\, d\mu(w) \\ &= \begin{pmatrix} ik & 0 \\ 0 & -ik \end{pmatrix}^{-1} \Lambda_k^{-1}(z) F_s(z) \\ &\quad - \frac{1}{\pi}\int \begin{pmatrix} z-w & 0 \\ 0 & \bar z - \bar w \end{pmatrix}^{-1} \begin{pmatrix} -ik & 0 \\ 0 & ik \end{pmatrix}^{-1} \Lambda_k^{-1}(w) DF_s(w)\, d\mu(w) \\ &= I + II. \end{aligned}$$

By (3.5) and the fact that $\|F_s\|_{C^\epsilon} \le \|F\|_{C^\epsilon}$, we have that

$$\|I\|_{C^\epsilon} \le \frac{C}{|k|}(1+|k|)^\epsilon \|F\|_{C^\epsilon}.$$

Using the support restrictions on F_s, we also have that

$$\|II\|_{C^\epsilon} \;\le\; \frac{C}{|k|}\|\Lambda_k DF_s\|_{L^\infty} \;\le\; \frac{C}{|k|} t^{\epsilon-1}\|F\|_{C^\epsilon}.$$

Combining these observations gives

$$\|D^{-1}E_k F\|_{C^\epsilon} \le C\|F\|_{C^\epsilon}\left(t^\epsilon + \frac{(1+|k|)^\epsilon}{|k|} + \frac{t^{\epsilon-1}}{|k|}\right).$$

Choosing $t = 1/|k|$ gives

$$\|D^{-1}E_k F\|_{C^\epsilon} \le C(1+|k|)^{-\epsilon}\|F\|_{C^\epsilon}$$

for $0 < \epsilon \le 1/2$. The estimate (3.3) follows immediately. The estimate (3.2) follows from (3.5). □

One reason it is convenient to study the map $f \to D_k^{-1}Qf$ on the spaces C^ϵ is that these spaces form an algebra under pointwise multiplication. The following elementary inequality

$$\|QF\|_{C^\epsilon} \le C\|F\|_{C^\epsilon}\|Q\|_{C^\epsilon},\ 0 < \epsilon < 1$$

will be used several times below.

Proposition 3.3. *Suppose Q is C^ϵ, compactly supported and $Q^d = 0$. Then there is a constant C_0 so that if*

$$C_0(1+|k|)^{-2}\|Q\|^2_{C^\epsilon} = \theta < 1, \qquad \textit{then} \qquad m(\cdot,k) = \sum_{j=0}^{\infty}(D_k^{-1}Q)^j(I)$$

and the series converges in $C^\epsilon(\mathbf{R}^2)$ and $\|m(\cdot,k)\|_{C^\epsilon} \le \frac{C}{1-\theta}$.

Proof. We will show that

$$\|(D_k^{-1}Q)^{2j}(1)\|_{C^\epsilon} \le C(1+|k|)^{-\epsilon}\|Q\|^{2j}_{C^\epsilon}. \tag{3.6}$$

We note that if $F = F^d$, then

$$(D_k^{-1}Q)^2F = D^{-1}QD_k^{-1}QF.$$

Now by (3.4), we have that

$$\|D_k^{-1}QF\|_{L^\infty} \le C(1+|k|)^{-\epsilon}\|Q\|_{C^\epsilon}\|F\|_{C^\epsilon}.$$

Since Q is compactly supported, we have

$$\|D^{-1}QF\|_{C^\epsilon} \le C(R)\|F\|_{L^\infty}\|Q\|_{L^\infty}.$$

Combining these estimates gives

$$\|(D_k^{-1}Q)^2F\|_{C^\epsilon} \le C(1+|k|)^{-\epsilon}\|F\|_{C^\epsilon}\|Q\|^2_{C^\epsilon}.$$

Our claim (3.6) and hence the proposition follow from this estimate and (3.3). □

Corollary 3.4. *Suppose Q is in C^ϵ, compactly supported and $Q^d = 0$. Then there is a constant C so that*

$$\|m(\cdot,k)\|_{C^\epsilon} \le C.$$

Here, C depends on Q.

Proof. For k large, this follows from the previous proposition. To see that we also have a bound for small k, we observe that since Q is compactly supported, $k \to (I - D_k^{-1}Q)^{-1}(I)$ is a continuous map into $L^r_{-\beta}$ and thus on each compact subset $K \subset \mathbf{C}$,

$$\sup_{k\in K}\|m(\cdot,k)\|_{L^r_{-\beta}} < \infty. \tag{3.7}$$

Using that Q is bounded and compactly supported and the equation

$$m = 1 + D_k^{-1}Qm$$

we may use (3.1) to improve (3.7) and obtain

$$\sup_{k\in K}\|m(\cdot,k)\|_{C^\epsilon} < \infty.$$

Combining this with our result for large k gives the Corollary. □

Our next step is to observe that the diagonal part of m is smoother. This is an easy consequence of the integral equation.

Proposition 3.5. *If $Q \in C^\epsilon(\mathbb{R}^2)$ and is compactly supported, then m^d is in $C^{1+\epsilon}(\mathbb{R}^2)$ and we have*

$$\|m^d(\cdot, k)\|_{C^{1+\epsilon}} \leq C.$$

Proof. We have that $m - D_k^{-1}Qm \equiv 1$ and that $m \in C^\epsilon(\mathbf{R}^2)$. In particular,

$$m^d = \begin{pmatrix} \bar{\partial}^{-1}q_{12}m_{21} & 0 \\ 0 & \partial^{-1}q_{21}m_{12} \end{pmatrix}.$$

We have $\bar{\partial}^{-1}, \partial^{-1} : C^\epsilon(\mathbf{R}^2) \to C^{1+\epsilon}(\mathbf{R}^2)$. Thus the entries of m^d are $C^{1+\epsilon}(\mathbf{R}^2)$ since Qm is $C^\epsilon(\mathbf{R}^2)$. We obtain the L^∞-estimates for m since Qm is compactly supported. □

Corollary 3.6. *Let Q be in $C^\epsilon(\mathbb{R}^2)$ and compactly supported. Then the special solutions constructed in Theorem 2.2 and Proposition 3.3 satisfy*

$$\|m(\cdot, k) - 1\|_{L^\infty} \leq C|k|^{-\epsilon}.$$

Proof. We have $m(\cdot, k) \in C^\epsilon(\mathbf{R}^2)$ from Corollary 3.4. We observe that

$$m^{off} = D_k^{-1}Qm^d,$$

then the estimate (3.3) implies $\|m^{off}\|_{L^\infty} \leq C|k|^{-\epsilon}$. Now we have

$$m^d = 1 + D^{-1}Qm^{off}$$

and since Q is compactly supported, we have

$$\|m^d - 1\|_{L^\infty} \ \leq \ C(Q)\|m^{off}\|_{L^\infty} \ \leq \ C|k|^{-\epsilon}.$$

□

Using these estimates and L^2 estimates for pseudodifferential operators with non-smooth symbols proven in [CM78] one can easily conclude the proof of Lemma 3.1. For more details see [BBR01].

4. Stability estimates and reconstruction

Conditional stability estimate for Nachman's approach for conductivities with one derivatives was proved in [Liu97], while stability estimate using the $\overline{\partial}\partial$ approach for conductivities in $C^{1+\epsilon}$, $\epsilon > 0$, was shown in [BBR01]. It is the latter result that is stated below.

Theorem 4.1. *Let $\Omega \subset \mathbb{R}^2$ be a bounded domain with Lipschitz boundary and assume that γ_1 and γ_2 are two conductivities in $\overline{\Omega}$ such that for $i = 1, 2$,*

(i) *There exists a constant $C > 0$ such that for every $x \in \Omega$*

$$\frac{1}{C} < \gamma_i < C, \tag{4.1}$$

(ii) *$\gamma_i \in C^{1+\epsilon}(\overline{\Omega})$ for some $\epsilon > 0$ and there exists $M > 0$ such that*

$$\|\gamma_i\|_{C^{1+\epsilon}} \leq M. \tag{4.2}$$

If $\|\Lambda_{\gamma_1} - \Lambda_{\gamma_2}\|_{\frac{1}{2},-\frac{1}{2}}$ *is small enough, then*

$$\|\gamma_1 - \gamma_2\|_{L^\infty(\Omega)} \leq C\omega\left(\|\Lambda_{\gamma_1} - \Lambda_{\gamma_2}\|_{\frac{1}{2},-\frac{1}{2}}\right), \tag{4.3}$$

where $\|\cdot\|_{\frac{1}{2},-\frac{1}{2}}$ *denotes the operator norm as operators from* $H^{\frac{1}{2}}(\partial\Omega)$ *to* $H^{-\frac{1}{2}}(\partial\Omega)$ *and* $\omega : [0,\delta] \to \mathbb{R}$ *is such that*

$$\omega(t) \leq |\log t|^{-\alpha}, \tag{4.4}$$

for some $\alpha > 0$.

For details on the proof we refer to [BBR01].

Next we focus our attention on the reconstruction method. In [Nac96] a reconstruction method is proposed for conductivities in $W^{2,p}(\Omega)$, $p > 2$. In the remaining of the section we outline the reconstruction method using the $\overline{\partial}\partial$ approach developed in [KT01].

With $Q \in C^\epsilon(\overline{\Omega})$, the natural Cauchy data for the system (2.5) is

$$\mathcal{C}_Q = \{(v|_{\partial\Omega}, w|_{\partial\Omega}) : (v,w) \in C^{1+\epsilon}(\overline{\Omega}) \times C^{1+\epsilon}(\overline{\Omega}),\ (D-Q)(v,w)^T = 0\}.$$

We remark that the method in [KT01] also shows that Q is uniquely determined by $\mathcal{C}_Q$. However, it is still unclear how to reconstruct it. In the case when Q comes from a conductivity (2.2) $\mathcal{C}_Q$ is determined by Λ_γ.

Theorem 4.2. *Let* $\Omega \subset \mathbb{R}^2$ *be bounded with smooth boundary and the conductivity* $\gamma \in C^{1+\epsilon}(\overline{\Omega})$ *be such that* $0 < c \leq \gamma$ *for some constant c. Then* γ *can be reconstructed from* Λ_γ

The reconstruction method is done in three steps.

Step 1. Determine ψ on $\partial\Omega$. There are two ideas involved. The first consists of a complete characterization of $\mathcal{C}_Q$ in terms of Λ_γ as follows. Let $\dot{C}^{1+\epsilon}(\overline{\Omega})$ be the space of functions in $C^{1+\epsilon}(\overline{\Omega})$ which integrate to zero along the boundary. Let ∂_s^{-1} denote an inverse to the tangential field ∂_s along the boundary. Notice that it is well defined when acting on boundary maps which integrate to zero. The complexified normal ν was defined in (2.14).

Theorem 4.3. *Let* $Q \in C^\epsilon(\overline{\Omega})$ *compactly supported in* Ω *be as in (2.5). Then*

$$\begin{aligned}\mathcal{C}_Q = \{(h_1,h_2) \in \dot{C}^{1+\epsilon}(\partial\Omega) \times \dot{C}^{1+\epsilon}(\partial\Omega) :\\ i\Lambda_\gamma\partial_s^{-1}(\nu h_1 - \overline{\nu}h_2) = (\nu h_1 + \overline{\nu}h_2)\}.\end{aligned} \tag{4.5}$$

The proof relies on the equivalence between the boundary value problem for the conductivity equation (1.2) with $f = i\partial_s^{-1}(\nu h_1 - \overline{\nu}h_2)$ and the boundary value problem for the equation (2.5) with the boundary data $(u,v)|_{\partial\Omega} = (h_1,h_2)$ and on the following straightforward relation on the boundary $\partial\Omega$,

$$\begin{pmatrix} v \\ w \end{pmatrix}\Bigg|_{\partial\Omega} = \frac{1}{2}\begin{pmatrix} \overline{\nu} & -i\overline{\nu} \\ \nu & i\nu \end{pmatrix}\begin{pmatrix} \Lambda_\gamma(f) \\ \partial_s(f) \end{pmatrix}. \tag{4.6}$$

The second idea takes into account the behaviour of the complex geometrical optics ψ for z near infinity. Notice that the first row has entries which are analytic outside Ω, while the second row has entries which are anti-analytic outside Ω.

Due to the symmetries

$$m_{11}(z,k) = \overline{m_{22}(z,\overline{k})}, \qquad m_{21}(z,k) = \overline{m_{12}(z,\overline{k})}, \tag{4.7}$$

which follows from the differential equations, the asymptotic for the columns of m and the uniqueness in Theorem 2.2, it suffices to reconstruct the first column $(\psi_{11}, \psi_{21})^T$ of $\psi(z,k)$.

For every $(\zeta, z) \in \mathbb{C}$ with $\zeta \neq z$ we introduce $g_k(\zeta, z) = \frac{1}{\pi}\frac{e^{-ik(\zeta-z)}}{\zeta - z}$, a Green kernel for $\overline{\partial}$ which also takes into account exponential growth at infinity. Using $g_k(\cdot, z)$ with $z \in \partial\Omega$, we define the single layer potentials $\mathcal{S}_k$ and $\overline{\mathcal{S}}_k$ as boundary integral operators by

$$\mathcal{S}_k f(z) = \int_{\partial\Omega} f(\zeta) g_k(\zeta, z) d\zeta, \ \overline{\mathcal{S}}_k f(z) = \int_{\partial\Omega} f(\zeta) \overline{g}_k(\zeta, z) d\zeta, \tag{4.8}$$

where the integrals are understood in the sense of principal value. It is a classical result in singular integral theory that these operators are well defined bounded operators from $C^{1+\epsilon}(\partial\Omega)$ to $C^{1+\epsilon}(\partial\Omega)$, see [Musk53].

The following result completely characterizes the traces on $\partial\Omega$.

Theorem 4.4. *The only pair* $(h_1, h_2) \in \dot{C}^{1+\epsilon}(\partial\Omega) \times \dot{C}^{1+\epsilon}(\partial\Omega)$ *which satisfies*

$$\begin{pmatrix} I - i\mathcal{S}_k & 0 \\ 0 & I - i\overline{\mathcal{S}}_k \end{pmatrix} \begin{pmatrix} h_1 \\ h_2 \end{pmatrix} = \begin{pmatrix} 2e^{izk} \\ 0 \end{pmatrix} \tag{4.9}$$

together with

$$(I - i\Lambda_\gamma \partial_s^{-1})(\nu h_1)(z) = (I + i\Lambda_\gamma \partial_s^{-1})(\overline{\nu} h_2)(z), \tag{4.10}$$

is $(\psi_{11}(\cdot, k), \psi_{21}(\cdot, k))|_{\partial\Omega}$.

The idea of the proof is based on the uniqueness of direct scattering. The pair (h_1, h_2) may be extended inside Ω since they belong to the Cauchy data $\mathcal{C}_Q$, see Theorem 4.3. Moreover, this extension is unique due to the ellipticity of the system. Outside Ω they can be explicitly extended by

$$v(z) = -\frac{1}{2i} \int_{\partial\Omega} h_1(\zeta) g_k(\zeta, z) d\zeta + e^{izk}, \tag{4.11}$$

respectively

$$w(z) = -\frac{1}{2i} \int_{\partial\Omega} h_2(\zeta) \overline{g}_k(\zeta, z) d\zeta. \tag{4.12}$$

It is easy to see that they are analytic respectively anti-analytic and that they satisfy the decay condition from Theorem 2.2. The system (2.5) is also satisfied across the boundary $\partial\Omega$, since Plemelj's formula [Musk53] ensures continuity. The analyticity, respectively anti-analyticity is obtained using Morera's theorem.

Step 3. Using the formula (2.13) we find the scattering matrix S_Q.

Next we solve the $\overline{\partial}$-equation (2.11). In [BU97] the fact that $S_Q \in L^2(\mathbb{R}^2)$ suffices. Here we will use the further decay on S_Q given by Lemma 3.1.

To simplify notations we introduce the operator $(\partial/\partial\overline{k})^{-1} = \partial_{\overline{k}}^{-1}$ defined by

$$\partial_{\overline{k}}^{-1} f(k) = \frac{1}{\pi}\int_{\mathbb{C}} \frac{f(k')}{k-k'}dk'.$$

We can now write the integral equations for m and prove unique solvability of these. We denote below by S_{ij} the components of the matrix S_Q.

Lemma 4.5. *Let $Q \in C^{\epsilon}(\mathbb{R}^2)$ be compactly supported, and let $z \in \mathbb{C}$ be fixed. Then for $q > 4/\epsilon$ the equation*

$$(I - \partial_{\overline{k}}^{-1}(e(z,-k)S_{21}\cdot))(m-I) = \partial_{\overline{k}}^{-1}(e(z,-k)S_{21}), \tag{4.13}$$

has a unique solution $m(z,\cdot)$ such that $m(z,\cdot) - I \in L^q(\mathbb{R}^2)$ given by

$$m(z,k) = I + (I - \partial_{\overline{k}}^{-1}(e(z,-k)S_{21}\cdot))^{-1}\partial_{\overline{k}}^{-1}(e(z,-k)S_{21}). \tag{4.14}$$

Moreover, $m(z,\cdot) - I \in C^{\alpha}(\mathbb{R}^2)$, for $\alpha < (1+\epsilon)/2$.

Proof. Since $S_{21} \in L^2(\mathbb{R}^2)$ and $|e(z,-k)| = 1$ we know (see [Nac96]) that $\partial_{\overline{k}}^{-1}(e(z,-k)S_{21}\cdot)$ is a compact operator in $L^s(\mathbb{R}^2)$ for $2 < s < \infty$. Furthermore, by Lemma 3.1, since $S_{21} \in L^r(\mathbb{R}^2)$ for some $4/(2+\epsilon) < r < 2$, it follows by using the Hardy-Littlewood-Sobolev inequality that $\partial_{\overline{k}}^{-1}(e(z,-k)S_{21}) \in L^q(\mathbb{R}^2)$ for $q > \epsilon/2$. The unique solvability of (4.13) in $L^q(\mathbb{R}^2)$ follows from the Fredholm alternative. The fact that $(I + \partial_{\overline{k}}^{-1}(e(z,-k)S_{21}\cdot))$ has trivial kernel in $L^q(\mathbb{R}^2)$ is a consequence of Liouville's theorem for pseudoanalytic functions with coefficients in $L^r(\mathbb{R}^2) \cap L^{r'}(\mathbb{R}^2)$ (see [Vek62]).

To prove the Hölder continuity of $m(z,\cdot) - I$ we use the fact, that convolution by $1/z$ maps $L^p(\mathbb{R}^2)$ into $C^{\alpha}(\mathbb{R}^2)$ for $1 < p < \infty$, $\alpha = 1 - 1/p_0$ and $\max(2,p) < p_0 < \infty$, see [SuU93]. □

Step 3. Recover γ in Ω. If we solve (2.11) for $m(z,k)$ for each fixed $z \in \mathbb{C}$, then Q can be found by

$$Q(z) = \lim_{k_0\to\infty} \mu(B_r(0))^{-1}\int_{\{k:|k-k_0|<r\}} D_k m(z,k)d\mu(k).$$

This would solve the inverse problem for (2.5).

However, as observed in [BBR01] we can reconstruct γ directly. First solve (2.11) with $S_Q(k)$ replaced by $S_Q(-\overline{k})^T$ and call the solution $\tilde{m}(z,k)$. Then define

$$\tilde{m}_+(z,k) = \tilde{m}_{11}(z,k) + \overline{\tilde{m}_{12}(z,\overline{k})}.$$

The conductivity γ is given by the formula [KT01]

$$\gamma^{1/2}(z) = \mathrm{Re}(\tilde{m}_+(z,0)). \tag{4.15}$$

5. The Pauli Hamiltonian

The Pauli Hamiltonian describes particles in a magnetic field with spin. In two dimensions it is a direct sum of the pair of operators

$$H_{\vec{A},q}u := \sum_{j=1}^{2}\left(\frac{1}{i}\frac{\partial}{\partial x_j} - A_j\right)^2 u \pm Bu - qu \tag{5.1}$$

where $\vec{A}$ denotes the magnetic potential, $B = \operatorname{rot}\vec{A}$ is the magnetic field, and q is the electrical potential (see for instance Chapter 6 of [CFKS87]). Thus both direct and inverse problems for the Pauli Hamiltonian consider separately both signs in B in (5.1). For simplicity we will choose the plus sign. All of the results below are also valid for the minus sign in B with minor changes in the arguments.

We first describe the inverse boundary problem we consider. Let $\Omega \subset \mathbb{R}^2$ be a bounded domain with smooth boundary. We consider the Pauli Hamiltonian given by a real-valued vector field $\vec{A} = (A_1, A_2) \in W^{1,p}(\Omega)$ and an electric potential $q \in L^p(\Omega)$, $p > 2$,

$$H_{\vec{A},q}u := \sum_{j=1}^{2}\left(\frac{1}{i}\frac{\partial}{\partial x_j} - A_j\right)^2 u + Bu - qu = 0 \quad \text{in } \Omega \tag{5.2}$$

where $B = \operatorname{rot}\vec{A}$. Fix $\alpha = \frac{p-2}{p}$ throughout this section. The set of Cauchy data of the solutions of (5.2) is given by

$$\begin{aligned} \mathcal{C}_{\vec{A},q} := \{(f,g) \in C^{1,\alpha}(\partial\Omega) \times C^{\alpha}(\partial\Omega) :\ & \text{there exists } u \in C^{1,\alpha}(\overline{\Omega}) \\ \text{such that } H_{\vec{A},q}u = 0,\ u|_{\partial\Omega} = f,\ & ((\nabla - i\vec{A})u)|_{\partial\Omega}\cdot\nu = g\}. \end{aligned} \tag{5.3}$$

Here ν denotes the unit normal to $\partial\Omega$. In the case that 0 is not a Dirichlet eigenvalue for $H_{\vec{A},q}$, $\mathcal{C}_{\vec{A},q}$ is the graph of the Dirichlet-to-Neumann (DN) map

$$\Lambda_{\vec{A},q} : C^{1,\alpha}(\partial\Omega) \to C^{\alpha}(\partial\Omega). \tag{5.4}$$

The inverse boundary value problem we consider in this section is whether we can determine $\vec{A}$ and q from $\mathcal{C}_{\vec{A},q}$.

It was observed in [Sun93a, Sun93b] that there is a gauge invariance in the problem. That is, if $\varphi \in C^1(\overline{\Omega})$ with $\varphi|_{\partial\Omega} = 1$, $\nabla\varphi|_{\partial\Omega} = 0$, then

$$\mathcal{C}_{\vec{A}+\nabla\varphi,q} = \mathcal{C}_{\vec{A},q}.$$

Therefore we can recover at best the magnetic field, $\operatorname{rot}\vec{A}$, and q from the DN map.

In [KU02] it was proven the following semiglobal identifiability result:

Theorem 5.1. *Let $\vec{A}_j \in W_0^{1,p}(\Omega)$, $j = 1,2$, $\operatorname{rot}\vec{A}_1 \in W_0^{1,p}(\Omega)$, $q_1 \in W^{1,p}(\Omega)$, $q_2 \in L^p(\Omega)$, $p > 2$. For each $M > 0$, there exists $\epsilon(M,\Omega,p) > 0$ such that if $\|\vec{A}_1\|_{L^p(\Omega)} \le M$ and $\|q_1\|_{W^{1,p}(\Omega)} \le \epsilon$ and*

$$\mathcal{C}_{\vec{A}_1,q_1} = \mathcal{C}_{\vec{A}_2,q_2}, \tag{5.5}$$

we conclude

$$\operatorname{rot}\vec{A}_1 = \operatorname{rot}\vec{A}_2 \quad \text{and} \quad q_1 = q_2 \quad \text{in } \Omega. \tag{5.6}$$

$W_0^{1,p}(\Omega)$ denotes the space of $W^{1,p}(\Omega)$-functions whose boundary traces are zero.

Observe that no smallness condition is assumed on the electric potential q_2. We also remark that the only place where we need that the magnetic potential has boundary trace zero is in the proof of Lemma 5.6. If we assume further regularity in the magnetic and electrical potentials then Theorem D of [NSU] (which is also valid in two dimensions) allows to extend the magnetic and electrical potentials to $\mathbb{R}^2$ with compact support. More precisely we have:

Theorem 5.2. *Let $\vec{A}_j$, $q_j \in C^\infty(\overline{\Omega})$, $j = 1, 2$, and $p > 2$. There exists $\epsilon(\Omega, p) > 0$ such that if $\|q_1\|_{W^{1,p}(\Omega)} \leq \epsilon$ and*

$$\mathcal{C}_{\vec{A}_1,q_1} = \mathcal{C}_{\vec{A}_2,q_2}, \tag{5.7}$$

we conclude

$$\operatorname{rot}\vec{A}_1 = \operatorname{rot}\vec{A}_2 \quad \text{and} \quad q_1 = q_2 \quad \text{in } \Omega. \tag{5.8}$$

In [Sun93b] Sun proved in two dimensions for the Schrödinger equation in a magnetic field that if $\|\operatorname{rot}\vec{A}_j\|_{W^{1,\infty}(\Omega)}$ $(j = 1, 2)$ is small enough and q_j $(j = 1, 2)$ are in an open and dense set in an appropriate topology, then we can determine uniquely $\operatorname{rot}\vec{A}_j$ and q_j from the DN map associated to the magnetic potentials and electrical potentials.

We remark that in dimensions $n \geq 3$ a global identifiability result of the magnetic field and electrical potential was proven in [NSU95] for the Schrödinger equation in a magnetic field assuming some smoothness conditions on the coefficients. We also note that even if we assume that $\vec{A}_1 = 0$ in Theorem 5.1, it is unknown whether we can recover q from $\mathcal{C}_q := \mathcal{C}_{\vec{0},q}$, the set of Cauchy data for the standard Schrödinger equation, although several partial results have been proven ([BU97], [Kan00], [Nac96], [SuU91], [SuU93], [SyU86]).

A particular case of Theorem 5.1 is when the electrical potential in (5.1) is zero. Thus we obtain the following global uniqueness result:

Corollary 5.3. *Let $\vec{A}_j \in W_0^{1,p}(\Omega)$, $j = 1, 2$, $\operatorname{rot}\vec{A}_1 \in W_0^{1,p}(\Omega)$, $q_1 = 0$, $q_2 \in L^p(\Omega)$, $p > 2$. For each $M > 0$, there exists $\epsilon(M, \Omega, p) > 0$ such that if $\|\vec{A}_1\|_{L^p(\Omega)} \leq M$*

$$\mathcal{C}_{\vec{A}_1,0} = \mathcal{C}_{\vec{A}_2,q_2}, \tag{5.9}$$

we conclude

$$\operatorname{rot} A_1 = \operatorname{rot} A_2 \quad \text{and} \quad q_1 = q_2 = 0 \quad \text{in } \Omega. \tag{5.10}$$

As a consequence of Theorem 5.2, a similar result to Corollary 5.3 holds with $\vec{A}_j \in C^\infty(\overline{\Omega})$, $j = 1, 2$, without the assumption that the magnetic potentials have zero boundary trace.

In the case that the magnetic potentials are zero, we conclude the following semiglobal identifiability result:

Corollary 5.4. *Let $q_1 \in W^{1,p}(\Omega)$, $q_2 \in L^p(\Omega)$, $p > 2$. There exists $\epsilon(\Omega, p) > 0$ such that if $\|q_1\|_{W^{1,p}(\Omega)} \leq \epsilon$ and $\mathcal{C}_{q_1} = \mathcal{C}_{q_2}$, then $q_1 = q_2$.*

Corollary 5.4 was known previously if q_1 and q_2 are both a priori close to constant [SyU86]. As far as we know, even the case $q_1 = 0$ was previously unknown.

The method of proof of Theorem 5.1 is by reducing the problem to a similar one for a second order equation which can be factored in terms of $\overline{\partial}$ and ∂.

We rewrite (5.2) in the form

$$(\overline{\partial} + \bar{a})(\partial - a)u - \tilde{q}u = 0 \quad \text{in } \Omega \tag{5.11}$$

where

$$a := \frac{1}{2}(A_2 + iA_1), \quad \tilde{q} = \frac{1}{4}q. \tag{5.12}$$

We define the set of Cauchy data associated to (5.11) by

$$\begin{aligned} \mathcal{C}_{a,\tilde{q}} := \{(f, g) \in C^{1,\alpha}(\partial\Omega) \times C^{\alpha}(\partial\Omega) : u|_{\partial\Omega} = f, \\ ((\partial - a)u)|_{\partial\Omega} = g,\ u \in C^{1,\alpha}(\overline{\Omega}) \text{ a solution of (5.11) } \}. \end{aligned} \tag{5.13}$$

Theorem 5.1 is then a consequence of

Theorem 5.5. *Let $a_j \in W^{1,p}(\Omega)$, $\tilde{q}_1 \in W^{1,p}(\Omega)$, $\tilde{q}_2 \in L^p(\Omega)$, $p > 2$, $j = 1, 2$. For each $M > 0$, there exists $\epsilon(M, \Omega, p) > 0$ such that if $\|a_1\|_{L^p(\Omega)} \leq M$ and $\|\tilde{q}_1\|_{W^{1,p}(\Omega)} \leq \epsilon$ and*

$$\mathcal{C}_{a_1,\tilde{q}_1} = \mathcal{C}_{a_2,\tilde{q}_2}, \tag{5.14}$$

then

$$\tilde{q}_1 = \tilde{q}_2 \quad \textit{and} \quad \overline{\partial}^{-1}\bar{a}_1 + \partial^{-1}a_1 = \overline{\partial}^{-1}\bar{a}_2 + \partial^{-1}a_2 \quad \textit{in } \Omega. \tag{5.15}$$

Of course (5.15) implies that

$$\operatorname{rot} a_1 = \frac{1}{2}(\overline{\partial}\bar{a}_1 + \partial a_1) = \operatorname{rot} a_2 = \frac{1}{2}(\overline{\partial}\bar{a}_2 + \partial a_2) \quad \text{in } \Omega. \tag{5.16}$$

Sketch of proof of Theorem 5.5

The method of proof of Theorem 5.5 reduces (5.11) to a first order system and follows the outline of Section 2.

Consider the equation

$$(\overline{\partial} + \bar{a})(\partial - a)u - qu = 0 \quad \text{in } \Omega\,. \tag{5.17}$$

Let u be a solution of (5.17) and set $w := (\partial - a)u$. Then the equation (5.17) takes the form

$$\left[\begin{pmatrix} \overline{\partial} + \bar{a} & 0 \\ 0 & \partial - a \end{pmatrix} - \begin{pmatrix} 0 & q \\ 1 & 0 \end{pmatrix}\right]\begin{pmatrix} w \\ u \end{pmatrix} = 0. \tag{5.18}$$

By conjugating the equation (5.18), we have

$$\begin{pmatrix} e^{-\overline{\partial}^{-1}\bar{a}} & 0 \\ 0 & e^{\partial^{-1}a} \end{pmatrix}\left[\begin{pmatrix} \overline{\partial} & 0 \\ 0 & \partial \end{pmatrix} - \begin{pmatrix} 0 & e^{Ta}q \\ e^{-Ta} & 0 \end{pmatrix}\right]\begin{pmatrix} e^{\overline{\partial}^{-1}\bar{a}} & 0 \\ 0 & e^{-\partial^{-1}a} \end{pmatrix}\begin{pmatrix} w \\ u \end{pmatrix} = 0 \tag{5.19}$$

where

$$Ta(z) := \overline{\partial}^{-1}\bar{a} + \partial^{-1}a.$$

Set

$$D = \begin{pmatrix} \overline{\partial} & 0 \\ 0 & \partial \end{pmatrix}$$

and

$$\widetilde{Q} = \begin{pmatrix} 0 & e^{Ta}q \\ e^{-Ta} & 0 \end{pmatrix}.$$

We are seeking special solutions of the system

$$(D - \widetilde{Q})\psi = 0 \quad \text{in } \Omega \tag{5.20}$$

in the form

$$\psi = \begin{pmatrix} e^{-\overline{\partial}^{-1}\bar{a}} & 0 \\ 0 & e^{\partial^{-1}a} \end{pmatrix} m(z,k) \begin{pmatrix} e^{izk} & 0 \\ 0 & e^{-i\bar{z}k} \end{pmatrix} \tag{5.21}$$

where $m(z,k)$ is a 2×2 matrix-valued function in Ω.

We will not repeat all the arguments of the proof that follows pretty much the steps of Section 2. We mention, though, that in order to prove that the boundary value of the solutions ψ_j, $j = 1,2$ in Step 1 one needs the following Lemma due to Sun (equation (3.44) in [Sun93b]).

Lemma 5.6. *If $\mathcal{C}_{a_1,q_1} = \mathcal{C}_{a_2,q_2}$, then $\partial^{-1}a_1 = \partial^{-1}a_2$ on $\partial\Omega$.*

Z. Sun proved this lemma under the assumption that the DN maps are the same. However, exactly the same argument works with the assumption $\mathcal{C}_{a_1,q_1} = \mathcal{C}_{a_2,q_2}$. Also Sun's proof works under the weaker regularity assumptions assumed in this section on the electrical and magnetic potentials.

6. Determination of convection terms

In this section we sketch a proof, using the method of Sections 2 and 4, of a Theorem of Cheng and Yamamoto [CY98] (for an announcement see [CY00]) regarding the unique identification of the convection term $B = (B_1, B_2) \in L^p(\Omega), p > 2$ entering

$$\Delta u + B \cdot \nabla u = 0 \quad \text{in } \Omega \tag{6.1}$$

in terms of the Dirichlet-to-Neumann map Λ_B defined by

$$\Lambda_B(u) = \frac{\partial u}{\partial \nu} \tag{6.2}$$

with u solution to (6.1).

We define

$$b := \frac{1}{4}(B_1 + iB_2).$$

Then, (6.1) becomes

$$\overline{\partial}\partial u + \bar{b}\partial u + b\overline{\partial} u = 0 \quad \text{in } \Omega.$$

Put $v := \partial u$ and $w := \overline{\partial} u$. We obtain

$$\left[\begin{pmatrix} \overline{\partial} & 0 \\ 0 & \partial \end{pmatrix} + \begin{pmatrix} \bar{b} & b \\ \bar{b} & b \end{pmatrix}\right]\begin{pmatrix} v \\ w \end{pmatrix} = 0. \tag{6.3}$$

Conjugating (6.3) we have

$$\begin{pmatrix} e^{-\overline{\partial}^{-1}\bar{b}} & 0 \\ 0 & e^{-\partial^{-1}b} \end{pmatrix}\left[\begin{pmatrix} \overline{\partial} & 0 \\ 0 & \partial \end{pmatrix} - \begin{pmatrix} 0 & e^{Tb}b \\ e^{\overline{Tb}}\bar{b} & 0 \end{pmatrix}\right]\begin{pmatrix} e^{\overline{\partial}^{-1}\bar{b}} & 0 \\ 0 & e^{\partial^{-1}b} \end{pmatrix}\begin{pmatrix} v \\ w \end{pmatrix} = 0$$

where

$$Tb(z) := \overline{\partial}^{-1}\bar{b} - \partial^{-1}b.$$

Put $q := e^{Tb}b$. Then we have the following system of equations:

$$\left[\begin{pmatrix} \overline{\partial} & 0 \\ 0 & \partial \end{pmatrix} - \begin{pmatrix} 0 & q \\ \bar{q} & 0 \end{pmatrix}\right]\begin{pmatrix} v_1 \\ v_2 \end{pmatrix} = 0 \quad \text{in } \Omega. \tag{6.4}$$

By Theorem 2.2, for each $k \in \mathbb{C}$, there exists a solution of (6.4) of the form

$$\psi(z,k) = m(z,k)\begin{pmatrix} e^{izk} & 0 \\ 0 & e^{-i\bar{z}k} \end{pmatrix}.$$

If $B^{(1)}$ and $B^{(2)}$ are two convection coefficients and satisfy $\Lambda_{B^{(1)}} = \Lambda_{B^{(2)}}$, then by using arguments similar to Lemma 5.6, $\overline{\partial}^{-1}B^{(1)} = \overline{\partial}^{-1}B^{(2)}$ on $\partial\Omega$. As in Step 1 of the proof of Theorem 2.1 outlined in Section 2 one gets that the scattering matrix $S_{Q_j}(k)$ corresponding to (6.4) coincide. It then follows from the Liouville Theorem for pseudo-analytic functions, Theorem 3.1 of [BU97], that

$$m^{(1)}(z,k) = m^{(2)}(z,k), \quad k \in \mathbb{C}.$$

Therefore, we have

$$e^{\overline{\partial}^{-1}\bar{b}^{(1)} - \partial^{-1}b^{(1)}}b^{(1)} = e^{\overline{\partial}^{-1}\bar{b}^{(2)} - \partial^{-1}b^{(2)}}b^{(2)}$$

and hence

$$b^{(1)} = b^{(2)}$$

concluding the proof of the Theorem.

7. Open problems

In this section we mention some open problems directly related to the results of this paper.

- **Bounded measurable conductivities**
  The Dirichlet-to-Neumann map is well defined for L^∞ conductivities. Is it possible to extend Theorem 2.1, Theorem 4.1 and Theorem 4.2 to this case? A less ambitious project would be to extend these results to conductivities $\gamma \in W^{1,2}(\Omega)$. If this is possible then it is very likely that one can show that the inequality in Step 3 of Section 2 is an equality providing a Plancherel formula for the map $Q \longrightarrow S_Q$. This was done by Brown [B01] for the case

that $\|Q\|_{L^2}$ is small enough. It is known [BC88] that there is a Plancherel identity for Schwartz potentials.

- **Numerical algorithm**
  Develop a numerical algorithm based on the reconstruction method proposed in [KT01].
- **Complex conductivities**
  Show that Theorem 2.1, Theorem 4.1 and Theorem 4.2 extend to complex conductivities, without the assumption of smallness in the imaginary part.
- **The Schrödinger equation**
  Extend Theorem 5.1 (or Theorem 5.2) to a global uniqueness result, without the smallness assumption on the potential.
- **Other elliptic systems**
  Can the $\overline{\partial}\partial$ method used in this paper be applied to other elliptic systems? One important case to consider is the isotropic elasticity system in two dimensions. Up to the present time only a local uniqueness result is known [NU93].

References

[Ale88] G. Alessandrini, Stable determination of conductivity by boundary measurements, Appl. Anal. 27 (1988), no. 1–3, 153–172.

[BBR01] J.A. Barceló, T. Barceló, and A. Ruiz, Stability of the inverse conductivity problem in the plane for less regular conductivities, J. Differential Equations 173 (2001), no. 2, 231–270.

[BC88] R. Beals and R. Coifman, The spectral problem for the Davey-Stewartson and Ishimori hierarchies in *Nonlinear evolution equations: Integrability and spectral methods*, Manchester University Press (1988), 15–23.

[B01] R.M. Brown, Estimates for a scattering map associated to a two-dimensional first order system, to appear in J. Nonlinear Science.

[BU96] R.M. Brown and G. Uhlmann, Uniqueness in the inverse conductivity problem with less regular conductivities in two dimensions, unpublished, 1996.

[BU97] R.M. Brown and G. Uhlmann, Uniqueness in the inverse conductivity problem for nonsmooth conductivities in two dimensions, Comm. in PDE 22 (1997), 1009–1027.

[Cal80] A. Calderón, On an inverse boundary value problem, Seminar on Numerical Analysis and its Applications to Continuum Physics (Rio de Janeiro, 1980), Soc. Brasil. Mat., Rio de Janeiro, 1980, 65–73.

[CIN99] M. Cheney, D. Isaacson, and J. Newell, Electrical impedance tomography, SIAM Rev. 41 (1999), 85–101.

[CY98] J. Cheng and M. Yamamoto, Determination of two convection terms from Dirichlet to Neumann map in two-dimensional case, University of Tokyo, Graduate School of Mathematical Sciences Technical Note UTMS, 98-31.

[CY00] J. Cheng and M. Yamamoto, The global uniqueness for determining two convection coefficients from Dirichlet to Neumann map in two dimensions, Inverse Problems 16 (2000), L25–L30.

[CM78] R. Coifman and Y. Meyer, Au delà des Opérateurs pseudo-différentiels, Vol. 57 of Astérisque, Société Mathématique de France, 1978.

[CFKS87] H.L. Cycon, R. Froese, W. Kirsch, and B. Simon, Schrödinger operators with applications to quantum mechanics and global geometry, Texts and Monographs in Physics, Springer-Verlag, 1987.

[Fra00] E. Francini, Recovering a complex coefficient in a planar domain from the Dirichlet-to-Neumann map, Inverse Problems 16 (2000), 107–119.

[Kan00] H. Kang, A uniqueness theorem for an inverse boundary value problem in two dimensions, to appear in J. Math. Anal. Appl.

[KU02] H. Kang and G. Uhlmann, Inverse problems for the Pauli Hamiltonian in two dimensions, preprint.

[KV84] R. Kohn and M. Vogelius, Determining conductivity by boundary measurements, Comm. Pure Appl. Math. 37 (1984), 289–298.

[KT01] K. Knudsen, A. Tamasan, Reconstruction of less regular conductivities in the plane, MSRI preprint series, Berkeley, 2001.

[Liu97] L. Liu, Stability estimates for the two-dimensional inverse conductivity problem, Ph.D. thesis, Department of Mathematics, University of Rochester, New York, 1997.

[Musk53] N.I. Muskhelishvili, Singular integral equations. Boundary problems of function theory and their application to mathematical physics, P. Noordhoff N. V., Groningen, 1953.

[Nac88] A.I. Nachman, Reconstructions from boundary measurements, Ann. of Math. 128 (1988), 531–576.

[Nac96] A.I. Nachman, Global uniqueness for a two-dimensional inverse boundary value problem, Ann. of Math. 143 (1996), 71–96.

[NSU95] G. Nakamura, Z. Sun, and G. Uhlmann, Global Identifiability for an inverse problem for the Schrödinger equation in a magnetic field, Math. Ann. 303 (1995), 377–388.

[NU93] G. Nakamura and G. Uhlmann, Identification of Lamé parameters by boundary observations, American J. of Math. 115 (1993), 1161–1187.

[SMI00] S. Siltanen, J. Mueller, and D. Isaacson, An implementation of the reconstruction algorithm of A. Nachman for the 2D inverse conductivity problem, Inverse Problems 16 (2000), 681–699.

[Sun93a] Z. Sun, An inverse boundary value problem for Schrödinger operator with vector potentials, Trans. of AMS 338 (1993), 953–971.

[Sun93b] ——, An inverse boundary value problem for Schrödinger operator with vector potential in two dimensions, Comm. in PDE 18 (1993), 83–124.

[SyU86] J. Sylvester and G. Uhlmann, A uniqueness theorem for an inverse boundary value problem in electrical prospection, Comm. Pure Appl. Math. 39 (1986), 91–112.

[SyU88] J. Sylvester and G. Uhlmann, Inverse boundary value problems at the boundary-continuous dependence, Comm. Pure Appl. Math. 41 (1988), 197–219.

[SuU91] Z. Sun and G. Uhlmann, Generic uniqueness for an inverse boundary value problem, Duke Math. J. 62 (1991), 131–155.

[SuU93] ——, Recovery of singularities for formally determined inverse problems, Comm. Math. Phys. 153 (1993), 431–445.

[Sung94a] L. Sung, An inverse scattering transform for the Davey-Stewartson II equations. I, J. Math. Anal. Appl. 183 (1994), 121–154.

[Sung94b] ——, An inverse scattering transform for the Davey-Stewartson II equations. II, J. Math. Anal. Appl. 183 (1994), 289–325.

[Sung94c] ——, An inverse scattering transform for the Davey-Stewartson II equations. III, J. Math. Anal. Appl. 183 (1994), 477–494.

[Uhl99] G. Uhlmann, Developments in inverse problems since Calderón's foundational paper in *Harmonic analysis and partial differential equations* (Chicago, IL, 1996), Univ. Chicago Press, Chicago, IL, 1999, pp. 295–345.

[Vek62] I.N. Vekua, Generalized analytic functions, Pergamon Press, London, 1962.

[ZK94] M.S. Zhdanov and G.V. Keller, The geolectric methods in geophysical exploration, Methods in Geochemistry and Geophysics, 31, Elsevier (1994).

Gunther Uhlmann
Department of Mathematics
University of Washington
Box 354350
Seattle, WA 98195, USA
E-mail address: gunther@math.washington.edu

Part III

D. Haroske, T. Runst, H.-J. Schmeisser (eds.): Function Spaces, Differential Operators and Nonlinear Analysis. The Hans Triebel Anniversary Volume.

On the Regularity of Weak Solutions of Elliptic Systems in Banach Spaces

Marina Borovikova and Rüdiger Landes

Dedicated to Prof. Hans Triebel on the occasion of his 65th birthday

Introduction

On a domain $\Omega \subset \mathbb{R}^N$ we consider weak solutions $u : \Omega \to \mathbb{R}^M$ for elliptic systems of the kind

$$A(u) + B(u) = f. \tag{E}$$

Here $A(u)$ is a quasilinear elliptic operator of second order satisfying certain structure conditions and $B(u)$ is a perturbation with critical growth in the gradient, i.e. the growth exponent for the gradient p, $1 < p < \infty$, is the same as the integration exponent of the Sobolev Space $W^{1,p}(\Omega)$ for which (the Nemitzky operator of) $A(u)$ is coercive.

We use specially constructed test functions recently introduced by the second author to prove a Caccioppoli-inequality for bounded weak solutions with the L^∞-bound depending on the maximal angle γ between the direction vector of the perturbation and direction vector of the solution. Then we discuss how the classical approach of Giaquinta and Guisti for $p = 2$, can be modified to obtain higher integrability properties and local C^α-regularity for such a geometric condition on the perturbation. It generalizes the well-known smallness condition, c.f.: [G,HW,W].

1. Preliminaries and main result

We consider elliptic operators $A(u)$ of the form

$$\big(A(u)\big)^k = -\sum_{i=1}^{N} D_i A_i^k\big(x, u, D(u)\big) , \qquad k = 1, \dots, M$$

where the coefficient functions $A_i^k(x, \eta, \zeta)$ are subject to the hypothesis (A) consisting of

$$\text{(A, i)} \qquad \sum_{i=1}^{N} \sum_{k=1}^{M} A_i^k(x, \eta, \zeta)\, \zeta_i^k \geq \lambda\, |\zeta|^p ;$$

(A, ii) $$\sum_{i=1}^{N} \Big(\sum_{k=1}^{M} A_i^k(x,\eta,\zeta)\, \mu^k \Big) \Big(\sum_{k=1}^{M} \mu^k\, \zeta_i^k \Big) \ \geq\ 0\,;$$

(A, iii) $$|A_i^k(x,\eta,\zeta)| \ \leq\ C|\zeta|^{p-1}.$$

Condition (A, i) is the usual ellipticity condition. The structure condition (A, ii) is satisfied by systems in "strict diagonal form" such as the p-Laplacian for instance. Elliptic operators with coefficient functions of the form

$$A_i^k(x,u,Du) \ = \ \sum_{j=1}^{N} a_{ij}(x,u,|Du|) D_j u^k\,,$$

are of this form. For further discussion on this structure conditions c.f.: [L], Section 5.

The perturbation $B(u) = b(x,u,Du) = \big(b^k(x,u,Du)\big)_{k=1}^{M}$ is subject to the growth condition

(B) $\big|b(x,\eta,\zeta)\big| \ \leq\ a\ (\ |\zeta|^p\ +\ 1\ )\,.$

Here C, a and λ are constants. The inhomogeneity f is always at least in $L^1(\Omega)$.

Strictly speaking the above hypotheses only need to be satisfied for the actual range of $\eta = u(x) \in \mathbb{R}^N$, $\zeta = Du(x) \in \mathbb{R}^M \times \mathbb{R}^N$ and $\mu = \mu(u(x)) \in \mathbb{R}^M$, $x \in \Omega \subset \mathbb{R}^N$.

Since we are considering interior regularity we define a weak solution u of (E) to be a function $u \in W^{1,p}(\Omega)$ with the properties:

1) $A_i^k\ (x,u,Du) \in L^{p-1}(\Omega)$, $\ b^k(x,u,Du) \in L^1(\Omega)$,
2) $(A(u),\varphi) +\ (B(u),\varphi) = f(\varphi)$, for all φ in $W_o^{1,p}(\Omega) \cap L^\infty(\Omega)$, with

$$(A(u),\varphi) = \int_\Omega \sum_{k=1}^{M}\ \sum_{i=1}^{N}\ A_i^k(x,u,Du)\ D_i\varphi^k dx\,, \text{ etc..}$$

To state our main result we introduce some notations: The function M is defined by

$$M(\gamma) \ = \ \frac{\lambda}{a} \begin{cases} \big(\exp(-\gamma\ \cot\gamma)\ \sin\gamma\big)^{-1}, & if\ \gamma\ <\ \frac{\pi}{2}\,, \\ 1, & if\ \gamma\ \geq\ \frac{\pi}{2}\,. \end{cases}$$

With $B_\rho = B_\rho(x_o)$ we denote the ball in $\mathbb{R}^N$ with radius ρ centered at x_o, and $\overline{u}$ stands for the average of u over B_{2R}, i.e.:

$$\overline{u} = \fint_{B_{2R}} u\, dx = \frac{1}{|B_{2R}(x_o)|} \int_{B_{2R}(x_o)} u\, dx.$$

We want to show the following

Theorem. Suppose that the hypotheses (A) and (B) are valid and suppose that a weak solution u is subject to the estimate

$$\|u\|_\infty \;<\; M(\gamma),$$

where γ is the maximal angle between the direction vectors of the solution and the perturbation i.e.: $\gamma = \sup\{\sphericalangle\, \big(u(x), b(x,u,D(u))\big) \mid x \in \Omega\}$.

Then for $x_0 \in \Omega$, $B_R = B_R(x_o)$ and for some $R_o(x_o) > 0$ we have the Caccioppoli-inequality

$$\int\limits_{B_R} |Du|^p \, dx \;\le\; \mathcal{K}_1 \, R^{-p} \int\limits_{B_{2R}} |u-\overline{u}|^p \, dx + \mathcal{K}_2 \int\limits_{B_{2R}} (|f|+a)\,|u-\overline{u}| dx, \tag{1}$$

where the constants $\mathcal{K}_1$ and $\mathcal{K}_2$ do not depend on R, $0 < R \le R_o$.

For the proof we need test functions constructed by projection onto convex sets. If K is a convex set of class C^2, then for a given function u we define a modified function $u^{[K]}$ by

$$u^{[K]}(x) \;=\; \begin{cases} P(u(x)) & , \quad if \;\; u(x) \;\notin\; \overline{K}\;, \\ \;u(x) & , \quad if \;\; u(x) \;\in\; \overline{K}\;, \end{cases}$$

where $P(u)$ is the nearest point of $\overline{K}$ to u. Even though we do not know an explicit formula for $P(u)$ the derivative of P as a mapping from $\mathbb{R}^M \to \mathbb{R}^M$ can be determined in terms of u, $P(u)$, and the principal curvatures of the boundary ∂K at $P(u)$. For sets K with the property

K$_1$) The boundary ∂K is a smooth manifold of class C^2 such that the minimal principal curvature is positive,

it is shown in [L1] that $u^{[K]}(x)$ is in $W^{1,2}(\Omega)$ and satisfies

Lemma 1.1. If $x \in \{y \in \Omega \;\mid\; u(y) \notin \overline{K}\}$ then we have the estimate

$$\sum_{i=1}^{N} \sum_{k=1}^{M} A_i^k\big(x,u,D(u)\big)\big(Du(x) - Du^{[k]}(x)\big) \ge \lambda\tau(x)\; |Du(x)|^p \;,$$

where

$$\tau(x) \;=\; 1 \;-\; \frac{1}{1+\big|u(x) \;-\; P\big(u(x)\big)\big|\;\mu(x)} \;=\; \frac{\big|u(x) \;-\; P\big(u(x)\big)\big|\;\mu(x)}{1\;+\;\big|u(x) \;-\; P\big(u(x)\big)\big|\;\mu(x)}\,,$$

and $\mu(x)$ is the minimal principal curvature of ∂K at $P(u(x))$.

In the first step of our proof we need test functions obtained by projections onto sets K_γ, for $\gamma < \frac{\pi}{2}$, with the following property:

K$_2$) The angle between the position vector of a point v of the boundary and the outer normal at this point is less or equal to $\frac{\pi}{2} - \gamma$.

In order to choose these sets as in the best manner possible we note that the elliptic spiral in the plane is the locus for which the position vector of the points of the curve has a constant angle with the normal direction at the points. Hence, in the x_1x_2-plane, say, we consider the curve $\mathcal{L}$, where $\mathcal{L}$ is given for nonnegative values of x_2 by two connected curves $\mathcal{L}_1$ and $\mathcal{L}_2$. The curve $\mathcal{L}_1$ is part of the logarithmic spiral

$$\mathcal{L}_1(t) = ||u||_\infty \, e^{-t \cot\gamma} \, (\cos t \, , \, \sin t) \, , \quad \text{for} \quad 0 \le t \le \tfrac{\pi}{2} + \gamma \, , \text{ and}$$

$\mathcal{L}_2$ is the vertical line connecting

$$P_1 = ||u||_\infty e^{-\left(\frac{\pi}{2}+\gamma\right)\cot\gamma}\left(\cos\left(\frac{\pi}{2}+\gamma\right), 0\right) \quad \text{with}$$

$$P_2 = ||u||_\infty e^{-\left(\frac{\pi}{2}+\gamma\right)\cot\gamma}\left(\cos\left(\frac{\pi}{2}+\gamma\right), \sin\left(\frac{\pi}{2}+\gamma\right)\right).$$

Then rotating $\mathcal{L}$ around the x_1-axis we obtain the boundary of a convex set. We rotate this set around the origin until its axis is parallel to $\overline{u}$ and denote it $\mathcal{S}$. It is elementary to see (but to verify the details is quite cumbersome) that there are sets K_γ containing $\mathcal{S}$ and satisfying K_1 and K_2, in any neighborhood of $\mathcal{S}$. In Fig. 1 below the inner curve is an example for $\mathcal{L}$ with $\gamma = \frac{1}{5}\pi$.

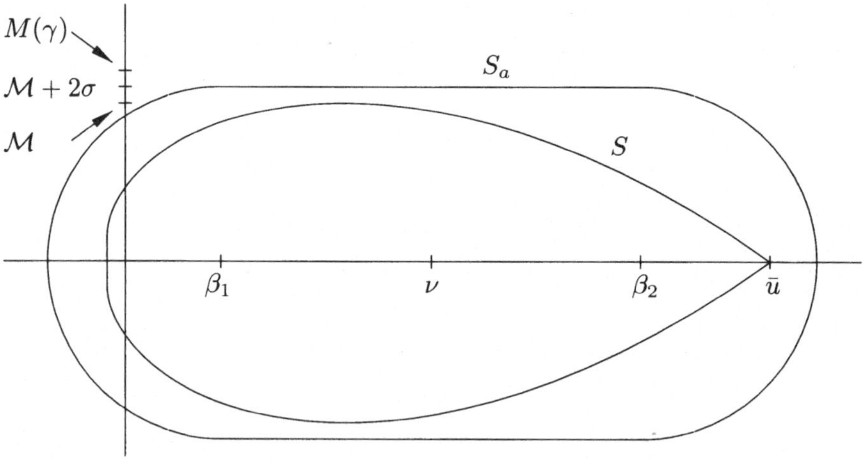

FIGURE 1

We further note that the maximal x_2-value $\mathcal{M}$ of $\mathcal{L}$ is given by

$$\mathcal{M} = ||u||_\infty e^{-\gamma\cot\gamma} \sin\gamma \; < \; M(\gamma) e^{-\gamma\cot\gamma} \sin\gamma,$$

and $\mathcal{M} \; < \; \frac{\lambda}{a}$.

2. Proof of the Caccioppoli-inequality

In case $\gamma < \frac{\pi}{2}$ we obtain a first estimate for sets $B_{2R} \cap \Omega_\gamma$ where $\Omega_\gamma = \{x \in \Omega \mid u(x) \notin \overline{K}_\gamma\}$; we have

Proposition 2.1. For every $\delta > 0$ there are constants $\mathcal{K}_\delta$ not depending on R such that

$$\int_{\Omega_\gamma \cap B_{2R}} |Du|^p \eta^p \frac{|u - u^{[K_\gamma]}|\,\mu}{1 + |u - u^{[K_\gamma]}|\mu}\, dx$$
$$\leq \delta \int_{B_{2R}} \eta^p |Du|^p dx + \mathcal{K}_\delta \frac{1}{R^p} \int_{B_{2R}} |u - \overline{u}|^p dx + \frac{1}{\lambda} \int_{B_{2R}} |f||u - \overline{u}|\, dx,$$

where η is a standard smooth cut-off function with support in B_{2R}, i.e.: $0 \leq \eta \leq 1$, $\eta\big|_{B_R} \equiv 1$, and $|D\eta| \leq C_1/R$, for some constant C_1.

Proof. Because of the assumption on the angle between the perturbation $b(x, u, Du)$ and the solution u we have $\big(B(u), u - u^{[K_\gamma]}\big) \geq 0$; further $|u - u^{[K_\gamma]}| \leq |u - \overline{u}|$, since $\overline{u} \in \overline{K}_\gamma$. Using Lemma 1.1 and the Sobolev-Poincaré-inequality the desired estimate follows as in [L2] replacing 2 by p.

Proposition 2.1 is not yet useful for our purpose since $|u - P(u)|$ is not bounded away from zero on Ω_γ. But we can choose sets K_γ such that $\operatorname{dist}\{\partial K_\gamma, \mathcal{S}\}$ becomes small enough, without μ going to zero: Let $\Omega_o = \{x \in \Omega \mid \operatorname{dist}\{u(x), \mathcal{S}\} > \sigma\}$, with $\sigma = \frac{1}{4}\big(\frac{\lambda}{a} - \mathcal{M}\big)$, say, and choose K_γ such that $\operatorname{dist}\{\partial K_\gamma, \mathcal{S}\} < \frac{\sigma}{2}$, and there is a number $\epsilon > 0$ such that for $x \in \Omega_o$ we have $\dfrac{|u - u^{[K_\gamma]}|\,\mu}{1 + |u - u^{[K_\gamma]}|\mu} > \epsilon$. Hence Proposition 2.1 implies

Lemma 2.1.

$$\int_{\Omega_o \cap B_{2R}} \eta^p |Du|^p\, dx$$
$$\leq \delta \int_{B_{2R}} |Du|^p \eta^p dx + \frac{1}{R^p} \mathcal{K}_{1,\delta} \int_{B_{2R}} |u - \overline{u}|^p\, dx + \mathcal{K}_{2,\delta} \int_{B_{2R}} |f||u - \overline{u}|\, dx$$

for every $\delta > 0$. The constants $\mathcal{K}_{1,\delta}$ and $\mathcal{K}_{2,\delta}$ do not depend on R, $0 < R < R_o$.

In order to set up the finite induction we define sets $Z(r, \nu) \subset \mathbb{R}^M$ as cylinders of radius r with a half ball of the same radius attached to their faces. The rotation axis of the cylinder is on the line through the origin with the direction of $\overline{u}$. The center of the cylinder is at $\frac{\nu}{|\overline{u}|}\overline{u}$, and the centers of the half ball are at $\frac{\beta_1}{|\overline{u}|}\overline{u}$, and $\frac{\beta_2}{|\overline{u}|}\overline{u}$, respectively, where $\beta_1 = m(\gamma) + r$ and $\beta_2 = M(\gamma) - r$ with $M(\gamma)$ as defined above and $m(\gamma)$ some number less than $||u||_\infty e^{-\left(\frac{\pi}{2}+\gamma\right)\cot\gamma} \cos\!\big(\frac{\pi}{2} + \gamma\big)$. (In case

$|\overline{u}| = 0$ the direction of the axis can be chosen arbitrarily.) Fig. 1 shows $\mathcal{S}$ and S_α. For $\gamma \geq \frac{\pi}{2}$, we set $m(\gamma) = -M(\gamma) = -\dfrac{\lambda}{a}$ and define $\Omega_o = \phi$, then we have for all γ and $\alpha = \frac{m(\gamma)+M(\gamma)}{2}$ that $(\Omega \setminus u^{-1}(Z(\mathcal{M}+\sigma,\alpha))) \subset \Omega_o$. Since the sets $Z(r,\nu)$ are not of class C^2 we cannot use them for the construction of the test functions directly. Instead we use convex sets S_ν of class C^2 containing $Z(\mathcal{M}+2\sigma,\nu)$ such that the boundary of the half ball coincides with the boundary of S_ν for those points which have a distance of $(\mathcal{M}+\sigma)$ or less to the axis of $Z(\mathcal{M}+2\sigma,\nu)$.

The idea of the proof now is to construct test functions with S_α moving it step by step up and down with the step length β, say. Roughly speaking, we will estimate $|Du|^p$ on the preimages of the sets cut successively from $\mathcal{S}$ by $S_{\alpha\pm j\beta}$ as long as $\overline{u}$ is in $S_{\alpha\pm(j+1)\beta}$. Adjusting if necessary the step length in the last steps the remaining set will be so small that the usual smallness argument can be applied. We define $\Omega_j = \Omega_o \cup \{x \in \Omega \mid u(x) \notin S_{\alpha+j\beta}\}$ for all γ, where β is some fixed number with $0 < \beta < \sigma$, and get

Proposition 2.2. For every $\delta > 0$ we have the estimate

$$\int\limits_{B_{2R}\cap(\Omega_2\setminus\Omega_0)} \frac{|u-u^{[S_{\alpha+2\beta}]}|\,(\mathcal{M}+2\sigma)^{-1}}{1+|u-u^{[S_{\alpha+2\beta}]}|\,(\mathcal{M}+2\sigma)^{-1}}|Du|^p\eta^p dx$$

$$\leq \delta \int\limits_{B_{2R}} |Du|^p\eta^p dx + \frac{1}{R^p}\mathcal{K}_{1,\delta}\int\limits_{B_{2R}} |u-\overline{u}|^p\,dx + \mathcal{K}_{2,\delta}\int\limits_{B_{2R}} (|f|+a)\,|u-\overline{u}|\,dx.$$

Proof. We test the equation (E) with $\eta^p(u-u^{[S_{\alpha+2\beta}]})$ and obtain the estimate observing that on $\Omega_2 \setminus \Omega_o$ we have $|u(x)-u^{[S_{\alpha+2\beta}]}(x)| < 2\beta$, hence and the inequality

$$\frac{(\mathcal{M}+2\sigma)^{-1}}{1+|u-u^{[S_{\alpha+2\beta}]}|\,(\mathcal{M}+2\sigma)^{-1}} > \frac{1}{\mathcal{M}+4\sigma} = \frac{a}{\lambda},$$

is providing the result in the analogous manner as in [L].

Proof of the Theorem. First we note that there is a positive constant $\epsilon > 0$ such that for all $x \in \Omega_1 \setminus \Omega_0$ we have $\dfrac{|u-u^{[S_{\alpha+2\beta}]}|\,(\mathcal{M}+2\sigma)^{-1}}{1+|u-u^{[S_{\alpha+2\beta}]}|\,(\mathcal{M}+2\sigma)^{-1}} \geq \epsilon$, therefore Proposition 2.1 yields the Caccioppoli estimate for this set. For every $\delta > 0$ we have

$$\int\limits_{B_{2R}\cap(\Omega_1\,\cup\,\Omega_0)} \eta^p|Du|^p dx$$

$$\leq \delta \int\limits_{B_{2R}} |Du|^p\eta^p dx + \frac{1}{R^p}\mathcal{K}_{1,\delta}\int\limits_{B_{2R}} |u-\overline{u}|^p\,dx + \mathcal{K}_{2,\delta}\int\limits_{B_{2R}} (|f|+a)\,|u-\overline{u}|\,dx.$$

The estimate for the whole set B_{2R} now proceeds with a finite induction. Suppose that the Caccioppoli estimate holds for $(\Omega_k \cup \Omega_0) \cap B_{2R}$. Then testing with $(u-u^{[S_{\alpha+(k+2)\beta}]})$ yields the estimate for $(\Omega_{k+1} \cup \Omega_0) \cap B_{2R}$. Likewise we obtain the estimate for $(\Omega_{-(k+1)} \cup \Omega_0) \cap B_{2R}$ from the one for $(\Omega_{-k} \cup \Omega_0) \cap B_{2R}$. After finitely many steps, m_+ and m_- say, we have $u((\Omega_{m_+} \cup \Omega_{m_-} \cup \Omega_0) \cap B_{2R}) \subset B_\epsilon(\overline{u})$, for some arbitrarily small ϵ, adjusting the step length β in the last steps, if necessary.

Finally, we use $\eta|u - \overline{u}|$ as test function to estimate $\eta^p |Du|^p$ on $u^{-1}(B_\epsilon(\overline{u}))$ with the usual smallness argument.

3. Regularity results

We get the higher integrability result without further restriction on the structure conditions on $A(u)$ and $B(u)$.

Corollary 3.1. Let $f \in L^l(\Omega)$ with $l > p/t$, $t = p - 1 + p/n$ and $n \geq 3$. If u satisfies the hypothesis of the theorem, then there are positive constants ϵ and $\mathcal{K}$ not depending on R such that $u \in W^{1,q}_{loc}(\Omega)$ for $q \in [p, p + \epsilon)$ and

$$\Big(\fint_{B_{R/2}} |Du|^q dx \Big)^{\frac{1}{q}} \leq \mathcal{K} \Big\{ \Big(\fint_{B_R} |Du|^p dx \Big)^{\frac{1}{p}} + \Big(R \Big[\fint_{B_R} (|f| + a)^{\frac{q}{t}}\, dx \Big]^{\frac{t}{q}} \Big)^{\frac{1}{p-1}} \Big\}.$$

Proof. The proof is based on application of the reverse Hölder inequality, cf.: [G,p.122]. From (1) and Sobolev-Poincaré inequality it follows that

$$\fint_{B_{R/2}} |Du|^p\, dx \leq \mathcal{K}_3 \Big(\fint_{B_R} |Du|^{\frac{np}{n+p}}\, dx \Big)^{\frac{n+p}{n}} + \mathcal{K}_2 \fint_{B_R} (|f| + a)|u - \bar{u}|\, dx\,.$$

Applying at first the Hölder inequality, then the Sobolev - Poincaré inequality and Young's inequality to the last integral and we get for $r = np/(n - p)$

$$\begin{aligned}
\mathcal{K}_2 \int_{B_R} (|f| + a)|u - \bar{u}|\, dx &\leq \mathcal{K}_2 \Big(\int_{B_R} |u - \bar{u}|^r\, dx \Big)^{\frac{1}{r}} \Big(\int_{B_R} (|f| + a)^{\frac{r}{r-1}}\, dx \Big)^{\frac{r-1}{r}} \\
&\leq \theta \int_{B_R} |Du|^p\, dx + \mathcal{K}_5(\theta) \Big(\int_{B_R} (|f| + a)^{\frac{r}{r-1}}\, dx \Big)^{\frac{r-1}{r} \frac{p}{p-1}} \\
&= \theta \int_{B_R} g^{\frac{n+p}{n}}\, dx + \int_{B_R} F^{\frac{n+p}{n}}\, dx,
\end{aligned}$$

where $0 < \theta < 1$, $g = |Du|^{\frac{np}{n+p}}$ and the function F is given by

$$F = \mathcal{K}_6 (|f| + a)^{\frac{q}{st}} \Big(\int_{B_R} (|f| + a)^{\frac{p}{t}}\, dx \Big)^{\frac{n}{n+p} (\frac{t}{p-1} - 1)}$$

for s and q with $q = \dfrac{nps}{n+p}$. Consequently

$$\fint_{B_{R/2}} g^{\frac{n+p}{n}}\,dx \le \mathcal{K}_3 \Big(\fint_{B_R} g\,dx\Big)^{\frac{n+p}{n}} + \theta \fint_{B_R} g^{\frac{n+p}{n}}\,dx + \fint_{B_R} F^{\frac{n+p}{n}}\,dx.$$

By Proposition 1.1 [G.,p.122], there is an $\epsilon > 0$ not depending on R such that $g \in L^s_{loc}(\Omega)$ for $s \in \left[\frac{n+p}{n}, \frac{n+p}{n} + \epsilon\right)$ and

$$\Big(\fint_{B_{R/2}} |Du|^q\,dx\Big)^{\frac{1}{q}} \le \mathcal{K}_8 \Big\{ \Big(\fint_{B_R} |Du|^p\,dx\Big)^{\frac{1}{p}} + \Big(\fint_{B_R} F^s\,dx\Big)^{\frac{1}{q}} \Big\}.$$

Let's estimate the last integral using the Hölder inequality:

$$\begin{aligned}
\fint_{B_R} F^s\,dx &= \mathcal{K}_6^s \fint_{B_R} (|f|+a)^{\frac{q}{t}}\,dx \Big(\int_{B_R} (|f|+a)^{\frac{p}{t}}\,dx\Big)^{\frac{q}{p}(\frac{t}{p-1}-1)} \\
&\le \mathcal{K}_6^s \fint_{B_R} (|f|+a)^{\frac{q}{t}}\,dx \Big\{ \Big[\int_{B_R} (|f|+a)^{\frac{q}{t}}\,dx\Big]^{\frac{p}{q}} \Big(\frac{R^n}{R^n}\Big)^{\frac{p}{q}} \Big(\int_{B_R} dx\Big)^{1-\frac{p}{q}} \Big\}^{\frac{q}{p}(\frac{t}{p-1}-1)} \\
&\le \mathcal{K}_9 \Big\{ \Big[\int_{B_R} (|f|+a)^{\frac{q}{t}}\,dx\Big]^{\frac{t}{q}} R \Big\}^{\frac{q}{p-1}},
\end{aligned}$$

and the corollary follows.

In order to obtain the Hölder continuity for $p = 2$ the solution is compared locally to the solution of the unperturbed system with constant coefficients for which regularity properties are known from the classical theory, cf.: [G, p.167, ff.]. The argument is based on the fact that for systems with constant coefficients the ellipticity implies an inequality of the type

$$|D(u-v)|^p \le \sum_{i=1}^{N} \sum_{k=1}^{M} \big(A_i^k(x,u,D(u)) - A_i^k(x,v,D(v))\big)\big(Dv(x) - Dv(x)(x)\big),$$

with $p = 2$. Such a condition is often referred to as a strict monotonicity condition. However, it is not satisfied even for the p-Laplacian, if $p < 2$. Instead of employing such a strict monotonicity condition we have an argument based on the assumption of the convexity of potential of the elliptic operator. That, of course, is satisfied by the p-Laplacian for all $p > 1$. In order not to introduce further technical details we assume in the following that $A(u)$ actually is the p-Laplacian. Our argument immediately applies to more general operators as long as the regularity result of Tolksdorf is available, cf.: [T]. We have

Proposition 3.1. Let u be a weak solution of the p-Laplace system:

$$-\operatorname{div}(|Du|^{p-2}Du) + B = f$$

with $u, B,$ and f as in Corollary 3.1. If, moreover, $f \in L^s(\Omega)$ for $s > n/p$, then for every $0 < \rho < R_o(x_o)$ for some $R_o(x_o)$ we have

$$\int_{B_\rho} (1+|Du|^p)\,dx \le C\{[(\frac{\rho}{R})^n + \chi(x_o,R)] \int_{B_R} (1+|Du|^p)dx + \|f\|_{L^\sigma(\Omega)} R^{n-p+p\gamma}[1+\chi(x_o,R)]\} \tag{2}$$

with $\sigma = \max\{\frac{q}{t}, s\}$, $\gamma = \min\{1 - \frac{1}{p-1}(\frac{n}{\sigma}-1), 1-\frac{n}{\sigma p}\} > 0$ and

$$\chi(x_o,R) = (R^{p-n}\int_{B_\rho} |Du|^p\,dx)^{\frac{q-p}{pq}}.$$

Proof. Since (2) is obvious for $\rho \ge R/2$, we assume $\rho < R/2$ and consider a weak solution of the unperturbed system

$$\begin{aligned} \operatorname{div}(|Dv|^{p-2}Dv) &= 0, \quad \text{on } B_{R/2}, \\ v &= u, \quad \text{on } \partial B_{R/2}. \end{aligned}$$

For the p-Laplacian we have the maximum principle that the values of the solution are in the convex hull of its boundary values, hence $\|v\|_{L^\infty(B_{R/2})} \le \|u\|_{L^\infty(\Omega)}$, cf. eg.: [L]. Since a weak solution of the homogeneous problem is the unique minimum of the associated functional we have $\int_{B_{R/2}} |Dv|^p\,dx \le \int_{B_{R/2}} |Du|^p\,dx$;

and from [T] we infer $\int_{B_\rho} |Dv|^p\,dx \le \text{const}\,(\rho/R)^n \int_{B_{R/2}} (1+|Dv|^p)\,dx$.

For $w = u - v$, we have $w \in W_0^{1,p}(B_{R/2})$, and

$$\int_{B_{R/2}} |Dw|^p\,dx \le \text{const} \int_{B_{R/2}} (|Du|^p + |Dv|^p)\,dx \le \text{const} \int_{B_{R/2}} |Du|^p\,dx.$$

Setting $\tilde\gamma = 1 - \frac{1}{p-1}(\frac{n}{\sigma}-1)$ we want to show now that

$$\int_{B_{R/2}} (1+|Du|^p)|w|\,dx \le \text{const}\,\chi(x_0,R)[\int_{B_R} (1+|Du|^p)\,dx + R^{n-p+p\tilde\gamma}\|f\|_{L^\sigma(\Omega)}]. \tag{3}$$

Indeed, applying at first Hölder inequality and then Corollary 3.1, we get

$$\begin{aligned} \fint_{B_{R/2}} (1+|Du|^p)|w|\,dx &\le (\fint_{B_{R/2}} |w|^{\frac{q}{q-p}}\,dx)^{\frac{q-p}{q}} [(\fint_{B_{R/2}} dx)^{\frac{p}{q}} + (\int_{B_{R/2}} |Du|^q\,dx)^{\frac{p}{q}}] \\ &\le c_1(\fint_{B_{R/2}} |w|^{\frac{q}{q-p}}\,dx)^{\frac{q-p}{q}} [1 + \fint_{B_R} |Du|^p\,dx + \{R(\fint_{B_R} (|f|+a)^{\frac{q}{t}}\,dx)^{\frac{t}{q}}\}^{\frac{p}{p-1}}]. \end{aligned}$$

Since $\|w\|_{L^\infty(B_{R/2})} \leq 2\|u\|_{L^\infty(\Omega)}$, we can write $|w|^{q/q-p} = |w|^{p/q-p}|w| \leq \text{const}|w|$ and obtain (3) with Poincare's inequality:

$$\Big(\fint_{B_{R/2}} |w|^{\frac{q}{q-p}}\,dx\Big)^{\frac{q-p}{q}} \leq c_2\Big(\fint_{B_{R/2}} |w|^p dx\Big)^{\frac{q-p}{pq}} \leq c_3\Big(R^p \fint_{B_{R/2}} |Du|^p\,dx\Big)^{\frac{q-p}{pq}}.$$

To prove (2) we use the convexity of the map $\xi \to |\xi|^p$, as in [LM], to get

$$\int_{B_\rho} |Du|^p\,dx \leq \frac{1}{2}\int_{B_\rho} |Du|^p\,dx + c\int_{B_\rho} |Dv|^p\,dx$$
$$+\,p\sum_{\alpha=1}^{n}\sum_{i=1}^{m}\int_{B_{R/2}} |Du|^{p-2}\,D_\alpha u^i D_\alpha (u-v)^i\,dx$$

Since the trivial extension of $w = u - v$ also can be used as a test function for the original equation, that yields

$$\int_{B_\rho} |Du|^p\,dx \leq \frac{1}{2}\int_{B_\rho} |Du|^p\,dx + c\int_{B_\rho} |Dv|^p\,dx + ap\int_{B_{R/2}} (1+|Du|^p)|w|\,dx + \int_{B_{R/2}} |f||w|\,dx.$$

We further note

$$\int_{B_{R/2}} |f||w|\,dx \leq \|f\|_{L^\sigma(\Omega)}\,R^{n(1-1/\sigma)} = \|f\|_{L^\sigma(\Omega)}\,R^{n-p+p(1-n/\sigma p)}.$$

Using the facts gathered at the beginning of the proof we conclude

$$\int_{B_\rho} |Du|^p\,dx \leq \text{const}\Big\{(\frac{\rho}{R})^n\int_{B_R}(1+|Du|^p)\,dx + \chi(x_o,R)\Big[\int_{B_R}(1+|Du|^p)\,dx$$
$$+\,R^{n-p+p\gamma}\|f\|_{L^\sigma(\Omega)}\Big] + \|f\|_{L^\sigma(\Omega)}R^{n-p+p\gamma}\Big\}.$$

Now from Proposition 3.1 we obtain the local Hölder continuity

Corollary 3.2. With the assumptions of Proposition 3.1, there is an open set $\Omega_o \subset \Omega$ such that $u \in C^{o,\gamma}(\Omega_o)$, where γ is the same as in Proposition 3.1 and the $(n-q)$-dimensional Hausdorff measure $H^{n-q}(\Omega \setminus \Omega_o) = 0$ for some $q > p$.

Proof. Let

$$\phi(R) = R^{p-n}\int_{B_R}(1+|Du|^p)\,dx \quad \text{and} \quad \rho = \tau R \quad \text{with} \quad 0 < \tau < 1.$$

It follows from (3) that

$$\phi(\tau R) \leq C(\tau R)^{p-n}\{[\tau^n + \chi(R)]\,R^{n-p}\phi(R) + \|f\|_{L^\sigma(\Omega)}(1+\chi(R))\,R^{n-p+p\gamma}\}$$
$$= C\{(\tau^p + \chi(R)\,\tau^{p-n})\,\phi(R) + \tau^{p-n}\|f\|_{L^\sigma(\Omega)}(1+\chi(R))\,R^{p\gamma}\}$$

Let now $\gamma < \beta < 1$ and choose τ in such a way that $2\mathcal{C}\tau^{p-p\beta} = 1$ (we may assume $2\mathcal{C} > 1$ and so $\tau < 1$). Defining

$$\Omega_0 = \{x_0 \in \Omega \mid \exists R_0 < \min\{1, \operatorname{dist}(x_0, \partial\Omega)\} \text{ s. t. } \sup_{R<R_0} \chi(x_o, R) < \tau^n\},$$

we get for $x_0 \in \Omega_0$ and $R < R_0$, with the arguments analogous to those of [G.p170] the estimate

$$\phi(\tau^k R) \leq \operatorname{const}(\tau^k)^{p\gamma},$$

and hence for any $\rho < R$, we have $\phi(x_0, \rho) \leq \operatorname{const}(\rho/R)^{p\gamma}$, yielding

$$\int_{B_\rho} |Du|^p \, dx \leq \operatorname{const} \rho^{n-p+p\gamma},$$

Morrey's classical criterion providing the local Hölder continuity with exponent γ, c.f.e.g.: [G, p. 64].

References

[G] Giaquinta, M.: *Multiple Integrals in the Calculus of Variations and Nonlinear Elliptic Systems.* Princeton University Press, Princeton, NJ; 1983

[HW] Hildebrandt, S., Widman, K.-0.: *On the Hölder continuity of weak solutions of quasilinear elliptic systems of second order.* Ann. Sc. Norm. Sup. Pisa, **4** (1977), 145–178.

[L1] Landes, R.: *Test functions for elliptic systems and maximum principles.* Forum Math., **12** (2000), 23–52.

[L2] Landes, R.: *On the regularity of weak solutions of certain elliptic system.* Submitted.

[LM] Landes, R., Mirafzali, A.: *Some Regularity Results for Elliptic Systems.* Submitted.

[T] Tolksdorf, P.: *Everywhere regularity for some quasilinear systems with lack of ellipticity.* Ann. Math. P. Appl. (1983), 241–266.

[W1] Wiegner, M. *Ein optimaler Regularitätssatz für schwache Lösungen gewisser elliptischer Systeme.* Math. Z. **147** (1976), 21–28.

Marina Borovikova
Department of Mathematics
The University of Oklahoma
Norman, OK 73019, USA

Rüdiger Landes
Department of Mathematics
The University of Oklahoma
Norman, OK 73019, USA
E-mail address: `rlandes@AFTERMATH.math.ou.edu`

D. Haroske, T. Runst, H.-J. Schmeisser (eds.): Function Spaces, Differential Operators and Nonlinear Analysis. The Hans Triebel Anniversary Volume.

Complements and Results on h-sets

Michele Bricchi

Dedicated to Prof. Hans Triebel on the occasion of his 65th birthday

1. How to measure smoothness

This is the celebrated enticing title of Chapter 1 of the monograph *Theory of function spaces II*, by Hans Triebel ([Tri92]). On this occasion we could not withstand the temptation to quote such an impressive heading title. But instead of considering function spaces we shall take into consideration geometrical sets, i.e., smoothness is to be referred to the geometry of some sets in $\mathbb{R}^n$. To be honest, our efforts to describe smoothness (or beautiful badness, as it shall be clear) of some irregular sets in $\mathbb{R}^n$ are motivated by the desire to define function spaces of Besov and Triebel-Lizorkin type *on* these sets. We shall not treat this aspect here and we rather refer to [Bri02] and to forthcoming papers for a complete discussion on this subject.

1.1. d-sets

So, leaving the realm of function spaces (which, nonetheless, lurk eagerly behind the scene), we precise with some rigour the measure-geometrical problem we have in mind, beginning with some heuristical considerations.

It is known that if $\Gamma \subset \mathbb{R}^n$ is a self-similar fractal generated by the system of contractive similarities $\mathfrak{S} = \{S_i\}_{i=1}^N$ (say, with the Open Set Condition), then there is a (uniquely determined normalised) finite Radon measure μ supported by Γ with the property

$$\mu = \sum_{i=1}^{N} r_i^d \mu \circ S_i^{-1}, \tag{1}$$

where $\mu \circ S_i^{-1}(\cdot)$ is the image measure $\mu(S_i^{-1}(\cdot))$, $r_i = \mathrm{Lip}(S_i)$, for $i = 1, \ldots, N$ and $d \in [0, n]$ is the similarity dimension of the system $\mathfrak{S}$ (which, by the way, turns out to be also the Hausdorff dimension of the set Γ). See for instance [Hut81, Fal85].

From (1) and some additional considerations one arrives at the estimation

$$c_1 r^d \le \mu(B(\gamma, r)) \le c_2 r^d, \qquad r \in (0,1],\ \gamma \in \Gamma, \tag{2}$$

for some positive constants c_1 and c_2 independent of r and γ as above.

In sharp contrast to (1), in (2) the similarities related to Γ do not appear. Indeed, those sets satisfying the last assertion have got a proper name.

Definition 1.1. *Let Γ be a non-empty closed set in $\mathbb{R}^n$. Let d be a real number in $[0,n]$. Then Γ is a* d-set *if there exists a Radon measure μ with*

$$\operatorname{supp}\mu = \Gamma; \tag{3}$$

$$\mu(B(\gamma,r)) \sim r^d, \qquad r \in (0,1],\ \gamma \in \Gamma. \tag{4}$$

Here and in the sequel we use $\sim$ in (4) as an abbreviation of (2).

Remark 1.2. In complete analogy, we shall also use the same notation for positive sequences $a_k \sim b_k$, $k \in \mathbb{N}_0$, positive functions $f(r) \sim g(r)$, $0 < r \leq 1$ and Radon measures $\mu(A) \sim \nu(A)$, A Borel set.

Definition 1.1 (actually, even a more general version) has been given by A. Jonsson and H. Wallin in [JW84, Ch. 8, p. 205]. As we have already remarked d-sets need not to show any self-similarity behaviour: for instance d-sets are stable under bi-Lipschitz maps, whereas the image of a self-similar fractal under such a map is no more (in general) a self-similar fractal.

Remark 1.3. The relevant role played by the measure μ in the above definition is only apparent: if Γ is a d-set, then any admitted measure μ with (3) and (4) is equivalent to $\mathcal{H}^d|\Gamma$ (d-dimensional Hausdorff measure restricted to Γ) which therefore can be regarded as the (essentially) uniquely determined measure verifying the properties (3) and (4).

Any d-set Γ has Hausdorff dimension $\dim_{\mathcal{H}}\Gamma$ equal to d. One has even the following stronger assertion (which is obvious for self-similar fractals):

$$\dim_{\mathcal{H}}(\Gamma \cap B(\gamma,r)) = d, \qquad \gamma \in \Gamma,\ r \in (0,1]. \tag{5}$$

Remark 1.4. For any given $d \in [0,n]$ there exists a self-similar fractal with Hausdorff dimension d. The question of the *existence* of d-sets is then easily solved. In contrast to this case, the proof of the existence of more general sets defined in some formal way is one of the central difficulties we shall face later on.

1.2. A refined construction: (d,Ψ)-sets

The definition of d-sets is admittedly handsome: they contain self-similar fractals as outstanding examples, their geometrical structure (which can be very nasty and highly non-rectifiable) is described only in terms of the local behaviour of some Radon measures (whose exact nature is immaterial by Remark 1.3) and still one can study successfully function spaces defined on these sets, such as the Besov spaces $B^s_{pq}(\Gamma)$. The works [JW84] and [Tri97] are mostly dedicated to this subject.

Lately, D. Edmunds and H. Triebel have considered in [ET98, ET99] a refined variant of d-sets. We quote the necessary definitions.

Definition 1.5. *A positive monotone function* $\Psi : (0,1] \to \mathbb{R}$ *is called* admissible *if*

$$\Psi(2^{-2j}) \sim \Psi(2^{-j}), \qquad j \in \mathbb{N}_0. \tag{6}$$

Roughly speaking, admissible functions have at most a logarithmic growth or decay near zero. Typical examples are in fact ($c > 0$ appropriately small) $\Psi(r) = |\log cr|^b$ and $\Psi(r) = (\log|\log cr|)^b$, $b \in \mathbb{R}$.

The definition of (d, Ψ)-sets reads then as follows.

Definition 1.6. *Let Γ be a closed non-empty set in $\mathbb{R}^n$ and let $d \in (0, n]$. Then Γ is a (d, Ψ)-set if either $d < n$ and there exists a Radon measure μ with*

$$\operatorname{supp} \mu = \Gamma, \tag{7}$$

$$\mu(B(\gamma, r)) \sim r^d \Psi(r), \qquad r \in (0,1],\ \gamma \in \Gamma, \tag{8}$$

or $d = n$, Ψ is unbounded and (7) and (8) hold true (with $d = n$) for some Radon measure μ.

Remark 1.7. Actually in [ET98] the definition of (d, Ψ)-sets embraces only compact sets but there is no problem to consider here also unbounded variants.

The difficult problem concerning (d, Ψ)-sets is to assert their existence. Indeed, if one knew that a (d, Ψ) exists, then one could easily extract properties analogous to those stated for d-sets: for instance the Hausdorff dimension would be (even locally) again d (in accord to the intuitive idea that Ψ is only a minor perturbation of the main term r^d) and, up to equivalence, there is only one measure satisfying (7) and (8), constructed as a refinement of the usual d-dimensional Hausdorff measure restricted to Γ.

The main proposition proved by D. Edmunds and H. Triebel concerning the geometry of (d, Ψ) is actually an existence assertion:

Proposition 1.8. *Given any admissible function Ψ and given any $d \in (0, n]$ (Ψ unbounded if $d = n$) then there exists an example of a compact (d, Ψ)-set in $\mathbb{R}^n$. Moreover, any such example can be provided as an attractor of (possibly) infinitely many similarities (pseudo self-similar fractal).*

Remark 1.9. The suggestive second part of the above proposition is the counterpart to the case of d-sets: as self-similar fractals (attractors of finitely many similarities) are special cases of d-sets, so pseudo self-similar fractals are special cases of (d, Ψ)-sets. We cannot go into details. We refer to the quoted papers and specially to [Tri01, Sections 22.1–22.8, pp. 329–339], [Bri99, Bri00] and [dM01] where this subject is discussed in great generality.

2. h-sets

Taking into account Definitions 1.1 and 1.6 (and the results obtained starting with these definitions, both from a geometrical point of view and – what's more – from the point of view of function-space theory) one cannot withstand the temptation to replace the special functions r^d and $r^d \Psi(r)$ by a general function $h(r)$, trying to unify sparse examples in a unique language (h-sets). This is our point of view and we follow here our results contained in [Bri01] and [Bri02]. We shall also cover the small gap between necessary and sufficient conditions for the existence of h-sets stated in [Bri02, Theorem 1.7.6], ending up with the complete description of these sets.

Now we quote the necessary rigorous definitions.

Definition 2.1. *Let* $\mathbb{H}$ *denote the class of all positive continuous and non-decreasing functions* $h\colon (0,1] \to \mathbb{R}$ *(*gauge functions*).*

Definition 2.2. *Let* $h \in \mathbb{H}$. *Then a closed non-empty compact set* $\Gamma \subset \mathbb{R}^n$ *is called* h-set *if there exists a Radon measure* μ *with*

$$\operatorname{supp}\mu = \Gamma, \tag{9}$$

$$\mu(B(\gamma, r)) \sim h(r), \qquad r \in (0,1],\ \gamma \in \Gamma, \tag{10}$$

If for a given $h \in \mathbb{H}$ *there exists an* h*-set* $\Gamma \subset \mathbb{R}^n$, *we call* h *a* measure function *(*in $\mathbb{R}^n$*) and any related measure* μ *with* (9) *and* (10) *will be called* h-measure *(*related to Γ*).*

Our definition contains as special cases d-sets and (d, Ψ)-sets (after the immaterial replacement, if it is the case, of Ψ with a continuous version).

Also in this case the difficult part is to characterise the functions $h \in \mathbb{H}$ for which there actually exists an h-set. In other words, our task reads as follows:

$$\textit{characterise all measure functions in } \mathbb{R}^n. \tag{11}$$

It is easier to provide necessary conditions. We quote two results in this direction concerning the local Hausdorff (and packing) dimension of an h-set and the existence of a unique h-measure (up to equivalence), before the characterisation theorem 2.7. Both are stated and proved in [Bri01, Bri02]. In these works one can also find a more detailed account on geometric properties of h-sets (ball condition, local Markov's inequality, Minkowski's contents and others).

Theorem 2.3. *Let* Γ *be an* h*-set in* $\mathbb{R}^n$. *Then one has for any* $r \in (0,1]$ *and any* $\gamma \in \Gamma$

$$\dim_{\mathcal{H}}(\Gamma \cap B(\gamma, r)) = \liminf_{t \to 0} \frac{\log h(t)}{\log t} := \underline{\omega}(h) \tag{12}$$

and

$$\dim_{\mathcal{P}}(\Gamma \cap B(\gamma, r)) = \limsup_{t \to 0} \frac{\log h(t)}{\log t} := \overline{\omega}(h). \tag{13}$$

Remark 2.4. $\dim_{\mathcal{H}} A$ and $\dim_{\mathcal{P}} A$ stand for the Hausdorff and the packing dimension of $A \subset \mathbb{R}^n$, respectively. The expressions $\underline{\omega}(h)$ and $\overline{\omega}(h)$ are sometimes called *lower order* and *upper order* of the function h, respectively. Of course Theorem 2.3 implies the weaker global assertion $\dim_{\mathcal{H}} \Gamma = \underline{\omega}(h)$ and $\dim_{\mathcal{P}} \Gamma = \overline{\omega}(h)$.
Since one always has $0 \le \dim_{\mathcal{H}} A \le \dim_{\mathcal{P}} A \le n$ for any subset A of $\mathbb{R}^n$ (see for instance [Mat95]), by virtue of the above theorem one has some information on the decay of $h(r)$, as $r \to 0$.

This observation rules out functions like $h(r) = \exp\{-r^{\varkappa}\}$, $\varkappa < 0$, (extreme case: $\underline{\omega}(h) = \overline{\omega}(h) = \infty$) but also functions as $h(r) = r^\delta$, $\delta > n$ ($\underline{\omega}(h) = \overline{\omega}(h) = \delta$) from the list of measure functions (in $\mathbb{R}^n$). On the other hand, the function $h(r) = r^n |\log cr|^{-1}$ ($c > 0$ suitably small) has lower and upper order both equal to n and up to now we cannot neither accept it nor reject it from the list, though one might guess that the decay near zero of this function exceeds the "limit of tolerance". This limit is represented, in vague terms, by the function $h(r) = r^n$ which is surely a measure function (even related to an n-set). We shall be more precise in the sequel, however, we warn here that the attractive feeling that $0 < \underline{\omega}(h) \leq \overline{\omega}(h) < n$ might be a sufficient condition for a gauge function to be a measure function is totally misleading. See Remark 2.10 for a discussion.

The second result we want to quote before the main assertion concerns the uniqueness of the h-measure related to an h-set.

Theorem 2.5. *Let $\Gamma \subset \mathbb{R}^n$ be an h-set. Then all h-measures μ related to Γ are equivalent to $\mathcal{H}^h|\Gamma$, where the latter stands for the restriction to Γ of the generalised Hausdorff measure with respect to the gauge function h.*

Remark 2.6. Also for this assertion we wish to shift details to the above quoted papers and to [Rog70] for the theory of generalised Hausdorff measures. Here we remark that for any gauge function h one can define the measure $\mathcal{H}^h$: in particular, if $h(r) = r^d$ then $\mathcal{H}^h$ coincides with the usual d-dimensional Hausdorff measure, whereas with $h(r) = r^d\Psi(r)$, where $d \in (0, n]$ and Ψ is a (continuous) admissible function, then the measure $\mathcal{H}^h$ restricted to a (d, Ψ)-set Γ satisfies (7) and (8).

Now we can state the main assertion of this note.

Theorem 2.7. *Let h be a gauge function. Then h is a measure function in $\mathbb{R}^n$ if, and only if,*

$$\frac{h(2^{-k}r)}{h(r)} \gtrsim 2^{-kn}, \qquad k \in \mathbb{N}_0,\ r \in (0,1]. \tag{14}$$

Remark 2.8. We have used a compact notation here. The meaning of (14) is the following: there exists a gauge function $\widetilde{h} \sim h$ for which $\widetilde{h}(2^{-k}r)/h(r) \geq 2^{-kn}$, $k \in \mathbb{N}_0$, $r \in (0, 1]$.

We shall prove also the following discrete version of the above theorem, whose "if"-part is occasionally easier to be checked.

Theorem 2.9. *Let h be a gauge function. Then h is a measure function in $\mathbb{R}^n$ if, and only if,*

$$\frac{h(2^{-k-j})}{h(2^{-j})} \gtrsim 2^{-kn}, \qquad j, k \in \mathbb{N}_0. \tag{15}$$

Proof of Theorems 2.7 and 2.9. We consider in detail the proof of 2.7 and at the end we comment on the necessary changes to deal with the proof of 2.9.

STEP 1. Let $h \in \mathbb{H}$ be a given gauge function. If $h \sim 1$ then h is obviously a measure function (corresponding to a 0-set) and Condition (14) is clearly fulfilled. So we can assume without loss of generality $h(0+) = 0$, which was the condition we systematically considered in [Bri01] and [Bri02].

STEP 2. In [Bri02, Theorem 1.7.6] we have proved that if $h(\lambda r)/h(r) \gtrsim \lambda^n$, for $0 < \lambda \le \lambda_0 \le 1$ and $r \in (0,1]$ then there exists an example of a compact h-set in $\mathbb{R}^n$. The basic step in order to prove this assertion was [Bri02, Lemma 1.7.7]. Examining the proof of this lemma it is evident that one can reach the same conclusion only with the weaker assumption (14) (it suffices to know that there exists a gauge function $g \sim h$ satisfying (14) or (15) with the usual $\ge$-symbol only definitively in k). Then the rest of the proof of the sufficiency follows literally from that given in [Bri02] and one ends up with a compact h-set which is even the attractor of infinitely many similarities. We shall come back in Remark 2.11 to this point. Of course the existence of a compact h-set immediately implies the existence of an unbounded h-set.

STEP 3. The less trivial part of the proof is to show that (14) is necessary. Let us suppose to have an h-set Γ. We have to find a gauge function $\widetilde{h} \sim h$ with

$$\frac{\widetilde{h}(2^{-k}r)}{\widetilde{h}(r)} \ge 2^{-kn}. \tag{16}$$

In order to get this, let us consider an h-measure μ related to Γ. Let

$$Q := \{x = (x_1, \dots, x_n) \in \mathbb{R}^n : 0 < x_i \le 1,\ i = 1, \dots, n\} \tag{17}$$

be the unit cube. Then define, for $j \in \mathbb{N}_0$ and $m \in \mathbb{Z}^n$,

$$Q_{j,m} := 2^{-j}Q + 2^{-j}m. \tag{18}$$

Suppose without loss of generality $\Gamma \cap 1/2Q \neq \emptyset$ and define, for $j \in \mathbb{N}_0$,

$$\widetilde{h}(2^{-j}) := \max_{Q_{j,m} \subset Q} \mu(Q_{j,m}) = \mu(Q_j^*). \tag{19}$$

Then the other values of $\widetilde{h}(r)$ for $r \in (0,1]$ are defined by piecewise linear interpolation: if $r \in (2^{-j-1}, 2^{-j})$, for some $j \in \mathbb{N}_0$, we let

$$\widetilde{h}(r) := (2^{j+1}r - 1)(\widetilde{h}(2^{-j}) - \widetilde{h}(2^{-j-1})) + \widetilde{h}(2^{-j-1}). \tag{20}$$

Now, notice that

$$\frac{\widetilde{h}(2^{-k-j})}{\widetilde{h}(2^{-j})} \ge 2^{-kn}, \qquad j, k \in \mathbb{N}_0. \tag{21}$$

As a matter of fact, for $j, k \in \mathbb{N}_0$,

$$\widetilde{h}(2^{-j}) = \mu(Q_j^*) = \sum_{Q_{j+k,m} \subset Q_j^*} \mu(Q_{j+k,m}) \le 2^{kn}\mu(Q_{j+k}^*) = 2^{kn}\widetilde{h}(2^{-j-k}), \tag{22}$$

which gives (21). In general, if $r \in [2^{-j-1}, 2^{-j}]$, for some $j \in \mathbb{N}_0$, the expression $\widetilde{h}(2^{-k}r)/\widetilde{h}(r)$ is of the form

$$\frac{ar + b}{a'r + b'} := F(r), \qquad r \in [2^{-j-1}, 2^{-j}], \tag{23}$$

where a, a', b and b' are numbers depending on j and k which can be calculated with some patience. By elementary arguments F attains its minimum at 2^{-j} or at 2^{-j-1}. In other words,

$$\frac{\widetilde{h}(2^{-k}r)}{\widetilde{h}(r)} \geq \frac{\widetilde{h}(2^{-k-j'})}{\widetilde{h}(2^{-j'})}, \tag{24}$$

where j' is either j or $j+1$. In any case, by virtue of (21),

$$\frac{\widetilde{h}(2^{-k}r)}{\widetilde{h}(r)} \geq 2^{-kn}, \tag{25}$$

which is what we claimed and therefore the proof of 2.7 is concluded.

STEP 4. The proof of 2.9 is practically contained in the above lines: surely, (15) follows as a necessary condition from (14). So it remains to show that if (15) holds, then h is a measure function. In order to accomplish this task one considers a function $\widetilde{h} \sim h$ with $\widetilde{h}(2^{-j-k})/\widetilde{h}(2^{-j}) \geq 2^{-kn}$, which exists by assumption. Afterwards one consider the function h^* which interpolates in a piecewise linear way the nodes $\widetilde{h}(2^{-j})$ as we did in the previous step. This gauge function h^* is equivalent to h and verifies $h^*(2^{-k}r)/h^*(r) \geq 2^{-kn}$, $k \in \mathbb{N}_0$, $0 < r \leq 1$. Then by what we have proved in the previous steps h is a measure function and this concludes the proof. □

Remark 2.10. One can regard (14) as refined doubling condition for the function h, in which the dimension n of the space $\mathbb{R}^n$, where we state the problem, is taken into account (as it should be). The usual doubling condition for a function, say, in $\mathbb{H}$ has the form

$$h(2^{-1}r) \geq ch(r), \qquad r \in (0,1]. \tag{26}$$

There is an abyss between gauge functions satisfying this condition and those gauge functions which are not doubling. The example of the function $h(r) = \exp\{-1/r\}$ is only the tip of the iceberg: given any positive number d, there are even gauge functions h with $\underline{\omega}(h) = \overline{\omega}(h) = d$ but still not doubling (and, a fortiori, not measure functions). This means that there are gauge functions of the form $h(r) = r^d\Phi(r)$ with $\log\Phi(r) = o(\log r)$ which are not measure functions.

Remark 2.11. There is an additional assertion which can be appended to Theorem 2.7 (or 2.9). Only in the proof we have spent some words on the possibility to exhibit for any given measure function h a compact h-set which is the attractor of a system of (possibly) infinitely many similarities. This fact stresses the analogy to the case of d-sets and self-similar fractals and it is in accord with the analogous assertion obtained by D. Edmunds and H. Triebel for (d, Ψ)-sets. Here we cannot go into details and we refer to [Bri01] where we discussed carefully this aspect, quoting the results obtained by G. Follo in [Fol01] on attractors in a wider context.

As a conclusion of this section we reveal the answer concerning the function $h(r) = r^n |\log cr|^{-1}$, where c is chosen small enough to guarantee $h \in \mathbb{H}$. We have

$$h(2^{-j-k})/h(2^{-j}) \sim 2^{-kn} k^{-1}, \qquad k, j \in \mathbb{N}_0, \tag{27}$$

and now it is evident that there cannot exist an equivalent function $\widetilde{h}$ to h with (15).

The next section will be dedicated entirely to the discussion of examples.

3. Examples

(i) As we already remarked, d-sets, $0 \le d \le n$ and (d, Ψ)-sets, $0 < d \le n$ and Ψ admissible, are particular cases of h sets, letting, respectively $h(r) = r^d$ and $h(r) = r^d \Psi(r)$.

(ii) The above examples are particular cases of a much wider situation. Let $\Phi(r)$ be a continuous function equivalent to a slowly varying function in the sense of [BGT87].
Then any function of the form $h(r) = r^d \Phi(r)$, $0 < d < n$, is a measure function ([Bri02, Theorem 1.9.15-(i)]), agreeing on the sloppy convention to pass to an equivalent gauge function if h is not monotone (which is possible). Celebrated examples of slowly varying functions are (r suitably small)

$$\begin{aligned}
&\Phi_1(r) = |\log r|^b, && b \in \mathbb{R};\\
&\Phi_2(r) = (\log r |\log r|)^b, && b \in \mathbb{R};\\
&\Phi_3(r) \sim \Psi(r), && \Psi \text{ admissible function};\\
&\Phi_4(r) = \exp\{b |\log r|^{\varkappa}\}, && b \in \mathbb{R},\ 0 < \varkappa < 1;\\
&\Phi_5(r) = \exp\{\log r / \log|\log r|\};\\
&\Phi_6(r) = \exp\{-\textstyle\int_r^1 \varepsilon(s)\, ds/s\}, && \varepsilon(s) = o(1) \text{ and measurable.}
\end{aligned}$$

The first two examples are a special case of the third one. Here we must add a comment: an admissible function Ψ needs not always to be slowly varying, but there is always an equivalent representative which is slowly varying (see [Bri02, Proposition 1.9.7]). Φ_4 and Φ_5 are slowly varying functions and they are not equivalent to any admissible function (despite of their definition, they are not "logarithmic functions"). Finally, Φ_6 is the most general form of a slowly varying function (actually it is a characterisation: any slowly varying function is equivalent to Φ_6 for an appropriate function $\varepsilon(s)$) and hence all previous cases are subsumed in the latter.

(iii) We can also consider limiting cases (r suitably small): the function $h(r) = |\log r|^b$, $b < 0$, is tout court a measure function. The same holds for $h(r) = \exp\{b |\log r|^{\varkappa}\}$, with $b < 0$ and $0 < \varkappa < 1$. The corresponding sets have Hausdorff dimension and packing zero (by Theorem 2.3). More generally any continuous and monotone increasing slowly varying function is a measure function ([Bri02, Theorem 1.9.15-(iii)]).

Conversely, $h(r) = r^n \exp\{b|\log r|^{\varkappa}\}$, $b > 0$, $0 < \varkappa < 1$ is an example of a measure function corresponding to an n-dimensional h-set, which is not an (n, Ψ)-set, for any admissible Ψ. More generally, $r^n\Phi(r)$ is a measure function in $\mathbb{R}^n$, provided Φ is a decreasing slowly varying function ([Bri02, Theorem 1.9.15-(ii)]).

(iv) All these examples share the nice property that all involved functions h have lower and upper order equal. We want to show that it may also happen that $\underline{\omega}(h) < \overline{\omega}(h)$, for some measure function h. The following example is taken from [Bri02, Example 1.10.5].

Let $\{k_l\}_{l\in\mathbb{N}}$ be a strictly increasing sequence of positive integers.

Let $\xi\colon (0,1] \to \mathbb{R}$ be defined as follows.

$$\xi(s) = \begin{cases} 1, & \text{if } s \in (2^{-k_l-1}, 2^{-k_l}), \\ 0, & \text{otherwise.} \end{cases} \tag{28}$$

Now define

$$h(r) = \exp\Big\{-\int_r^1 \xi(s)\,\frac{ds}{s}\Big\}, \qquad r \in (0,1]. \tag{29}$$

h is readily a gauge function and one can prove that h is a measure function in, say, $\mathbb{R}^2$.

Let us evaluate the expression $\log h(r)/\log r$, as $r \to 0$. Suppose $r \in (2^{-k_l-1}, 2^{-k_l}]$. Then

$$\frac{\log h(r)}{\log r} = \frac{\int_r^1 \xi(s)\,ds/s}{|\log r|} \leq \frac{\int_{2^{-k_l-1}}^1 \xi(s)\,ds/s}{k_l} = \frac{l}{k_l}. \tag{30}$$

Analogously,

$$\frac{\log h(r)}{\log r} \geq \frac{\int_{2^{-k_l}}^1 \xi(s)\,ds/s}{k_l+1} = \frac{l-1}{k_l+1}. \tag{31}$$

Therefore,

$$\overline{\omega}(h) = \limsup_{r\to 0} \frac{\log h(r)}{\log r} = \limsup_{l\to\infty} \frac{l}{k_l} \tag{32}$$

and, similarly,

$$\underline{\omega}(h) = \liminf_{r\to 0} \frac{\log h(r)}{\log r} = \liminf_{l\to\infty} \frac{l}{k_l}. \tag{33}$$

So, it remains to choose an appropriate sequence $\{k_l\}_{l\in\mathbb{N}}$ in order to have the different lim inf and lim sup in the above expression to conclude, and this can be done easily.

Final remarks

In view of Theorem 2.7 we can assert that our Task (11) is now completely solved. In the above section we provided a considerable class of explicit examples and in [Bri02] we have considered even more measure functions, relying on the beautiful theory of O-regular functions developed originally by W. Feller (see [Fel69]) and then considered in full generality in [BGT87]. In particular, one could show that any measure function h is of the form

$$h(r) \sim \exp\left\{-\int_r^1 \xi(s)\,\frac{ds}{s}\right\}, \tag{34}$$

for some measurable and bounded function ξ. This sheds some light on the peculiarity of these functions. We have examined these aspects with some care in [Bri02, Chapter 1], so we refer to this work for further observations.

I wish to conclude this note by expressing my gratitude to Prof. Hans Triebel: thanks to his hints I could eventually prove the main results 2.7 and 2.9 in full generality.

References

[BGT87] N.H. Bingham, C.M. Goldie, and J.L. Teugels. *Regular variation.* Cambridge Univ. Press, 1987.

[Bri99] M. Bricchi. On some properties of (d,Ψ)-sets and related Besov spaces. *Jenaer Schriften zur Mathematik und Informatik*, 99/31, 1999.

[Bri00] M. Bricchi. On the relationship between Besov spaces $B^{(s,\Psi)}_{p,q}(\mathbb{R}^n)$ and L_p spaces on a (d,Ψ)-set. *Jenaer Schriften zur Mathematik und Informatik*, 00/13, 2000.

[Bri01] M. Bricchi. Existence and properties of h-sets. *Georgian Math. J.*, 2002

[Bri02] M. Bricchi. *Tailored function spaces and h-sets.* PhD thesis, University of Jena, 2002.

[dM01] S. Domingues de Moura. Function spaces of generalised smoothness. *Dissertationes Math.*, CCCXCVIII, 2001.

[ET98] D. Edmunds and H. Triebel. Spectral theory for isotropic fractal drums. *C. R. Acad. Sci. Paris*, 326, série I:1269–1274, 1998.

[ET99] D. Edmunds and H. Triebel. Eigenfrequencies of isotropic fractal drums. *Operator Theory: Advances and Applications*, 110:81–102, 1999.

[Fal85] K. J. Falconer. *The Geometry of fractal sets.* Cambridge Univ. Press, 1985.

[Fel69] W. Feller. One-sided analogues of Karamata's regular variation. *Enseign. Math.*, 15:107–121, 1969.

[Fol01] G. Follo. Some remarks on fractals generated by a sequence of finite systems of contractions. *Georgian Math. J.*, 2001.

[Hut81] J. E. Hutchinson. Fractals and self similarity. *Indiana Univ. Math. J.*, 30:713–747, 1981.

[JW84] A. Jonsson and H. Wallin. *Function spaces on subsets of* $\mathbb{R}^n$. Math. Reports 2, part 1. Harwood Acad. Publ., 1984.

[Mat95] P. Mattila. *Geometry of sets and measures in euclidean spaces.* Cambridge Univ. Press, 1995.

[Rog70] C. A. Rogers. *Hausdorff measure.* Cambridge Univ. Press, 1970.

[Tri92] H. Triebel. *Theory of function spaces II.* Birkhäuser, Basel, 1992.

[Tri97] H. Triebel. *Fractals and spectra.* Birkhäuser, Basel, 1997.

[Tri01] H. Triebel. *The structure of functions.* Birkhäuser, Basel, 2001.

Michele Bricchi
Dipartimento di Matematica
Università degli Studi di Pavia
Strada Ferrata 1
I-27100 Pavia (PV), Italy
E-mail address: bricchi@unipv.it

D. Haroske, T. Runst, H.-J. Schmeisser (eds.): Function Spaces, Differential Operators and Nonlinear Analysis. The Hans Triebel Anniversary Volume.

Lifting Properties of Sobolev Spaces

Viktor I. Burenkov[1]

Dedicated to Prof. Hans Triebel on the occasion of his 65th birthday

Abstract. We prove that a Sobolev-type inequality for an arbitrary open set $\Omega \subset \mathbf{R}^N$ is equivalent to inequalities of such type with different exponents for $\Omega \times I$, where I is an interval, or $\Omega \times \Omega$. The proofs are based on the factorisation property of the appropriate heat kernels.
AMS subject classifications: 35P15, 35J25, 47A75, 47B25
keywords: Sobolev inequalities, Neumann Laplacian, heat kernels.

Let, for a positive integer N and an open set $\Omega \subset \mathbf{R}^N$, $W^{1,2}(\Omega)$ denote the standard Sobolev space equipped with the norm

$$\|f\|_{W^{1,2}(\Omega)} = \left(\int_\Omega (|f|^2 + |\nabla f|^2)\, \mathrm{d}^N x \right)^{\frac{1}{2}}.$$

We shall prove some statements related to the validity of the embedding

$$W^{1,2}(\Omega) \subset L^q(\Omega) \tag{1}$$

for $q > 2$.

Embedding (1) is equivalent to existence of $c > 0$ such that

$$\|f\|_{L^q(\Omega)} \le c \|f\|_{W^{1,2}(\Omega)}$$

for all $f \in W^{1,2}(\Omega)$. (See, for example, [2, Section 4.1].) Recall that if $\Omega_1 \subset \mathbf{R}^N$ is a bounded open set such that $\overline{\Omega}_1 \subset \Omega$, then $W^{1,2}(\Omega) \subset L^q(\Omega_1)$ if, and only if, $q \le \infty$ for $N = 1$, $q < \infty$ for $N = 2$ and $q \le \frac{2N}{N-2}$ for $N \ge 3$.

Theorem 1. *Let $\Omega \subset \mathbf{R}^N$ be an open set and $I = (a, b)$, where $-\infty \le a < b \le \infty$. Moreover, let $2 < q < \infty$ for $N = 1, 2$ and $2 < q \le \frac{2N}{N-2}$ for $N \ge 3$, and let*

$$r = \frac{2(3q-2)}{q+2}.$$

Then embedding (1) *is equivalent to the embedding*

$$W^{1,2}(\Omega \times I) \subset L^r(\Omega \times I). \tag{2}$$

[1]Supported by the grants of INTAS (project 99-01080) and RFBR (project 99-01-00843).

The idea of the proof below and of the proof of Theorem 3 is actually a kind of indirect induction in dimension based on the fact that the heat kernel of the semigroup associated with the Neumann Laplacian, the domain of the square root of which coincides with $W^{1,2}(\Omega \times I)$, possesses the factorisation property making possible separation of variables.

Proof. 1. Embedding (2) is equivalent to the embedding

$$W^{1,2}(\Omega \times \mathbf{R}) \subset L^r(\Omega \times \mathbf{R}). \tag{3}$$

Indeed, let (2) be satisfied. Then for some $c_1 > 0$

$$\|f\|_{L^r(\Omega\times I)} \le c_1 \|f\|_{W^{1,2}(\Omega\times I)} \tag{4}$$

for all $f \in W^{1,2}(\Omega \times I)$. If $-\infty < a < b < \infty$, $I_k = I + k(b-a)$ for integer k, then $\mathbf{R} = \bigcup_{k=-\infty}^{\infty} \bar{I}_k$ and since $r > 2$

$$\begin{aligned} \|f\|_{L^r(\Omega\times\mathbf{R})} &= \left(\sum_{k=-\infty}^{\infty} \|f\|^r_{L^r(\Omega\times I_k)} \right)^{\frac{1}{r}} \\ &\le c_1 \left(\sum_{k=-\infty}^{\infty} \|f\|^r_{W^{1,2}(\Omega\times I_k)} \right)^{\frac{1}{r}} \\ &\le c_1 \left(\sum_{k=-\infty}^{\infty} \|f\|^2_{W^{1,2}(\Omega\times I_k)} \right)^{\frac{1}{2}} \\ &= c_1 \|f\|_{W^{1,2}(\Omega\times\mathbf{R})} \end{aligned}$$

for all $f \in W^{1,2}(\Omega \times \mathbf{R})$ which implies (3). If $a = -\infty$ and $b < \infty$ or $a > -\infty$ and $b = \infty$, (3) follows by a similar argument.

Next let (3) be satisfied. Then for some $c_2 > 0$

$$\|f\|_{L^r(\Omega\times\mathbf{R})} \le c_2 \|f\|_{W^{1,2}(\Omega\times\mathbf{R})}$$

for all $f \in W^{1,2}(\Omega \times \mathbf{R})$. Since $\Omega \times I$ is a cylinder, there exists a bounded extension operator $T : W^{1,2}(\Omega \times I) \to W^{1,2}(\Omega \times \mathbf{R})$. (See, for example [2, Section 6.1].) Consequently inequality (4) is valid with $c_1 = c_2 \|T\|$. Hence (2) follows.

2. Next we note that embedding (2) implies embedding (1) with r replacing q. Indeed, let $g \in C_0^\infty(I)$ be fixed and let $f \in W^{1,2}(\Omega)$. By applying (4) to $f(x)g(\xi), x \in \Omega, \xi \in I$, we get that for some $c_3 > 0$ independent of f

$$\|f\|_{L^r(\Omega)} \le c_3 \|f\|_{W^{1,2}(\Omega)} .$$

Hence (1) with r replacing q follows.

3. Let us define the Neumann Laplacian to be the non-negative self-adjoint operator $H = -\Delta_N$ acting in $L^2(\Omega)$ and associated with the quadratic form

$$Q(f) = \begin{cases} \int_\Omega |\nabla f|^2 \, \mathrm{d}^N x & \text{if } f \in W^{1,2}(\Omega) \\ +\infty & \text{otherwise} \end{cases}$$

as described in [5, Section 4.4]. Note that $\mathrm{Dom}\,(H^{1/2}) = W^{1,2}(\Omega)$ [4, page 19]. If either of embeddings (1) or (2) holds, then by step 2 $W^{1,2}(\Omega) \subset L^p(\Omega)$ for some $p > 2$. Consequently [4, Corollary 2.4.3, Lemma 2.1.2] it follows that the semigroup e^{-Ht} has a positive continuous integral kernel $K(t,x,y), t > 0, x, y, \in \Omega$.

Next we consider the operator

$$H' = H \otimes 1 + 1 \otimes \left(-\frac{\partial^2}{\partial \xi^2} \right)$$

on $L^2(\Omega \times \mathbf{R})$. Then $H' = -\Delta_N - \frac{\partial^2}{\partial \xi^2}$ is the Neumann Laplacian acting in $L^2(\Omega \times \mathbf{R})$ and $e^{-H't}$ has an integral kernel of the form

$$K'(t,x,y,x'y') = K(t,x,y) \frac{1}{2\sqrt{\pi t}} \, e^{-\frac{(x'-y')^2}{4t}}, \tag{5}$$

where $t > 0, x, y \in \Omega, x', y' \in \mathbf{R}$.

4. If (2) holds, then it follows [4, Corollary 2.4.3, Lemma 2.1.2] that for some $c_4 > 0$

$$0 < K'(t,x,y,x,'y') \le c_4 \, t^{-\frac{\mu}{2}}$$

for all $0 < t \le 1, x, y, \in \Omega, x', y' \in \mathbf{R}$ where $\mu = \frac{2r}{r-2}$. By putting $x' = y'$ we get

$$0 < K(t,x,y) \le 2c_4 \sqrt{\pi} \, t^{-\frac{\nu}{2}},$$

where $\nu = \mu - 1 = \frac{r+2}{r-2}$. Since $\mu > 2$ it follows [4, Lemma 2.1.2, Corollary 2.4.3] that (1) is satisfied because $\frac{2\nu}{\nu-2} = \frac{2(r+2)}{6-r} = q$.

Similarly, if (1) holds, then it follows that for some $c_5 > 0$

$$0 < K(t,x,y) \le c_5 \, t^{-\frac{\nu}{2}},$$

for all $0 < t \le 1, x, y \in \Omega$, where $\nu = \frac{2q}{q-2}$. Hence by (5)

$$0 < K(t,x,y,x',y') \le \frac{c_5}{2\sqrt{\pi}} \, t^{-\frac{\mu}{2}},$$

where $\mu = \nu + 1 = \frac{3q-2}{q-2}$, and (2) follows since $\frac{2\mu}{\mu-2} = \frac{2(3q-2)}{q+2} = r$.

Corollary 2. *Let k be a positive integer and let*

$$r_k = \frac{2(k(q-2) + 2q)}{k(q-2) + 4}.$$

Then the embedding (1) is equivalent to the embedding

$$W^{1,2}(\Omega \times I^k) \subset L^{r_k}(\Omega \times I^k),$$

where $I^k = \underbrace{I \times \cdots \times I}_{k \text{ times}}$.

Proof. This could be easily proved by induction if we take in account that

$$\frac{1}{r_{k+1} - 2} = \frac{1}{r_k - 2} + \frac{1}{4}.$$

Theorem 3. *Let $\Omega \subset \mathbf{R}^N$ be an open set, $2 < q < \infty$ for $N = 1, 2$ and $2 < q \leq \frac{2N}{N-2}$ for $N \geq 3$, and let*

$$s = \frac{4q}{q+2}.$$

Then embedding (1) *is equivalent to the embedding*

$$W^{1,2}(\Omega \times \Omega) \subset L^s(\Omega \times \Omega). \tag{6}$$

Proof. The proof is similar to the proof of Theorem 1 (steps 2–4). In this case

$$H' = H \otimes 1 + 1 \otimes H,$$

on $L^2(\Omega \times \Omega)$ and $e^{-H't}$ has an integral kernel of the form

$$K'(t, x, y, x', y') = K(t, x, y)K(t, x', y'), \tag{7}$$

where now $t > 0, x, y, x', y' \in \Omega$. If (6) is satisfied, then for some $c_6 > 0$

$$0 < K'(t, x, y, x', y') \leq c_6 t^{-\frac{\mu}{2}}$$

for all $0 < t \leq 1, x, y, x', y' \in \Omega$, where $\mu = \frac{2s}{s-2}$. By putting $x = x', y = y'$ we get

$$0 < K(t, x, y) \leq \sqrt{c_6} t^{-\frac{\nu}{2}},$$

where now $\nu = \frac{\mu}{2} = \frac{s}{s-2}$. Hence (1) is satisfied since $\frac{2\nu}{\nu-2} = \frac{2s}{4-s} = q$. Similarly, if (1) holds, then for some $c_4 > 0$

$$0 < K(t, x, y) \leq c_4 t^{-\frac{\nu}{2}},$$

for all $0 < t \leq 1, x, y \in \Omega$, where $\nu = \frac{2q}{q-2}$. Hence by (7)

$$0 < K'(t, x, y, x', y') \leq c_4^2 t^{-\frac{\mu}{2}}$$

where $\mu = 2\nu = \frac{4q}{q-2}$ and (6) holds since $\frac{2\mu}{\mu-2} = \frac{4q}{q+2} = s$.

Corollary 4. *Let $k \geq 2$ be an integer and let*

$$s_k = \frac{2kq}{(k-1)q + 2}.$$

Then the embedding (1) *is equivalent to the embedding*

$$W^{1,2}(\Omega^k) \subset L^{s_k}(\Omega^k),$$

where $\Omega^k = \underbrace{\Omega \times \cdots \times \Omega}_{k \text{ times}}$.

Example 5. *Let in Corollary 2 $\Omega = \mathbf{R}^3, I = \mathbf{R}$ and $k = N - 3 \geq 1$. Then it follows that the embedding*

$$W^{1,2}(\mathbf{R}^N) \subset L^{\frac{2N}{N-2}}(\mathbf{R}^N) \tag{8}$$

is equivalent to the embedding

$$W^{1,2}(\mathbf{R}^3) \subset L^6(\mathbf{R}^3) \tag{9}$$

for any $N \geq 4$. *Moreover, all embeddings (8) are equivalent for different* $N \geq 3$. *By Theorem 1 it also follows that (9) implies the embedding*

$$W^{1,2}(\mathbf{R}^2) \subset L^q(\mathbf{R}^2) \tag{10}$$

for any $2 < q < \infty$. *In its turn, (10) implies the embedding*

$$W^{1,2}(\mathbf{R}^N) \subset L^r(\mathbf{R}^N) \tag{11}$$

for any $N \geq 3, r < \frac{2N}{N-2}$.

Example 6. *Let* $N \geq 2, 0 < \gamma \leq 1$ *and* $\Omega \subset \mathbf{R}^N$ *be an open set such that* $\partial\Omega \in \operatorname{Lip}\gamma$. *It is well known* [6], [7], [8] *that (1) holds with* $q = \frac{2(N-1+\gamma)}{N-1-\gamma}$. *By Theorems 1 and 2, (1) is equivalent to (2), where* $r = \frac{2(N-1+2\gamma)}{N-1-\gamma}$, *and to (6), where* $s = \frac{2(N-1-\gamma)}{N-1}$. *Note that if* $\gamma = 1$, *then* $r = \frac{2(N+1)}{(N+1)-2}$ *and* $s = \frac{2(2N)}{(2N)-2}$. *However, if* $\gamma < 1$, *then* $r > \frac{2((N+1)-1+\gamma)}{(N+1)-1-\gamma}$ *and* $s > \frac{2((2N)-1+\gamma)}{(2N)-1-\gamma}$. *The reason for this is that, in contrast to the case* $\gamma = 1$, *open sets* $\Omega \times I$ *and* $\Omega \times \Omega$, *for which* $\partial(\Omega \times I) \in \operatorname{Lip}\gamma$ *and* $\partial(\Omega \times \Omega) \in \operatorname{Lip}\gamma$ *are "better" than arbitrary open sets* $G_1 \subset \mathbf{R}^{N+1}$ *and* $G_2 \subset \mathbf{R}^{2N}$ *satisfying* $\partial G_1, \partial G_2 \in \operatorname{Lip}\gamma$. *This conforms with Theorem 12.1 in* [1]. *Since* $\Omega \times I$ *and* $\Omega \times \Omega$ *satisfy the so-called horn condition* [1] *with the parameters* $(1, \underbrace{\gamma, \ldots, \gamma}_{N-1 \text{ times}}, 1), (1, \underbrace{\gamma, \ldots, \gamma}_{N-1 \text{ times}}, 1, \underbrace{\gamma, \ldots, \gamma}_{N-1 \text{ times}})$ *respectively, embeddings (2) and (6) follow by the afore mentioned theorem with* $r = \frac{2(N-1+2\gamma)}{N-1}$, $s = \frac{2(N-1-\gamma)}{N-1}$ *respectively.*

Acknowledgement. I would like to thank Professor E.B. Davies, FRS, for fruitful discussions during the work on our joint paper [3] due to which this paper has arisen.

References

[1] Besov, O.V., Il'in, V.P., Nikol'skiĭ, S.M.: Integral Representations of Functions and Embedding Theorems. Nauka, Moscow, 1966.

[2] Burenkov, V.I.: Sobolev Spaces on Domains. Teubner-Texte Math. 137, B. G. Teubner, Stuttgart-Leipzig, 1998.

[3] Burenkov, V.I., Davies, E.B.: Spectral stability of the Neumann Laplacian (to appear in J. Diff. Equations).

[4] Davies, E.B.: Heat Kernels and Spectral Theory. Cambridge Univ. Press, Cambridge, 1989.

[5] Davies, E.B.: Spectral Theory and Differential Operators. Cambridge Univ. Press, Cambridge, 1995.

[6] Globenko, I.G.: Embedding theorems for a region with null corner points. Dokl. Akad. Nauk USSR 132(2) (1960) 251–253.

[7] Maz'ya, V.G.: Classes of regions and embedding theorems for function spaces. Dokl. Akad. Nauk. USSR 133(3) (1960) 527–530.

[8] Maz'ya, V.G.: Sobolev Spaces. Springer-Verlag, Berlin, 1985.

Viktor I. Burenkov
School of Mathematics
Cardiff University
23 Senghennydd Road
Cardiff CF24 4YH, Wales, UK
E-mail address: Burenkov@Cardiff.ac.uk

D. Haroske, T. Runst, H.-J. Schmeisser (eds.): Function Spaces, Differential Operators and Nonlinear Analysis. The Hans Triebel Anniversary Volume.

Sharp Estimates of Approximation Numbers via Growth Envelopes

António M. Caetano[1] and Dorothee D. Haroske[1]

Dedicated to Prof. Hans Triebel on the occasion of his 65th birthday

Abstract. We give sharp asymptotic estimates for the approximation numbers of the compact embedding $id : L_p(\log L)_a(\Omega) \longrightarrow B^{-1}_{\infty,\infty}(\Omega)$, $a > 0$, $n < p < \infty$, applying the newly developed tool of (growth) envelopes for function spaces.

Introduction

We present an application of our recently developed concept of envelopes in function spaces: this is a relatively simple tool for the study of rather complicated spaces, say, of Sobolev type H^s_p, or Besov type $B^s_{p,q}$, in '*limiting*' situations. Arising from the famous Sobolev embedding theorem it is, for instance, well known that $B^{n/p}_{p,q} \hookrightarrow L_\infty$ if, and only if, $0 < p \le \infty$, $0 < q \le 1$; thus a natural question is in what sense the unboundedness of functions belonging to $B^{n/p}_{p,q}$ with $1 < q \le \infty$ (or $H^{n/p}_p$ with $1 < p < \infty$) can be qualified. Concentrating on this particular feature only, the concept of growth envelope functions $\mathcal{E}^X_{\mathsf{G}}$ 'measuring' the unboundedness of functions f belonging to some function space X of regular distributions is introduced by means of their non-increasing rearrangement $f^*(t)$,

$$\mathcal{E}^X_{\mathsf{G}}(t) = \sup_{\|f|X\|\le 1} f^*(t) , \quad t > 0.$$

We found rather simple and final answers characterising the unboundedness of functions in spaces like $B^s_{p,q}$ and H^s_p; in fact, the results contain an even finer description of this feature than measured by $\mathcal{E}^X_{\mathsf{G}}$ merely and cover far more settings than the above-described limiting ones; see [Har02], [Tri01], [CaM01]. There are some direct connections with classical results, for instance with the so-called fundamental function φ_X when X is rearrangement-invariant. More interesting, however, are the new and, in our opinion, elegant results for spaces like $B^s_{p,q}$ or H^s_p.

[1]The research was partially supported by the program 'Projektbezogener Personenaustausch mit Portugal – Acções Integradas Luso-Alemãs / DAAD-ICCTI'.

Analogously one can investigate limiting situations when questions of (un)boundedness of functions are replaced by inquiries about (almost) Lipschitz continuity; though there are immediate counterparts we do not pursue this point here further. Our main goal in [Har02] (as well as in [Tri01], [CaM01]) was to obtain precise but simple characterisations of technically rather complicated (scales of) spaces in the sense described above. In the course of our studies it turned out that this also leads to a lot of interesting consequences. Naturally one arrives at Hardy-type inequalities, but there is also an interplay between envelope functions and lift operators in function spaces as well as applications to related questions of compactness; this latter subject will be explained in the present paper in a little detail. Our idea is twofold: on the one hand we give an immediate (though not obvious) application of our envelope results to estimates for approximation numbers which are surprisingly sharp. Secondly we close a gap concerning asymptotic estimates for approximation numbers in function spaces (characterising in that way the compactness of the corresponding natural embedding) which was not covered by earlier results so far. Roughly speaking, the upper estimate rests upon an envelope function argument whereas the corresponding lower one is obtained by a careful combination of earlier approximation number estimates and complex interpolation.

We collect the necessary background material first, present our main result in Section 2, whereas its proof is postponed to Section 3. Finally, we end this paper with a small collection of related results.

1. Preliminaries

Let $\Omega \subset \mathbb{R}^n$ be a bounded domain with C^∞ boundary $\partial\Omega$. Recall that for $1 < p < \infty$, $a \in \mathbb{R}$, the space $L_p(\log L)_a(\Omega)$ consists of all measurable functions $f : \Omega \longrightarrow \mathbb{C}$ such that

$$\left(\int_0^{|\Omega|} \left[(1 + |\log t|)^a \, f^*(t) \right]^p \, \mathrm{d}t \right)^{\frac{1}{p}} \tag{1}$$

is finite, where $f^*(t)$ denotes the non-increasing rearrangement of f, as usual. For our later purpose of estimating approximation numbers, however, we prefer a characterisation of $L_p(\log L)_a(\Omega)$ by extrapolation techniques; this was obtained by Edmunds and Triebel in [ET96, Thm. 2.6.2, p. 69]. For convenience we adopt the notation

$$\frac{1}{p^\sigma} \equiv \frac{1}{p} + \frac{\sigma}{n}, \tag{2}$$

where $0 < \sigma < \varepsilon$ and we shall always assume in the sequel $\varepsilon > 0$ to be sufficiently small such that $p^\sigma > 1$. Then

$$\left(\int_0^{\varepsilon} \left[\sigma^a \| f | L_{p^\sigma}(\Omega) \| \right]^p \frac{\mathrm{d}\sigma}{\sigma} \right)^{1/p} \tag{3}$$

defines an equivalent norm on $L_p(\log L)_{-a}(\Omega)$, $1 < p < \infty$, $a > 0$; cf. [ET96, Thm. 2.6.2, p. 69].

Recall that $B^s_{p,q}(\Omega)$ stand for the Besov spaces, where $s \in \mathbb{R}$ and $0 < p, q \leq \infty$; see, for example, Chapter 2 of [ET96] for definitions and properties, in particular for embedding relations between such spaces.

Recall also the notion of approximation numbers $a_k(T)$, $k \in \mathbb{N}$, of $T \in L(E,F)$, where $L(E,F)$ stands for the collection of all bounded linear operators acting from the quasi-Banach space E into the quasi-Banach space F:

$$a_k(T) \equiv \inf\{\|T - S\| : S \in L(E,F),\ \operatorname{rank} S < k\},$$

where $\operatorname{rank} S$ is the dimension of the range of S. There are two properties of these numbers which will be useful in the sequel:

(i) (monotonicity)

$$\|T\| = a_1(T) \geq a_2(T) \geq \cdots \geq 0;$$

(ii) (multiplicativity)

$$a_{k+l-1}(R \circ S) \leq a_k(R) a_l(S),$$

where $S \in L(E,F)$ and $R \in L(F,G)$, with G another quasi-Banach space.

We shall also need the notion of the (local) growth envelope function applied to the spaces $L_p(\log L)_a(\Omega)$, which, for simplicity, we define just as

$$\mathcal{E}_{\mathsf{G}}^{L_p(\log L)_a}(t) \equiv \sup\{f^*(t) : \|f|L_p(\log L)_a(\Omega)\| \leq 1\}$$

for all small positive numbers t; for general considerations concerning growth envelope functions, see [Har02], [Tri01].

The behaviour of such functions will be important to us only up to equivalence $\sim$, where $g(t) \sim h(t)$ means the existence of two positive constants c_1 and c_2 such that $c_1\, g(t) \leq h(t) \leq c_2\, g(t)$ for all t under consideration. By the way, the notion of equivalence $\sim$ will also be applied to sequences, that is, to functions of the natural variable k instead of the continuous variable t, and in whatever situation where the existence of constants as c_1 and c_2 independent of the variable in the formula is implied.

Finally, let us recall that p' stands for the conjugate of p (that is, satisfies $\frac{1}{p} + \frac{1}{p'} = 1$) and that, for convenience, log will be used instead of $\log_2$.

2. The main result

Theorem 1. *Let $\Omega \subset \mathbb{R}^n$ be a bounded domain with C^∞ boundary $\partial\Omega$.*

(i) *Assume $a > 0$, $n < p < \infty$ for $n \geq 2$, and $2 \leq p < \infty$ when $n = 1$. Then for all $k \in \mathbb{N}$,*

$$a_k\left(id : L_p(\log L)_a(\Omega) \longrightarrow B^{-1}_{\infty,\infty}(\Omega)\right) \sim k^{-\left(\frac{1}{n}-\frac{1}{p}\right)} (1 + \log k)^{-a}\ . \tag{4}$$

(ii) *Assume* $a > 0$, $1 < p < \frac{n}{n-1}$ *for* $n \geq 2$, *and* $1 < p \leq 2$ *when* $n = 1$. *Then for all* $k \in \mathbb{N}$,

$$a_k \left(id : B^1_{1,1}(\Omega) \longrightarrow L_p(\log L)_{-a}(\Omega) \right) \sim k^{-\left(\frac{1}{n}-\frac{1}{p'}\right)} (1+\log k)^{-a} . \tag{5}$$

3. Proofs

We start with three auxiliary results which will be used in the proof of the theorem; we thus split the (otherwise rather long) argument for our main result in more handy pieces. Moreover, it becomes more obvious that way which different tools will be applied afterwards. Let all function spaces be defined on $\Omega \subset \mathbb{R}^n$ in the sequel unless otherwise stated; here $\Omega \subset \mathbb{R}^n$ stands for a bounded domain with C^∞ boundary $\partial\Omega$.

Proposition 2. [Har02, Prop. 5.1.2] *Let* $1 < p < \infty$, *and* $a \in \mathbb{R}$. *Then*

$$\mathcal{E}_{\mathsf{G}}^{L_p(\log L)_a}(t) \sim t^{-\frac{1}{p}} \, |\log t|^{-a}, \quad 0 < t < \frac{1}{2} . \tag{6}$$

Let $X(\Omega)$ be some function space of regular distributions with $X(\Omega) \not\hookrightarrow L_\infty(\Omega)$. Denote by $X^\nabla(\Omega) \subset X(\Omega)$ the subspace

$$X^\nabla(\Omega) = \{ g \in D'(\Omega) \ : \ D^\alpha g \in X(\Omega), \ |\alpha| \leq 1 \} \tag{7}$$

normed by

$$\| g | X^\nabla(\Omega) \| = \sum_{|\alpha| \leq 1} \| D^\alpha g | X(\Omega) \| . \tag{8}$$

Let C stand for the space of all complex-valued bounded uniformly continuous functions on $\overline{\Omega}$, equipped with the sup-norm as usual.

Proposition 3. *Let* $X(\Omega)$, $X^\nabla(\Omega)$ *be the spaces given by* (7), (8). *Assume* $X(\Omega) \hookrightarrow B^{-1}_{\infty,\infty}(\Omega)$ *and* $X^\nabla(\Omega) \hookrightarrow C(\Omega)$. *Let* $\mathcal{E}^X_{\mathsf{G}}$ *satisfy*

$$\sum_{k=0}^{\infty} 2^{-k} \, \frac{\mathcal{E}^X_{\mathsf{G}}\left(2^{-(k+J)n}\right)}{\mathcal{E}^X_{\mathsf{G}}\left(2^{-Jn}\right)} \leq c , \tag{9}$$

for some number $C > 0$ *and for all large* $J \in \mathbb{N}$, *and assume that there is a bounded (linear) lift operator* L *mapping* $X(\Omega)$ *into* $X^\nabla(\Omega)$ *such that its inverse* L^{-1} *exists and maps* $C(\Omega)$ *into* $B^{-1}_{\infty,\infty}(\Omega)$. *Then*

$$a_k \left(id : X(\Omega) \longrightarrow B^{-1}_{\infty,\infty}(\Omega) \right) \leq c \, k^{-\frac{1}{n}} \, \mathcal{E}^X_{\mathsf{G}}\left(k^{-1}\right) . \tag{10}$$

Proof of Proposition 3. The proof essentially relies on a result of Carl and Stephani [CaS90, Thm. 5.6.1, p. 178] together with the definition of envelope functions. It can be obtained parallel to [Har02, Cor. 7.2.3] (dealing with entropy numbers), see also [Har01, 6.15]. One simply combines results on approximation numbers for compact embeddings in $C(\Omega)$ (as target space) with the properties of the operator L and its inverse L^{-1}, a lift argument for envelope functions (which requires (9)), and, finally, the multiplicativity of approximation numbers.

Proposition 4. *Let $1 < r_0 < r_1 \leq 2$ be such that $n < r_1'$. Then*

$$a_k(B^1_{1,1}(\Omega) \hookrightarrow L_r(\Omega)) \geq c\, k^{-(\frac{1}{n}-\frac{1}{r'})} \tag{11}$$

for $r \in [r_0, r_1]$, with $c > 0$ independent of r and k.

Proof of Proposition 4. It follows by inspection of Step 3 of the proof of Theorem 3.3.4 in [ET96, pp. 123–125], and taking care of $L_r(\Omega) \hookrightarrow B^0_{r,2}(\Omega)$, that $id^l : l_1^{N_j} \to l_r^{N_j}$ can be decomposed (maybe after some translation and rescaling arguments) as

$$l_1^{N_j} \xrightarrow{A} B^1_{1,1}(\Omega) \xrightarrow{id^L} L_r(\Omega) \xrightarrow{B} l_r^{N_j},$$

i.e.,

$$id^l = B \circ id^L \circ A, \tag{12}$$

where $N_j = 2^{jn}$, $j \in \mathbb{N}$, and where the bounded linear operators A and B can be chosen independently of $r \in [r_0, r_1]$. Furthermore, [ET96, pp. 123-125] gives us the estimates

$$\|A\| \leq c_1 2^{j(1-n)} \quad \text{and} \quad \|B\| \leq c_2\, 2^{j\frac{n}{r}},$$

where the positive constants c_1 and c_2 are independent of j, though c_2 can depend on r. However, writing $\frac{1}{r} = \frac{1-\theta}{r_0} + \frac{\theta}{r_1}$, for some $\theta \in]0,1[$, one has the complex interpolation formulae

$$[L_{r_0}(\Omega), L_{r_1}(\Omega)]_\theta = L_r(\Omega) \quad \text{and} \quad [l_{r_0}^{N_j}, l_{r_1}^{N_j}]_\theta = l_r^{N_j},$$

with equality of norms too (cf. [Tri78, 1.18.1,1.18.4]), so that

$$\|B : L_r(\Omega) \to l_r^{N_j}\| \leq c_2^{1-\theta}(r_0)\, 2^{(1-\theta)j\frac{n}{r_0}}\, c_2^{\theta}(r_1)\, 2^{\theta j \frac{n}{r_1}} = c_3 2^{j\frac{n}{r}},$$

where now $c_3 > 0$ is independent of j and r.

The multiplicativity of the approximation numbers applied to (12) then gives

$$a_k(id^L) \geq \|A\|^{-1}\|B\|^{-1} a_k(id^l) \geq c_1^{-1} c_3^{-1} 2^{-j(1-\frac{n}{r'})} a_k(id^l)$$

for all $k \in \mathbb{N}$. Choosing $k \leq \frac{N_j}{4} = 2^{jn-2}$, one can write

$$a_k(id^L) \geq c_4 2^{-j(1-\frac{n}{r'})}$$

where c_4 is independent of k, j and r (cf. [ET96, Corol. 3.2.3]).

Since for each $k \in \mathbb{N}$ there is (exactly) one $j \in \mathbb{N}$ such that $2^{(j-1)n-2} < k \leq 2^{jn-2}$, (11) now follows easily by standard arguments.

Proof of Theorem 1.

STEP 1. We first verify that (5) is an immediate consequence of (4) and duality arguments; note that $[L_p(\log L)_a(\Omega)]' = L_{p'}(\log L)_{-a}(\Omega)$, $1 < p < \infty$, $a \in \mathbb{R}$, see [BR80, Thm. 8.4]. On the other hand, $\left[B^1_{1,1}(\mathbb{R}^n)\right]' = B^{-1}_{\infty,\infty}(\mathbb{R}^n)$, cf. [Tri83, Thm. 2.11.2 (i), p. 178]; for the counterpart on Ω, one could start with spaces $\widetilde{B^s_{p,q}}$ – containing functions $f \in B^s_{p,q}$ with supp $f \subset \overline{\Omega}$ – first and extend afterwards.

Thus taking (i) for granted at the moment, the duality result for approximation numbers, see [CaS90, Prop. 2.5.4, p. 80] and [ETy86], yields (ii),

$$a_k\left(L_p(\log L)_a(\Omega) \hookrightarrow B^{-1}_{\infty,\infty}(\Omega)\right) \sim a_k\left(B^1_{1,1}(\Omega) \hookrightarrow L_{p'}(\log L)_{-a}(\Omega)\right). \qquad (13)$$

STEP 2. We show the upper estimate in (4) and benefit from the preceding Propositions 2, 3. As $\mathcal{E}_{\mathsf{G}}^{L_p(\log L)_a}(t)$ given by (6) satisfies (9),

$$\sum_{k=0}^{\infty} 2^{-k} \frac{\mathcal{E}_{\mathsf{G}}^X\left(2^{-(k+J)n}\right)}{\mathcal{E}_{\mathsf{G}}^X\left(2^{-Jn}\right)} \sim \sum_{k=0}^{\infty} 2^{-k(1-\frac{n}{p})} \left(\frac{k+J}{J}\right)^{-a} \leq c$$

for $p > n$, $a \in \mathbb{R}$, we can apply Proposition 3 with $X = L_p(\log L)_a$, $X^\nabla = H^1_p(\log L)_a$. The existence of the bounded linear lift operator $L : L_p(\log L)_a(\Omega) \longrightarrow H^1_p(\log L)_a(\Omega)$ is covered by [ET96, Thm. 2.6.3, p. 79], whereas the additional assumption on L^{-1} to map $C(\Omega)$ into $B^{-1}_{\infty,\infty}(\Omega)$ is a consequence of restriction-extension procedures and the usual lift operator $I_\sigma : B^s_{p,q} \longrightarrow B^{s-\sigma}_{p,q}$ in $\mathbb{R}^n$. Alternatively one can use regular elliptic differential operators adapted to Ω; see [Tri78, Thm. 4.9.2, p. 335] for the case $1 < p < \infty$, $1 \leq q \leq \infty$, and [Tri83, Thm. 4.3.4, p. 235] for the extensions to $0 < p, q \leq \infty$. Hence (6) and (10) imply

$$a_k\left(id : L_p(\log L)_a(\Omega) \longrightarrow B^{-1}_{\infty,\infty}(\Omega)\right) \leq c\, k^{-\left(\frac{1}{n}-\frac{1}{p}\right)} (1+\log k)^{-a}. \qquad (14)$$

STEP 3. We prove the inequality converse to (14), i.e., the existence of some $c > 0$ such that for all $k \in \mathbb{N}$,

$$a_k\left(id : L_p(\log L)_a(\Omega) \longrightarrow B^{-1}_{\infty,\infty}(\Omega)\right) \geq c\, k^{-\left(\frac{1}{n}-\frac{1}{p}\right)} (1+\log k)^{-a}. \qquad (15)$$

By the same duality argument as stressed above it is sufficient to show

$$a_k\left(id : B^1_{1,1}(\Omega) \longrightarrow L_{p'}(\log L)_{-a}(\Omega)\right) \geq c_1\, k^{-\left(\frac{1}{n}-\frac{1}{p}\right)} (1+\log k)^{-a}. \qquad (16)$$

Note that the embedding $id^L : B^1_{1,1}(\Omega) \to L_{(p')^\sigma}(\Omega)$ can be decomposed as

$$B^1_{1,1}(\Omega) \stackrel{id}{\longrightarrow} L_{p'}(\log L)_{-a}(\Omega) \stackrel{id_\sigma}{\longrightarrow} L_{(p')^\sigma}(\Omega),$$

for each $\sigma > 0$ such that $(p')^\sigma > p'_0 > 1$, where p'_0 is fixed and less than p' (therefore, $0 < \sigma < \varepsilon \equiv n(\frac{1}{p'_0} - \frac{1}{p'})$). Since from (3) it follows that $\|id_\sigma\| \leq c_1 \sigma^{-a}$, with $c_1 > 0$ independent of σ, and from Proposition 4 we can write

$$a_k(id^L) \geq c_2\, k^{-\left(\frac{1}{n}-\frac{1}{p}\right)} k^{-\frac{\sigma}{n}},$$

also with $c_2 > 0$ independent of σ, then the multiplicativity of the approximation numbers gives

$$a_k(id) \geq a_k(id^L)\|id_\sigma\|^{-1} \geq c_2\, c_1^{-1}\, k^{-\left(\frac{1}{n}-\frac{1}{p}\right)} k^{-\frac{\sigma}{n}} \sigma^a.$$

For large k we can now choose $\sigma = an(\log k)^{-1}$ and obtain

$$a_k(id) \geq c\, k^{-\left(\frac{1}{n}-\frac{1}{p}\right)} (1+\log k)^{-a},$$

with $c > 0$ independent of k. Of course this also holds for small $k \in \mathbb{N}$.

The proof is complete.

4. Related results

We collect a few closely related results and report, in particular, what was already known before.

(i) The 'non-logarithmic' case $a = 0$ is covered by [ET96, Thm. 3.3.4 (i), p. 119]; further improvements of this result can be found in [Cae98]. Note that the case $n = 1$, $1 < p < 2$, is not yet completely solved in this case, too.

(ii) The situation with $p = \infty$ was studied by Triebel in [Tri93]; then

$$a_k \left(id : H_r^{n/r}(\Omega) \longrightarrow L_\infty(\log L)_{-a}(\Omega) \right) \sim (1 + \log k)^{-\left(a - \frac{1}{r'}\right)}, \qquad (17)$$

where $1 < r < \infty$, $a > \frac{1}{r'}$, see [ET96, Thm. 3.4.2, p. 129]. Here H_r^s, $s \in \mathbb{R}$, $1 < r < \infty$, stand for the Bessel potential spaces (*fractional Sobolev spaces*).

(iii) In [EH00, Thm. 3.13] we studied the counterpart 'lifted' by smoothness 1, that is

$$a_k \left(id : B_{r,q}^{1+n/r}(\Omega) \longrightarrow \mathrm{Lip}^{(1,-\alpha)}(\Omega) \right) \sim (1 + \log k)^{-\alpha},$$

assuming $1 < r \le \infty$, $0 < q \le 1$, $\alpha > 0$; here $\mathrm{Lip}^{(1,-\alpha)}(\Omega)$ are (logarithmically) refined Lipschitz spaces. Similarly we obtained in [EH00, Cor. 3.19] estimates of the type

$$a_k \left(id : \mathrm{Lip}^{(1,-\alpha)}(\Omega) \longrightarrow B_{\infty,q}^{1-s}(\Omega) \right) \sim k^{-\frac{s}{n}} (1 + \log k)^{\alpha},$$

where $\alpha \ge 0$, $s > 0$, and $0 < q \le \infty$, and

$$a_k \left(id : B_{p,q}^{s}(\Omega) \longrightarrow \mathrm{Lip}^{(1,-\alpha)}(\Omega) \right) \sim k^{-\frac{s-1}{n} + \frac{1}{p}} (1 + \log k)^{-\alpha}$$

for $\alpha \ge 0$, $2 \le p \le \infty$, $0 < q \le \infty$ and $s > 1 + \frac{n}{p}$. In the last assertion $B_{p,q}^s(\Omega)$ can be replaced by $H_p^s(\Omega)$, $2 \le p < \infty$.

Parallel studies dealing with entropy numbers instead of approximation numbers can be found in [ET96], [EN98], [Cae00] (dealing with embeddings into spaces $L_p(\log L)_a$), and in [EH99], [EH00], [CoK01] (related to spaces $\mathrm{Lip}^{(1,-\alpha)}$).

Finally, we refer to the book [Tri01], [Har01], [Har02], and to [CaM01] for further details and results on envelopes.

References

[BR80] C. Bennett and K. Rudnick. On Lorentz-Zygmund spaces. *Dissertationes Math.*, 175:1–72, 1980.

[Cae98] A. Caetano. About approximation numbers in function spaces. *J. Approx. Theory*, 94:383–395, 1998.

[Cae00] A. Caetano. Entropy numbers of embeddings between logarithmic Sobolev spaces. *Port. Math.*, 57(3):355–379, 2000.

[CaM01] A. Caetano and S.D. Moura. Local growth envelopes of spaces of generalized smoothness : the sub-critical case. Preprint 01-23, University Coimbra, 2001.

[CaS90] B. Carl and I. Stephani. *Entropy, compactness and the approximation of operators*. Cambridge Univ. Press, Cambridge, 1990.

[CoK01] F. Cobos and Th. Kühn. Entropy numbers of embeddings of Besov spaces in generalized Lipschitz spaces. *J. Approx. Theory*, 112:73–92, 2001.

[EH99] D.E. Edmunds and D.D. Haroske. Spaces of Lipschitz type, embeddings and entropy numbers. *Dissertationes Math.*, 380:1–43, 1999.

[EH00] D.E. Edmunds and D.D. Haroske. Embeddings in spaces of Lipschitz type, entropy and approximation numbers, and applications. *J. Approx. Theory*, 104(2):226–271, 2000.

[EN98] D.E. Edmunds and Yu. Netrusov. Entropy numbers of embeddings of Sobolev spaces in Zygmund spaces. *Studia Math.*, 128(1):71–102, 1998.

[ET96] D.E. Edmunds and H. Triebel. *Function Spaces, Entropy Numbers, Differential Operators*. Cambridge Univ. Press, Cambridge, 1996.

[ETy86] D.E. Edmunds and H.-O. Tylli. On the entropy numbers of an operator and its adjoint. *Math. Nachr.*, 126:231–239, 1986.

[Har01] D.D. Haroske. Envelopes in function spaces – a first approach. Jenaer Schriften zur Mathematik und Informatik Math/Inf/16/01, p. 1–72, Universität Jena, Germany, 2001.

[Har02] D.D. Haroske. Limiting embeddings, entropy numbers and envelopes in function spaces. Habilitationsschrift, Friedrich-Schiller-Universität Jena, Germany, 2002.

[Tri78] H. Triebel. *Interpolation Theory, Function Spaces, Differential Operators*. North-Holland, Amsterdam, 1978.

[Tri83] H. Triebel. *Theory of Function Spaces*. Birkhäuser, Basel, 1983.

[Tri93] H. Triebel. Approximation numbers and entropy numbers of embeddings of fractional Besov-Sobolev spaces in Orlicz spaces. *Proc. London Math. Soc.*, 66(3):589–618, 1993.

[Tri01] H. Triebel. *The structure of functions*. Birkhäuser, Basel, 2001.

António M. Caetano
Departamento de Matemática
Universidade de Aveiro
3810-193 Aveiro, Portugal
E-mail address: acaetano@mat.ua.pt

Dorothee D. Haroske
Mathematisches Institut
Friedrich-Schiller-Universität Jena
D-07740 Jena, Germany
E-mail address: haroske@minet.uni-jena.de

D. Haroske, T. Runst, H.-J. Schmeisser (eds.): Function Spaces, Differential Operators and Nonlinear Analysis. The Hans Triebel Anniversary Volume.

Sharp Summability of Functions from Orlicz-Sobolev Spaces

Andrea Cianchi

Dedicated to Prof. Hans Triebel on the occasion of his 65th birthday

1. Introduction

Recent years have witnessed an increasing interest of researchers working on function spaces, and on related fields, about optimal embeddings of Sobolev type, as demonstrated by a number of papers on this topic. One of the ancestors of these kind of results may be considered a sharpened version of the classical Sobolev inequality, independently proved by O'Neil [23] and by Peetre [25], which can be stated as follows. Let G be an open subset of $\mathbb{R}^n$, $n \geq 2$, and let $W_0^{1,p}(G)$, $1 \leq p \leq \infty$, be the first order Sobolev space of those real-valued weakly differentiable functions in G, decaying to 0 on ∂G, whose gradient belongs to $L^p(G)$. If $1 \leq p < n$, then a constant C exists such that

$$\|u\|_{L^{p^*,p}(G)} \leq C\| |\nabla u| \|_{L^p(G)} \tag{1}$$

for every $u \in W_0^{1,p}(G)$. Here, $p^* = \frac{np}{n-p}$, the Sobolev conjugate of p, ∇ stands for gradient and $L^{p^*,p}(G)$ is a Lorentz space. Lorentz spaces, as well as Lebesgue spaces, are rearrangement invariant Banach function spaces (briefly, r.i. spaces); namely, Banach function spaces (in the sense of Luxemburg – see e.g. [4]) where the norm of a function depends only on its decreasing rearrangement. Recall that, given any measurable subset G of $\mathbb{R}^n$ and any real-valued measurable function u in G, the decreasing rearrangement u^* of u is the unique non-increasing, right-continuous function from $[0, \infty)$ into $[0, \infty]$ which is equidistributed with u. In formulas,

$$u^*(s) = \sup\{t \geq 0 : |\{x \in G : |u(x)| > t\}| > s\} \qquad \text{for } s \geq 0, \tag{2}$$

where $|\ |$ denotes the Lebesgue measure. Thus, a Banach function space $X(G)$, endowed with a norm $\|\cdot\|_{X(G)}$, is an r.i. space if

$$\|v\|_{X(G)} = \|u\|_{X(G)} \quad \text{whenever} \quad v^* = u^*.$$

In a sense, r.i. spaces can be regarded as those Banach function spaces whose norm involves only summability properties of functions.

When $p \geq q$ (and hence also in the case involved in inequality (1)), the quantity

$$\|u\|_{L^{p,q}(G)} = \|s^{\frac{1}{p}-\frac{1}{q}} u^*(s)\|_{L^q(0,|G|)} \tag{3}$$

is a (r.i.) norm in the Lorentz space $L^{p,q}(G)$. In general, for $1 < p < \infty$ and $1 \leq q \leq \infty$, or $p = q = \infty$, $L^{p,q}(G)$ is equipped with a norm, equivalent to the right-hand side of (3), obtained on replacing u^* by $u^{**}(s) = \frac{1}{s}\int_0^s u^*(r)\,dr$. Since $L^{p^*,p}(G)$ is (strictly) contained in $L^{p^*}(G)$, inequality (1) recovers, and is indeed stronger, then the standard Sobolev embedding of $W_0^{1,p}(G)$ into $L^{p^*}(G)$. In fact, it has been recently shown in [14] that $L^{p^*,p}(G)$ is the best possible range space for embeddings of $W_0^{1,p}(G)$ among all r.i. spaces.

A counterpart of inequality (1) in the limiting case where $p = n$ was obtained by Brezis and Wainger [6] and by Hansson [19] (and also follows via certain capacitary estimates contained in [21]). The result can be stated in terms of Lorentz-Zygmund spaces, a refinement, in the logarithmic scale, of the Lorentz spaces, defined as follows. Given any measurable subset G of $\mathbb{R}^n$, the Lorentz-Zygmund space $L^{p,q}(\log L)^\alpha(G)$ is defined, for $\alpha \in \mathbb{R}$ and either $1 < p < \infty$ and $1 \leq q \leq \infty$, or $p = q = \infty$, as the space of all real-valued measurable functions u in G for which the expression

$$\|u\|_{L^{p,q}(\log L)^\alpha(G)} = \|s^{\frac{1}{p}-\frac{1}{q}}(1+\log(|G|/s))^\alpha u^*(s)\|_{L^q(0,|G|)} \tag{4}$$

is finite. As for plain Lorentz spaces, this is actually a norm in the r.i. space $L^{p,q}(\log L)^\alpha(G)$ for those values of p, q, α such that the function $s^{\frac{1}{p}-\frac{1}{q}}(1+\log(|G|/s))^\alpha$ is non-increasing; otherwise, the right-hand side of (4) can be turned into a norm on replacing u^* by u^{**}. With this notation in place, the Sobolev type inequality in question for $W_0^{1,n}(G)$ tells us that, if $|G| < \infty$, then there exists a positive constant C such that

$$\|u\|_{L^{\infty,n}(\log L)^{-1}(G)} \leq C\||\nabla u|\|_{L^n(G)} \tag{5}$$

for every $u \in W_0^{1,n}(G)$. Inequality (5) implies, in particular, the classical embedding of $W_0^{1,n}(G)$ into $L^q(G)$ for every $q < \infty$; it also reproduces, and improves, Trudinger's embedding of $W_0^{1,n}(G)$ into $L^{exp^{n'}}(G)$, the Orlicz space associated with the Young function $\exp^{n'}(s) = e^{s^{n'}} - 1$, where $n' = \frac{n}{n-1}$, the Hölder conjugate of n ([30]; see also [26, 31]). Actually, $L^{\infty,n}(\log L)^{-1}(G)$ turns out to be the optimal r.i. range space for embeddings of $W_0^{1,n}(G)$ ([11, 14]).

In the case where $p > n$ and $|G| < \infty$,

$$\|u\|_{L^\infty(G)} \leq C\||\nabla u|\|_{L^p(G)} \tag{6}$$

for some positive constant C and for every $u \in W_0^{1,p}(G)$. This is trivially an optimal result as far as r.i. range spaces are concerned, since $L^\infty(G)$ is contained in every other r.i. space when $|G| < \infty$.

Besides (1), (5) and (6), other Sobolev embeddings with sharp r.i. spaces are known; for instance, for Sobolev type spaces modelled on generalizations of the Lorentz and Lorentz-Zygmund spaces ([14]), and for Sobolev spaces close to the borderline space $W_0^{1,n}(G)$ ([11]). Related results are contained in [24] and [15]. Purpose of the present note is to report on a recent result, established in [10], concerning the solution of the optimal r.i. range space for embeddings of Orlicz-Sobolev spaces.

The notion of Orlicz space rests on that of Young function. Here, a continuous function $A : [0, \infty) \to [0, \infty]$ will be called a Young function if it has the form

$$A(s) = \int_0^s a(r)\, dr \qquad \text{for } s \geq 0, \tag{7}$$

for some non-decreasing, left-continuous function $a : [0, \infty) \to [0, \infty]$, which is neither identically equal to 0 nor to ∞. Hence, A is convex and vanishes at 0. The Orlicz space $L^A(G)$ associated with A is the Banach space of those real-valued measurable functions u in G for which the Luxemburg norm

$$\|u\|_{L^A(G)} = \inf\left\{\lambda > 0 : \int_G A\left(\frac{|u|}{\lambda}\right) dx \leq 1\right\} \tag{8}$$

is finite. $L^A(G)$, equipped with this norm, is an r.i. space. Note that $L^A(G) = L^p(G)$ if $A(s) = s^p$ for some $p \in [1, \infty)$; moreover, $L^A(G) = L^\infty(G)$ (up to equivalent norms) if $A(s)$ is any Young function which vanishes for small s and equals ∞ for large s. The Orlicz-Sobolev space $W_0^{1,A}(G)$ built upon $L^A(G)$ is the space of those weakly differentiable functions in G satisfying $\||\nabla u|\|_{L^A(G)} < \infty$ and decaying to 0 on ∂G, in the sense that the continuation of u outside G is a weakly differentiable function in $\mathbb{R}^n$ and $|\{x \in G : |u(x)| > t\}| < \infty$ for every $t > 0$.

Embedding theorems for Orlicz-Sobolev spaces were established by Donaldson and Trudinger [12] and by R.A. Adams [1]; a sharp embedding theorem for $W_0^{1,A}(G)$ into $L^\infty(G)$ is due to Talenti [28]. In [8] we proved a general result concerning optimal embeddings of $W_0^{1,A}(G)$ into Orlicz spaces. An equivalent version of that result, appearing in [9], tells us the following. Let G be either $\mathbb{R}^n$, or any open subset of $\mathbb{R}^n$ having finite measure. Given any Young function A such that

$$\int_0 \left(\frac{r}{A(r)}\right)^{\frac{1}{n-1}} dr < \infty, \tag{9}$$

define $A_n : [0, \infty) \to [0, \infty]$ as

$$A_n(s) = A \circ {H_n}^{-1}(s) \quad \text{for } s \geq 0, \tag{10}$$

where ${H_n}^{-1}$ is the (generalized) right-continuous inverse of the function $H_n : [0, \infty) \to [0, \infty)$ given by

$$H_n(r) = \left(\int_0^r \left(\frac{t}{A(t)}\right)^{\frac{1}{n-1}} dt\right)^{\frac{1}{n'}} \qquad \text{for } r \geq 0 \tag{11}$$

(and $A(\infty)$ is taken equal to ∞). Then A_n is a Young function and there exists a constant C such that

$$\|u\|_{L^{A_n}(G)} \leq C\||\nabla u|\|_{L^A(G)} \tag{12}$$

for every $u \in W_0^{1,A}(G)$. Moreover, $L^{A_n}(G)$ is the optimal Orlicz range space in (12), in the sense that, if an inequality of type (12) holds with $L^{A_n}(G)$ replaced by some other Orlicz space, then the latter must contain $L^{A_n}(G)$.

The Sobolev type inequality which will be exhibited in the next section yields, instead, the best norm (up to equivalence) on the left-hand side of inequalities having the form (12) in the larger class of r.i. spaces. Such a norm belongs to a family of Orlicz-Lorentz norms, containing those appearing on the left-hand sides of (1) and (5) as special cases. Let us mention that various notions of Orlicz-Lorentz spaces appear in the literature, whose properties have been studied by several authors – see e.g. [17], [22], [29]. However, the spaces coming into play in the present setting are somewhat different and are introduced in the next section.

2. Optimal Orlicz-Sobolev embeddings

Let G be a measurable subset of $\mathbb{R}^n$, $n \geq 1$. Let $q \in (1, \infty)$ and let D be any Young function such that

$$\int^{\infty} \frac{D(r)}{r^{1+q}} dr \ < \ \infty. \tag{13}$$

Then we define $L(q, D)(G)$ as the space of real-valued measurable functions u on G for which the quantity

$$\|u\|_{L(q,D)(G)} = \|s^{-\frac{1}{q}} u^*(s)\|_{L^D(0,|G|)} \tag{14}$$

is finite. We have

Proposition 1. *Let G be any measurable subset of $\mathbb{R}^n$, $n \geq 1$. Let $q \in (1, \infty)$ and let D be a Young function satisfying (13). Then the space $L(q, D)(G)$, equipped with the norm (14), and the space $L(q, D)(G) \cap L^\infty(G)$, equipped with the norm*

$$\|u\|_{L(q,D)(G) \cap L^\infty(G)} = \|s^{-\frac{1}{q}} u^*(s)\|_{L^D(0,|G|)} + \|u\|_{L^\infty(G)},$$

are r.i. spaces.

Notice that the Lorentz space $L^{p^*,p}(G)$ and the Lorentz-Zygmund space $L^{\infty,n}(\log L)^{-1}(G)$ can be identified (up to equivalent norms) as special instances of the spaces $L(q, D)(G)$. Indeed, when $D(s) = s^p$, with $1 \leq p < n$, then $L(n, D)(G) = L^{p^*,p}(G)$. If $D(s) = s^n \log^{-n}(1 + s)$ for large s, and $|G| < \infty$, then [4, Lemma 6.12, Chapter 4] ensures that $L(n, D)(G) = L^{\infty,n}(\log L)^{-1}(G)$, up to equivalent norms. Thus, the optimal r.i. spaces for embeddings of $W_0^{1,p}(G)$ have the form $L(n, D)(G)$ for a suitable choice of the Young function D. It turns out that this is always the case even in the more general framework of Orlicz-Sobolev spaces $W_0^{1,A}(G)$. Namely, the optimal r.i. space for embeddings of $W_0^{1,A}(G)$ is a space $L(n, D)(G)$ (or its intersection with $L^\infty(G)$) for some Young function D. A general recipe for constructing such a function, called $B_{A,n}$ in what follows, from any Young function A satisfying (9), is provided by the formula

$$B_{A,n}(s) = \int_0^s b(r)\, dr,$$

where b is the left-continuous function in $[0,\infty)$ such that

$$b^{-1}(s) = \left(\int_{a^{-1}(s)}^{\infty}\left(\int_0^t\left(\frac{1}{a(r)}\right)^{\frac{1}{n-1}}dr\right)^{-n}\frac{dt}{a(t)^{n'}}\right)^{\frac{1}{1-n}} \qquad \text{for } s\geq 0. \tag{15}$$

Here, a is the function appearing in (7), and a^{-1} and b^{-1} are the (generalized) left-continuous inverses of a and b.

The function $B_{A,n}$ is always an admissible function in the definition of the space $L(n,B_{A,n})(G)$, as shown by the following proposition.

Proposition 2. *Let $n\geq 2$ and let A be any Young function satisfying (9). Then the function $B_{A,n}$ is a finite-valued Young function and*

$$\int^{\infty}\frac{B_{A,n}(r)}{r^{1+n}}dr<\infty. \tag{16}$$

We are now in a position to state our embedding theorem.

Theorem 1. *Let $n\geq 2$ and let G be either $\mathbb{R}^n$, or an open subset of $\mathbb{R}^n$ having finite measure. Let A be any Young function satisfying (9).*

I. *If*

$$\int^{\infty}\left(\frac{r}{A(r)}\right)^{\frac{1}{n-1}}dr=\infty, \tag{17}$$

then there exists a constant C_1, depending only on n, such that

$$\|u\|_{L(n,B_{A,n})(G)}\leq C_1\||\nabla u|\|_{L^A(G)} \tag{18}$$

for every $u\in W_0^{1,A}(G)$.

II. *If*

$$\int^{\infty}\left(\frac{r}{A(r)}\right)^{\frac{1}{n-1}}dr<\infty, \tag{19}$$

then there exists a constant C_2, depending only on n and $\int_0^{\infty}\frac{\widetilde{A}(r)}{r^{1+n'}}dr$, such that

$$\|u\|_{L(n,B_{A,n})(G)\cap L^{\infty}(G)}\leq C_2\||\nabla u|\|_{L^A(G)} \tag{20}$$

for every $u\in W_0^{1,A}(G)$. Here, $\widetilde{A}$ denotes the Young conjugate of A.

Moreover, $L(n,B_{A,n})(G)$ and $L(n,B_{A,n})(G)\cap L^{\infty}(G)$ are the optimal r.i. range spaces in inequalities (18) and (20), respectively, in the sense that if (18) [resp. (20)] holds with $L(n,B_{A,n})(G)$ $[L(n,B_{A,n})(G)\cap L^{\infty}(G)]$ replaced by some other r.i. space $X(G)$, then $L(n,B_{A,n})(G)\subseteq X(G)$ $[L(n,B_{A,n})(G)\cap L^{\infty}(G)\subseteq X(G)]$, with continuous inclusion.

Let us make a few comments on Theorem 1.

To begin with, let us stress that assumption (9) is necessary for the theorem to hold when $G = \mathbb{R}^n$. In fact, one can show that such an assumption is indispensable for any inequality of the type

$$\|u\|_{X(\mathbb{R}^n)} \leq C\||\nabla u|\|_{L^A(\mathbb{R}^n)}$$

to hold for some r.i. space $X(\mathbb{R}^n)$, for some constant C and for every $u \in W_0^{1,A}(\mathbb{R}^n)$. On the other hand, assumption (9) is irrelevant in Theorem 1 when $|G| < \infty$. Actually, A can be replaced, if necessary, by another Young function which is equivalent to A near infinity and makes the integral in (9) converge. Such a replacement turns $\|\cdot\|_{L^A(G)}$ into an equivalent norm, up to multiplicative constants depending on A and $|G|$, and hence leaves $W_0^{1,A}(G)$ unchanged.

In some applications, especially to partial differential equations, inequalities in integral form are more useful than norm inequalities. In this regard, observe that inequality (18) is equivalent to the integral inequality

$$\int_0^{|G|} B_{A,n}\big(C_1^{-1}s^{-1/n}u^*(s)\big)ds \leq \int_G A(|\nabla u|)\,dx \tag{21}$$

for every $u \in W_0^{1,A}(G)$. Indeed, (21) implies (18) by the very definition of Luxemburg norm in Orlicz spaces. Conversely, on applying (18) with $A(s)$ replaced by $A(s)/M$, where $M = \int_G A(|\nabla u|)\,dx$, and observing that $B_{A,n}(s)$ is transformed into $B_{A,n}(s)/M$ after this replacement, one easily obtains (21).

If $|G| < \infty$ and (19) is in force, then Theorem 1, Part II, reproduces the fact that $L^\infty(G)$ is the optimal r.i. range space for embeddings of $W_0^{1,A}(G)$ ([8, 28]). Actually, $L(n, B_{A,n})(G) \cap L^\infty(G)$ trivially equals $L^\infty(G)$ when $|G| < \infty$.

Theorem 1 and inequality (12) tell us that either $L(n, B_{A,n})(G)$ or $L(n, B_{A,n})(G) \cap L^\infty(G)$ is (continuously) included in $L^{A_n}(G)$, according to whether (17) or (19) is fulfilled. A direct proof of these inclusions can be accomplished as well, thus enabling us to recover inequality (12) via Theorem 1.

A version of Theorem 1 holds for anisotropic Orlicz-Sobolev spaces, involving Orlicz type norms of the full gradient of test functions, not necessarily depending on its modulus (see [10]).

Theorem 1 includes inequalities (1) and (5) as special cases. Indeed, $B_{A,n}(s)$ is equivalent to $A(s)$ when $A(s) = s^p$, with $1 \leq p < n$; hence, $L(n, B_{A,n})(G) = L(n, A)(G) = L^{p^*,p}(G)$. If $A(s) = s^n$ for large s and satisfies (9), then $B_{A,n}(s)$ is equivalent to $s^n \log^{-n}(1+s)$ near infinity; thus, as observed above, $L(n, B_{A,n})(G) = L^{\infty,n}(\log L)^{-1}(G)$, up to equivalent norms, provided that $|G| < \infty$.

In view of these examples, one is led to expect that $B_{A,n}$ is equivalent to A in sub-limiting cases, but that $B_{A,n}$ grows more slowly than A in borderline situations. This is actually the case. Indeed, one can show that A always dominates $B_{A,n}$ and that A and $B_{A,n}$ are equivalent if and only if $A(s)$ is dominated, in a strong sense, by s^n. The latter assertion can be made precise in terms of the Matuszewska-Orlicz indices. Recall that if A is a real-valued and strictly positive

Young function in $(0,\infty)$, the upper index $I(A)$ of A is defined as

$$I(A) = \lim_{\lambda\to\infty} \frac{\log\left(\sup_{s>0} \frac{A(\lambda s)}{A(s)}\right)}{\log \lambda}. \tag{22}$$

The upper index at infinity $I_\infty(A)$ and the upper index at zero $I_0(A)$ are defined analogously, with $\sup_{s>0}$ replaced by $\limsup_{s\to\infty}$ and by $\limsup_{s\to 0+}$, respectively. Note that $I_0(A)$ is well defined even if A takes the value ∞. From Theorem 1, via some extra work, one obtains the following characterization of non-limiting optimal embeddings.

Corollary 1. *Let $n \geq 2$ and let A be any Young function satisfying (9).*

I. *Assume that (17) holds. Then:*

(i) *$L(n,A)(\mathbb{R}^n)$ is the optimal r.i. range space for embeddings of $W_0^{1,A}(\mathbb{R}^n)$ if and only if A is finite-valued and $I(A) < n$;*

(ii) *Given any open subset $G \subset \mathbb{R}^n$ having finite measure, $L(n,A)(G)$ is the optimal r.i. range space for embeddings of $W_0^{1,A}(G)$ into r.i. spaces if and only if A is finite-valued and $I_\infty(A) < n$.*

II. *Assume that (19) holds. Let $\widehat{A}$ be any Young function which agrees with A near 0 and satisfies $\int^\infty \frac{\widehat{A}(r)}{r^{1+n}} dr < \infty$. Then $L(n,\widehat{A})(\mathbb{R}^n) \cap L^\infty(\mathbb{R}^n)$ is the optimal r.i. range space for embeddings of $W_0^{1,A}(\mathbb{R}^n)$ if and only if $I_0(A) < n$.*

Example 1. Let G be an open subset of $\mathbb{R}^n$ having finite measure. We shall exhibit the optimal Orlicz range space and the optimal r.i. range space for embeddings of $W_0^{1,A}(G)$ when A is a Young function which equals $s^p \log^\alpha(1+s)$ for large s, where either $p = 1$ and $\alpha \geq 0$, or $p > 1$ and $\alpha \in \mathbb{R}$. Let us mention that special instances of this example overlap with results scattered in various papers, including [2, 3, 6, 8, 11, 13, 14, 16, 18, 19, 23, 25]

Case $1 \leq p < n$. Condition (17) is fulfilled. The Young function $A_n(s)$ is equivalent to $s^{p^*} \log^{\frac{n\alpha}{n-p}}(1+s)$ near infinity; thus the Orlicz space associated with this Young function agrees with the optimal Orlicz range space $L^{A_n}(G)$. Since $I_\infty(A) = p < n$, then by Corollary 1 the optimal r.i. space is $L(n,A)(G)$. By [4, Lemma 6.12, Chap. 4], the latter space agrees (up to equivalent norms) with the Lorentz-Zygmund space $L^{p^*,p}(\log L)^{\alpha/p}(G)$. Notice that this reproduces (1) when $\alpha = 0$.

Case $p = n$, $\alpha < n-1$. Condition (17) is in force. The optimal Orlicz range space $L^{A_n}(G)$ equals the Orlicz space associated with the Young function $e^{s^{\frac{n}{n-1-\alpha}}} - 1$. The Young function $B_{A,n}(s)$ is equivalent to $s^n(\log(1+s))^{\alpha-n}$ near infinity. Thus, by [4, Lemma 6.12, Chap. 4], the optimal r.i. range space $L(n, B_{A,n})(G)$ agrees with the Lorentz-Zygmund space $L^{\infty,n}(\log L)^{\alpha/n-1}(G)$. In the special case where $\alpha = 0$, this recovers (5).

Case $p = n$, $\alpha = n-1$. Condition (17) is satisfied. The Orlicz space associated with the Young function $e^{e^{s^{n'}}} - e$ is the optimal Orlicz range space. The Young function $B_{A,n}(s)$ is equivalent to $s^n(\log(1+s))^{-1}(\log(1+\log(1+s)))^{-n}$ near infinity. Hence, by a simple extension of [4, Lemma 6.12, Chap. 4], we deduce that the optimal r.i. range space $L(n, B_{A,n})(G)$ agrees with the generalized Lorentz-Zygmund space $L^{\infty,n}(\log L)^{-1/n}(\log\log L)^{-1}(G)$ equipped with the norm

$$\|u\|_{L^{\infty,n}(\log L)^{-1/n}(\log\log L)^{-1}(G)} = \|s^{-1/n}(1+\log(|G|/s))^{-1/n}(1+\log(1+\log(|G|/s)))^{-1}u^*(s)\|_{L^n(0,|G|)}.$$

Cases $p > n$, or $p = n$ and $\alpha > n-1$. Condition (19) is now fulfilled. The Young function $A_n(s)$ equals ∞ for large s. Thus, $L^\infty(G)$ is the optimal Orlicz range space, and also the optimal r.i. range space for embeddings of $W_0^{1,A}(G)$.

We refer to [10] for proofs of Theorem 1 and Corollary 1. Here, we limit ourselves to outlining our approach to Theorem 1. The first step reduces the problem of the optimal r.i. range space for embeddings of $W_0^{1,A}(G)$ to a one-dimensional problem, via symmetrization. The key result involved at this stage is the Pólya-Szegő principle on the decrease of gradient norms under rearrangement. In its Orlicz space version, such a principle tells us that if A is any Young function and $u \in W_0^{1,A}(G)$, then u^* is locally absolutely continuous in $(0,\infty)$, and

$$\left\| n\omega_n{}^{1/n} r^{1/n'}\left(-\frac{du^*}{dr}\right)\right\|_{L^A(0,|G|)} \le \| |\nabla u| \|_{L^A(G)}, \tag{23}$$

where ω_n denotes the measure of the unit ball in $\mathbb{R}^n$ (see e.g. [7]). After applying this result, inequalities (18) and (20) are shown to be equivalent to certain Hardy type inequalities in Orlicz spaces. The proof of these inequalities is a crucial point, and can be accomplished by interpolation. The optimality of the spaces $L(n, B_{A,n})(G)$ and $L(n, B_{A,n})(G) \cap L^\infty(G)$ on the left-hand sides of inequalities (18) and (20), respectively, relies, instead, on a Hölder type inequality in Orlicz spaces for non-increasing functions. Such inequality is established via discretization and truncation arguments.

References

[1] R.A. Adams, On the Orlicz-Sobolev imbedding theorem, *J. Funct. Anal.* **24** (1977), 241–257.

[2] A. Alvino, V. Ferone, and G. Trombetti, Moser-type inequalities in Lorentz spaces, *Potential Anal.* **5** (1996), 273–299.

[3] C. Bennett and K. Rudnick, On Lorentz-Zygmund spaces, *Dissert. Math.* **175** (1980), 1–72.

[4] C. Bennett and R. Sharpley, "Interpolation of operators", Academic Press, Boston, 1988.

[5] O.V. Besov, V.P. Il'in and S.M. Nikolskii, 'P Integral representations of functions and embedding theorems", Nauka, Moscow, 1975; English translation: Wiley, 1979.

[6] H. Brezis and S. Wainger, A note on limiting cases of Sobolev embeddings and convolution inequalities, *Comm. Part. Diff. Eq.* **5** (1980), 773–789.

[7] J.E. Brothers and W.P. Ziemer, Minimal rearrangements of Sobolev functions, *J. Reine Angew. Math* **384** (1988), 153–179.

[8] A. Cianchi, A sharp embedding theorem for Orlicz-Sobolev spaces, *Indiana Univ. Math. J.* **45** (1996), 39–65.

[9] A. Cianchi, Boundedness of solutions to variational problems under general growth conditions, *Comm. Part. Diff. Eq.* **22** (1997), 1629–1646.

[10] A. Cianchi, Optimal Orlicz-Sobolev embeddings, preprint.

[11] M. Cwikel and E. Pustylnik, Sobolev type embeddings in the limiting case, *J. Fourier Anal. Appl.* **4** (1998), 433–446.

[12] D.T. Donaldson and N.S. Trudinger, Orlicz-Sobolev spaces and embedding theorems, *J. Funct. Anal.* **8** (1971), 52–75.

[13] D.E. Edmunds, P. Gurka, and B. Opic, Double exponential integrability of convolution operators in generalized Lorentz-Zygmund spaces, *Indiana Univ. Math. J.* **44** (1995), 19–43.

[14] D.E. Edmunds, R.A. Kerman and L. Pick, Optimal Sobolev embeddings involving rearrangement invariant quasi-norms, *J. Funct. Anal.* **170** (2000), 307–355.

[15] D.E. Edmunds and H. Triebel, Sharp Sobolev embeddings and related Hardy inequalities, *Math. Nachr.* **207** (1999), 79–92.

[16] N. Fusco, P.L. Lions and C. Sbordone, Sobolev embedding theorems in borderline cases, *Proc. Amer. Math. Soc.* **124** (1996), 561–565.

[17] M.L. Goldman and R. Kerman, The dual for the cone of decreasing functions in a weighted Orlicz class and the associate of an Orlicz-Lorentz space, *Proc. Amer. Math. Soc.*, to appear.

[18] L. Greco and G. Moscariello, An embedding theorem in Lorentz-Zygmund spaces, *Potential Anal.* **5** (1996), 581–590.

[19] K. Hansson, Imbedding theorems of Sobolev type in potential theory, *Math. Scand.* **45** (1979), 77–102.

[20] V.I. Kolyada, On the differential properties of the rearrangements of functions, *in* "Progress in Approximation Theory" (A.A. Gonchar and E.B. Saff, Eds.), pp. 333–352, Springer-Verlag, Berlin, 1992.

[21] V.M. Maz'ya, "Sobolev spaces", Springer-Verlag, Berlin, 1985.

[22] S.J. Montgomery-Smith, Comparison of Orlicz-Lorentz spaces, *Studia Math.* **103** (1992), 161–189.

[23] R. O'Neil, Convolution operators in $L(p,q)$ spaces, *Duke Math. J.* **30** (1963), 129–142.

[24] Yu.V. Netrusov, Embedding theorems for Lizorkin-Triebel spaces, *Notes of Scientific Seminar LOMI* **159** (1987), 103–112. (Russian)

[25] J. Peetre, Espaces d'interpolation et théorème de Soboleff, *Ann. Inst. Fourier* **16** (1966), 279–317.

[26] S.I. Pohozhaev, On the embedding Sobolev theorem for $pl = n$, *Doklady Conference, Section Math. Moscow Power Inst.* (1965), 158–170 (Russian).

[27] L. Tartar, Imbedding theorems of Sobolev spaces into Lorentz spaces, *Boll. Un. Mat. Ital.* (8) **1-B** (1998), 473–500.

[28] G. Talenti, An embedding theorem, *in* "Partial differential equations and the calculus of variations, Vol. II", pp. 919–924, Progr. Nonlinear Differential Equations Appl., 2, Birkhäuser, Boston, 1989.

[29] A. Torchinsky, Interpolation of operators and Orlicz classes, *Studia Math.* **59** (1976), 177–207.

[30] N.S. Trudinger, On imbeddings into Orlicz spaces and some applications, *J. Math. Mech.* **17** (1967), 473–483.

[31] V.I. Yudovich, Some estimates connected with integral operators and with solutions of elliptic equations, *Soviet Math. Doklady* **2** (1961), 746–749.

[32] W.P. Ziemer, "Weakly differentiable functions", Springer-Verlag, New York, 1989.

Andrea Cianchi
Dipartimento di Matematica e Applicazioni per l'Architettura
Università di Firenze
Piazza Ghiberti 27
50122 Firenze, Italy
E-mail address: cianchi@unifi.it

D. Haroske, T. Runst, H.-J. Schmeisser (eds.): Function Spaces, Differential Operators and Nonlinear Analysis. The Hans Triebel Anniversary Volume.

Regularity Problems for Some Semi-linear Problems

Serguei Dachkovski

Dedicated to Prof. Hans Triebel on the occasion of his 65th birthday

1. Introduction

In this article we consider the regularity problems for some partial differential and integral equations (also and specially related to fractals) containing the semilinearity of the following type

$$T^+f(x) = f_+(x) = \max(0, f(x)). \tag{1}$$

T^+ is defined on real functions and it is usually called *truncation operator*. Its properties were studied by many authors, see for example [Osw92], [Tri01b], [Zie89]. We will consider this operator in (real) Besov $B^s_{pq}(\mathbb{R}^n)$ and Hardy-Sobolev $H^s_p(\mathbb{R}^n)$ spaces and we look in these spaces for the best regularity of the solutions of semilinear equations below. The final statement on the boundedness and Lipschitz continuity of the operator T^+ in these spaces can be found in [Tri00] or [Tri01b], §25. We will see that the operator T^+ has rather bad continuity properties in the spaces involved. To overcome the difficulties caused by this circumstance one can try to apply bootstrapping arguments using the lifting property of the corresponding (hypo)elliptic operators as we do it in Section 2. But it may happen (as we see in Sections 4 and 5) that there is no space to start from. An efficient tool to obtain the optimal smoothness also in such case is the so-called Q-method recently invented by H. Triebel in [Tri01a] (see also [Tri01b], §27). We describe it briefly in Section 3.

To handle the equations with the hypoelliptic operators we will also need anisotropic (real) spaces $B^{s,a}_{pq}(\mathbb{R}^n)$. The properties of T^+ in these spaces were studied in [Dac02a], see also [Dac02b], where the truncation property of $B^{s,a}_{pq}(\mathbb{R}^n)$ (i.e. boundedness of the operator T^+ in $B^{s,a}_{pq}(\mathbb{R}^n)$) was obtained. We also have proved there the Fubini and Fatou properties of the anisotropic Besov and Triebel-Lizorkin spaces. First we consider the regularity problem for the semi-linear hypoelliptic equation where we use the bootstrapping arguments and the properties of the spaces mentioned above. Then we take under consideration the equation related to a fractal set Γ, where these arguments do not work. To handle this problem we use the Q-method.

In the following sections we always deal with the real parts of Besov and Hardy-Sobolev spaces.

2. Hypo-elliptic equation

Let us consider the following hypo-elliptic operator

$$A = \mathrm{id} - \frac{\partial^2}{\partial x_1^2} + \frac{\partial^4}{\partial x_2^4}. \tag{2}$$

In view of the structure of this operator we consider the following anisotropy vector $a = (\frac{4}{3}, \frac{2}{3})$ for the real anisotropic Besov spaces $B^{s,a}_{pq}(\mathbb{R}^n)$, $s \in \mathbb{R}$, $0 < p \leq \infty$, $0 < q \leq \infty$. We consider the following semilinear model equation

$$Au(x) = \varepsilon u_+(x) + h(x), \quad x \in \mathbb{R}^2, \tag{3}$$

where $h \in B^{s,a}_{pq}(\mathbb{R}^n)$ is a given function. We are looking for the best possible space (i.e. for the maximal s) for the solution of this equation. The answer is the following.

Theorem 2.1. *Let*

$$0 < p \leq \infty, \quad 0 < q \leq \infty, \tag{4}$$

$$2\left(\frac{1}{p} - 1\right)_+ < s + \lambda < \frac{2}{3}(1 + \frac{1}{p}) \tag{5}$$

for some $\lambda \in [0, \frac{8}{3}]$. *Then there exists a positive number* $\varepsilon_0 > 0$ *such that, if* $0 < \varepsilon \leq \varepsilon_0$, *then for any* $h \in B^{s,a}_{pq}(\mathbb{R}^2)$ *the equation (3) has a uniquely determined solution in* $B^{s+\frac{8}{3},a}_{pq}(\mathbb{R}^2)$.

Proof. We give only the scheme. Let $h(x) \in B^{s,a}_{pq}(\mathbb{R}^2)$. Let $u^0(x) = 0$ and by iteration

$$u^{j+1} = \varepsilon A^{-1} T^+ u^j + A^{-1} h, \quad j = 0, 1, \dots . \tag{6}$$

Then by the mapping properties of A and the truncation property of $B^{s,a}_{pq}(\mathbb{R}^n)$ it follows $u^1(x) = A^{-1}h(x) \in B^{s+\lambda,a}_{pq}(\mathbb{R}^2)$ and by the same reasoning one can prove that

$$u^j \in B^{s+\lambda,a}_{pq}(\mathbb{R}^2), \quad j = 1, 2, \dots . \tag{7}$$

Then for ε small enough it follows by induction that

$$\|u^j \mid B^{s+\lambda,a}_{pq}(\mathbb{R}^2)\| \leq 2\|A^{-1}h \mid B^{s+\lambda,a}_{pq}(\mathbb{R}^2)\|. \tag{8}$$

Now one can prove that this sequence converges in $L_r(\mathbb{R}^2)$ (for some $r > 1$) $\{u^j(x)\} \to u$. Applying the Fatou property of the anisotropic Besov spaces it follows $u \in B^{s+\lambda,a}_{pq}(\mathbb{R}^2)$. By the lifting property of the operator A^{-1} we finally obtain $u \in B^{s+\frac{8}{3},a}_{pq}(\mathbb{R}^2)$. For details of the proof and remarks see [Dac02a]. □

The region of admissible parameters s and p is depicted in Figure 1.

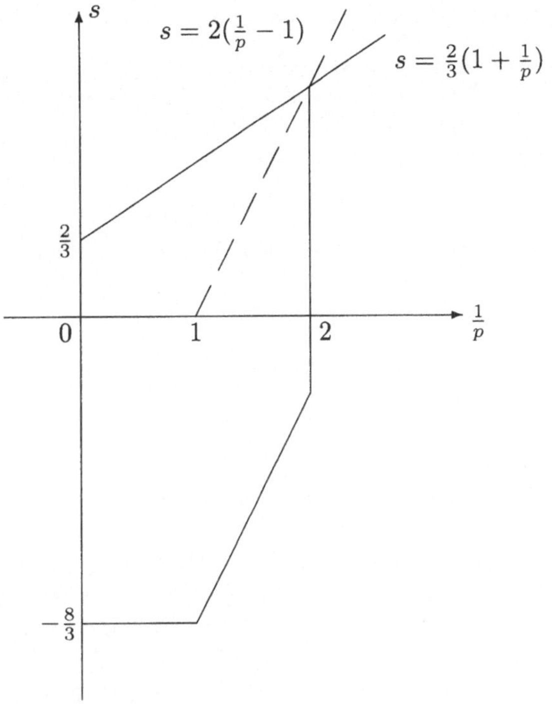

FIGURE 1. $n = 2$

3. On the Q-method

A very elegant method of gaining the optimal regularity for the semilinear equations was invented by H. Triebel, see [Tri01a] or [Tri01b]. It is based on the nice properties of the so-called quarkonial decomposition of function spaces and probably for this reason it was named Q-method. We give briefly the idea how it works. Under some reasonable assumptions on parameters s, p and q it is possible to decompose any function f of $B^s_{pq}(\mathbb{R}^n)$ or $H^s_p(\mathbb{R}^n)$ (also for the anisotropic spaces with some modifications) in the following sum

$$f(x) = \sum_{\gamma\in\mathbb{N}_0^n} \sum_{j=0}^{\infty} \sum_{m\in\mathbb{Z}^n} \lambda^\gamma_{jm} (\gamma qu)_{jm}(x), \tag{9}$$

where the real coefficients

$$\lambda^\gamma = \{\lambda^\gamma_{jm} \in \mathbb{R} : \quad j \in \mathbb{N}_0, \quad m \in \mathbb{Z}^n\}, \quad \gamma \in \mathbb{N}_0^n, \tag{10}$$

can be defined in such way that they depend linearly on f and the sum (9) converges absolutely and unconditionally in S'; and the quarks $(\gamma qu)_{jm}(x)$ are smooth and nonnegative functions. This allows us to define the following operator

$$(Qf)(x) = \sum_{\gamma \in \mathbb{N}_0^n} \sum_{j=0}^{\infty} \sum_{m \in \mathbb{Z}^n} |\lambda_{jm}^{\gamma}(f)| (\gamma qu)_{jm}(x), \quad x \in \mathbb{R}^n. \tag{11}$$

It follows from the definition that

$$f_+(x) \leq |f(x)| \leq (Qf)(x), \quad x \in \mathbb{R}^n. \tag{12}$$

It was proved in [Tri01a] (see also [Tri01b]) that for

$$0 < p \leq \infty, \quad 0 < q \leq \infty, \quad s > n\left(\frac{1}{p} - 1\right)_+ \tag{13}$$

the operator Q is not only bounded in $B^s_{pq}(\mathbb{R}^n)$ (or $H^s_p(\mathbb{R}^n)$) but also Lipschitz continuous. This allows to apply the Banach contraction theorem in this spaces. We explain the idea in the following equation

$$u(x) = \varepsilon \int_{\mathbb{R}^n} K(y)(T^+u)(x-y)\, dy + h(x), \quad x \in \mathbb{R}^n, \tag{14}$$

with an integrable nonnegative kernel $K(y)$ and a given function $h \in B^s_{pq}(\mathbb{R}^n)$. We can consider the right-hand side of this equation as an operator, say B, acting on a function u. We would like to prove that for some small $\varepsilon > 0$ it yields a contraction in $B^s_{pq}(\mathbb{R}^n)$, but it turns out that T^+ is not Lipschitz continuous in those spaces where it is bounded [Tri00]. To overcome this H. Triebel replaced T^+ by Q in B and considered the following operator

$$(B^Q u)(x) = \varepsilon \int_{\mathbb{R}^n} K(y)(Qu)(x-y)\, dy + h(x), \quad x \in \mathbb{R}^n, \tag{15}$$

using the better continuity properties of the operator Q he proved by the Banach contraction theorem that B^Q has a fixed point $u_0(x)$ in $B^s_{pq}(\mathbb{R}^n)$ which is a supersolution of the equation (14). Then one can use the supersolution technique (which requires $K(y) \geq 0$) to obtain the unique solution $u(x)$ in $B^s_{pq}(\mathbb{R}^n)$ which is optimal for $h \in B^s_{pq}(\mathbb{R}^n)$. For details we refer to [Tri01a] and [Tri01b], where also another model equation

$$(\mathrm{id} - \Delta)u(x) = \varepsilon(T^+u)(x) + h(x), \quad x \in \mathbb{R}^n \tag{16}$$

was considered. In [Dac00] we have applied this method to some more general operators, for example we considered the uniformly elliptic operator in place of $\mathrm{id} - \Delta$, the Dirichlet problem in domain Ω and the semilinear Volterra equation of the second kind

$$u(x) = \lambda \int_0^x K(x-y) u_+(y) dy + h(x), \quad x \in \mathbb{R}_+, \tag{17}$$

which plays an important role in the visco-elasticity theory. For details, discussion and connection to visco-elasticity see [Dac00]. We just quote two theorems from there. Let A be a uniformly elliptic operator of the second order.

Theorem 3.1. *Let* $\mathbb{N} \ni n \geq 2$, $0 < p \leq \infty$, $n(\frac{1}{p} - 1)_+ < s + \lambda < 1 + \frac{1}{p}$ *for some* $\lambda \in [0, 2]$. *There is a positive number* $\varepsilon_0 > 0$ *with the following properties:*
(i) if $0 < \varepsilon < \varepsilon_0$, *then for any* $h \in B^s_{pq}(\mathbb{R}^n)$ *the equation*

$$(A + \mathrm{id})u(x) = \varepsilon u_+(x) + h(x), \qquad x \in \mathbb{R}^n \tag{18}$$

has a uniquely determined solution $u \in B^{s+2}_{pq}(\mathbb{R}^n)$.
(ii) if $p < \infty$ *and* $0 < \varepsilon < \varepsilon_0$, *then for any* $h \in H^s_p(\mathbb{R}^n)$ *the equation (18) has a uniquely determined solution* $u \in H^{s+2}_p(\mathbb{R}^n)$.

Theorem 3.2. *Let*

$$1 \leq p \leq \infty, \quad 1 \leq q \leq \infty, \quad 0 < s < \frac{1}{p}. \tag{19}$$

Let $K(y) \geq 0$ *in* $\mathbb{R}_+$ *with* $K \in L_1(\mathbb{R}_+)$. *There is a positive number* λ_0 *such that*
(i) If $0 < \lambda < \lambda_0$, *then for any* $h \in B^s_{pq}(\mathbb{R}_+)$ *the equation (17) has a uniquely determined solution* $u \in B^s_{pq}(\mathbb{R}_+)$.
(ii) Let, in addition, $p < \infty$. *If* $0 < \lambda < \lambda_0$, *then for any* $h \in H^s_p(\mathbb{R}_+)$ *the equation (17) has a uniquely determined solution* $u(x) \in H^s_p(\mathbb{R}_+)$.

Actually for such equations like (14) and (16) the Q-method is not the only way to find a solution with the optimal smoothness. One can even get the so-called refined solutions of such equations, i.e. to build such solutions which make sense not only almost everywhere (in the Lebesgue sense), but also μ-a.e., for some Radon measure μ (singular with respect to the Lebesgue one). It means that we can consider these equations on some sets of the Lebesgue measure zero, for example on a d-set Γ corresponding to the above Radon measure μ. For details see [Tri02].

4. Equations related to fractals

In this section we are going to consider a fractal set Γ and an equation including the trace of the unknown function on Γ. In contrast to the previous section the Q-method seems to be necessary, at least we do not see any other way to get the solution with maximal regularity. However it does not solve the question of uniqueness of the solution.

Definition 4.1. *Let* $n \in \mathbb{N}$ *and* Γ *be a compact set in* $\mathbb{R}^n$ *and let* $0 \leq d \leq n$. *Then* Γ *is called a d-set if there exists a Borel measure* μ *in* $\mathbb{R}^n$ *such that* $\operatorname{supp} \mu = \Gamma$ *and there are two positive constants* c_1, c_2 *such that for all* $\gamma \in \Gamma$ *and all* r *with* $0 < r < 1$,

$$c_1 r^d \leq \mu(B(\gamma, r) \cap \Gamma) \leq c_2 r^d. \tag{20}$$

If $0 \leq d < n$, then by [Tri97], Corollary 3.6, $|\Gamma| = 0$, where $|\Gamma|$ stands for the Lebesgue measure, and μ in (20) is a Radon measure. There are different ways how to define the trace of f say belonging to $B^s_{pq}(\mathbb{R}^n)$ on such sets. We refer to [Tri97], §18 and [Tri01b], Section 1.9. The idea is the following. For any smooth function $\phi(x) \in S(\mathbb{R}^n)$ one can define $tr_\Gamma \phi = \phi|_\Gamma$ as a pointwise restriction of ϕ on Γ. Let us suppose, that for some $s > 0, 0 < p < \infty, 0 < q < \infty$ the inequality

$$\|tr_\Gamma \phi \mid L_p(\Gamma)\| \le c\|\phi \mid B^s_{pq}(\mathbb{R}^n)\| \tag{21}$$

holds for all $\phi \in S(\mathbb{R}^n)$ (which is dense in $B^s_{pq}(\mathbb{R}^n)$ for $0 < p < \infty$ and $0 < q < \infty$), where $c > 0$ does not depend on ϕ. Then, by completion, one can also define the trace of any function f of $B^s_{pq}(\mathbb{R}^n)$ and it is denoted by $tr_\Gamma f$. It was proved in [Tri97] that if $0 < p < \infty$ and $0 < q \le \min(1, p)$, then

$$tr_\Gamma B_{pq}^{\frac{n-d}{p}}(\mathbb{R}^n) = L_p(\Gamma). \tag{22}$$

Another approach, which uses so-called strictly defined functions can be found for example in [Bri00], §6, but it works only for $p > 1$ and $s > \frac{n-d}{p}$.

The above optimal assertion (22) allows us to give the following definition (see [Tri97], §20.2), if $0 < d < n,\ s > 0,\ 0 < p \le \infty$ and $0 < q \le \infty$, then

$$B^s_{pq}(\Gamma) = tr_\Gamma B_{pq}^{s+\frac{n-d}{p}}(\mathbb{R}^n) \tag{23}$$

quasi-normed by

$$\|f \mid B^s_{pq}(\Gamma)\| = \inf \left\| g \mid B_{pq}^{s+\frac{n-d}{p}}(\mathbb{R}^n)\right\|, \tag{24}$$

where the infimum is taken over all $g \in B_{pq}^{s+\frac{n-d}{p}}(\mathbb{R}^n)$ with $tr_\Gamma g = f$. By [Tri97], Theorem 20.6, $B^s_{pq}(\Gamma)$ is embedded in $L_p(\Gamma)$ if $s > 0$. If $p > 1$, then by [Tri97], Theorem 18.2, any $f^\Gamma \in L_p(\Gamma)$ can be identified with the singular distribution

$$f(\phi) = \int_\Gamma f^\Gamma(\gamma)(\phi|\Gamma)(\gamma)\mu(d\gamma), \quad \phi \in S(\mathbb{R}^n), \tag{25}$$

which belongs to $B_{p,\infty}^{-\frac{n-d}{p'}}(\mathbb{R}^n)$ if $1 < p < \infty$, where $\frac{1}{p} + \frac{1}{p'} = 1$. We denote this identification by

$$id_\Gamma: \quad L_p(\Gamma) \mapsto B_{p,\infty}^{-\frac{n-d}{p'}}(\mathbb{R}^n). \tag{26}$$

The operator tr^Γ is defined as (see [Tri97], Chapter IV)

$$tr^\Gamma = id_\Gamma \circ tr_\Gamma, \tag{27}$$

which maps

$$B_{pq}^{s+\frac{n-d}{p}}(\mathbb{R}^n) \mapsto B_{p,\infty}^{-\frac{n-d}{p'}}(\mathbb{R}^n) \quad 1 < p < \infty, \quad 0 < q \le \infty, \quad s > 0. \tag{28}$$

Let $p > 1$, Γ be a compact d-set as above and μ is a corresponding Radon measure. Let tr_Γ and id_Γ be the operators defined as above. The truncation operator can be considered also on $L_p(\Gamma)$:

$$(T^+_\Gamma f^\Gamma)(\gamma) = \max(f^\Gamma(\gamma), 0), \quad \gamma \in \Gamma. \tag{29}$$

Remark 4.2. We preserve the same notation $T^+_\Gamma f = f_+$ also in this case. Obviously T^+_Γ is bounded and a Lipschitz-continuous operator on $L_p(\Gamma)$.

Consider the equation

$$(-\Delta + \mathrm{id})u(x) = \varepsilon\, \mathrm{id}_\Gamma\, T_\Gamma^+ tr_\Gamma u(x) + h(x), \quad x \in \mathbb{R}^n, \tag{30}$$

in some $B_{pq}^{s+\frac{n-d}{p}}(\mathbb{R}^n)$ with $s > 0$. It is known, that the Green's function $G(x)$ of the operator $-\Delta + \mathrm{id}$ is a positive and exponentially decreasing function. It allows us to rewrite (30) as follows

$$\begin{aligned} u(x) &= \varepsilon(-\Delta + \mathrm{id})^{-1}(\mathrm{id}_\Gamma\, T_\Gamma^+ tr_\Gamma u(x) + h(x)) \\ &= \varepsilon(G * \mathrm{id}_\Gamma\,(tr_\Gamma u)_+)(x) + H(x), \quad x \in \mathbb{R}^n, \end{aligned} \tag{31}$$

where $H(x) = (-\Delta + \mathrm{id})^{-1}h(x)$. Using the interpretation (25–27) we get

$$u(x) = \varepsilon \int_\Gamma (tr_\Gamma u(\gamma))_+ G(x-\gamma)\, \mu(d\gamma) + H(x), \quad x \in \mathbb{R}^n. \tag{32}$$

We generalize this equation inserting an additional volume term

$$u(x) = \varepsilon \int_\Gamma G(x-\gamma)(tr_\Gamma u)_+(\gamma)\, \mu(d\gamma) + \varepsilon \int_{\mathbb{R}^n} K(x-y) u_+(y)\, dy + H(x), \tag{33}$$

$x \in \mathbb{R}^n$, where $K(y)$ is supposed to be a non-negative and integrable on $\mathbb{R}^n$.

For the equation (33) we ask again the following question: if we know that the given function H belongs to a given space $B_{pq}^s(\mathbb{R}^n)$, then what can we conclude about the solution of this equation. We give the answer on the regularity in the next theorem, but we still do not know whether the solution is unique.

Remark 4.3. In all theorems above the uniqueness followed from the uniqueness of the solution in $L_p(\mathbb{R}^n)$. We stress that we cannot consider the equation (33) in $L_p(\mathbb{R}^n)$ since tr_Γ makes no sense there.

Theorem 4.4. *Let $1 < p < \infty$ and $s < 1 + \frac{1}{p}$,*

$$0 < s < 2 - n + d, \quad n - 2 < d < n \quad \text{and} \quad 1 < q \leq \infty. \tag{34}$$

Let Γ be a compact d-set as above. If $\varepsilon > 0$ is small enough, then for any $H(x) \in B_{pq}^{s+\frac{n-d}{p}}(\mathbb{R}^n)$ there exists at least one solution of the equation (33) belonging to $B_{pq}^{s+\frac{n-d}{p}}(\mathbb{R}^n)$.

For proof and remarks we refer to [Dac02a], Section 2.5.

5. Heat equation with fractal source of heat

5.1. Parabolic equation with fractional inhomogeneity

In this section we are going to demonstrate that equations of fractal type can appear in some mechanical models.

Let Ω be a subset of $\mathbb{R}^3$. Let us suppose that Ω is filled with a heat-conducting material and the distribution of the sources of heat is given by the function $f(x,t)$.

It is known that the behavior of this system can be described with the help of the following heat equation.

$$\frac{\partial u}{\partial t}(x,t) - \Delta u(x,t) = f(x,t), \quad x \in \Omega, \quad t \in (0,T). \tag{35}$$

If we suppose that the temperature on the boundary $\partial\Omega$ is always zero and that initial distribution of the temperature in Ω is $u_0(x)$ then we have the following initial and boundary conditions:

$$u\Big|_{t=0} = u_0(x), \quad x \in \Omega; \tag{36}$$

$$u(x,t) = 0, \quad x \in \partial\Omega, \quad t \in (0,T). \tag{37}$$

This is a well-known problem of the heat conduction which is studied in many classical monographs (see e.g. [LSU67]). We will not discuss the history of this problem. Later we will give only those known facts that are interesting for us here.

Now let us additionally suppose that, say thanks to some chemical changes, each point of material in Ω is a source of heat (if temperature overbalance some level, say zero centigrade degree) with intensity proportional to the value of temperature. Then an additional term should appear on the right-hand side of the equation (36), namely $k_1(T^+u)(x,t)$, where k_1 is a positive constant and T^+ is the truncation operator defined above.

Also we suppose that we deal with a composite material, such that its matrix is a conductor, which can conduct the current and also can conduct chemical changes with the filling (say if the current is switched on), yielding the heat evolution. We suppose that this interaction starts beginning from a certain level of the temperature (say again zero centigrade degree), and that the matrix has a very small rate in comparison with the filling, such that its volume is negligible in comparison with the volume of the filling.

We are going to model the matrix as a d-set γ with the Lebesgue measure zero. The current is not always switched on, but only at some discrete periods of time of the set τ, such that the set $\Gamma = (\gamma, \tau) \subset \Omega \times (0,T)$ forms a regular anisotropic d-set with respect to the anisotropy $a = (\sigma, \ldots \sigma, \sigma/2)$ according to the definition given in [Far98], §3.1, p. 27. In this case the additional term in the right-hand side of (35) looks as follows: $(T_\Gamma^+ tr^\Gamma u)(x,t)$. Hence for such a problem we have the following equation of state

$$\frac{\partial u}{\partial t}(x,t) - \Delta u(x,t) = k_1(T^+u)(x,t) + k_2(T^+tr^\Gamma u)(x,t) + f(x,t), \tag{38}$$

where $x \in \Omega$ and $t \in (0,T)$.

From now on we will consider this equation in $\mathbb{R}^n$. However it is not our aim to consider the problems of such type in full generality. We are going to study the regularity properties of solutions of such problems. For this purpose we will consider only the model equation (38) with $\Omega = \mathbb{R}^n$ and homogeneous initial condition:

$$u(x,t)\Big|_{t=0} = 0, \quad x \in \mathbb{R}^n. \tag{39}$$

In the following three subsections we introduce some notions and collect some facts that will be useful for us in the sequel.

5.2. Related spaces

Let us denote $\mathbb{R}^{n+1}_T = \mathbb{R}^n \times (0,T)$, $T \in (0,\infty]$ and let $L_p(\mathbb{R}^{n+1}_T)$ be the usual Lebesgue space of measurable and p-power integrable functions on $\mathbb{R}^{n+1}_T$. We confine ourselves to the Hilbert case and set $p = q = 2$. In this section we will need anisotropic spaces.

Let $a = (a_1, \dots a_n)$ be a given anisotropy. It is known that

$$B^{s,a}_{2,2}(\mathbb{R}^n) = F^{s,a}_{2,2}(\mathbb{R}^n) = H^{s,a}_2(\mathbb{R}^n), \quad s \in \mathbb{R}, \tag{40}$$

where $H^{s,a}_2(\mathbb{R}^n)$ are the anisotropic Bessel potential spaces and, if $s_1 = \frac{s}{a_1} \in \mathbb{N}$, $\dots$, $s_n = \frac{s}{a_n} \in \mathbb{N}$, then we have $B^{s,a}_{2,2}(\mathbb{R}^n) = W^{s,a}_2(\mathbb{R}^n)$ are the classical anisotropic Sobolev spaces.

For the heat equation considered above we will need the special case of the anisotropy: $W^{s,\dots,s,s/2}_2(\mathbb{R}^{n+1}_T) = W^{s,s/2}_2(\mathbb{R}^{n+1}_T)$. It is more convenient for us to use another notation of these spaces (as in Section 2) with the help of the anisotropy vector a and the mean smoothness σ. In our case $\sigma = \frac{n+1}{n+2}s$ and $a = (\frac{n+1}{n+2}, \dots, \frac{n+1}{n+2}, 2\frac{n+1}{n+2})$.

5.3. Regularity theorem for a linear equation

Before we can study the regularity of solutions of the equation (38) we need to know what happens in the linear case. We formulate the simplified version of Theorem 9.1 from Chapter IV, [LSU67] for the initial value problem (35), (39).

Theorem 5.1. *Let $n \in \mathbb{N}$. Then for any $f \in L_2(\mathbb{R}^{n+1}_T)$ there exists a unique solution $u \in H^{2,1}_2(\mathbb{R}^{n+1}_T)$ of (35), (39). (In our notation $u \in H^{2\frac{n+1}{n+2},a}_2(\mathbb{R}^{n+1}_T)$ with a is as above.)*

The solution of this problem can be presented with the help of the Green's function as (see [LSU67], p. 369)

$$u(x,t) = \int_0^t d\tau \int_{\mathbb{R}^n} K(x-y, t-\tau) f(y,\tau)\, dy, \quad x \in \mathbb{R}^n, \quad 0 < t < T, \tag{41}$$

where

$$K(x,t) = \frac{1}{(4\pi t)^{\frac{n}{2}}} e^{-\frac{x^2}{4t}}, \quad t > 0, \tag{42}$$

and $K(x,t) = 0$ if $t < 0$. Of course $K(x,t) \geq 0$ on $\mathbb{R}^{n+1}_T$ and $K \in L_1(\mathbb{R}^{n+1}_T)$. There are similar results for $\Omega = \mathbb{R}^n_+ = \{x \in \mathbb{R}^n, x_n > 0\}$ or a rectangle, see [Shi92], p. 40–43.

With the help of the lifting operator (see for example [Far98])

$$I_\sigma(f) = \left(\left(\sum_{k=1}^n (1+\xi_k^2)^{\frac{1}{2a_k}} \right)^\sigma \widehat{f} \right)^\vee, \quad \sigma \in \mathbb{R}, \tag{43}$$

which maps $H_2^{s,a}(\mathbb{R}^{n+1})$, $s \in \mathbb{R}$ isomorphically onto $H_2^{s-\sigma,a}(\mathbb{R}^{n+1})$, we can extend the above theorem:

Theorem 5.2. *Let $n \in \mathbb{N}$. Then for any $f \in H_2^{s,s/2}(\mathbb{R}_T^{n+1})$ there exists a unique solution $u \in H_2^{s+2,s/2+1}(\mathbb{R}_T^{n+1})$ of (35), (39). (In our notation $f \in H_2^{s\frac{n+1}{n+2},a}(\mathbb{R}_T^{n+1})$ and $u \in H_2^{(s+2)\frac{n+1}{n+2},a}(\mathbb{R}_T^{n+1})$, where a as above.)*

5.4. Anisotropic traces

Let Γ be a regular anisotropic d-set with respect to the given anisotropy $a = (a_1, \dots, a_n)$. We refer to [Far98], Section 3 for the description of these sets. Let $s \in \mathbb{R}$, $0 < p \leq \infty$ and $0 < q \leq \infty$. By definition

$$B_{pq}^{s,a;\Gamma}(\mathbb{R}^n) = \{f \in B_{pq}^{s,a}(\mathbb{R}^n) : \ f(\phi) = 0 \quad \text{if} \quad \phi \in S(\mathbb{R}^n), \quad \phi|_\Gamma = 0\}, \tag{44}$$

where $\phi|_\Gamma$ is the pointwise restriction of ϕ to Γ (see [Far98], §3.2). We quote Theorem 3.11 from [Far98].

Theorem 5.3. *Let $0 < d < n$ and let Γ be a regular anisotropic d-set in $\mathbb{R}^n$ with respect to the anisotropy $a = (a_1, \dots, a_n)$. If $1 < p \leq \infty$ and $\frac{1}{p} + \frac{1}{p'} = 1$, then (in the sense of identification (25))*

$$L_p(\Gamma) = B_{p,\infty}^{-\frac{n-d}{p'},a;\Gamma}(\mathbb{R}^n), \quad \|f^\Gamma \mid L_p(\Gamma)\| \sim \|f \mid B_{p,\infty}^{-\frac{n-d}{p'},a}(\mathbb{R}^n)\|. \tag{45}$$

Let us consider the initial value problem (38–39). With the help of the Green's function this is equivalent to the equation

$$u(x,t) = \int_0^t d\tau \int_{\mathbb{R}^n} K(x-y,t-\tau)\Big[k_1 u_+(y,\tau) + k_2(tr^\Gamma u)_+(y,\tau) + f(y,\tau)\Big]dy, \tag{46}$$

$x \in \mathbb{R}^n$, $0 < t < T$. We denote the third term of the right-hand side of (46) by

$$H(x,t) = \int_0^t d\tau \int_{\mathbb{R}^n} K(x-y,t-\tau) f(y,\tau)\, dy. \tag{47}$$

This function is known, and by Theorem 5.2, for any $f \in H_2^{s\frac{n+1}{n+2},a}(\mathbb{R}_T^{n+1})$ it belongs to $u \in H_2^{(s+2)\frac{n+1}{n+2},a}(\mathbb{R}_T^{n+1})$. We introduce new variables $z = (x,t)$ and $\xi = (y,\tau)$ and $\mathbf{I}_\Omega$ the characteristic function of the set Ω Then (46) can be written as ($\mathbf{I}_{\mathbb{R}\times[0,t]} = \mathbf{I}_{[0,t]}$ for brevity).

$$u(z) = \int_{\mathbb{R}_+^{n+1}} \mathbf{I}_{\times[0,t]}(z-\xi)K(z)\Big[k_1(T^+u)(z-\xi) + k_2(\mathrm{id}_\Gamma T^+ tr_\Gamma u)(z-\xi)\Big]d\xi + H(z). \tag{48}$$

This equation looks like equation (33). Since K is a positive kernel, integrable on $\mathbb{R}_+^{n+1}$, then the Q-method is applicable to this problem as for (33) and (14). It enables us to obtain the following

5.5. Result

Theorem 5.4. *Let* $p = 2, \quad s < \frac{1}{2}$, *and*

$$0 < s < 1 - n + d, \quad n - 1 < d < n + 1. \tag{49}$$

Let Γ *be a compact regular anisotropic* d*-set as above. If* k_1 *and* k_2 *are small enough, then for any* $f \in H_2^{s\frac{n+1}{n+2},a}(\mathbb{R}_T^{n+1})$ *there exists at least one solution of the equation (48) belonging to* $H_2^{(s+2)\frac{n+1}{n+2},a}(\mathbb{R}_T^{n+1})$, *where* a *has the same meaning as above.*

Remark 5.5. The idea of the proof is the same as in Theorem 4.4, however now we deal with the anisotropic spaces. The counterparts of the needed properties such as Fatou and truncation properties of these spaces can be found in [Dac02b]. To prove Theorem 3.2 we used the property of the characteristic function of the halfspace. Namely, that it is a pointwise multiplier in $B^s_{pq}(\mathbb{R}^n)$ under certain restrictions on s, p and q. We can use the Fubini property of the anisotropic spaces to extend this result to our case.

I would like to thank Prof. Hans Triebel for the supervision of my PhD thesis, where we have obtained these results.

References

[Bri00] M. Bricchi. On the relationship between Besov spaces $B^{s,\Psi}_{pq}(\mathbb{R}^n)$ and L_p-spaces defined on a (d, Ψ)-set. *Jenaer Schriften zur Mathematik und Informatik*, 00/13, 2000.

[Dac00] S. Dachkovski. Application of the Q-method to some semilinear differential and integral equations. *Jenaer Schriften zur Mathematik und Informatik*, 00/15, 2000.

[Dac02a] S. Dachkovski. *The Q-method and regularity problems of some integral equations and PDE's.* Dissertation, Jena, 2002.

[Dac02b] S. Dachkovski. Anisotropic function spaces and related hypoelliptic equations. *Math. Nachr.*, (to appear) 2002.

[Far98] W. Farkas. *Anisotropic function spaces, fractals, and spectra of some semi-elliptic differential operators.* Dissertation, Jena, 1998.

[LSU67] O. Ladizhinskaya, V. Solonnikov, and N. Uraltseva. *Linear and quasilinear parabolic equations (Russian).* Nauka, Moscow, 1967.

[Osw92] P. Oswald. On the boundedness of the mapping $f \to |f|$ in Besov spaces. *Comment. Univ. Carolina 33*, 57–66, 1992.

[Shi92] N. Shimakura. *Partial differential operators of elliptic type.* American Math. Soc., 1992.

[Tri97] H. Triebel. *Fractals and Spectra.* Birkhäuser, Basel, 1997.

[Tri00] H. Triebel. Truncations of functions. *Forum Math*, 12:731–756, 2000.

[Tri01a] H. Triebel. Regularity theory for some semi-linear equations: the Q-method. *Forum Math.*, 13:1–19, 2001.

[Tri01b] H. Triebel. *The structure of functions.* Birkhäuser, Basel, 2001.

[Tri02] H. Triebel. Refined solutions of some integral equations. *Functiones et Approximatio*, to appear, 2002.

[Zie89] W. P. Ziemer. *Weakly differentiable functions.* Springer, New York, 1989.

Serguei Dachkovski
MZH 2300
Zentrum für Technomathematik
Universität Bremen
D-28334 Bremen, Germany
E-mail address: dsn@math.uni-bremen.de

D. Haroske, T. Runst, H.-J. Schmeisser (eds.): Function Spaces, Differential Operators and Nonlinear Analysis. The Hans Triebel Anniversary Volume.

Besov Regularity for the Neumann Problem

Stephan Dahlke

Dedicated to Prof. Hans Triebel on the occasion of his 65th birthday

Abstract. This paper is concerned with the regularity of the solutions to the Neumann problem in Lipschitz domains Ω contained in $\mathbf{R}^d$. Especially, we consider the specific scale $B^s_\tau(L_\tau(\Omega))$, $1/\tau = s/d + 1/p$, of Besov spaces. The regularity of the variational solution in these Besov spaces determines the order of approximation that can be achieved by adaptive and nonlinear numerical schemes. We show that the solution to the Neumann problem is much smoother in the specific Besov scale than in the usual L_p-Sobolev scale which justifies the use of adaptive schemes. The proofs are performed by combining some recent regularity results derived by Zanger [23] with some specific properties of harmonic Besov spaces.

1. Introduction

Quite recently, the regularity of the solutions to second order elliptic boundary value problems

$$\begin{aligned} Lu &= F \quad \text{in} \quad \Omega \subset \mathbf{R}^d, \\ u &= g \quad \text{on} \quad \partial\Omega, \end{aligned} \tag{1}$$

where Ω is a Lipschitz domain, in specific Besov spaces has been investigated, see, e.g., [6, 7, 8, 11]. The aim was to provide some theoretical foundations for the use of adaptive schemes for the numerical treatment of (1). The order of convergence as measured in L_p of usual (linear) Galerkin schemes obtained, e.g., by finite element spaces based on uniform grid refinement, is determined by the regularity of the variational solution u to (1) in the Sobolev scale $W^s(L_p(\Omega))$. Unfortunately, on a general Lipschitz domain, this Sobolev regularity may not be very high, even if the right-hand side F is sufficiently smooth. This fact is caused by singularities near the boundary. Therefore, to increase efficiency, one often uses *adaptive* methods, i.e., the underlying grid is only refined in regions where the solution lacks smoothness. In this case, one does not use the whole linear spaces, hence an adaptive scheme can be interpreted as some kind of *nonlinear approximation*. Then the question arises if nonlinear methods indeed provide some gain of efficiency when compared with linear schemes. So far, the problem is best understood for numerical schemes based on a *wavelet basis* $\Psi = \{\psi_\lambda,\ \lambda \in \mathcal{J}\}$. (We refer to one of the textbooks [2, 12, 19, 22] for the definition and the basic properties of wavelets.) An adaptive wavelet scheme approximates the solution u to (1) by a linear combination of n wavelets. Therefore a natural benchmark for its performance is given by the *best n-term approximation*. Then one approximates a function $f \in L_p(\mathbf{R}^d)$ by the nonlinear manifolds $\mathcal{M}_n$ of all functions

$$S = \sum_{\lambda \in \Gamma} a_\lambda \psi_\lambda$$

with $\Gamma \subset \mathcal{J}$ of cardinality n and studies the error

$$\sigma_n(f)_{L_p(\mathbf{R}^d)} := \inf_{S \in \mathcal{M}_n} \|f - S\|_{L_p(\mathbf{R}^d)}. \tag{2}$$

For the L_2-metric and an orthonormal wavelet basis, the approximation problem (2) has a simple solution. We order the wavelet coefficients by their absolute values and choose Γ corresponding to the n largest values. (Similar results also hold for other values of p, see, e.g., [13] for details.) In contrary to linear schemes, the order of approximation that can be achieved by best n-term approximation is not determined by the Sobolev regularity but by certain *non-classical* scales of function spaces. Indeed, the following characterization has been derived in [14]

$$\sum_{n=1}^{\infty} [n^{s/d} \sigma_n(f)_{L_p(\mathbf{R}^d)}]^\tau \frac{1}{n} < \infty \text{ if and only if } f \in B^s_\tau(L_\tau(\mathbf{R}^d)), \ \frac{1}{\tau} = \frac{s}{d} + \frac{1}{p}, \tag{3}$$

where the $B^s_\tau(L_\tau(\mathbf{R}^d))$ are the *Besov* spaces (see, e.g., [15, 20] for the definition and the main properties of Besov spaces).

Of course, best n-term approximation is not directly applicable in our setting for catching the n biggest wavelet coefficients requires knowing *all* coefficients of the unknown solution u. Nevertheless, quite recently, an *implementable* adaptive wavelet scheme has been developed which produces asymptotically the same rate of convergence as the best n-term approximation [3], see also [4, 9, 10]. Having these results and the characterization (3) in mind, it is therefore natural to ask the following question: what is the regularity of the solution u to (1) as measured in the scale $B^s_\tau(L_\tau(\Omega))$, $\tau = (s/d + 1/p)^{-1}$? Especially, does the solution have a higher smoothness order in these spaces compared to the usual Sobolev scale? For then, adaptive wavelet methods would definitely perform better than linear schemes and the use of adaptive schemes is completely justified. The results in [6, 7, 8, 11] indicate that this is indeed the case for many problems. So far, the deepest results were obtained for the classical Dirichlet problem for harmonic functions:

$$\begin{aligned} \Delta v &= 0 \text{ in } \Omega, \\ v &= g \text{ on } \partial\Omega, \mathcal{N}, \end{aligned} \tag{4}$$

see [11] for details. Once these results are established, it is clearly desirable to generalize them also to the Neumann problem

$$\begin{aligned} \Delta v &= 0 \text{ in } \Omega, \\ \frac{\partial v}{\partial n} &= g \text{ on } \partial\Omega. \end{aligned} \tag{5}$$

However, the proofs in [11] made heavy use of a very systematic study of the homogeneous and inhomogeneous Dirichlet problem presented by Jerison and Kenig [17]. Unfortunately, for the Neumann problem, such a systematic study was an open problem for a long time. Consequently, in his famous book [18], C. Kenig presented this problem as a suggestion for further researches. Soon afterwards, D. Jerison gave this problem to his Ph.D. student D. Zanger who solved it completely [23, 24]. Therefore we can now use his results to establish Besov regularity for both, the homogeneous and the inhomogeneous Neumann problem, and this is the main objective of this note.

This paper is organized as follows. In Section 2, we recall Zanger's developments as far as they are needed for our purposes. Then, in Section 3, we explain how these results can be exploited to establish nonclassical Besov regularity.

2. Classical regularity results

In this section, we want to summarize some of Zanger's results as far as they are needed for our purposes. The first step is to introduce the space

$$B^s_p(L_p(\partial\Omega))_{1^\perp} := \{h \in B^s_p(L_p(\partial\Omega)) \mid h(1) = 0\}. \tag{6}$$

Then the main result for the Neumann problem reads as follows.

Theorem 2.1. *Consider ϵ such that $0 < \epsilon \leq 1$. Define p_0 and p_0' by $1/p_0 = (1+\epsilon)/2$ and $1/p_0' = (1-\epsilon)/2$. Let α and p be numbers satisfying one of the following:*

(a) $p_0 < p < p_0'$ *and* $0 < \alpha < 1$

(b) $1 < p \leq p_0$ *and* $2/p - 1 - \epsilon < \alpha < 1$

(c) $p_0' \leq p < \infty$ *and* $0 < \alpha < 2/p + \epsilon$.

Let Ω be a bounded Lipschitz domain in $\mathbf{R}^d$ for some $d \geq 3$ whose complement is connected. There exists ϵ depending only on the Lipschitz constant of Ω such that for $g \in B^{\alpha-1}_p(L_p(\partial\Omega))_{1^\perp}$ there exists a unique solution to the Neumann problem

$$\begin{aligned} \Delta v &= 0 \quad \text{in} \quad \Omega, \\ \frac{\partial v}{\partial n} &= g \quad \text{on} \quad \partial\Omega, \end{aligned} \tag{7}$$

which satisfies $v \in B^{\alpha+1/p}_p(L_p(\Omega))$.

For later use, let us briefly sketch the idea of the proof. It can be performed by combining estimates for the homogeneous Dirichlet problem with those for the operator sending Neumann boundary values to the Dirichlet boundary values of the harmonic function exhibiting those Neumann boundary values, loosely speaking the inverse of the Calderón operator. We start by recalling the classical method of layer potentials to solve the Neumann problem. We define the *nontangential cone* $\Gamma_a(Q)$ for $a > 0$ via

$$\Gamma_a(Q) := \{X \in \Omega \mid |X - Q| < (1+a)\mathrm{dist}(X, \partial\Omega)\}. \tag{8}$$

If u is a function on Ω we may define its *nontangential maximal function* $M(u)$ by setting

$$M(u)(Q) := \sup\{|u(P)| \mid P \in \Gamma_1(Q)\}. \tag{9}$$

We say that u has a *nontangential limit* at $Q \in \partial\Omega$ if there is a finite, well-defined limit as $P \longrightarrow Q$ from within $\Gamma_a(Q)$ for all $a > 0$. Furthermore, given $h \in L_1(\partial\Omega)$, its *single layer potential* is the function defined via

$$Sh(X) := \frac{-1}{\omega_d(d-2)} \int_{\partial\Omega} \frac{h(Q)}{|X-Q|^{(d-2)}} d\sigma(Q), \tag{10}$$

where ω_d is the surface area of the unit sphere in $\mathbf{R}^d$. We also need the operators

$$K^*h(P) \;:=\; \text{p.v.}\frac{1}{\omega_d} \int_{\partial\Omega} \frac{\langle P-Q, n(P)\rangle}{|P-Q|^d} h(Q) d\sigma(Q), \tag{11}$$

$$T \;:=\; \frac{1}{2}I - K^*. \tag{12}$$

Here $n(P)$ clearly denotes the outward unit normal vector on $\partial\Omega$. Then the solutions to the Neumann problem can be constructed as follows, see Dahlberg and Kenig [5] and Verchota [21] for details.

Theorem 2.2. *Let $\Omega \subseteq \mathbf{R}^d$ be a bounded Lipschitz domain whose complement is connected. Then there is $\epsilon = \epsilon(\Omega) > 0$ such that, whenever $1 < p < 2 + \delta$, T is an invertible mapping from $L_p(\partial\Omega)_{1^\perp}$ onto $L_p(\partial\Omega)_{1^\perp}$, and S is an invertible mapping from $L_p(\partial\Omega)$ onto $W^1(L_p(\Omega))$. Moreover, given $g \in L_p(\partial\Omega)_{1^\perp}$ with $1 < p < 2 + \delta$ and writing $v = ST^{-1}g$, i.e.,*

$$v(X) = \frac{1}{\omega_d(d-2)} \int_{\partial\Omega} |X-Q|^{2-d} (\frac{1}{2}I - K^*)^{-1}(g)(Q) d\sigma(Q), \tag{13}$$

it follows that v is the unique (modulo constants) harmonic function on Ω such that the nontangential maximal function $M(\nabla v)$ is bounded in $L_p(\partial\Omega)$ and $\frac{\partial v}{\partial n} = g$ nontangentially a.e. on $\partial\Omega$. Finally, we have

$$\|M(\nabla u)\|_{L_p(\partial\Omega)} \leq C\|g\|_{L_p(\partial\Omega)}. \tag{14}$$

In order to determine the Dirichlet boundary values of our single layer potentials one has

Proposition 2.1. *If $h \in L_1(\partial\Omega)$ then $Sh(X) \longrightarrow Sh(Q)$ as $X \longrightarrow Q$ nontangentially for a.e. $Q \in \partial\Omega$. In particular, if $1 < p < 2 + \delta$, then for all $g \in L_p(\partial\Omega)_{1^\perp}$, $v(X) = ST^{-1}g(X) \longrightarrow ST^{-1}g(Q)$ for a.e. Q.*

For the proof, we refer to [24]. Consequently, for $1 < p < 2 + \delta$ we may define the *inverse Calderón or Neumann to Dirichlet operator* $\Upsilon : L_p(\partial\Omega)_{1^\perp} \longrightarrow W^1(L_p(\partial\Omega))$ by setting

$$\Upsilon(g) := (ST^{-1}(g))|_{\partial\Omega}. \tag{15}$$

One of the main results in [23] states that the inverse Calderón operator moreover acts as a bounded operator on a whole scale of Besov spaces.

Theorem 2.3. *There exists ϵ with $0 < \epsilon \le 1$ so that the inverse Calderón operator Υ introduced in* (15) *satisfies*

$$\|\Upsilon g\|_{B_p^\alpha(L_p(\partial\Omega))} \le C\|g\|_{B_p^{\alpha-1}(L_p(\partial\Omega))}, \tag{16}$$

provided

(a) $p_0 < p < p_0'$ *and* $0 < \alpha < 1$

(b) $1 < p \le p_0$ *and* $2/p - 1 - \epsilon < \alpha < 1$

(c) $p_0' < p < \infty$ *and* $0 < \alpha < 2/p + \epsilon$,

wherein $1/p_0 = (1+\epsilon)/2$, $1/p_0' = (1-\epsilon)/2$.

The proof of Theorem 2.1 now follows by combining Theorem 2.3 with the following fundamental result for the Dirichlet problem which was proved by Jerison and Kenig [17].

Theorem 2.4. *Consider ϵ such that $0 < \epsilon \le 1$. Define p_0 and p_0' by $1/p_0 = (1+\epsilon)/2$ and $1/p_0' = (1-\epsilon)/2$. Let α and p be numbers satisfying one of the following:*

(a) $p_0 < p < p_0'$ *and* $0 < \alpha < 1$

(b) $1 < p \le p_0$ *and* $2/p - 1 - \epsilon < \alpha < 1$

(c) $p_0' \le p < \infty$ *and* $0 < \alpha < 2/p + \epsilon$.

Let Ω be a bounded Lipschitz domain in $\mathbf{R}^d$ for some $d \ge 3$. There exists ϵ depending only on the Lipschitz constant of Ω such that for every $g \in B_p^\alpha(L_p(\partial\Omega))$ there exists a unique harmonic function v such that $Tr\ v = g$ and $v \in B_p^{\alpha+1/p}(L_p(\Omega))$. Moreover,

$$\|v\|_{B_p^{\alpha+1/p}(L_p(\Omega))} \le C\|g\|_{B_p^\alpha(L_p(\partial\Omega))}. \tag{17}$$

Theorems 2.1 and 2.3 can also be used to derive a regularity result for the inhomogeneous Neumann problem

$$\begin{aligned} \Delta w &= F \quad \text{in} \quad \Omega, \\ \frac{\partial w}{\partial n} &= 0 \quad \text{on} \quad \partial\Omega. \end{aligned} \tag{18}$$

Indeed, by a judicious homogenization procedure, (18) can be reduced to a problem of the form (7). For a detailed elaboration of these ideas, we refer again to [23] where the following fundamental result is proved, see also Section 3.

Theorem 2.5. *Let Ω be a bounded Lipschitz domain in $\mathbf{R}^d$, $d \ge 3$, and let $1 < p < \infty$ and $1/p + 1/p' = 1$. There is ϵ, $0 < \epsilon \le 1$, depending only on the Lipschitz constant of Ω, such that, for every $F \in (W^{2-\alpha}(L_{p'}(\Omega)))^*_{1\perp}$, there exists a solution $w \in W^\alpha(L_p(\Omega))$ to the inhomogeneous Neumann problem* (18) *provided one of the following holds:*

(a) $p_0 < p < p_0'$ *and* $1/p < \alpha < 1 + 1/p$

(b) $1 < p \le p_0'$ *and* $3/p - 1 - \epsilon < \alpha < 1 + 1/p$

(c) $p_0' \le p < \infty$ *and* $1/p < \alpha < 3/p + \epsilon$

wherein $1/p_0 = 1/2 + \epsilon/2$ *and* $1/p_0' = 1/2 - \epsilon/2$.

Moreover, for all $F \in (W^{2-\alpha}(L_{p'}(\Omega)))^*_{1\perp}$ *we have the estimate*

$$\|w\|_{W^\alpha(L_p(\Omega))} \leq C\|F\|_{(W^{2-\alpha}(L_{p'}(\Omega)))^*}. \tag{19}$$

Finally, modulo constants, this solution is unique.

Remark 2.1.

i) *The space* $(W^{2-\alpha}(L_{p'}(\Omega)))^*_{1\perp}$ *is clearly defined analogously to* (6).

ii) *Quite recently, similar results were also derived by Fabes, Mendez, and Mitrea* [16].

3. Nonclassical regularity results

In this section, we want to derive some nonclassical regularity results for the Neumann problem, i.e., we want to estimate the regularity of the solution as measured in the specific Besov scale $B^s_\tau(L_\tau(\Omega))$, $1/\tau = s/d + 1/p$, which determines the approximation order of adaptive numerical schemes. Let us first discuss the homogeneous case.

Theorem 3.1. *Let* Ω *be a bounded Lipschitz domain in* $\mathbf{R}^d$, $d \geq 3$, *whose complement is connected. Let* v *be the solution to the Neumann problem*

$$\begin{aligned} \triangle v &= 0 \quad in \quad \Omega \subset \mathbf{R}^d, \\ \frac{\partial v}{\partial n} &= g \quad on \quad \partial\Omega, \end{aligned} \tag{20}$$

where $g \in B^{\alpha-1}_p(L_p(\partial\Omega))$ *and* α *and* p *satisfy the conditions of Theorem* 2.1. *Then*

$$v \in B^s_\tau(L_\tau(\Omega)), \quad \tau = \left(\frac{s}{d} + \frac{1}{p}\right)^{-1}, \quad 0 < s < \frac{(\alpha + 1/p)d}{(d-1)}. \tag{21}$$

Proof. The proof can be performed by combining Theorem 2.1 with the following nonclassical regularity result proved in [11] which states a specific property of harmonic Besov spaces.

Theorem 3.2. *Let* Ω *be a bounded Lipschitz domain in* $\mathbf{R}^d$. *If* v *is an harmonic function on* Ω *which is in the Besov class* $B^\mu_p(L_p(\Omega))$, *for some* $1 < p < \infty$ *and* $\mu > 0$, *then*

$$v \in B^s_\tau(L_\tau(\Omega)), \quad \tau = \left(\frac{s}{d} + \frac{1}{p}\right)^{-1}, \quad 0 < s < \frac{\mu d}{(d-1)}. \tag{22}$$

Now Theorem 2.1 implies that $v \in B^{\alpha+1/p}_p(L_p(\Omega))$. However, the solution v to (20) is clearly an harmonic function. Therefore an application of Theorem 3.2 proves the assertion. □

Theorem 3.1 says that we indeed gain regularity in the scale $B^s_\tau(L_\tau(\Omega))$, $1/\tau = s/d + 1/p$, compared with the usual scale $B^s_p(L_p(\Omega))$, $s > 0$, since the maximal smoothness parameter according to Theorem 2.1 is multiplied by $d/(d-1)$. Consequently, the use of adaptive schemes is completely justified. By interpolation and embeddings for Besov spaces, we can moreover conclude that v is in a family of Besov spaces $B^{\tilde{s}}_{\tilde{\tau}}(L_{\tilde{\tau}}(\Omega))$ for a certain range of the parameters $\tilde{\tau}$ and $\tilde{s}$. This is depicted in Figure 1 for the special case $p = 2$, $d = 3$. If $g \in L_2(\partial\Omega)$, then v is in $B^{\tilde{s}}_{\tilde{\tau}}(L_{\tilde{\tau}}(\Omega))$ whenever $(1/\tilde{\tau}, \tilde{s})$ is in the interior of the quadrilateral with vertices $(1/2, 0)$, $(1/2, 3/2)$, $(1.25, 0)$, $(1.25, 2.25)$. The heavy line connecting $(1/2, 0)$ to $(1.25, 2.25)$ corresponds to the spaces $B^s_\tau(L_\tau(\Omega))$ of Theorem 3.1.

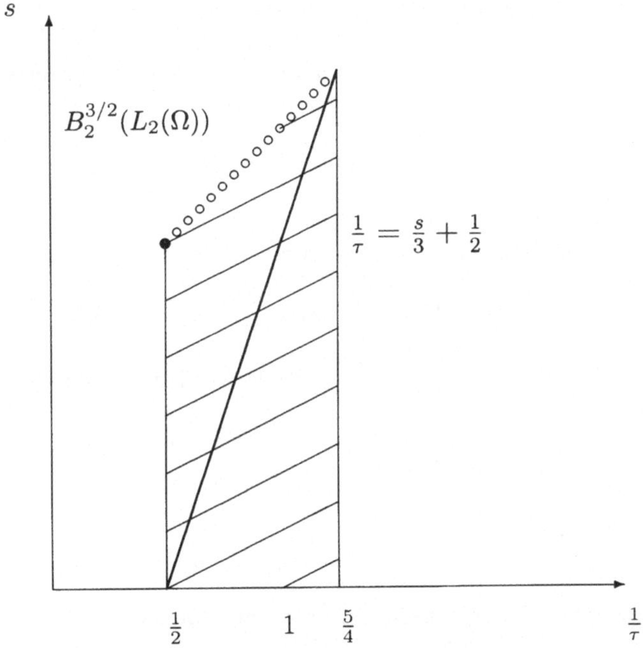

FIGURE 1. Regularity spaces according to Theorem 3.1, $g \in L_2(\partial\Omega)$.

A further improvement of the smoothness index for v can be obtained by repeatedly applying Theorem 3.2. Indeed, if α and p satisfy the conditions of Theorem 2.1, then we know that

$$v \in B^s_\tau(L_\tau(\Omega)), \quad \tau = \left(\frac{s}{d} + \frac{1}{p}\right)^{-1}, \quad 0 < s < \frac{(\alpha + 1/p)d}{d-1},$$

see (21). If

$$\frac{1}{\tau_1} := \frac{\alpha}{d-1} + \frac{1}{p}\left(\frac{d}{d-1}\right) < 1,$$

we may apply Theorem 3.2 for another time which yields

$$v \in B^s_\tau(L_\tau(\Omega)), \quad \tau = \left(\frac{s}{d} + \frac{1}{\tau_1}\right)^{-1}, \quad 0 < s < \frac{(\alpha + 1/p)d^2}{(d-1)^2}.$$

We can always keep on going until $\tau \leq 1$. By this kind of bootstrapping arguments, we always obtain a regularity result for $\tau < 1$. A slightly more sophisticated version of the bootstrapping strategy yields the following result.

Theorem 3.3. *Let Ω be a bounded Lipschitz domain. If v is the solution to* (20) *where $g \in B_p^{\alpha-1}(L_p(\Omega))$ and α and p satisfy the conditions of Theorem* 2.1, *then*

$$v \in B_{\tilde{\tau}}^{\tilde{s}}(L_{\tilde{\tau}}(\Omega)), \quad 0 < \tilde{s} < \alpha + \frac{1}{\tilde{\tau}}, \quad p \geq \tilde{\tau} > \tau^*, \quad \tau^* := \left(\frac{\alpha+1}{d-1} + 1\right)^{-1}. \tag{23}$$

Proof. We first observe that the critical value for s in (21) is exactly given by the intersection of the lines

$$s = \alpha + \frac{1}{\tau} \qquad \text{and} \qquad s = \frac{d}{\tau} - \frac{d}{p}.$$

Consequently, if we apply Theorem 3.2 repeatedly and use interpolation and embedding theorems for Besov spaces, we can conclude that

$$v \in B_{\tilde{\tau}}^{\tilde{s}}(L_{\tilde{\tau}}(\Omega)), \quad \tilde{s} < \alpha + \frac{1}{\tilde{\tau}}, \quad p \geq \tilde{\tau} > 1. \tag{24}$$

For any $\tilde{\tau}$ satisfying the condition in (24), we can apply Theorem 3.2 for another time and obtain

$$v \in B_{\tau}^{s}(L_{\tau}(\Omega)), \quad s < \frac{d(\alpha + 1/\tilde{\tau})}{d-1}, \quad \frac{1}{\tau} = \frac{s}{d} + \frac{1}{\tilde{\tau}}.$$

Since

$$\alpha + \frac{1}{\tau^*} = \frac{d}{d-1}(\alpha+1),$$

the result follows again by interpolation and embeddings. □

As an example, let us again consider the case $p = 2$, $d = 3$, and $g \in L_2(\partial\Omega)$. Theorem 3.3 gives that v is in $B_{\tilde{\tau}}^{\tilde{s}}(L_{\tilde{\tau}}(\Omega))$ for all $\tilde{s}$ and $\tilde{\tau}$ such that $(1/\tilde{\tau}, \tilde{s})$ is in the shaded region of Figure 2.

The results stated in the Theorems 3.1 and 3.3 can also be used to establish Besov regularity for the inhomogeneous Neumann problem.

Theorem 3.4. *Let Ω be a bounded Lipschitz domain in $\mathbf{R}^d$. Let ϵ and p_0' be defined as in Theorem* 2.1. *Let w be the solution to*

$$\begin{aligned} \Delta w &= F \quad in \quad \Omega \subset \mathbf{R}^d, \\ \frac{\partial w}{\partial n} &= 0 \quad on \quad \partial\Omega, \end{aligned} \tag{25}$$

with $F \in (W^{2-\mu}(L_{p'}(\Omega)))^*_{1^\perp}$ for some non-integer $\mu > 1/p$.

(a) *Suppose that $p_0' > p > 1$. If $\mu \geq 1 + 1/p$, then*

$$w \in B_{\tilde{\tau}}^{\tilde{s}}(L_{\tilde{\tau}}(\Omega)), \quad 0 < \tilde{s} < \min\left(\mu, 1 + \frac{1}{\tilde{\tau}}\right), \quad p \geq \tilde{\tau} > \frac{d-1}{d+1}.$$

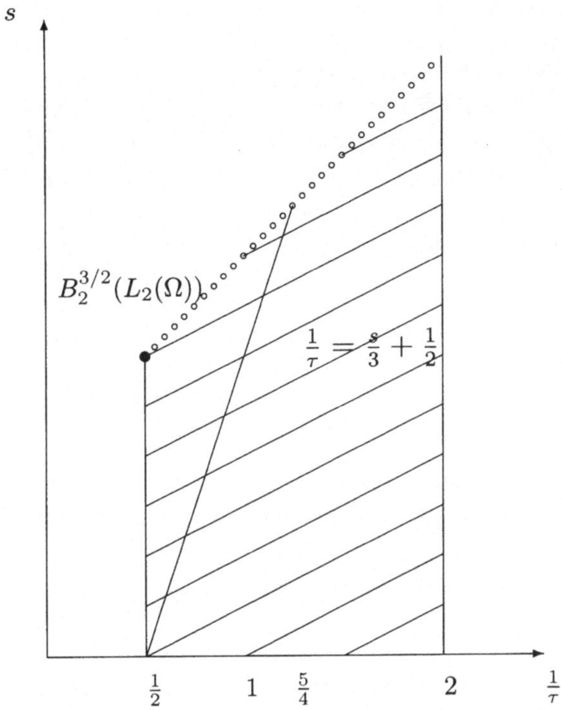

FIGURE 2. Regularity spaces according to Theorem 3.3, $g \in L_2(\partial\Omega)$.

(b) *Suppose that* $p \geq p_0'$. *If* $\mu \geq 3/p + \epsilon$, *then*

$$w \in B^{\tilde{s}}_{\tilde{\tau}}(L_{\tilde{\tau}}(\Omega)), \quad 0 < \tilde{s} < \min\left(\mu, 2/p + \epsilon + \frac{1}{\tilde{\tau}}\right), \quad p \geq \tilde{\tau} > \frac{d-1}{2/p + \epsilon + d}.$$

Proof. We shall only prove the first case in detail. The second case can be studied analogously.

Let us first assume that $2 > \mu \geq 1 + 1/p$. Let $R_\Omega(f)$ denote the restriction of a function f on $\mathbf{R}^d$ to Ω. By using the Newtonian potential $N(x) := C_d|x|^{2-d}$, we define

$$\tilde{w} = N * (R^*_\Omega F). \tag{26}$$

Then

$$\triangle\tilde{w} = \triangle N * (R^*_\Omega F) = R^*_\Omega F. \tag{27}$$

Therefore we can write the solution w to (25) as

$$w = \tilde{w} - v \qquad \text{on} \quad \Omega, \tag{28}$$

where v is the solution to the homogeneous Neumann problem

$$\begin{aligned} \triangle v &= 0 \quad \text{in} \quad \Omega \subset \mathbf{R}^d, \\ \frac{\partial v}{\partial n} &= \frac{\partial \tilde{w}}{\partial n} := g \quad \text{on} \quad \partial\Omega. \end{aligned} \tag{29}$$

Therefore we have to establish Besov regularity for both, $\tilde{w}$ and v. Let us start with $\tilde{w}$. It can be shown that $R_\Omega^* F \in W^{\mu-2}(L_p(\mathbf{R}^d))$, see [23] for details. Hence the classical elliptic regularity theory implies that $\tilde{w} \in W^\mu(L_p(\mathbf{R}^d)) = B_p^\mu(L_p(\mathbf{R}^d))$ (since $\mu \notin \mathbf{N}$). We refer to [1] for further information. Hence $\tilde{w}|_\Omega \in B_p^\mu(L_p(\Omega))$, and by the embeddings of Besov spaces: $B_p^\mu(L_p(\Omega)) \hookrightarrow B_p^\mu(L_{\tilde{\tau}}(\Omega)) \hookrightarrow B_{\tilde{\tau}}^{\tilde{s}}(L_{\tilde{\tau}}(\Omega))$, we have $\tilde{w}|_\Omega \in B_{\tilde{\tau}}^{\tilde{s}}(L_{\tilde{\tau}}(\Omega))$ for any $\tilde{s}, \tilde{\tau}$ as in the statement (a). It remains to establish Besov regularity for v. It can be shown that $g \in B_p^{\beta-1}(L_p(\Omega))$ for all $\beta < 1$, compare again with [23]. Therefore an application of Theorem 3.3 implies that

$$v \in B_{\tilde{\tau}}^{\tilde{s}}(L_{\tilde{\tau}}(\Omega)), \qquad 0 < \tilde{s} < 1 + \frac{1}{\tilde{\tau}}, \qquad p > \tilde{\tau} > \left(\frac{2}{d-1} + 1\right)^{-1} = \frac{d-1}{d+1}$$

and the result follows. The case $\mu > 2$ can be treated analogously by employing a classical extension technique as, e.g., outlined in [11]. □

Remark 3.1. *We have formulated Theorem* 3.4 *only for the 'interesting' case of a sufficiently smooth right-hand side. For smaller values of* μ, *our theory is consistent with Theorem* 2.5.

References

[1] S. Agmon, A. Douglis, and L. Nierenberg, Estimates near the boundary for solutions to elliptic partial differential equations satisfying general boundary conditions, I., *Comm. Pure Appl. Math.* **12** (1959), 623–727.

[2] C.K. Chui, *An Introduction to Wavelets*, Academic Press, Boston, 1992.

[3] A. Cohen, W. Dahmen, and R. DeVore, Adaptive wavelet methods for elliptic operator equations – Convergence rates, *Math. Comp.* **70** (2001), 27–75.

[4] A. Cohen, W. Dahmen, and R. DeVore, Adaptive wavelet methods II: Beyond the elliptic case, IGPM-Preprint No. 199, RWTH Aachen, 2000.

[5] B.E.J. Dahlberg and C.E. Kenig, Hardy spaces and the Neumann problem in L^p for Laplace's equation in Lipschitz domains, *Annals of Math.* **125** (1987), 437–466.

[6] S. Dahlke, *Wavelets: Construction Principles and Applications to the Numerical Treatment of Operator Equations*, Shaker Verlag, Aachen, 1997.

[7] S. Dahlke, Besov regularity for second order elliptic boundary value problems with variable coefficients, *Manuscripta Math.* **95** (1998), 59–77.

[8] S. Dahlke, Besov regularity for elliptic boundary value problems in polygonal domains, *Appl. Math. Letters* **12** (1999), 31–36.

[9] S. Dahlke, W. Dahmen, R. Hochmuth, and R. Schneider, Stable multiscale bases and local error estimation for elliptic problems, *Appl. Numer. Math.* **8** (1997), 21–47.

[10] S. Dahlke, W. Dahmen, and K. Urban, Adaptive wavelet methods for saddle point problems – Optimal convergence rates, Report 01–07, Zentrum für Technomathematik, Universität Bremen, 2001.

[11] S. Dahlke and R. DeVore, Besov regularity for elliptic boundary value problems, *Comm. Partial Differential Equations* **22(1&2)** (1997), 1–16.

[12] I. Daubechies, *Ten Lectures on Wavelets*, CBMS–NSF Regional Conference Series in Applied Math. **61**, SIAM, Philadelphia, 1992.

[13] R. DeVore, Nonlinear approximation, *Acta Numerica* **7** (1998), 51–150.

[14] R. DeVore, B. Jawerth, and V. Popov, Compression of wavelet decompositions, *Amer. J. Math.* **114** (1992), 737–785.

[15] R. DeVore and V. Popov, Interpolation of Besov spaces, *Trans. Amer. Math. Soc.* **305** (1988), 397–414.

[16] E. Fabes, O. Mendez, and M. Mitrea, Boundary layers on Sobolev-Besov spaces and Poisson's equation for the Laplacian in Lipschitz domains, *J. of Funct. Anal.* **159** (1998), 323–368.

[17] D. Jerison and C.E. Kenig, The inhomogeneous Dirichlet problem in Lipschitz domains, *J. of Funct. Anal.* **130** (1995), 161–219.

[18] C.E. Kenig, *Harmonic Analysis Techniques for Second Order Elliptic Boundary Value Problems,* CBMS Regional Conference Series in Math. **83**, AMS, Providence, Rhode Island, 1994.

[19] Y. Meyer, *Wavelets and Operators*, Cambridge Studies in Advanced Mathematics **vol. 37**, Cambridge, 1992.

[20] H. Triebel, *Interpolation Theory, Function Spaces, Differential Operators*, North-Holland, Amsterdam, 1978.

[21] G. Verchota, Layer potentials and regularity of the Dirichlet problem for Laplace's equation, *J. of Funct. Anal.* **59** (1984), 572–611.

[22] P. Wojtaszczyk, *A Mathematical Introduction to Wavelets,* Cambridge University Press, 1997.

[23] D. Zanger, The inhomogeneous Neumann problem in Lipschitz domains, *Comm. Partial Differential Equations* **25(9&10)** (2000), 1771–1808.

[24] D. Zanger, *Regularity and Boundary Variations for the Neumann Problem,* Ph.D. thesis, Massachusetts Institute of Technology, 1997.

Stephan Dahlke
Fachbereich 3, ZeTeM
Universität Bremen
Postfach 33 04 40
D-28334 Bremen, Germany
E-mail address: dahlke@math.uni-bremen.de

D. Haroske, T. Runst, H.-J. Schmeisser (eds.): Function Spaces, Differential Operators and Nonlinear Analysis. The Hans Triebel Anniversary Volume.

Intrinsic Descriptions Using Means of Differences for Besov Spaces on Lipschitz Domains

Sophie Dispa

Dedicated to Prof. Hans Triebel on the occasion of his 65th birthday

Abstract. The aim of this paper is to study the equivalence between quasi-norms of Besov spaces on bounded Lipschitz open subsets in $\mathbb{R}^n$. We define Besov spaces on Ω as the restrictions of the corresponding Besov spaces on $\mathbb{R}^n$. Then we extend the well-known characterization of Besov spaces on $\mathbb{R}^n$ described in Theorem 2.4 to the case of Lipschitz domains. So, we obtain an equivalent and intrinsic quasi-norm using generalized differences and moduli of smoothness.

1. Introduction

This paper deals with the obtention of intrinsic characterizations of Besov spaces on Lipschitz domains using moduli of smoothness. As explained in [T-92], Sobolev, Hölder-Zygmund and Besov spaces on domains are as old as the corresponding spaces on $\mathbb{R}^n$. For spaces with indices $1 < p < \infty$, $1 \leq q \leq +\infty$, $s > 0$, we refer to the works of H. Triebel [Tri-78], T. Muramuta [Mur-71] and of the Russian school, see [BIN-75], [Nik-75], [BKLN-90] and especially G.A. Kaljabin [Kal-85], [Kal-88]. For a wider range of spaces, i.e. indices satisfying $0 < p, q \leq +\infty$, $s > \sigma_p := n(\frac{1}{p} - 1)_+$ and bounded C^∞ domains, descriptions in terms of moduli of continuity, though not formulated explicitly, are given in Theorem 1.10.4, p. 74 in [T-92].

As far as numerical applications are concerned, Besov spaces on Lipschitz domain are particularly efficient to solve significant pde problems. Defining these spaces as a restriction (see Definition 3.3 and [R-99]) of the elements of $B^s_{pq}(\mathbb{R}^n)$ (see Definition 2.2) to Ω, we wonder if it is possible to give intrinsic descriptions and, better, intrinsic characterizations based on generalized differences and moduli of smoothness as in Theorem 2.4? A first answer was given by V. Rychkov [R-99] who found equivalent intrinsic quasi-norms using maximal function and convolutions (see Theorem 3.9 and 3.12). Using these descriptions obtained by V. Rychkov, we get Theorem 3.17, the aim of this paper.

Notations 1.1. *The symbol* $\mathbb{N}$ *denotes the set of positive integers* $\{1, 2, 3, \dots\}$ *and* $\mathbb{N}_0 := \{0, 1, 2, 3, \dots\}$. *Let* $D(\mathbb{R}^n)$ *denote the set of functions* $f \in C^\infty(\mathbb{R}^n)$ *which have a compact support in* $\mathbb{R}^n$ *and* $S(\mathbb{R}^n)$ *the usual set of functions* $f \in C^\infty(\mathbb{R}^n)$ *which are rapidly decreasing in* $\mathbb{R}^n$. *If* $f \in S(\mathbb{R}^n)$, *then* $\mathcal{F}^\pm f$ *denotes the usual Fourier transform of function* f. *Let* $D'(\mathbb{R}^n)$ *denote the set of distributions and* $S'(\mathbb{R}^n)$ *the set of tempered distributions on* $\mathbb{R}^n$.

2. Besov spaces on $\mathbb{R}^n$

Let us remember a few results concerning Besov spaces on $\mathbb{R}^n$.

2.1. General definition

The general definition given by H. Triebel [T-83] uses balls and differences of balls as underlying decomposition of $\mathbb{R}^n$ and a sequence of C^∞ functions with compact support included in the sets of this covering.

Definition 2.1. *Let us denote by* $\Phi(\mathbb{R}^n)$ *the set of sequences*

$$\varphi := (\varphi_j)_{j \in \mathbb{N}_0} \subset S(\mathbb{R}^n)$$

such that

1) $\operatorname{supp} \varphi_0 \subset \{x \in \mathbb{R}^n : |x| \leq 2\}$, $\operatorname{supp} \varphi_j \subset \{x \in \mathbb{R}^n : 2^{j-1} \leq |x| \leq 2^{j+1}\}$, $j = 1, 2, 3 \dots,$
2) $\forall \alpha \in \mathbb{N}^n$, $\exists C_\alpha > 0 : 2^{j|\alpha|} |D^\alpha \varphi_j(x)| \leq C_\alpha, \quad j \in \mathbb{N}_0, x \in \mathbb{R}^n,$
3) $\sum_{j=0}^{+\infty} \varphi_j(x) = 1, \quad x \in \mathbb{R}^n.$

Let us denote $L^p(\mathbb{R}^n)$ $(0 < p \leq +\infty)$ the set of functions $f : \mathbb{R}^n \to \mathbb{C}$ such that

$$\|f\|_p := \left(\int_{\mathbb{R}^n} |f(x)|^p dx \right)^{1/p} < +\infty, \quad \text{if } 0 < p < +\infty$$

and

$$\|f\|_\infty := \sup_{\text{pp}} |f(.)| < +\infty, \quad \text{if } p = +\infty.$$

Here follows the general definition of $B^s_{pq}(\mathbb{R}^n)$.

Definition 2.2. *If* $s \in \mathbb{R}$, $0 < p, q \leq +\infty$ *and* $\varphi := (\varphi_j)_{j \in \mathbb{N}_0} \subset \Phi(\mathbb{R}^n)$ *then*

$$\mathcal{B}^s_{pq}(\mathbb{R}^n) := \left\{ f \in S'(\mathbb{R}^n) : \|f\|_{\mathcal{B}^s_{pq}(\mathbb{R}^n)} := \left(\sum_{k=0}^{+\infty} 2^{skq} \parallel \mathcal{F}^+[\varphi_k \mathcal{F}^- f] \parallel_p^q \right)^{1/q} < +\infty \right\}.$$

This definition does not depend on the choice of φ (see [T-83], 2.3.2, p. 46). One can find other definitions using differences.

Definition 2.3. *For $M \in \mathbb{N}$, $f : \mathbb{R}^n \to \mathbb{C}$, the difference of order M of f is defined by*

$$(\Delta_h^M f)(.) := \sum_{j=0}^{M} C_M^j (-1)^{M-j} f(. + jh), \qquad h \in \mathbb{R}^n.$$

The following theorem, proved in [T-83], 2.5.12, p. 110, shows the equivalence between some definitions.

Theorem 2.4. *If $0 < p, q \leq +\infty$, $s > \sigma_p := n(\frac{1}{p} - 1)_+$, $M \in \mathbb{N}$ such that $M > s$, let*

$$\begin{aligned} B_{pq}^s(\mathbb{R}^n) &:= \left\{ f \in S'(\mathbb{R}^n) : \|f\|^{(1)}_{B_{pq}^s(\mathbb{R}^n),M} < +\infty \right\} \\ &= \left\{ f \in S'(\mathbb{R}^n) : \|f\|^{(2)}_{B_{pq}^s(\mathbb{R}^n),M} < +\infty \right\} \end{aligned}$$

where

$$\|f\|^{(1)}_{B_{pq}^s(\mathbb{R}^n),M} := \|f\|_p + \left(\int_{\mathbb{R}^n} |h|^{-sq} \sup_{\rho \in \mathbb{R}^n : |\rho| \leq |h|} \|\Delta_\rho^M f\|_p^q \frac{dh}{|h|^n} \right)^{1/q}$$

and

$$\|f\|^{(2)}_{B_{pq}^s(\mathbb{R}^n),M} := \|f\|_p + \left(\int_{\mathbb{R}^n} |h|^{-sq} \|\Delta_h^M f\|_p^q \frac{dh}{|h|^n} \right)^{1/q}.$$

In these conditions,

$$\| f \|^{(1)}_{B_{pq}^s(\mathbb{R}^n),M} \text{ and } \| f \|^{(2)}_{B_{pq}^s(\mathbb{R}^n),M}$$

are equivalent quasi-norms to

$$\|f\|_{\mathcal{B}_{pq}^s(\mathbb{R}^n)}$$

and

$$B_{pq}^s(\mathbb{R}^n) = \mathcal{B}_{pq}^s(\mathbb{R}^n).$$

2.2. Other formulation

To define $B_{pq}^s(\mathbb{R}^n)$, we may also take $L_\varphi \in \mathbb{N}_0$ and a function $\varphi_0 \in S(\mathbb{R}^n)$ such that

$$\left\{ \begin{array}{ll} \text{(i)} & \int_{\mathbb{R}^n} \varphi_0(x) dx \neq 0 \\ \text{(ii)} & \int_{\mathbb{R}^n} x^\alpha \varphi(x) dx = 0 \text{ if } |\alpha| \leq L_\varphi, L_\varphi \geq [s], \varphi(x) := \varphi_0(x) - 2^{-n}\varphi_0(\frac{x}{2}). \end{array} \right. \tag{1}$$

For every $s \in \mathbb{R}$, we can find functions φ_0 having such properties (see [R-99], 1.1, p. 241).

Proposition 2.5 ([R-99]). *Suppose that $\varphi_0 \in S(\mathbb{R}^n)$ satisfies the conditions (1). If $s \in \mathbb{R}$ and $0 < p, q \leq +\infty$ then*

$$B_{pq}^s(\mathbb{R}^n) = \left\{ f \in S'(\mathbb{R}^n) : \|f\|^{(\varphi_0)}_{B_{pq}^s(\mathbb{R}^n)} := \left(\sum_{j=0}^{+\infty} 2^{jsq} \|\varphi_j \star f\|^q_{L^p(\mathbb{R}^n)} \right)^{1/q} < +\infty \right\}.$$

Here again, the space $B^s_{pq}(\mathbb{R}^n)$ is independent of the particular choice of $\varphi_0 \in S(\mathbb{R}^n)$ satisfying (1). The quasi-norms arising for different φ_0 are equivalent as shown for example in [BPT-96]. For a complete proof of Proposition 2.5, we refer to [R-99].

3. Besov spaces on domains

Let us introduce a few definitions.

3.1. Definitions

By a Lipschitz domain, we mean either a special or a bounded Lipschitz domain (see [ST-70] and [R-99]).

Definition 3.1. *A special Lipschitz domain is an open set $\Omega \subset \mathbb{R}^n$ lying above the graph of a Lipschitz function $\omega : \mathbb{R}^{n-1} \to \mathbb{R}$. More precisely,*

$$\Omega := \{(x', x_n) \in \mathbb{R}^n : x_n > \omega(x')\},$$

where there exists $A > 0$ such that

$$|\omega(x') - \omega(y')| \leq A|x' - y'|, \qquad x', y' \in \mathbb{R}^{n-1}.$$

Definition 3.2. *A bounded Lipschitz domain is a bounded domain Ω whose boundary $\partial\Omega$ can be covered by a finite number of open balls B_k so that, possibly after a proper rotation, $\partial\Omega \cap B_k$ for each k is a part of the graph of a Lipschitz function.*

For a Lipschitz domain Ω, the space $\mathcal{B}^s_{pq}(\Omega)$ is defined as the restrictions of the corresponding space from $\mathbb{R}^n$ to Ω.

Definition 3.3. *If $s \in \mathbb{R}$ and $0 < p, q \leq +\infty$ and if Ω is a Lipschitz domain in $\mathbb{R}^n$ then $\mathcal{B}^s_{pq}(\Omega)$ is the set of functions $f : \Omega \to \mathbb{C}$ such that there exists $g \in B^s_{pq}(\mathbb{R}^n)$: $g_{|\Omega} = f$ in $D'(\Omega)$. It is endowed with the quasi-norm quotient:*

$$\|f\|_{\mathcal{B}^s_{pq}(\Omega)} := \inf\left\{\|g\|_{B^s_{pq}(\mathbb{R}^n)} : g \in B^s_{pq}(\mathbb{R}^n), g_{|\Omega} = f \quad in \quad D'(\Omega)\right\}.$$

Definition 3.4. *A function $\psi \in D(\mathbb{R}^n)$ has vanishing moments up to order $L_\psi \in \mathbb{N}_0$ if*

$$\int_{\mathbb{R}^n} x^\alpha \psi(x) dx = 0$$

for all $\alpha \in \mathbb{N}^n$ with $|\alpha| \leq L_\psi$.

3.2. Problem

A natural question arises then: is it possible to obtain an equivalent intrinsic characterization using differences and moduli of smoothness on Ω as in Theorem 2.4?

To solve this problem, we will use the extension theorem and the intrinsic characterizations obtained by V. Rychkov [R-99].

3.3. Extension theorem and intrinsic characterizations

We rely on V. RYCHKOV [R-99] and collect some of main results of intrinsic characterizations already proved there.

First of all, by the arguments used by V. RYCHKOV (see [R-99], 1.2, p. 244 for details and [D-01]), we see that it is possible to reduce the problem of finding an intrinsic characterization with general differences to the case of Ω being a special Lipschitz domain.

Then let Ω be a special Lipschitz domain. We verify immediately that the cone K with vertex at the origin given by

$$K := \{(x', x_n) : |x'| < A^{-1}x_n\}$$

has the property that its shifts satisfy $x + K \subset \Omega$ for each $x \in \Omega$. Let $-K := \{-x : x \in K\}$ be the reflected cone. Then for every $\gamma \in D(-K)$ and $f \in D'(\Omega)$, this allows us to define the convolution product $\gamma \star f(x) = f(\gamma(x - .))$ in Ω since $\operatorname{supp}\gamma(x - .) \subset \Omega$ for $x \in \Omega$.

Notations 3.5. *The dyadic dilates of a given function g on $\mathbb{R}^n$ are defined by $g_j(x) := 2^{jn}g(2^jx)$ for $j \in \mathbb{N}_0$. Note that $g_j(x)$ for $j = 0$ is not covered by the last formula, but is just a value of a function g_0.*

The following particular decomposition of distributions on Ω is one of the keys of our problem.

Proposition 3.6 ([R-99]). *Let $\varphi_0 \in D(-K)$ have non zero integral, and let $\varphi(x) = \varphi_0(x) - 2^{-n}\varphi_0(x/2)$. Then for any given $L \in \mathbb{N}_0$, there exist functions $\psi_0, \psi \in D(-K)$ such that $L_\psi \geq L$ and*

$$f = \sum_{j=0}^{+\infty} \psi_j \star \varphi_j \star f \text{ in } D'(\Omega) \tag{2}$$

for all $f \in D'(\Omega)$.

For every $s \in \mathbb{R}$, one can choose such a $\varphi_0 \in D(-K)$ satisfying (1) (see [R-99], 2, p. 246).

Theorem 3.7 ([R-99]). *Suppose that $\varphi_0 \in D(-K)$ satisfies (1) and that $\psi_0, \psi \in D(-K)$ are given by Proposition 3.6 with $L_\psi > \frac{n}{\inf\{p,q\}}$ and that (2) holds. For any $g : \Omega \to \mathbb{R}$, denote by g_Ω its extension from Ω to $\mathbb{R}^n$ by zero. Then the map $\mathcal{E}$ defined by*

$$D'(\Omega) \ni f \mapsto \sum_{j=0}^{+\infty} \psi_j \star (\varphi_j \star f)_\Omega \tag{3}$$

yields a linear continuous extension operator from $\mathcal{B}^s_{pq}(\Omega)$ into the corresponding space on $\mathbb{R}^n$.

More precisely, the theorem claims that for each $f \in \mathcal{B}^s_{pq}(\Omega)$, the series in the right-hand side of (3) converges in $S'(\mathbb{R}^n)$ to an $\mathcal{E}f$ such that $\mathcal{E}f_{|\Omega} = f$ in the sense of $D'(\Omega)$ and that the estimate $\|\mathcal{E}f\| \leq C\|f\|_{\mathcal{B}^s_{pq}(\Omega)}$ is true.

Definition 3.8. *Let $\varphi_0 \in D(-K)$ satisfy (1) and let $N > n/\inf\{p,q\}$ be a real number. For every $f \in D'(\Omega)$ define the sequence of its Peetre-type maximal functions by*

$$\varphi_{j,N}^{\Omega} f(x) = \sup_{y\in\Omega} \frac{|\varphi_j \star f(y)|}{(1+2^j|x-y|)^N} \qquad x \in \Omega, j \in \mathbb{N}.$$

Theorem 3.9 (Peetre-type characterization [R-99]). *Let $\varphi_0 \in D(-K)$ satisfy (1) and let $N > n/\inf\{p,q\}$ be a real number. Then for all $f \in D'(\Omega)$*

$$\left(\sum_{j=0}^{+\infty} 2^{jsq} \|\varphi_{j,N}^{\Omega} f\|_{L^p(\Omega)}^q\right)^{1/q} \tag{4}$$

and $\|f\|_{\mathcal{B}^s_{pq}(\Omega)}$ are equivalent quasi-norms.

In the following theorem, V. Rychkov obtains a refinement of Theorem 3.9 involving "pure" convolutions $\varphi_j \star f$ instead of the maximal functions $\varphi_{j,N}^{\Omega} f$. He calls this type of description a "characterization via local means", adopting the terminology used by H. Triebel [T-92] in the case of $\mathbb{R}^n$. This is the counterpart of Proposition 2.5 for domains.

Definition 3.10. *A distribution $f \in D'(\Omega)$ belongs to $S'(\Omega)$ if there exist $C > 0, M, k \in \mathbb{N}_0$ such that*

$$f(\varphi) \leq C \sup_{x\in\Omega, |\alpha|\leq k} (1+|x|)^M)|D^\alpha \varphi(x)|, \quad \forall \varphi \in D(\Omega).$$

Proposition 3.11 ([R-99]). *For every $f \in S'(\Omega)$, there exists a $g \in S'(\mathbb{R}^n)$ so that $f = g_{|\Omega}$.*

Theorem 3.12 (Characterization via local means [R-99]). *Let $\varphi_0 \in D(-K)$ satisfy (1). Then for all $f \in S'(\Omega)$*

$$\left(\sum_{j=0}^{+\infty} 2^{jsq} \|\varphi_j \star f\|_{L^p(\Omega)}^q\right)^{1/q}$$

and $\|f\|_{\mathcal{B}^s_{pq}(\Omega)}$ are equivalent quasi-norms.

3.4. Intrinsic characterization using differences

We need a first lemma.

Lemma 3.13. *If $1 \leq p \leq +\infty, 0 < q \leq +\infty$ and $s > 0$ then the following continuous embedding holds:*

$$\mathcal{B}^s_{pq}(\Omega) \subset L^p(\Omega).$$

If $0 < p < 1, 0 < q \leq +\infty$ and $s > n(\frac{1}{p} - 1)$ then the following continuous embeddings hold:

$$\mathcal{B}^s_{pq}(\Omega) \subset \mathcal{B}^{s-\sigma_p}_{1q}(\Omega) \subset L^1(\Omega).$$

Furthermore, suppose that $\varphi_0, \varphi, \psi_0, \psi \in D(-K)$ *are given by Proposition 3.6. If* $0 < p \leq +\infty$ *and* $s > \sigma_p := n(\frac{1}{p} - 1)_+$, *the equality*

$$f = \sum_{k=0}^{+\infty} \psi_k \star \varphi_k \star f$$

holds in $S'(\Omega)$ *and* $L^p(\Omega)$ *for all* $f \in \mathcal{B}^s_{pq}(\Omega)$.

Proof. The continuous inclusion between Besov spaces follows from Definition 3.3 and the similar lemma proved by H. Triebel ([T-83], Remark 1, p. 83) in the case of $\mathbb{R}^n$. For the other inclusions and the equality, details of the proof are given by S. Dispa in [D-01]. It uses essentially Theorem 3.9 and Proposition 3.6. □

Let us introduce an adapted definition of generalized differences.

Definition 3.14. *If* $f : \Omega \to \mathbb{C}$, $M \in \mathbb{N}$, $h \in \mathbb{R}^n$, *we define*

$$\Delta_h^M f(x) := \begin{cases} \sum_{j=0}^M C_M^j (-1)^{M-j} f(x+jh) \text{ if } x, x+h, \ldots, x+Mh \in \Omega \\ 0 \text{ otherwise }. \end{cases}$$

In order to solve our problem, we ask some more properties on the function $\varphi_0 \in D(-K)$. We will mention when it is really important to choose a particular definition (see (7)) for φ_0.

The following two lemmas are adapted from results used by H. Triebel ([T-92], 3.3.1, 3.3.2, p. 173–177) in the case of C^∞ domains. For complete proofs, we refer to [D-01].

Lemma 3.15. *Let* $M \in \mathbb{N}$. *If* $\alpha < 0$, *there exist functions* $g, h \in C^\infty(\mathbb{R})$: $\operatorname{supp} g, \operatorname{supp} h \subset]\alpha, 0[$ *and*

$$\int_{\mathbb{R}} g(x)dx = 1 \text{ and } g(t) - \frac{1}{2}g(\frac{t}{2}) = D_t^M h(t), \quad t \in \mathbb{R}. \tag{5}$$

The function g and the number $M \in \mathbb{N}$ have the same meaning as in preceding lemma. Let

$$\Phi_B(x) := \frac{1}{B} g(\frac{x_1}{B}) g(x_2) \ldots g(x_n), \quad x \in \mathbb{R}^n, B > 0 \tag{6}$$

$$\varphi_0(x) := \frac{(-1)^{M+1}}{M!} \sum_{r=1}^M \sum_{m=1}^M (-1)^{r+m} C_M^r C_M^m r^{-n} m^{M-n} \Phi_B(\frac{x}{rm}) \tag{7}$$

$$\varphi(x) := \varphi_0(x) - 2^{-n}\varphi_0(\frac{x}{2}) \quad \text{and} \quad \varphi_j(x) := 2^{nj}\varphi(2^j x), \quad j \in \mathbb{N}. \tag{8}$$

Next lemma shows that the functions φ_0, φ satisfy the condition (1).

Lemma 3.16.

(i) *The functions* φ_0 *and* φ *are compactly supported functions on* $\mathbb{R}^n$ *with*

$$\operatorname{supp} \varphi_0, \operatorname{supp} \varphi \subset \mathbb{R}^n_-$$

where

$$\mathbb{R}^n_- := \{x = (x_1, \ldots, x_n) \in \mathbb{R}^n : x_n < 0\}.$$

Furthermore,

$$\mathcal{F}_0^- \varphi_0 = 1 \quad \text{and} \quad \mathcal{F}_\xi^- \varphi = O(|\xi|^M) \quad \text{near the origin} . \tag{9}$$

(ii) *Let* $f \in S'(\Omega)$. *Then, for* $j \in \mathbb{N}$ *and* $x \in \Omega$,

$$(\varphi_j \star f)(x) = \frac{(-1)^{M+1}}{M!} \sum_{m=1}^{M} (-1)^{M-m} C_M^m m^{M-n}$$
$$\times \int_K [\Phi_B(-\frac{y}{m}) - 2^{-n}\Phi_B(-\frac{y}{2m})] \Delta^M_{2^{-j}y} f(x) dy,$$

and

$$(\varphi_0 \star f)(x) = \frac{(-1)^{M+1}}{M!} \sum_{m=1}^{M} (-1)^{M-m} C_M^m m^{M-n}$$
$$\times \int_K \Phi_B(-\frac{y}{m}) \sum_{r=1}^{M} (-1)^{M-r} C_M^r f(x + r2^{-j}y) dy.$$

Here follows the theorem which provides a solution to our question.

Theorem 3.17 (Characterization with differences). *If* $0 < p, q \leq +\infty$, $s > \sigma_p := n(\frac{1}{p} - 1)_+$ *and* $M \in \mathbb{N}$ *such that* $M > s$ *then*

$$\|f\|_{\mathcal{B}^s_{pq}(\Omega),M} := \|f\|_{L^p(\Omega)} + \left(\int_0^1 t^{-sq} \sup_{|\rho| \leq t} \|\Delta^M_\rho f\|^q_{L^p(\Omega)} \frac{dt}{t} \right)^{1/q}$$

is a quasi-norm equivalent to $\|f\|_{\mathcal{B}^s_{pq}(\Omega)}$ *and under these conditions,*

$$\mathcal{B}^s_{pq}(\Omega) = B^s_{pq}(\Omega).$$

Proof. We refer to [D-01] for details of demonstration. Let us simply give here a few ideas of it. In [D-01], we split the proof into two parts. The first step providing a first estimate

$$\exists C > 0 : \|f\|_{B^s_{pq}(\Omega),M} \leq C \|f\|_{\mathcal{B}^s_{pq}(\Omega)}, \quad \forall f \in \mathcal{B}^s_{pq}(\Omega)$$

between the two quasi-norms is adapted from the proof of Theorem 2.4 and uses Lemma 3.13, Proposition 3.6, and Theorem 3.9. In the second step, we get the reverse inequality using Definition 7, Lemma 3.16 and Theorem 3.12. □

References

[BIN-75] O.V. Besov, V.P. Il'in and S.M. Nikol'skij, *Integral Representations of Functions and Embedding Theorems*, Moskva, Nauka, 1975.

[BKLN-90] O.V. Besov, L.D. Kudrjavcev, P.I. Lizorkin and S.M. Nikol'skij, *Studies on the theory of spaces of differentiable functions of several variables*, Proc. Steklov Inst. Math. **182** (1990), 73–140.

[BPT-96] H.-Q. Bui, M. Paluszynski and M.H. Taibleson, *A maximal function characterization of weighted Besov-Lipschitz and Triebel-Lizorkin spaces*, Studia Math. **119** (1996), 219–246.

[BPT-97] H.-Q. Bui, M. Paluszynski and M.H. Taibleson, *Characterization of the Besov-Lipschitz and Triebel-Lizorkin spaces. The case $q < 1$*, J. Fourier Appl. **3** (1997), 837–846 (special issue).

[DD-97] S. Dahlke and R.A. DeVore, *Besov regularity for elliptic boundary value problems*, Comm. Partial Differential Equations **22** 1997, 1–16.

[D-01] S. Dispa, *Intrinsic characterizations of Besov spaces on Lipschitz domains*, to appear.

[Kal-85] G.A. Kaljabin, *A characterization of certain functions spaces by means of generalized differences*, Dokl. Akad. Nauk SSSR **284** (1985), 1305–1308.

[Kal-88] G.A. Kaljabin, *Characterizations of spaces of Besov-Lizorkin-Triebel type by generalized differences*, Trudy Mat. Inst. Steklov **181** (1988), 95–116.

[Kal-87] G.A. Kaljabin and P.I. Lizorkin, *Spaces of functions of generalized smoothness*, Math. Nachr. **133** (1987), 7–32.

[Mur-71] T. Muramatu, *On imbedding theorems for Besov spaces of functions defined in general regions*, Publ. Res. Inst. Math. Sci. Kyoto Univ. **7** (1971–72), 261–285.

[Nik-75] S.M. Nikol'skij, *Approximation of Functions of Several Variables and Imbedding Theorems*, Berlin, Springer 1975.

[R-99] V. Rychkov, *On restrictions and extensions of the Besov and Triebel-Lizorkin spaces with respect to Lipschitz domains*, J. London Math. Soc. **(2) 60**(1999), 237–257.

[ST-70] E.M. Stein, *Singular Integrals and Differentiability Properties of functions*, Princeton Univ. Press, 1970.

[Tri-78] H. Triebel, *Interpolation Theory, Function Spaces, Differential Operators*, Amsterdam, North-Holland, 1978.

[T-83] H. Triebel, *Theory of function spaces*, Birkhäuser Verlag, Basel, Boston, Stuttgart, 1983.

[T-92] H. Triebel, *Theory of function spaces II*, Birkhäuser Verlag, Basel, Boston, Berlin, 1992.

Sophie Dispa
Institut de Mathématique
University of Liège
Grande Traverse, 12
B-4000 Liège, Belgium
E-mail address: sdispa@ulg.ac.be

D. Haroske, T. Runst, H.-J. Schmeisser (eds.): Function Spaces, Differential Operators and Nonlinear Analysis. The Hans Triebel Anniversary Volume.

Landesman-Lazer Type like Results for the p-Laplacian

Pavel Drábek[1]

Dedicated to Prof. Hans Triebel on the occasion of his 65th birthday

Abstract. We study the Dirichlet boundary value problem for the p-Laplacian of the form

$$-\Delta_p u - \lambda |u|^{p-2}u + g(x,u) = f \text{ in } \Omega, \quad u = 0 \text{ on } \partial\Omega,$$

where $\Omega \subset \mathbb{R}^N$ is a bounded domain with smooth boundary $\partial\Omega$, $N \geq 1$, $p > 1$, $f \in C(\bar{\Omega})$, $\lambda > 0$ is a spectral parameter and g is a bounded function. We give the characterization of the right-hand sides f for which the Dirichlet problem above is solvable and has multiple solutions.

1. Statement of the results

Our aim is to study the solvability of the Dirichlet boundary value problem

$$\begin{cases} -\Delta_p u - \lambda |u|^{p-2}u + g(x,u) = f & \text{in } \Omega, \\ u = 0 & \text{on } \partial\Omega. \end{cases} \tag{1}$$

Here $p > 1$ is a real number, Ω is a bounded domain in $\mathbb{R}^N$ with sufficiently smooth boundary $\partial\Omega$ [2], $\Delta_p u = \operatorname{div}(|\nabla u|^{p-2}\nabla u)$ is the p-Laplacian, $\lambda \in \mathbb{R}$ is a spectral parameter, $g = g(x,u)$ is a bounded function and $f \in C(\bar{\Omega})$. In this paper, the function u is said to be a (*weak*) *solution* of (1) if $u \in W_0^{1,2}(\Omega)$ and the integral identity

$$\int_\Omega |\nabla u|^{p-2}\nabla u \cdot \nabla v - \lambda \int_\Omega |u|^{p-2}uv + \int_\Omega g(\cdot,u)v = \int_\Omega fv \tag{2}$$

holds for all $v \in W_0^{1,p}(\Omega)$.

Let us consider the homogeneous eigenvalue problem for the p-Laplacian,

$$\begin{cases} -\Delta_p u - \lambda |u|^{p-2}u = 0 & \text{in } \Omega, \\ u = 0 & \text{on } \partial\Omega. \end{cases} \tag{3}$$

[1]This research was partially supported by the Grant Agency of the Czech Republic, grant 201/00/0376. The results were presented in June–July 2001 at FSDONA - 01, Teistungen, Germany.

[2]We assume that if $N \geq 2$ then $\partial\Omega$ is a compact connected manifold of class C^2.

The structure of the set of all eigenvalues of (3) is not known if $N \geq 2$. On the other hand there are several possibilities how to find a sequence of so-called variational eigenvalues which tend to infinity. Here we present the method used in [10] which allows to find one such sequence and also to give a variational proof of the existence of at least one solution of (1) if g satisfies some additional conditions (of Landesman-Lazer type) and λ is any (even non variational) eigenvalue of (3).

Consider the even functional

$$I(u) = \frac{\int_\Omega |\nabla u|^p}{\int_\Omega |u|^p}$$

for $u \in W_0^{1,p}(\Omega) \setminus \{0\}$, and the manifold

$$\mathcal{S} \colon = \{u \in W_0^{1,p}(\Omega) \colon \|u\|_{L^p} = 1\}.$$

It is a straightforward task to verify that the eigenvalues and eigenfunctions of $-\Delta_p$ correspond to the critical values and critical points of $I|_\mathcal{S}$.

For any $k \in \mathbb{N}$ let

$$\mathcal{F}_k \colon = \{\mathcal{A} \subset \mathcal{S} \colon \exists \text{ continuous odd surjection } h \colon \mathcal{S}^{k-1} \to \mathcal{A}\},$$

where $\mathcal{S}^{k-1}$ represents the unit sphere in $\mathbb{R}^k$. Next define

$$\lambda_k \colon = \inf_{\mathcal{A} \in \mathcal{F}_k} \sup_{u \in \mathcal{A}} I(u).$$

It is proved in [10] that $\lambda_k, k = 1, 2, \ldots$, are the critical values of $I|_\mathcal{S}$, and hence the eigenvalues of $-\Delta_p$.

Let $\{\mu_k\}$ be the eigenvalues defined by the Ljusternik-Schnirelman characterization involving a minimax over sets of genus greater than k. Then $\lambda_1 = \mu_1, \lambda_2 = \mu_2$ and $\lambda_k \geq \mu_k, k = 3, 4, \ldots$. In particular,

$$\lambda_k \to \infty \text{ as } k \to \infty.$$

It is worth mentioning that whether equalities $\lambda_k = \mu_k, k = 3, 4, \ldots$, hold true is an open problem if $N \geq 2$. Also, as already pointed out, it is not clear if $\{\lambda_k\}_{k=1}^\infty$ forms the complete set of eigenvalues of $-\Delta_p$ if $N \geq 2$. On the other hand both problems are solved positively for $N = 1$ applying the shooting argument.

As for the properties of λ_1 (see e.g. [1], [15]), let us mention that λ_1 is positive, simple and isolated and the corresponding eigenfunction φ_1 (associated with λ_1) satisfies $\varphi_1 > 0$ in Ω, $\frac{\partial \varphi_1}{\partial n} < 0$ on $\partial\Omega$, where n denotes the exterior unit normal to $\partial\Omega$. One also has $\varphi_1 \in C^{1,\nu}(\bar{\Omega})$ with some $\nu \in (0,1)$ (see e.g. [8, Lemma 2.1, p. 115]). Moreover, λ_1 can be characterized as the best (the greatest) constant $C > 0$ in the Poincaré inequality

$$\int_\Omega |\nabla u|^p \geq C \int_\Omega |u|^p \tag{4}$$

for all $u \in W_0^{1,p}(\Omega)$, where identity

$$\int_\Omega |\nabla u|^p - \lambda_1 \int_\Omega |u|^p = 0$$

holds exactly for the multiples of the first eigenfunction φ_1.

In our further considerations we will use the standard spaces $W_0^{1,p}(\Omega)$, $L^p(\Omega)$, $C(\bar{\Omega})$ and $C^1(\bar{\Omega})$ (or $C_0^1(\bar{\Omega})$, respectively), with corresponding norms

$$\|u\| = \left(\int_\Omega |\nabla u|^p\right)^{\frac{1}{p}}, \qquad \|u\|_{L^p} = \left(\int_\Omega |u|^p\right)^{1/p},$$

$$\|u\|_C = \max_{x\in\Omega} |u(x)|, \qquad \|u\|_{C^1} = \|u\|_C + \max_{x\in\Omega} |\nabla u(x)|,$$

respectively, (here $|\cdot|$ denotes the Euclidean norm in $\mathbb{R}$ or $\mathbb{R}^N$). The subscript 0 indicates that the traces (or values) of functions are equal zero on $\partial\Omega$. Moreover, for the element h we use the following (L^2-non orthogonal) decomposition

$$h(x) = \tilde{h}(x) + \hat{h}, \quad \text{where} \quad \hat{h} \in \mathbb{R} \quad \text{and} \quad \int_\Omega \tilde{h}(x)\varphi_1(x)dx = 0.$$

The particular subspace formed by $\tilde{h}(x)$ will be denoted by $\tilde{C}(\bar{\Omega})$.

By $B_C(\tilde{f}, \rho)$ we denote the open ball in the space $C(\bar{\Omega})$ with the center $\tilde{f}$ and radius ρ.

In this paper we shall assume that $g = g(x,s)$ is a continuous function in both variables (this assumption can be relaxed to "a Carathéodory's function"), which is bounded and the limits

$$g^{\pm}(x) := \lim_{s\to\pm\infty} g(x,s)$$

exist finite for all $x \in \Omega$. The reader can have in mind e.g. the function

$$g(x,s) = \arctan s, \quad x \in \Omega, s \in \mathbb{R},$$

for which $g^-(x) \equiv -\frac{\pi}{2}$, $g^+(x) \equiv \frac{\pi}{2}$.

Our main results concern the solvability of (1). They are formulated in the theorems below.

Theorem 1.1. ([10]) *Assume that λ is an eigenvalue of $-\Delta_p$ (variational or non-variational) and either*

$$(LL)_\lambda^+ : \int_{v>0} g^+ v + \int_{v<0} g^- v > \int_\Omega fv, \quad \text{or} \quad (LL)_\lambda^- : \int_{v>0} g^+ v + \int_{v<0} g^- v < \int_\Omega fv$$

hold for any nonzero eigenfunction v associated with the eigenvalue λ. Then (1) has at least one solution.

Note that according to Theorem 1.1 the boundary value problem

$$\begin{cases} -\Delta_p u - \lambda_1 |u|^p u + \varepsilon \arctan u = f & \text{in } \Omega, \\ u = 0 & \text{on } \partial\Omega, \end{cases} \tag{5}$$

has at least one solution if

$$-\varepsilon\frac{\pi}{2} \;<\; \frac{1}{\|\varphi_1\|_{L^1}}\int_\Omega f\varphi_1 \;<\; \varepsilon\frac{\pi}{2}$$

(see Fig. 1). The inequalities above do not make any sense if $\varepsilon = 0$, i.e. Theorem 1.1 does not cover the solvability of (1) with $\lambda = \lambda_1$ and $g \equiv 0$, i.e. of

$$\begin{cases} -\Delta_p u - \lambda_1 |u|^{p-2}u = f & \text{in } \Omega, \\ \qquad\qquad u = 0 & \text{on } \partial\Omega. \end{cases} \tag{6}$$

But we have the following

Theorem 1.2. ([5]) *Let $p > 1$, $p \neq 2$, $\tilde{f} \in \tilde{C}(\bar{\Omega})$. Then the problem (6) has at least one solution if $f = \tilde{f}$. For $0 \neq \tilde{f} \in \tilde{C}(\bar{\Omega})$ there exists $\rho = \rho(\tilde{f}) > 0$ such that (6) has at least one solution for any $f \in B_C(\tilde{f}, \rho)$. Moreover, there exist real numbers $F_- < 0 < F_+$ (see Fig. 2) such that the problem (6) with $f = \tilde{f} + \hat{f}$ has*

(i) *no solution for $\hat{f} \notin [F_-, F_+]$;*

(ii) *at least two distinct solutions for $\hat{f} \in (F_-, 0) \cup (0, F_+)$;*

(iii) *at least one solution for $\hat{f} \in \{F_-, 0, F_+\}$.*

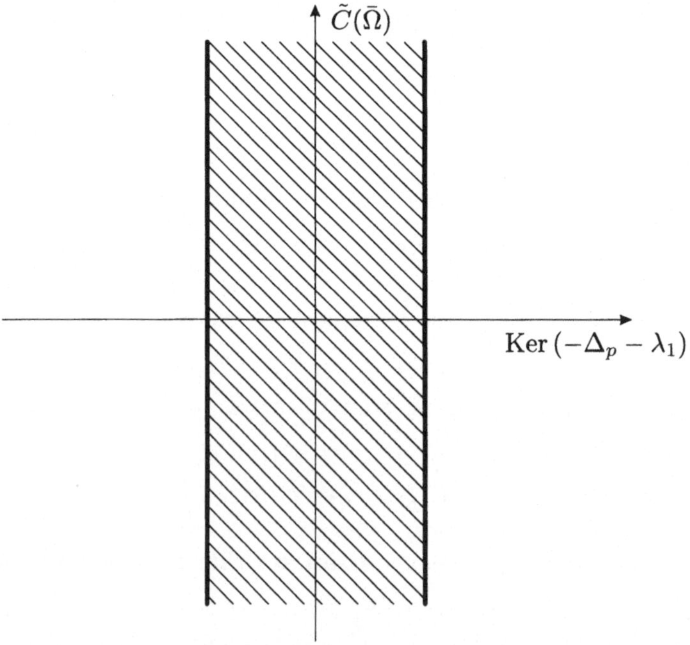

FIGURE 1. Illustration of Theorem 1.1

2. Remarks

Remark 2.1. Note that standard bootstrap regularity argument implies that any solution from Theorems 1.1–1.2 belongs to $L^\infty(\Omega)$ (cf. Drábek, Kufner, Nicolosi [9]). It follows then from the regularity results of Tolksdorf [19] (see also Di Benedetto [4] and Liebermann [14]) that it belongs to $C^{1,\nu}(\bar{\Omega})$ with some $\nu \in (0,1)$. In particular, our solution is an element of $C_0^1(\bar{\Omega})$.

Remark 2.2. We would like to emphasize that our Theorem 1.1 generalizes the classical result of Landesman and Lazer [13] and its proof can be found in [10]. However, we have to admit that both $(LL)_\lambda^\pm$ are sufficient conditions only and their necessity is an open problem even in the special case $g(x,s) = \arctan s$. The geometric interpretation of $(LL)_\lambda^\pm$ can be illustrated as in Fig. 1. One might get the impression that letting $\varepsilon \to 0$ in (5) we obtain that

$$\int_\Omega f\varphi_1 = 0$$

is a sufficient condition for the solvability of (6). Even if this is the case it cannot be proved passing to the limit for $\varepsilon \to 0$ in (5) because of the lack of a priori estimates of corresponding solutions.

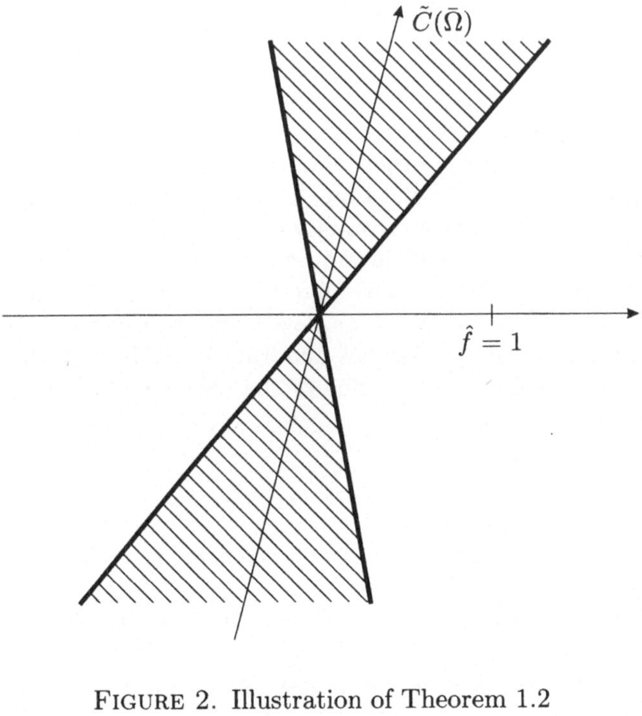

FIGURE 2. Illustration of Theorem 1.2

Remark 2.3. In particular, it follows from our Theorem 1.2 that the set of $f \in C(\bar{\Omega})$ for which (6) with $p \neq 2$ has at least one solution has a nonempty interior in $C(\bar{\Omega})$.

Remark 2.4. Note that Theorem 1.2 provides a necessary and sufficient condition for the solvability of the problem (6). This condition is in fact also of Landesman-Lazer type (see Theorem 1.1, [13], [10]). Indeed, given $\tilde{f} \in \tilde{C}(\bar{\Omega}), \tilde{f} \neq 0$, the problem (6) with the right-hand side $f(x) = \tilde{f}(x) + \hat{f}$ has a solution if and only if

$$F_-(\tilde{f}) \leq \frac{1}{\|\varphi_1\|_{L^1}} \int_\Omega f\varphi_1 \leq F_+(\tilde{f})$$

(see Fig. 2). However, it should be pointed out that this condition differs from the original condition of Landesman and Lazer due to the fact that F_- and F_+ depend on the component $\tilde{f}$ of the right-hand side f (and not on the perturbation term $g = g(x,u)$). By homogeneity we have that for any $t > 0$,

$$F_\pm(t\tilde{f}) = tF_\pm(\tilde{f}).$$

Our proof of Theorem 1.2 can be found in paper [5] and rely on the combination of the variational approach and the method of lower and upper solutions. One of the principal troubles is connected with the fact that usual a priori estimates and Palais-Smale condition fails. We also use essentially the results obtained by Drábek and Holubová [7], Takáč [17] and Fleckinger-Pellé and Takáč [12]. During the preparation of this manuscript the author received preprint of Takáč [18], where similar result to our Theorem 1.2 is proved. However, the approach used in [18] is very different from ours.

Our objective in this paper is to *avoid* complicated *technical assumptions.* For this reason we restrict to rather special domains Ω, nonlinear perturbations g and right-hand sides f. On the other hand, we believe that in our approach the main ideas appear more clearly and that possible generalization of Ω, g or f will not bring any new insight on the solvability of (1).

It should be mentioned that our approach covers also the case $N = 1$, and completes thus previous results in this direction proved by Del Pino, Drábek and Manásevich [3], Drábek, Girg and Manásevich [6], Manásevich and Takáč [16], Binding, Drábek and Huang [2], Drábek and Takáč [11]. In fact, the first relevant result which led to better understanding of the problem (1) with $\lambda = \lambda_1, p \neq 2$ and $g \equiv 0$ appeared in [3].

References

[1] A. Anane, *Etude des valeurs propres et de la résonance pour l'opérateur p-Laplacien,* Thèse de doctorat, U.L.B., 1987–1988.

[2] P.A. Binding, P. Drábek, Y.X. Huang, *On the Fredholm alternative for the p-Laplacian,* Proc. Amer. Math. Soc. 125 (1997), 3555–3559.

[3] M. Del Pino, P. Drábek, R. Manásevich, *The Fredholm alternative at the first eigenvalue for the one-dimensional p-Laplacian,* J. Differential Equations 151 (1999), 386–419.

[4] E. Di Benedetto, C^{1+d} *local regularity of weak solutions of degenerate elliptic equations*, Nonlinear Anal. 7 (8) (1983), 827–850.

[5] P. Drábek, *Geometry of the energy functional and the Fredholm alternative for the p-Laplacian in more dimensions*, To appear in Electronic J. Differential Equations.

[6] P. Drábek, P. Girg, R. Manásevich, *Generic Fredholm alternative for the one dimensional p-Laplacian*, Nonlinear Differential Equations and Applications 8 (2001), 285–298.

[7] P. Drábek, G. Holubová, *Fredholm alternative for the p-Laplacian in higher dimensions*, J. Math. Anal. Applications 263 (2001), 182–194.

[8] P. Drábek, P. Krejčí, P. Takáč, *Nonlinear Differential Equations*, Chapman & Hall/CRC, Boca Raton 1999.

[9] P. Drábek, A. Kufner, F. Nicolosi, *Quasilinear Elliptic Equations with Degenerations and Singularities*, De Gruyter Series in Nonlinear Anal. and Appl. 5, Walter de Gruyter, Berlin, New York 1997.

[10] P. Drábek, S.B. Robinson, *Resonance problems for the p-Laplacian*, J. Funct. Anal. 169 (1999), 189–200.

[11] P. Drábek, P. Takáč, *A counterexample to the Fredholm alternative for the p-Laplacian*, Proc. Amer. Math. Soc. 127 (1999), 1079–1087.

[12] J. Fleckinger-Pellé, P. Takáč, *An improved Poincaré inequality and the p-Laplacian at resonance for* $p > 2$, Preprint.

[13] E.M. Landesman, A.C. Lazer, *Nonlinear perturbations of linear elliptic boundary value problems at resonance*, J. Math. Mech. 19 (1970), 609–623.

[14] G. Liebermann, *Boundary regularity for solutions of degenerate elliptic equations*, Nonlinear Anal. 12 (11) (1998), 1203–1219.

[15] P. Lindqvist, *On the equation* $\operatorname{div}(|\nabla u|^{p-2}\nabla u) + \lambda|u|^{p-2}u = 0$, Proc. Amer. Math. Soc. 109 (1990), 157–164.

[16] R. Manásevich, P. Takáč, *On the Fredholm alternative for the p-Laplacian in one dimension*, Preprint.

[17] P. Takáč, *On the Fredholm alternative for the p-Laplacian at the first eigenvalue*, Preprint.

[18] P. Takáč, *On the number and structure of solutions for a Fredholm alternative with p-Laplacian*, Preprint.

[19] P. Tolksdorf, *Regularity for a more general class of quasilinear elliptic equations*, J. Differential Equations 51 (1984), 126–150.

Pavel Drábek
Centre of Applied Mathematics
University of West Bohemia
P. O. Box 314
306 14 Plzeň, Czech Republic
E-mail address: pdrabek@kma.zcu.cz

D. Haroske, T. Runst, H.-J. Schmeisser (eds.): Function Spaces, Differential Operators and Nonlinear Analysis. The Hans Triebel Anniversary Volume.

On the Sobolev, Hardy and CLR Inequalities Associated with Schrödinger Operators

W.D. Evans[1]

Dedicated to Prof. Hans Triebel on the occasion of his 65th birthday

Introduction

The following three inequalities have a central role in the spectral analysis of the Schrödinger operator $-\Delta - V$ in $L^2(\mathbb{R}^n)$ for $n \geq 3$.

(i) The Sobolev inequality: for all $u \in C_0^\infty(\mathbb{R}^n)$

$$\left(\int_{\mathbb{R}^n} |u(\mathbf{x})|^{\frac{2n}{n-2}} d\mathbf{x}\right)^{\frac{n-2}{n}} \leq c_n \int_{\mathbb{R}^n} |\nabla u(\mathbf{x})|^2 d\mathbf{x} \tag{1}$$

for some constant c_n independent of u; the best possible value of c_n is $\Gamma(\frac{n+1}{2})^{\frac{2}{n}} 2^{2(1-\frac{1}{n})}\{n(n-2)\pi^{1+\frac{1}{n}}\}^{-1}$, and there is equality if and only if $u(\mathbf{x})$ is a multiple of $[\mu^2 + (\mathbf{x}+\mathbf{a})^2]^{-\frac{(n-2)}{2}}$ with $\mu > 0$ and $\mathbf{a} \in \mathbb{R}^n$ arbitrary.

(ii) The Hardy inequality : for all $u \in C_0^\infty(\mathbb{R}^n \setminus \{0\})$

$$\int_{\mathbb{R}^n} \frac{|u(\mathbf{x})|^2}{|\mathbf{x}|^2}\, d\mathbf{x} \leq c_n \int_{\mathbb{R}^n} |\nabla u(\mathbf{x})|^2\, d\mathbf{x}; \tag{2}$$

the best possible value of this constant c_n is $4/(n-2)^2$, with equality if and only if $u = 0$.

(iii) The Cwikel-Lieb-Rosenblum (CLR) inequality : let $V \in L^{n/2}(\mathbb{R}^n)$, and let $-\Delta - V$ be defined in the quadratic form sense. Its spectrum is known to consist of the interval $[0, \infty)$ and a finite number of negative eigenvalues. Denoting the number of these negative eigenvalues by $N(-\Delta - V)$, the CLR inequality is

$$N(-\Delta - V) \leq c_n \int_{\mathbb{R}^n} V_+(\mathbf{x})\, d\mathbf{x}, \quad V_+ = \max(V, 0), \tag{3}$$

for some constant c_n. The evaluation of the sharp c_n is an interesting open problem.

[1]This work was partially supported by the European Union under the TMR grant FMRX-CT96-0001

The three inequalities are intimately connected. In [6, Theorem 3.1], Levin and Solomyak prove that the Sobolev inequality (1) implies the CLR inequality (3) on using the fact that $-\Delta$ generates a symmetric Markov semigroup, as required in their general abstract theorem. The importance of the Hardy inequality for the CLR inequality is demonstrated clearly in the proof of (3) given in [2]. An important feature of their proof is that, with $\mathcal{F}$ denoting the Fourier transform and $D^1(\mathbb{R}^n)$ the completion of $C_0^\infty(\mathbb{R}^n)$ with respect to the norm $\|\nabla \cdot \|_{L^2(\mathbb{R}^n)}$, the map $f \mapsto \mathcal{F}^{-1}(\frac{1}{|\cdot|}f)$ defines a unitary map U from $L^2(\mathbb{R}^n)$ to $D^1(\mathbb{R}^n)$ and $S(V) : f \mapsto V^{1/2}U$ is such that $S(V)S^*(V)$ is the Birman-Schwinger operator $V^{1/2}(-\Delta)V^{1/2}$. It can also be shown directly that (3) implies (1). By definition, the operator $-\Delta - V$ is the self-adjoint operator associated with the closure of the quadratic form

$$\int_{\mathbb{R}^n} \left(|\nabla u(\mathbf{x})|^2 - V(\mathbf{x})|u(\mathbf{x})|^2\right) d\mathbf{x}$$

on $C_0^\infty(\mathbb{R}^n)$. Suppose that $\varepsilon > 0$ is such that $c_n\varepsilon < 1$, where c_n is the optimal constant in (3), let $V \geq 0$ and

$$\int_{\mathbb{R}^n} V(\mathbf{x})^{n/2} d\mathbf{x} \leq \varepsilon. \tag{4}$$

Then, (3) implies that $N(-\Delta - V) = 0$ and hence that $-\Delta - V \geq 0$. This gives

$$\int_{\mathbb{R}^n} V(\mathbf{x})|u(\mathbf{x})|^2 d\mathbf{x} \leq \int_{\mathbb{R}^n} |\nabla u(\mathbf{x})|^2 d\mathbf{x}.$$

Since, for any $V \in L^{n/2}(\mathbb{R}^n)$, $\varepsilon^{2/n}|V|/\|V\|_{L^{n/2}(\mathbb{R}^n)}$ satisfies (4), we have

$$\int_{\mathbb{R}^n} |V(\mathbf{x})|\ |u(\mathbf{x})|^2 d\mathbf{x} \leq \varepsilon^{-2/n}\|V\|_{L^{n/2}(\mathbb{R}^n)} \int_{\mathbb{R}^n} |\nabla u(\mathbf{x})|^2 d\mathbf{x}.$$

As $L^{n/n-2}(\mathbb{R}^n)$ is the dual of $L^{n/2}(\mathbb{R}^n)$, it follows that $|u|^2 \in L^{n/n-2}(\mathbb{R}^n)$, which implies $u \in L^{2n/n-2}(\mathbb{R}^n)$, and (1) is satisfied.

When $n = 2$ each of the inequalities fails. In fact they fail even for spherically symmetric functions u in (1) and (2). For if $u(\mathbf{x}) = u(r), r = |\mathbf{x}|$, (1) and (2) become respectively

$$\|u\|^2_{L^\infty(0,\infty)} \leq C \int_0^\infty |u'(r)|^2 r dr, \tag{5}$$

$$\int_0^\infty |u(r)|^2 \frac{dr}{r} \leq C \int_0^\infty |u'(r)|^2 r dr. \tag{6}$$

But neither is satisfied, as is seen by taking a function $u \in C_0^\infty(0,1)$ with $u(r) = \ln(\ln r^{-1})$ for $0 < r < e^{-1}$.

In [7, Theorem XIII.II], it is proved that when $n = 1$ or 2, a function $V \in C_0^\infty(\mathbb{R}^n)$ which is non-positive everywhere and not identically zero is such that $-\Delta + \lambda V$ has a negative eigenvalue for all $\lambda > 0$. This implies that the CLR inequality does not hold for $n = 1$ or 2.

Our interest is with the case $n = 2$ and modified inequalities motivated by the Hardy-type inequality derived by Laptev and Weidl in [5].

1. The Laptev-Weidl inequality

In [5] Laptev and Weidl derived a Hardy-type inequality in the case $n = 2$ in which the gradient $\frac{1}{i}\nabla$ on the right-hand-side of (2) is replaced by a "magnetic gradient" $\frac{1}{i}\nabla + \vec{A}$, with *curl* $\vec{A}$ a magnetic field of Aharonov-Bohm type. Specifically, they prove, in particular, the following.

Theorem 1. *In polar co-ordinates* (r, θ), *let*

$$\vec{A}(r,\theta) = \frac{\Psi(\theta)}{r}\Big(\sin\theta, -\cos\theta\Big), \tag{7}$$

where $\Psi \in L^\infty(\mathbb{S}^1)$, *and set* $C = \Big(\min_{k\in\mathbb{Z}} |k + \tilde{\Psi}|\Big)^{-2}$, *where*

$$\tilde{\Psi} := \frac{1}{2\pi}\int_0^{2\pi} \Psi(\theta)d\theta. \tag{8}$$

Then, for all $u \in C_0^\infty(\mathbb{R}^2 \setminus \{0\})$,

$$\int_{\mathbb{R}^2} \frac{|u(\mathbf{x})|^2}{|\mathbf{x}|^2}\, d\mathbf{x} \;\le\; C \int_{\mathbb{R}^2} \left|\Big(\frac{1}{i}\vec{\nabla} + \vec{A}\Big)\, u(\mathbf{x})\right|^2 \, d\mathbf{x}. \tag{9}$$

The number $\tilde{\Psi}$ is the magnetic flux. Note that the radial component of the vector potential $\vec{A}$ is zero (Poincaré gauge), $B = curl\,\vec{A}$ is a scalar quantity and $B(r,\theta) = 0$ for $r \neq 0$. Aharonov-Bohm magnetic fields play a crucial role in the understanding of phenomena like high temperature super-conductivity and the quantum-Hall effect. These involve deep mathematical results for quantum mechanical systems in 2-dimensions which often do not have analogues in 3-dimensions.

Proof of Theorem 1. In polar co-ordinates (r, θ), we have for $\varphi \in C_0^\infty(\mathbb{R}^2 \setminus \{0\})$,

$$\begin{aligned} s_A^o[\varphi] &:= \int_{\mathbb{R}^2} \left|\Big(\frac{1}{i}\vec{\nabla} + \vec{A}\Big)\varphi\right|^2 d\mathbf{x} \\ &= \int_0^\infty \int_0^{2\pi} \Big(|\frac{\partial\varphi}{\partial r}|^2 + \frac{1}{r^2}\left|i\frac{\partial\varphi}{\partial\theta} + \Psi(\theta)\varphi\right|^2\Big) r dr d\theta \\ &= (S_A\varphi, \varphi) \end{aligned}$$

where

$$S_A = \Big(\frac{1}{i}\vec{\nabla} + \vec{A}\Big)^2 = -\frac{\partial^2}{\partial r^2} - \frac{1}{r}\frac{\partial}{\partial r} + \frac{1}{r^2}\Big(i\frac{\partial}{\partial\theta} + \Psi(\theta)\Big)^2 \tag{10}$$

and $(\cdot,\cdot)$ is the $L^2(\mathbb{R}^2)$ inner product. The operator S_A^o defined by S_A on $C_0^\infty(\mathbb{R}^2 \setminus \{0\})$ is non-negative and symmetric, and its Friedrichs extension, which we denote by S_A, is the self-adjoint operator associated with the closure s_A of s_A^o. The form domain H^1_A of S_A is the domain of s_A and this is the completion of $C_0^\infty(\mathbb{R}^2 \setminus \{0\})$ with respect to the norm

$$\|u\|_{H^1_A} := \Big(\|\Big(\frac{1}{i}\vec{\nabla} + \vec{A}\Big)u\|^2 + \|u\|^2\Big)^{1/2},$$

where $\|\cdot\|$ denotes the $L^2(\mathbb{R}^2)$ norm. Moreover, $H^1_A = \mathcal{D}(S_A^{1/2})$, the domain of $S_A^{1/2}$ with the graph norm.

The operator

$$K_\theta := i\frac{\partial}{\partial\theta} + \Psi(\theta), \quad 0 \le \theta < 2\pi,$$

with domain $H^1(\mathbb{S}^1)$ in $L^2(\mathbb{S}^1)$, where H^1 denotes the Sobolev space $H^{1,2} \equiv W^{1,2}$ and $\mathbb{S}^1$ is the unit circle, is self-adjoint in $L^2(\mathbb{S}^1)$ and has eigenvalues $\lambda_k = k+\tilde{\Psi}, k \in \mathbb{Z}$, and eigenfunctions

$$\varphi_k(\theta) = \frac{1}{\sqrt{2\pi}}\exp\Big(-i[\theta(k+\tilde{\Psi}) - \int_0^\theta \Psi(\eta)d\eta]\Big).$$

These eigenfunctions form an orthonormal basis of $L^2(\mathbb{S}^1)$, and so any $u \in L^2(\mathbb{R}^2)$ can be written

$$u(r,\theta) = \sum_{k\in\mathbb{Z}} u_k(r)\varphi_k(\theta),$$

where $u_k(r) = \int_0^{2\pi} u(r,\theta)\varphi_k(\theta)d\theta$. Furthermore, $u \in H^1_A$ if and only if

$$s_A[u] = \sum_{k\in\mathbb{Z}}\Big\{\int_0^\infty \Big(|u_k'(r)|^2 + \frac{\lambda_k^2}{r^2}|u_k(r)|^2\Big)rdr\Big\} < \infty. \tag{11}$$

Hence

$$\begin{aligned}\int_{\mathbb{R}^2}\frac{|u(\mathbf{x})|^2}{|\mathbf{x}|^2}d\mathbf{x} &= \int_0^\infty\int_0^{2\pi}|u(r,\theta)|^2\frac{dr}{r}\\ &\le \frac{1}{\min_{k\in\mathbb{Z}}\lambda_k^2}\int_0^\infty\sum_{k\in\mathbb{Z}}\lambda_k^2|u_k(r)|^2\frac{dr}{r}\\ &\le C\ s_A[u], \quad C = \Big(\min_{k\in\mathbb{Z}}|k+\tilde{\Psi}|\Big)^{-2}.\end{aligned}$$

Suppose C is attained at $k = k_0 : C = |k_0 + \tilde{\Psi}|^{-2} = \lambda_{k_0}^{-2}$, and let $u(r,\theta) = v(r)\exp\Big(-i[\theta(k_0+\tilde{\Psi}) - \int_0^\theta\Psi(\eta)d\eta]\Big)$. If C is not sharp, there exists $A \in (0, C)$ such that

$$\int_{\mathbb{R}^2}\frac{|u(\mathbf{x})|^2}{|\mathbf{x}|^2}d\mathbf{x} \le A\ s_A[u]$$

and this gives

$$C\int_0^\infty \lambda_{k_0}^2|v(r)|^2\frac{dr}{r} \le A\int_0^\infty\Big(|v'(r)|^2 + \frac{\lambda_{k_0}^2}{r^2}|v(r)|^2\Big)rdr$$

and

$$\frac{(C-A)}{A}\lambda_{k_0}^2 \int_0^\infty |v(r)|^2 \frac{dr}{r} \leq \int_0^\infty |v'(r)|^2 r dr.$$

This fails to hold for all $v \in C_0^\infty(0,\infty)$ (c.f.(6)) and hence C is sharp. The theorem is therefore proved.

2. Inequalities of Sobolev and CLR type

Since the Hardy inequality is intimately connected with (1) and (3) for $n \geq 3$, one is encouraged by the availability of Theorem 1 to seek analogues of (1) and (3) when $\frac{1}{i}\vec{\nabla}$ is replaced by $\frac{1}{i}\vec{\nabla} + \vec{A}$. We now proceed to confirm this. Further details may be found in [1]. The following two Banach spaces feature in the results:

$$X = L^\infty(\mathbb{R}^+; L^2(\mathbb{S}^1); rdr) \equiv L^\infty(\mathbb{R}^+; rdr) \bigotimes L^2(\mathbb{S}^1) \tag{12}$$

with norm

$$\|u\|_X = ess \sup_{r>0} \left\{ \left(\int_0^{2\pi} |u(r,\theta)|^2 d\theta \right)^{1/2} \right\}$$

and

$$Y = L^1(\mathbb{R}^+; L^\infty(\mathbb{S}^1); rdr) \equiv L^1(\mathbb{R}^+; rdr) \bigotimes L^\infty(\mathbb{S}^1) \tag{13}$$

with norm

$$\|u\|_Y = \int_0^\infty ess \sup_{\theta \in (0,2\pi)} |u(r,\theta)| rdr.$$

Theorem 2. *Suppose that the magnetic flux $\tilde{\Psi}$ is not an integer. Then, for all $u \in H_A^1$ and $\vec{A}$ defined by (7),*

$$\|u\|_X^2 \ \leq \ C\|(\frac{1}{i}\vec{\nabla} + \vec{A})u\|^2, \tag{14}$$

where $C = \left(\min_{k\in\mathbb{Z}} |k + \tilde{\Psi}|\right)^{-1}$ and $X = L^\infty(\mathbb{R}^+;\ L^2(\mathbb{S}^1);\ rdr)$.

Proof. For any $t \in (0,\infty)$ and $u_k(r) = \int_0^{2\pi} u(r,\theta)\varphi_k(\theta)d\theta$, in the notation of the proof of Theorem 1, we have

$$\begin{aligned} |u_k(t)|^2 &= 2 \text{ Re} \int_0^t \bar{u}_k(r)u_k'(r)\, dr \\ &\leq 2\left(\int_0^t |u_k'(r)|^2 \, rdr\right)^{\frac{1}{2}} \left(\int_0^t |u_k(r)|^2 \, \frac{dr}{r}\right)^{\frac{1}{2}} \end{aligned}$$

$$\leq \frac{2}{|\lambda_k|}\left(\int_0^t |u_k'(r)|^2 \, rdr\right)^{\frac{1}{2}}\left(\lambda_k^2\int_0^t |u_k(r)|^2 \,\frac{dr}{r}\right)^{\frac{1}{2}}$$
$$\leq \frac{1}{|\lambda_k|}\left\{\int_0^\infty \left(|u_k'(r)|^2 + \frac{\lambda_k^2}{r^2}|u_k(r)|^2\right) rdr\right\}.$$

Hence, on using (11),

$$\int_0^{2\pi} |u(t,\theta)|^2 \, d\theta = \sum_{k\in\mathbb{Z}} |u_k(t)|^2 \leq \frac{1}{\min\limits_{k\in\mathbb{Z}} |\lambda_k|}\left\|\left(\frac{1}{i}\nabla + \vec{A}\right)u\right\|^2,$$

whence (14).

The analogue of the CLR inequality will refer to the operator $T_A(V) = S_A - V$, which is defined in the following theorem.

Theorem 3. *Suppose that $\tilde{\Psi}$ is not an integer, and let*

$$V \in Y = L^1(\mathbb{R}^+;\ L^\infty(\mathbb{S}^1);\ rdr).$$

Then V is form bounded relative to S_A with S_A-bound zero, and hence $T_A(V) = S_A - V$ is defined as a form sum with form domain H_A^1. Moreover, the essential spectra of $S_A - V$ and S_A coincide and are equal to $[0,\infty)$.

Proof. It follows from (14) that S_A has no zero mode, and hence $S_A^{1/2}$ is injective, and has dense domain $\mathcal{D}(S_A^{1/2})$ and range $\mathcal{R}(S_A^{1/2})$ in $L^2(\mathbb{R}^2)$. Let D_A^1 denote the completion of $\mathcal{D}(S_A^{1/2})$ with respect to

$$\|\varphi\|_{D_A^1} := \|S_A^{1/2}\varphi\| = \|(\frac{1}{i}\vec{\nabla} + \vec{A})\varphi\|. \tag{15}$$

This is not a subspace of $L^2(\mathbb{R}^2)$, but, in view of the Laptev-Weidl inequality (9), it is a subspace of $L^2(\mathbb{R}^2; |\mathbf{x}|^{-2}d\mathbf{x})$.

We shall prove that $S_A^{-1/2}VS_A^{-1/2}$ is compact in $L^2(\mathbb{R}^2)$. It is sufficient to prove that $P \equiv S_A^{-1/2}|V|S_A^{-1/2}$ is compact and hence that $T = |V|^{1/2}S_A^{-1/2}$ is compact, since $P = T^*T$.

Given $\varepsilon > 0$, choose $W \in C_0^\infty(\mathbb{R}^+; L^\infty(\mathbb{S}^1))$ with support in $\Omega_\varepsilon = B(0,k_\varepsilon) \setminus B(0,1/k_\varepsilon)$ and such that $\|W\|_{L^\infty(\mathbb{R}^2)} \leq k_\varepsilon$, for some constant $k_\varepsilon > 1$, and $\|V - W\|_Y < \varepsilon$. Let $\varphi_n \rightharpoonup 0$ in $L^2(\mathbb{R}^2)$. Then with $\psi_n = S_A^{-1/2}\varphi_n$, $\psi_n \rightharpoonup 0$ in D_A^1 and

$$\|T\varphi_n\|^2 = \left\||V|^{\frac{1}{2}}\psi_n\right\|^2 \leq \left\||W|^{\frac{1}{2}}\psi_n\right\|^2 + \left\||V-W|^{\frac{1}{2}}\psi_n\right\|^2$$
$$\leq k_\varepsilon \int_{\Omega_\varepsilon} |\psi_n|^2 \, d\mathbf{x} + \|V-W\|_Y \|\psi_n\|_X^2$$
$$\leq k_\varepsilon \int_{\Omega_\varepsilon} |\psi_n|^2 \, d\mathbf{x} + C\varepsilon \|S_A^{\frac{1}{2}}\psi_n\|^2 \tag{16}$$

by (14). For any $\psi \in C_0^\infty(\mathbb{R}^2 \setminus \{0\})$, we have that there exists a constant $C(\varepsilon)$, depending on ε, such that

$$\|\psi\|^2_{L^2(\Omega_\varepsilon)} \leq C(\varepsilon)\|\psi\|^2_X \leq C(\varepsilon)\|(\frac{1}{i}\vec{\nabla} + \vec{A})\psi\|^2$$

by (14) and

$$\|\nabla\psi\|_{L^2(\Omega_\varepsilon)} \leq \|(\frac{1}{i}\vec{\nabla} + \vec{A})\psi\|^2 + C(\varepsilon)\|\psi\|^2_{L^2(\Omega_\varepsilon)} \leq C(\varepsilon)\|(\frac{1}{i}\vec{\nabla} + \vec{A})\psi\|^2.$$

Hence D_A^1 is continuously embedded in the standard Sobolev space $\mathrm{H}^1(\Omega_\varepsilon)$. Since $\mathrm{H}^1(\Omega_\varepsilon)$ is compactly embedded in $L^2(\Omega_\varepsilon)$ by Rellich's theorem, it follows that $T\varphi_n \to 0$ by (16) and the theorem is proved.

It follows from Theorem 3 that the spectrum of the operator $T_A(V)$ on the negative real axis consists only of eigenvalues. The next theorem gives bounds on their number.

Theorem 4. *Let $\vec{A}$ be given by (7), $V \in L^1_{loc}(\mathbb{R}^2 \setminus \{0\})$ and*

$$V_+ \in Y = L^1(\mathbb{R}^+,\ L^\infty(\mathbb{S}^1),\ rdr),$$

and suppose that the magnetic flux $\tilde{\Psi}$ is not an integer. Then $N(T_A(V))$, the number of negative eigenvalues of $T_A(V)$, is finite and the following inequalities are satisfied:

$$N(T_A(V)) \ \leq\ \sum{}' \frac{1}{2|m+\tilde{\Psi}|}\|V_+\|_Y\ , \tag{17}$$

where $\sum'$ indicates that all summands less than 1 are omitted;

$$N(T_A(V)) \ \leq\ c(\tilde{\Psi})\|V_+\|_Y, \tag{18}$$

where $c(\tilde{\Psi})$ is a constant depending only on $\tilde{\Psi}$.

Proof. Details of the proofs may be found in [1]. In the proofs of both inequalities it is noted that $T_A(V) \geq T_A(W)$, where

$$W(r) := \|V_+\|_{L^\infty(\mathbb{S}^1)},$$

and $T_A(W)$ is decomposed as

$$T_A(W) = \bigoplus_{m\in\mathbb{Z}} \Big\{(D_m - W)\otimes 1_m\Big\},$$

where

$$D_m = -\frac{1}{r}\frac{d}{dr}(r\frac{d}{dr}) + \frac{(m+\tilde{\Psi})^2}{r^2},$$

$D_m - W$ being defined in the quadratic form sense, and 1_m is the identity on the 1-dimensional space spanned by the eigenfunction $\varphi_m(\theta)$ (see the proof of Theorem 1). The inequality (17) follows from a celebrated result of Bargman in [3] which gives

$$N(D_m - W) \ \leq\ \frac{1}{2|m+\tilde{\Psi}|}\int_0^\infty W(r)\ rdr,$$

with sharp constant $(2[m+\tilde{\Psi}])^{-1}$. To prove (18) a result of Laptev and Netrusov in [4] for an operator of the form

$$-\Delta + \frac{b}{|\mathbf{x}|^2} - W$$

is used.

Also, it is proved in [1] that the $L^1(\mathbb{R}^+, L^\infty(\mathbb{S}^1, rdr)$ norm of V_+ in (18) cannot be replaced by the $L^1(\mathbb{R}^2)$ norm.

References

[1] A.A. Balinsky, W.D. Evans and R.T. Lewis. On the number of negative eigenvalues of Schrödinger operators with an Aharonov-Bohm magnetic field. *Proc. R. Soc. Lond. A* **457** (2001), 2481–2489.

[2] M.Sh. Birman and M. Solomyak. Estimates for the number of negative eigenvalues of the Schrödinger operator and its generalizations. *Advances in Soviet Math.* **7** (1991), 1–55.

[3] V. Bargmann. On the number of bound states in a central field of force. *Proc. Nat. Acad. Sci.* **38** (1952), 961.

[4] A. Laptev and Yu. Netrusov. On the negative eigenvalues of a class of Schrödinger operators. *Differential operators and spectral theory, 173–186,* Amer. Math. Soc. Transl. Ser. 2, 189, Amer. Math. Soc., Providence, RI, 1999.

[5] A. Laptev and T. Weidl. Hardy inequalities for magnetic Dirichlet forms. *In Mathematical results in quantum mechanics (Prague, 1998)*, 299–305, Oper. Theory Adv. Appl., 108, Birkhäuser, Basel, 1999.

[6] D. Levin and M. Solomyak. The Rozenblum-Lieb-Cwikel inequality for Markov generators. *J. d'Anal. Math.* **71** (1997), 173–193.

[7] M. Reed and B. Simon. *Methods of Modern Mathematical Physics; IV: Analysis of Operators.* Academic Press New York, San Francisco, London, 1978.

W.D. Evans
School of Mathematics
Cardiff University
Senghennydd Road
Cardiff, CF24 4YH, Wales, UK
E-mail address: EvansWD@cardiff.ac.uk

D. Haroske, T. Runst, H.-J. Schmeisser (eds.): Function Spaces, Differential Operators and Nonlinear Analysis. The Hans Triebel Anniversary Volume.

Mazur Distance and Normal Structure in Banach Spaces

Ji Gao

Dedicated to Prof. Hans Triebel on the occasion of his 65th birthday

Abstract. Let X be a Banach space and $\tilde{X}$ be class of spaces isomorphic to X. Using the concepts of supporting functionals in dual space X^* the condition on $\triangle(X,Y)$ is obtained, where $Y \in \tilde{X}$ and $\triangle(X,Y)$ denotes the distance between X and Y in $\tilde{X}$ which guarantees uniform normal structure.

Introduction

Let X be a normed linear space, and let

$$S(X) = \{x \in X : \|x\| = 1\} \qquad \text{and} \qquad B(X) = \{x \in X : \|x\| \leq 1\}$$

be the unit sphere and unit ball of X respectively.

Schäffer introduced two parameters, girth and perimeter, to measure the unit sphere $S(X)$ of X. These parameters, in particular

$$m(X) := \inf\{\delta(x, -x) : x \in S(X)\}$$

where the inner metric $\delta(x, -x) := \inf\{l(C) : C \text{ is a curve on } S(X) \text{ from } x \text{ to } -x\}$, were used to study reflexivity and isomorphism for some classical Banach spaces [8].

Gao and Lau considered a simplification of such a concept. Instead of curves they use an inscribed rhombus of $S(X_2)$, where X_2 denotes a two-dimensional subspace of X, to measure the unit sphere $S(X)$ of X. They defined four parameters. These parameters, in particular

$$J(X) := \sup\{\|x + y\| \wedge \|x - y\| : x, y \in S(X)\}$$

and

$$g(X) := \inf\{\|x + y\| \vee \|x - y\| : x, y \in S(X)\},$$

were used to study reflexivity, uniformly nonsquare, normal structure and isomorphism for some wider range of classical Banach spaces [5].

Recently Gao introduced two new parameters, $W(\epsilon)$ and $W_1(\epsilon)$, to measure the unit sphere $S(X)$ of X. These parameters were used to study uniformly nonsquare and normal structure for some classical Banach spaces, and some results in [6] were improved [4].

In this paper, further properties of $W(\epsilon)$ and $W_1(\epsilon)$ are studied; and by using the values of parameter $W_1(\epsilon)$, the Mazur distance between isomorphic Banach spaces is investigated which implies the uniformly nonsquare and uniform normal structure.

1. Preliminary

Let X, Y be two isomorphic normed linear spaces and $T: X \to Y$ be an isomorphism between them. Let X^* and Y^* be the dual spaces of X and Y respectively. For any $x \in X$ and $f \in X^*$, we use $\langle x, f\rangle$ to denote the value of the linear functional f at x. Let T^* be the conjugate mapping of T from Y^* to X^*: $\langle x, T^*g\rangle = \langle Tx, g\rangle$ for any $x \in X$ and $g \in Y^*$, then T^* is an isomorphism from Y^* to X^*, $(T^*)^{-1}$ exists and $(T^*)^{-1}$ is an isomorphism from X^* to Y^*.

Following the notation of Schäffer [8], we define $\partial T : S(X) \to S(Y)$ by $\partial T(x) = \frac{Tx}{\|Tx\|}$. It is clear that ∂T is a bijective mapping. Let $x \in S(X)$ we use $\nabla_x \subseteq S(X^*)$ to denote the set of norm 1 supporting functionals of $S(X)$ at x.

Lemma 1.1. *For any $x \in S(X)$, ∂T^* is a bijective mapping from $\nabla_{(\partial T)x}$ of $S(Y^*)$ to ∇_x of $S(X^*)$.*

Proof. Let $x \in S(X)$. For any $g_{(\partial T)x} \in \nabla_{(\partial T)x}$,

$$\langle x, T^*(g_{(\partial T)x})\rangle = \langle Tx, g_{(\partial T)x}\rangle = \|Tx\| .$$

So $\frac{T^*(g_{(\partial T)x})}{\|Tx\|} \in \nabla_x$, and hence $\partial T^* = \frac{T^*}{\|Tx\|}$ maps $\nabla_{(\partial T)x}$ to ∇_x. For any $f_x \in \nabla_x$, since $\|Tx\|(T^*)^{-1}(f_x) \in \nabla_{(\partial T)x}$ and $(\partial T^*)(\|Tx\|(T^*)^{-1}(f_x)) = f_x$, ∂T^* is surjective.

To prove ∂T^* is an injective mapping, let both $g^1_{(\partial T)x}, g^2_{(\partial T)x} \in \nabla_{(\partial T)x}$, but $g^1_{(\partial T)x} \neq g^2_{(\partial T)x}$. Then there exists an $y \in S(Y)$ such that

$$\langle y, g^1_{(\partial T)x}\rangle \neq \langle y, g^2_{(\partial T)x}\rangle.$$

Let $x \in S(X)$ such that $(\partial T)x = y$. Then the following chain of inequalities holds:

$$\begin{aligned}\langle(\partial T)x, g^1_{(\partial T)x}\rangle \neq \langle(\partial T)x, g^2_{(\partial T)x}\rangle &\Rightarrow \langle\frac{Tx}{\|Tx\|}, g^1_{(\partial T)x}\rangle \neq \langle\frac{Tx}{\|Tx\|}, g^2_{(\partial T)x}\rangle \\ &\Rightarrow \langle\frac{x}{\|Tx\|^2}, \frac{T^*g^1_{(\partial T)x}}{\|Tx\|}\rangle \neq \langle\frac{x}{\|Tx\|^2}, \frac{T^*g^2_{(\partial T)x}}{\|Tx\|}\rangle \\ &\Rightarrow \langle\frac{x}{\|Tx\|^2}, (\partial T^*)g^1_{(\partial T)x}\rangle \neq \langle\frac{x}{\|Tx\|^2}, (\partial T^*)g^2_{(\partial T)x}\rangle \\ &\Rightarrow (\partial T^*)g^1_{(\partial T)x} \neq (\partial T^*)g^2_{(\partial T)x}.\end{aligned}$$

For any $x, y \in S(X)$ and $f_x \in \nabla_x$, let $g_{(\partial T)x} = (\partial T^*)^{-1}(f_x)$. From Lemma 1.1,

$$g_{(\partial T)x} = \left(\frac{T^*}{\|Tx\|}\right)^{-1}(f_x) \in \nabla_{(\partial T)x}.$$

Furthermore, we have the following lemma:

Lemma 1.2. $\frac{1}{\|T\|\cdot\|T^{-1}\|}\langle y, f_x\rangle \leq \langle(\partial T)y, (\partial T^*)^{-1}(f_x)\rangle \leq \|T\|\cdot\|T^{-1}\|\langle y, f_x\rangle.$

Proof. $$\begin{aligned}\langle(\partial T)y, (\partial T^*)^{-1}(f_x)\rangle &= \langle \tfrac{Ty}{\|Ty\|}, g_{(\partial T)x}\rangle &&= \tfrac{\|Tx\|}{\|Ty\|}\langle Ty, \tfrac{g_{(\partial T)x}}{\|Tx\|}\rangle \\ &= \tfrac{\|Tx\|}{\|Ty\|}\langle y, \tfrac{T^*(g_{(\partial T)x})}{\|Tx\|}\rangle &&= \tfrac{\|Tx\|}{\|Ty\|}\langle y, f_x\rangle\ .\end{aligned}$$

Then Lemma 1.2 can be obtained by considering the chain of equations above and the following double inequality: $\frac{1}{\|T^{-1}\|} \leq \|Tx\|, \|Ty\| \leq \|T\|$.

Lemma 1.3. [5]. *For any* $x, y \in S(X)$,

$$\frac{1}{\|T\|\cdot\|T^{-1}\|}(\|x \pm y\| + 2) \leq \|(\partial T)y \pm (\partial T)x\| + 2 \leq \|T\|\cdot\|T^{-1}\|(\|x \pm y\| + 2)\ .$$

2. Main theorem

Definition 2.1. [7]. *A normed linear space X is called uniformly nonsquare if there exists a* $\delta > 0$ *such that*

$$\text{either}\quad \frac{\|x+y\|}{2} \leq 1-\delta \qquad \text{or}\quad \frac{\|x-y\|}{2} \leq 1-\delta.$$

Definition 2.2. *A normed linear space X is said to be uniformly convex if for any* $\epsilon > 0$ *there exists a* $\delta > 0$ *such that for any* $x, y \in S(X)$ *with* $\|\frac{x+y}{2}\| > 1-\delta$, *we have* $\|x-y\| < \epsilon$.

Definition 2.3. *For a normed linear space X,*

$$\delta(\epsilon) = \inf\{1 - \|\tfrac{x+y}{2}\| : \text{ for any } x, y \in S(X) \text{ with } \|x-y\| \geq \epsilon, 0 \leq \epsilon \leq 2\}$$

is called the modulus of convexity of X.

It is well known that X is uniformly convex if and only if for any $\epsilon > 0$, $\delta(\epsilon) > 0$.

Definition 2.4. [1]. *A bounded, convex subset K of a Banach space X is said to have normal structure if every convex subset H of K that contains more than one point contains a point* $x_0 \in H$, *such that* $\sup\{\|x_0 - y\|, y \in H\} < d(H)$, *where* $d(H) = \sup\{\|x-y\|, x, y \in H\}$ *denotes the diameter of H. A Banach space X is said to have normal structure if every bounded, convex subset of X has normal structure. A Banach space X is said to have weak normal structure if each weakly compact convex set K in X that contains more than one point has normal structure. X is said to have uniform normal structure if there exists* $0 < c < 1$ *such that for any subset K as above, there exists* $x_0 \in K$ *such that* $\sup\{\|x_0 - y\|, y \in K\} < c \cdot d(K)$.

For a reflexive Banach space X , the normal structure and weak normal structure coincide.

Definition 2.5. [8]. *Let X be a given normed linear space and $\tilde{X}$ be the class of spaces isomorphic to X. The Mazur distance (pseudo metric) D on $\tilde{X}$ is defined as:*

$$D(X,Y) = \inf\{\ln(\|T\| \cdot \|T^{-1}\|) : T : X \to Y \textit{ is an isomorphism}\}.$$

Recently, Gao introduced the parameters $W(\epsilon)$ and $W_1(\epsilon)$ for a Banach space:

Definition 2.6. [4]. *Let X be a given normed linear space, $x, y \in S(X)$, $\nabla_x \subseteq S(X^*)$ be the set of norm 1 supporting functionals of $S(X)$ at x, and $\gamma(x,y) = \sup\{\langle \frac{x-y}{2}, f_x\rangle : f_x \in \nabla_x\}$. Then*

$$W(\epsilon) = \inf\{\gamma(x,y), \|x-y\| \geq \epsilon\}, \quad 0 \leq \epsilon \leq 2$$

is called the modulus of W-convexity.

Definition 2.7. [4]. *Let X be a given normed linear space, $x, y \in S(X)$, $\nabla_x \subseteq S(X^*)$ be the set of norm 1 supporting functionals of $S(X)$ at x, and $\gamma_1(x,y) = \inf\{\langle \frac{x-y}{2}, f_x\rangle : f_x \in \nabla_x\}$. Then*

$$W_1(\epsilon) = \sup\{\gamma_1(x,y), \|x-y\| \leq \epsilon\}, \quad 0 \leq \epsilon \leq 2$$

is called the modulus of W_1-convexity.

Remarks.

1) In Definitions 2.6 and 2.7, $\|x-y\| \geq \epsilon$ ($\|x-y\| \leq \epsilon$) may be replaced by $\|x-y\| > \epsilon$ ($\|x-y\| < \epsilon$) or both replaced by $\|x-y\| = \epsilon$.
2) For a Hilbert space, $W(\epsilon) = W_1(\epsilon)$ for $0 \leq \epsilon \leq 2$.

Proposition 2.8. *For a Banach space X,*

(i) *$W(\epsilon)$ and $W_1(\epsilon)$ are increasing functions of ϵ, $0 \leq \epsilon \leq 2$.*
(ii) *$\delta(\epsilon) \leq W(\epsilon)$, $W_1(\epsilon) \leq \frac{\epsilon}{2}$ for any $0 \leq \epsilon \leq 2$.*

Proof. We only need to prove (ii). For any $x, y \in S(X)$ with $\|x-y\| = \epsilon$ and any $0 \leq \epsilon \leq 2$,

$$\delta(\epsilon) \leq 1 - \left\|\frac{x+y}{2}\right\| \leq 1 - \langle \frac{x+y}{2}, f_x\rangle = \langle x - \frac{x+y}{2}, f_x\rangle = \langle \frac{x-y}{2}, f_x\rangle \leq \|\frac{x-y}{2}\| = \frac{\epsilon}{2}.$$

We study further properties of $W(\epsilon)$ and $W_1(\epsilon)$.

Definition 2.9. *A normed linear space X is said to be weakly uniformly convex (WUC) if for any $\epsilon > 0$ and for any $f \in S(X^*)$ there exists a $\delta > 0$ such that for any $x, y \in S(X)$ with $\|\frac{x+y}{2}\| > 1 - \delta$, we have $|\langle x-y, f\rangle| < \epsilon$.*

Lemma 2.10. *If $x, y \in S(X)$, then $\|\frac{y-x}{\|y-x\|} - x\| \geq 1$.*

Proof.

$$\left\|\frac{y-x}{\|y-x\|} - x\right\| = \frac{\left\|y-x-\|y-x\|x\right\|}{\|y-x\|} = \frac{\left\|(\|y-x\|+1)x-y\right\|}{\|y-x\|}$$
$$\geq \frac{\left\|(\|y-x\|+1)x\right\| - \|y\|}{\|y-x\|} = \frac{\|y-x\|+1-1}{\|y-x\|} = 1 .$$

Let X be a normed linear space, X_2 be a two-dimensional subspace of X. For $x \in S(X_2)$, let k be one of the arcs of $S(X)$ from x to $-x$ counterclockwise, and let $g(s) : [0, L] \to k$ be the standard representation of k in terms of arc length with $x = g(0)$ and $-x = g(L)$, where L is the length of k.

Lemma 2.11. [6]. $\|g(s) - g(0)\|$ *is an increasing function of* s, $0 \le s \le L$.

Lemma 2.12. *If* x, y *and* $y_1 \in S(X_2)$ *with* $\|y - x\|, \|y_1 - x\| \le 1$, *then*

$$\min\{\|y - x\|, \|y_1 - x\|\} \le \|ty + (1-t)y_1 - x\| \le \max\{\|y - x\|, \|y_1 - x\|\}$$

for $0 \le t \le 1$.

Proof. Let $\|y - x\| = g(s_1)$, $\left\|\frac{y-x}{\|y-x\|} - x\right\| = g(s_2)$ and $\|y_1 - x\| = g(s_3)$. Without loss of generality we may assume $\|y - x\| \le \|y_1 - x\|$. From Lemma 2.11 we have $\left\|\frac{y+y_1}{2} - x\right\| \le \|y_1 - x\|$. Since $\left\|\frac{y-x}{\|y-x\|} - x\right\| \ge 1$, we have $0 \le s_1 \le s_3 \le s_2 \le L$. Then the convexity of $U(X) = \{x \in X : \|x\| = 1\}$ implies that for any $0 \le t \le 1$, $\left\|\frac{ty+(1-t)y_1-x}{\|y-x\|}\right\| = \left\|t\frac{y-x}{\|y-x\|} + (1-t)\frac{y_1-x}{\|y-x\|}\right\| \ge 1$.

Theorem 2.13. *If* X *is weakly uniformly convex, then both* $W(\epsilon)$ *and* $W_1(\epsilon)$ *are continuous for* $0 \le \epsilon < 1$.

Proof. Let $x, y \in S(X)$ with $\|y - x\| = a < 1$. For any $\epsilon > 0$ and $f \in \nabla_x$, let $\delta < \min\{\frac{a}{3}, \frac{\epsilon}{2}\}$ be in definition of WUC such that for any $x, y \in S(X)$ with $\left\|\frac{x+y}{2}\right\| > 1 - \frac{2\delta}{a}$, $|\langle x - y, f\rangle| < \frac{\epsilon}{2}$.

Let $y_1 \in S(X_2)$ such that $\big|\, \|y_1 - x\| - a\big| < \min\{\delta, 1 - a\}$, then

$$a - \delta \le \|y_1 - x\| \le \min\{a + \delta, 1\}.$$

If $\|y_1 - x\| \ge \|y - x\|$, there exists a $0 \le t \le 1$ such that $t\frac{y-x}{\|y-x\|} + (1-t)\frac{y_1-x}{\|y-x\|}$ is between

$$\frac{\frac{y-x}{\|y-x\|} + \frac{y_1-x}{\|y_1-x\|}}{2} \quad \text{and} \quad \frac{\frac{\|y_1-x\|}{\|y-x\|}\frac{y-x}{\|y-x\|} + \frac{y_1-x}{\|y-x\|}}{2}.$$

From Lemma 2.12,

$$\begin{aligned}
&\frac{1}{2}\left\|\frac{y-x}{\|y-x\|} + \frac{y_1-x}{\|y_1-x\|}\right\| \\
&\quad = \left\|t\frac{y-x}{\|y-x\|} + (1-t)\frac{y_1-x}{\|y-x\|}\right\| \\
&\qquad - \left\|t\frac{y-x}{\|y-x\|} + (1-t)\frac{y_1-x}{\|y-x\|} - \frac{1}{2}\left(\frac{y-x}{\|y-x\|} + \frac{y_1-x}{\|y_1-x\|}\right)\right\| \\
&\quad \ge 1 - \frac{1}{2}\left\|\frac{\|y_1-x\|}{\|y-x\|}\frac{y-x}{\|y-x\|} + \frac{y_1-x}{\|y-x\|} - \left(\frac{y-x}{\|y-x\|} + \frac{y_1-x}{\|y_1-x\|}\right)\right\| \\
&\quad \ge 1 - \left(\frac{\|y_1-x\|}{\|y-x\|} - 1\right) \ge 1 - \frac{\delta}{a}\,.
\end{aligned}$$

If $\|y_1-x\| \le \|y-x\|$, there exists a $0 \le t \le 1$ such that $t\frac{y-x}{\|y_1-x\|}+(1-t)\frac{y_1-x}{\|y_1-x\|}$ is between

$$\frac{\frac{y-x}{\|y-x\|}+\frac{y_1-x}{\|y_1-x\|}}{2} \quad \text{and} \quad \frac{\frac{y-x}{\|y_1-x\|}+\frac{\|y-x\|}{\|y_1-x\|}\frac{y_1-x}{\|y_1-x\|}}{2}.$$

From Lemma 2.12 again,

$$\begin{aligned}
&\frac{1}{2}\left\|\frac{y-x}{\|y-x\|}+\frac{y_1-x}{\|y_1-x\|}\right\| \\
&= \left\|t\frac{y-x}{\|y_1-x\|}+(1-t)\frac{y_1-x}{\|y_1-x\|}\right\| \\
&\quad - \left\|t\frac{y-x}{\|y_1-x\|}+(1-t)\frac{y_1-x}{\|y_1-x\|}-\frac{1}{2}\left(\frac{y-x}{\|y-x\|}+\frac{y_1-x}{\|y_1-x\|}\right)\right\| \\
&\ge 1-\frac{1}{2}\left\|\frac{y-x}{\|y_1-x\|}+\frac{\|y-x\|}{\|y_1-x\|}\frac{y_1-x}{\|y_1-x\|}-\left(\frac{y-x}{\|y-x\|}+\frac{y_1-x}{\|y_1-x\|}\right)\right\| \\
&\ge 1-\left(\frac{\|y-x\|}{\|y_1-x\|}-1\right) \ge 1-\frac{\delta}{a-\delta} \ge 1-\frac{3\delta}{2a}.
\end{aligned}$$

Hence

$$\begin{aligned}
&|\langle y_1-x, f\rangle - \langle y-x, f\rangle| \\
&= \|y-x\|\left|\langle\frac{y-x}{\|y-x\|}-\frac{y_1-x}{\|y-x\|}, f\rangle\right| \\
&\le \|y-x\|\left[\left|\langle\frac{y-x}{\|y-x\|}-\frac{y_1-x}{\|y_1-x\|}, f\rangle\right| + \left\|\frac{y_1-x}{\|y_1-x\|}-\frac{y_1-x}{\|y-x\|}\right\|\right] \\
&\le \frac{a\epsilon}{2}+\delta \le \frac{\epsilon}{2}+\delta \le \frac{\epsilon}{2}+\frac{\epsilon}{2} = \epsilon .
\end{aligned}$$

Theorem 2.14. *If X is weakly uniformly convex and uniformly nonsquare, then both $W(\epsilon)$ and $W_1(\epsilon)$ are continuous at 1.*

Proof. We first claim that if X is uniformly nonsquare then there exists an $\eta > 0$ such that for any $x, y \in S(X)$ with $\|y-x\| = 1$, $\|y-2x\| > 1+\eta$.

If not, for any $\eta > 0$ there exist $x, y \in S(X)$ with $\|y-x\| = 1$, but $\|y-2x\| \le 1+\eta$. Let $u = \frac{y-2x}{\|y-2x\|} \in S(X)$, $v = \frac{y}{\|y-2x\|}$, then $1-\eta \le \|v\| \le 1$. So,

$$\|u+y\|+\|u-y\| \ge \|u+v\|+\|u-v\| = \left\|\frac{2(y-x)}{\|y-2x\|}\right\|+\left\|\frac{-2x}{\|y-2x\|}\right\| = \frac{4}{\|y-2x\|} \ge 4-4\eta.$$

Hence $\min\{\|u+y\|, \|u-y\|\} \ge 2-4\eta$. Since η can be arbitrarily small, this is a contradiction with uniformly nonsquare.

The remainder of the proof is similar to the proof of Theorem 2.13.

Theorem 2.15. *If X is uniformly convex, then both $W(\epsilon)$ and $W_1(\epsilon)$ are continuous for $0 \le \epsilon \le 2$.*

Proof. We need only to prove this theorem for $1 \leq \epsilon \leq 2$.

We first claim that if X is uniformly convex then there exists an $\eta > 0$ such that for any $x, y \in S(X)$ with $\|y - x\| = \epsilon$, where $1 < \epsilon < 2$ and $\epsilon + \eta \leq 2$, $\|\frac{y-x}{\|y-x\|} - x\| \geq \epsilon + \eta$.

If not, for any $\eta > 0$ there exist $x, y \in S(X)$ with $\|y - x\| = \epsilon$, but $\epsilon \leq \|\frac{y-x}{\|y-x\|} - x\| < \epsilon + \eta$.

Let $z = \frac{y-x}{\|y-x\|} \in S(X)$, then $\epsilon \leq \|x - z\| \leq \epsilon + \eta$. Let $w, t \in [x, y]$, where $[x, y]$ denotes the line segment connecting x and y, such that $\|y - w\| = 1$, $\|(x - z) - t\| = \|x - z\|$, $u = \frac{w}{2}$ and $v = \frac{t}{2}$. Then $\|u - y\| \leq 1 - \delta(\|w\|)$, and $\|u - (x - z)\| \leq \|v - (x - z)\| + \|u - v\| \leq (\epsilon + \eta)(1 - \delta(\frac{\|t\|}{\epsilon+\eta})) + \frac{\eta}{2}$. This implies

$$1 + \epsilon \leq \|u - y\| + \|u - (x - z)\| \leq 1 + \epsilon + \eta + \frac{\eta}{2} - (\epsilon + \eta)\delta(\frac{\|t\|}{\epsilon + \eta}) - \delta(\|w\|).$$

We have $(\epsilon+\eta)\delta(\frac{\|t\|}{\epsilon+\eta}) + \delta(\|w\|) \leq \frac{3\eta}{2}$. So, $\delta(\|w\|)$ should be arbitrarily small if η is arbitrarily small. It is a contradiction with $\|w\| \geq \|x\| - \|x - w\| = 1 - (\epsilon - 1) = 2 - \epsilon$ and the uniform convexity.

The remainder of the proof is similar to the proof of Theorem 2.13 too.

Let X be a normed linear space, $x, y \in S(X)$. X is said to be Gateaux differentiable at $x \in S(X)$, if $\lim_{t\to 0} \frac{\|x+ty\|-\|x\|}{t}$ exists for each $y \in S(X)$. X is said to be Gateaux differentiable space, if the limit exists for each $x, y \in S(X)$. The normed linear space X is said to be Fréchet differentiable at $x \in S(X)$, if $\lim_{t\to 0} \frac{\|x+ty\|-\|x\|}{t}$ exists uniformly for $y \in S(X)$. X is said to be Fréchet differentiable space, if the limit exists uniformly for $y \in S(X)$ at each $x \in S(X)$. For any $x \in S(X)$, ∇_x is a w^*-compact convex subset of X^*. If X is Gateaux differentiable at $x \in S(X)$ then $S(X)$ is smooth at x, therefore ∇_x is a singleton. Let $K \subseteq X^*$ be a bounded closed subset, $f \in K$ is called a *weak** exposed point if there exists $x \in S(X)$ such that $\langle x, f\rangle = \sup\langle x, K\rangle > \langle x, g\rangle$ for any $f \neq g \in K$; $f \in K$ is called a w^*-strongly exposed point if there exists $x \in S(X)$ such that for any $\epsilon > 0$, there exists $\delta > 0$ satisfying for any $g \in K$, $\langle x, f\rangle < \langle x, g > +\delta$ implies $\|g - f\| < \epsilon$. It is clear that a w^*-strongly exposed point is a *weak** exposed point. (See [2] for a complete reference.)

A Banach space X is called a w-Asplund space (an Asplund space) if every continuous convex function on X is Gateaux (Fréchet) differentiable on a dense G_δ subset. It is well known that if X is a w-Asplund space then every bounded sequence in X^* has a w^*-convergent subsequence [9], and X is an Asplund space if and only if every w^*-compact convex subset K in X^* is the w^*-closed convex hull of its w^*-strongly exposed points [3].

Lemma 2.16. [2]. *Let X be a Banach space. For any $x, y \in S(X)$, let $u = \frac{x+\lambda y}{\|x+\lambda y\|}$ with $\lambda > 0$. Then for $f_x \in \nabla_x$, $f_u \in \nabla_u$, $\langle y, f_x\rangle \leq \frac{\|x+\lambda y\|-1}{\lambda} \leq \langle y, f_u\rangle$.*

Lemma 2.16 and the following Lemma 2.17 improve Lemma 3.3 and Lemma 3.6 in [6].

Lemma 2.17. *Let X be a Banach space, and let $x \in S(X), f, g \in \nabla_x$ and $z \in S(X)$ such that $\langle z, g\rangle < \langle z, f\rangle = 0$. Let X_2 be a two-dimensional subspace of X spanned by x and z, and let $A = \{y \in S(X_2) : \langle y, g\rangle < \langle y, f\rangle\}$ denote the half-sphere of $S(X_2)$. Suppose $\{y_n\} \subseteq A$ with $y_n = \frac{x+\frac{1}{n}z}{\|x+\frac{1}{n}z\|}$, and suppose $\{f_n\} \subseteq S(X^*)$ satisfies $\langle y_n, f_n\rangle \geq 1 - \frac{1}{n^2}$, for every n. Then $\langle z, f_n\rangle \geq 0$ for large n.*

Proof. From Lemma 2.16,

$$\underline{\lim}_{n\to\infty} \frac{\langle x - y_n, g\rangle}{\frac{1}{n}} = \underline{\lim}_{n\to\infty} \frac{\|x + \frac{1}{n}z\| - 1 - \frac{1}{n}\langle z, g\rangle}{\frac{1}{n}} \geq \langle z, f\rangle - \langle z, g\rangle = -\langle z, g\rangle.$$

Hence for large n,

$$\langle y_n, g\rangle = 1 - \langle x - y_n, g\rangle < 1 + \frac{1}{n}\langle z, g\rangle.$$

Let $\bar{f}_n$ be the normalization of f_n on X_2, and u_n be the support point of $\bar{f}_n$, then

$$\langle y_n, \bar{f}_n\rangle \geq \langle y_n, f_n\rangle > 1 - \frac{1}{n^2}.$$

We claim that $u_n \in A$ for large n. Suppose it is false. By passing to a subsequence we can assume that u_n are contained in the complementary half-sphere $A' = S(X)\backslash A$. Therefore x can be represented either $x = \frac{u_n+\lambda_n y_n}{\|u_n+\lambda_n y_n\|}, \lambda_n \geq 0$, or $-x = \frac{u_n+\lambda_n y_n}{\|u_n+\lambda_n y_n\|}, \lambda_n \geq 0$.

Apply Lemma 2.16 again, in the first case we have

$$\langle y_n, \bar{f}_n\rangle \leq \langle y_n, g\rangle, \quad \text{i.e.,} \quad \frac{1}{n} \geq -\langle z, g\rangle \quad \text{for large } n.$$

It is impossible. In the second case, take $-g$ as a supporting functional at $-x$, we have $\langle y_n, \bar{f}_n\rangle \leq \langle y_n, -g\rangle$. Consider $\langle y_n, -g\rangle \to -1$, it is impossible too.

Now for large n, $u_n \in A$, so either

$$u_n = \frac{x + \lambda_n z}{\|x + \lambda_n z\|}, \lambda_n \geq 0, \quad \text{or} \quad u_n = \frac{-x + \lambda_n z}{\| - x + \lambda_n z\|}, \lambda_n \geq 0.$$

In the first case Lemma 2.16 implies that $\langle z, \bar{f}_n\rangle \geq \langle z, f\rangle = 0$. In the second case, take $-f$ as a supporting functional at $-x$, Lemma 2.16 implies that $\langle z, \bar{f}_n\rangle \geq \langle z, -f\rangle = 0$. In both cases $\langle z, \bar{f}_n\rangle \geq 0$, therefore $\langle z, f_n\rangle \geq 0$.

Lemma 2.18. *Let X be a w-Asplund space, $x \in S(X)$, and let f be a weak* exposed point of ∇_x, then there exists a sequence $\{x_n\} \subseteq S(X)$, $\{f_n\} \subseteq S(X^*)$ with x_n converges in norm topology to x, f_n converges in weak* topology to f, where the x_n are Gateaux differentiable points of $S(X)$, and $f_n \in \nabla_{x_n}$ is the unique supporting functional of $S(X)$ at x_n.*

Proof. If ∇_x is a singleton, the assertion is clear. Suppose X contains more than one point, and let f be a *weak** exposed point of ∇_x. Then there exists a $z \in S(X)$ and a $g \in \nabla_x$ such that $\langle z, g\rangle < \langle z, f\rangle$, and for any $k \in \nabla_x \backslash f$, $\langle z, k\rangle < \langle z, f\rangle$. Without loss of generality, we may assume $\langle z, f\rangle = 0$. Let X_2 be a two-dimensional subspace spanned by x and z.

Let $A' = \{y \in S(X_2) : \langle y, g\rangle \leq \langle y, f\rangle\}$. Then A' is a half-sphere. Suppose $\{y_n\} \subseteq A'$ with $y_n = \frac{x+\frac{1}{n}z}{\|x+\frac{1}{n}z\|}$ as in Lemma 2.17. Let $\{\bar{y}_n\} \subseteq S(X)$ be the Gateaux differentiable points such that $\|y_n - \bar{y}_n\| < \frac{1}{n^2}$, and f_n be the unique supporting functional of $S(X)$ at $\bar{y}_n$. Then

$$1 \geq \langle y_n, f_n\rangle = \langle \bar{y}_n, f_n\rangle - \langle \bar{y}_n - y_n, f_n\rangle \geq 1 - \|\bar{y}_n - y_n\| > 1 - \frac{1}{n^2}.$$

Use Lemma 2.17 again, we have $\langle z, f_n\rangle > 0$.

Note that X is a w-Asplund space and $\{f_n\}$ are bounded, so there exists a w^*-convergent subsequence of $\{f_n\}$ which converges in *weak** topology to some $h \in X^*$. Without loss of generality we may assume that $\{f_n\}$ itself converges in *weak** topology to h. Since

$$1 \geq |\langle x, h\rangle| = \lim_{n\to\infty} |\langle x, f_n\rangle| = \lim_{n\to\infty} |\langle y_n, f_n\rangle + \langle x - y_n, f_n\rangle| \geq 1 - \lim_{n\to\infty} \|x - y_n\| = 1,$$

$h \in \nabla_x$, and $\langle z, h\rangle = \lim_{n\to\infty} \langle z, f_n\rangle \geq 0$. So, $h = f$. The proof is complete.

Theorem 2.19. *If X is an Asplund space, then $W(\epsilon) \leq W_1(\epsilon)$, $0 \leq \epsilon \leq 2$.*

Proof. Let $K \subseteq S(X)$ be the set of smooth points, then K is a dense G_δ subset of $S(X)$. Let $x \in S(X)\backslash K$, and P be the set of w^*-strongly exposed points of ∇_x, then $\nabla_x = \overline{co}^{w^*}(P)$ and for any $y \in S(X)$,

$$\sup\{\langle \frac{x-y}{2}, f_x\rangle, f_x \in \nabla_x\} = \sup\{\langle \frac{x-y}{2}, f_x\rangle, f_x \in P\}$$

and

$$\inf\{\langle \frac{x-y}{2}, f_x\rangle, f_x \in \nabla_x\} = \inf\{\langle \frac{x-y}{2}, f_x\rangle, f_x \in P\}.$$

From Lemma 2.18, for any $x, y \in S(X)$, $f_x \in P$, there exist $\{x_n\} \subseteq K$, $\{f_n\} \subseteq S(X^*)$, where f_n is the unique supporting functional of $S(X)$ at x_n such that x_n converges in norm topology to x, f_n converges in *weak** topology to f. Take $\{y_n\} \subseteq K$ such that y_n converges in norm topology to y. Then

$$0 \leq \lim_{n\to\infty} |\langle \frac{x_n - y_n}{2} - \frac{x-y}{2}, f_n\rangle| \leq \lim_{n\to\infty} (\|\frac{x - x_n}{2}\| + \|\frac{y - y_n}{2}\|) = 0.$$

So $\lim_{n\to\infty}\langle \frac{x_n-y_n}{2}, f_n\rangle = \lim_{n\to\infty}\langle \frac{x-y}{2}, f_n\rangle = \langle \frac{x-y}{2}, f_x\rangle$. Therefore

$$\begin{aligned}
W(\epsilon) &= \inf\{\sup\{\langle\frac{x-y}{2}, f_x\rangle : f_x \in \nabla_x\} : \|x-y\| = \epsilon, \text{ and } x, y \in S(X)\} \\
&= \inf\{\sup\{\langle\frac{x-y}{2}, f_x\rangle : f_x \in P\} : \|x-y\| = \epsilon, \text{ and } x, y \in S(X)\} \\
&= \inf\{\sup\{\langle\frac{x-y}{2}, f_x\rangle : f_x \in \nabla_x\} : \|x-y\| = \epsilon, \text{ and } x, y \in K\} \\
&= \inf\{\langle\frac{x-y}{2}, f_x\rangle : f_x \in \nabla_x, \|x-y\| = \epsilon, \text{ and } x, y \in K\} \\
&\le \sup\{\langle\frac{x-y}{2}, f_x\rangle : f_x \in \nabla_x, \|x-y\| = \epsilon, \text{ and } x, y \in K\} \\
&= \sup\{\inf\{\langle\frac{x-y}{2}, f_x\rangle : f_x \in \nabla_x\} : \|x-y\| = \epsilon, \text{ and } x, y \in K\} \\
&= \sup\{\inf\{\langle\frac{x-y}{2}, f_x\rangle : f_x \in P\} : \|x-y\| = \epsilon, \text{ and } x, y \in S(X)\} \\
&= \sup\{\inf\{\langle\frac{x-y}{2}, f_x\rangle : f_x \in \nabla_x\} : \|x-y\| = \epsilon, \text{ and } x, y \in S(X)\} \\
&= W_1(\epsilon)\ .
\end{aligned}$$

Theorem 2.20. *If X is an Asplund space, then $W(1) + W_1(1) = \frac{1}{2}$.*

Proof. Let $x \in K \subseteq S(X)$ and $y \in S(X)$ with $\|x-y\| = 1$, where K is the set of all smooth points of $S(X)$, and let $f_x \in \nabla_x$. Then

$$\langle\frac{x-y}{2}, f_x\rangle + \langle\frac{x-(x-y)}{2}, f_x\rangle = \langle\frac{x}{2}, f_x\rangle = \frac{1}{2},$$

so

$$\begin{aligned}
&\inf\{\langle\frac{x-y}{2}, f_x\rangle : f_x \in \nabla_x, \|x-y\| = 1, \text{ and } x \in K, y \in S(X)\} \\
&+\sup\{\langle\frac{x-y}{2}, f_x\rangle : f_x \in \nabla_x, \|x-y\| = 1, \text{ and } x \in K, y \in S(X)\} = \frac{1}{2}.
\end{aligned}$$

Therefore $W(1) + W_1(1) = \frac{1}{2}$.

Remark. If X is an Hilbert space, then $W(1) = W_1(1) = \frac{1}{4}$. It is also a special case of Proposition 2.23 below.

Theorem 2.21. [4]. *Let X be a Banach space. If $W_1(\epsilon) < \frac{\epsilon}{2}$, for any $0 \le \epsilon \le 2$ then X is uniformly nonsquare, and has uniform normal structure.*

We shall make use of the parameter $W_1(\epsilon)$ to obtain bounds of the pseudo-metric D for uniformly nonsquare and normal structure. To avoid confusion in the proof of next theorem, Theorem 2.22, we use $W_{1,X}(\epsilon)$ and $W_{1,Y}(\epsilon)$ to denote the parameter $W_1(\epsilon)$ of X and Y respectively.

Theorem 2.22. *If $\Delta(X, Y)$ satisfies the condition*

$$W_{1,X}((2+\delta)e^{\Delta(X,Y)} - 2) < \frac{1}{2}(\delta - 1)e^{\Delta(X,Y)} + \frac{1}{2},$$

for any $0 < \delta < 2$ *such that*

$$0 < (2+\delta)e^{\Delta(X,Y)} - 2 < 2,$$

then both X *with a continuous* $W_{1,X}(\epsilon)$ *and* Y *are uniformly nonsquare, and have uniform normal structure.*

Proof. Since $e^{\Delta(X,Y)} \geq 1$, $W_{1,X}((2+\delta)e^{\Delta(X,Y)} - 2) < \frac{1}{2}(\delta-1)e^{\Delta(X,Y)} + \frac{1}{2}$ implies $W_{1,X}((2+\delta)e^{\Delta(X,Y)} - 2) < \frac{(2+\delta)e^{\Delta(X,Y)}-2}{2}$. By Theorem 2.21, X is uniformly nonsquare, and has uniform normal structure.

We prove Y has uniform normal structure: If $\triangle(X,Y)$ satisfies $W_{1,X}((2+\delta)e^{\Delta(X,Y)} - 2) < \frac{1}{2}(\delta-1)e^{\Delta(X,Y)} + \frac{1}{2}$, we can take an isomorphism T such that

$$W_{1,X}((2+\delta)\|T\|\cdot\|T^{-1}\| - 2) < \frac{1}{2}(\delta-1)\|T\|\cdot\|T^{-1}\| + \frac{1}{2}$$

with

$$0 \leq (2+\delta)\|T\|\cdot\|T^{-1}\| - 2 \leq 2\,.$$

Let $\epsilon = (2+\delta)\|T\|\cdot\|T^{-1}\| - 2$, then $0 < \frac{2+\epsilon}{\|T\|\cdot\|T^{-1}\|} - 2 = \delta < 2$.

For any $(\partial T)x, (\partial T)y \in S(Y)$ with $\|(\partial T)x - (\partial T)y\| \leq \frac{2+\epsilon}{\|T\|\cdot\|T^{-1}\|} - 2$. From Lemma 1.3, we have $\|x-y\| \leq \epsilon$, hence $\langle\frac{x-y}{2}, f_x\rangle \leq W_{1,X}(\epsilon)$ for some $f_x \in \nabla_x$.

Let $g_{(\partial T)x} = (\partial T^*)^{-1}(f_x) \in \nabla_{(\partial T)x}$, then

$$\begin{aligned}\langle\frac{(\partial T)x - (\partial T)y}{2}, g_{(\partial T)x}\rangle &= \frac{1}{2} - \langle\frac{(\partial T)y}{2}, g_{(\partial T)x}\rangle \leq \frac{1}{2} - \frac{\langle\frac{y}{2}, f_x\rangle}{\|T\|\cdot\|T^{-1}\|}\\ &= \frac{1}{2} + \frac{\langle\frac{x-y}{2}, f_x\rangle}{\|T\|\cdot\|T^{-1}\|} - \frac{1}{2\|T\|\cdot\|T^{-1}\|}\\ &\leq \frac{W_{1,X}(\epsilon)}{\|T\|\cdot\|T^{-1}\|} + \frac{1}{2}(1 - \frac{1}{\|T\|\cdot\|T^{-1}\|})\,.\end{aligned}$$

Therefore, by definition of $W_{1,Y}$,

$$W_{1,Y}(\frac{2+\epsilon}{\|T\|\cdot\|T^{-1}\|} - 2) \leq \frac{W_{1,X}(\epsilon)}{\|T\|\cdot\|T^{-1}\|} + \frac{1}{2}(1 - \frac{1}{\|T\|\cdot\|T^{-1}\|}).$$

We have

$$\begin{aligned}W_{1,Y}(\delta) &\leq \frac{W_{1,X}((2+\delta)\|T\|\cdot\|T^{-1}\| - 2)}{\|T\|\cdot\|T^{-1}\|} + \frac{1}{2}\cdot(1 - \frac{1}{\|T\|\cdot\|T^{-1}\|})\\ &< \frac{1}{2}(\delta-1) + \frac{1}{2}\frac{1}{\|T\|\cdot\|T^{-1}\|} + \frac{1}{2}\cdot(1 - \frac{1}{\|T\|\cdot\|T^{-1}\|}) = \frac{\delta}{2}\,.\end{aligned}$$

From [4], Y is uniformly nonsquare and has uniform normal structure.

Proposition 2.23. *For a Hilbert space* H, $W_{1,H}(\epsilon) = \frac{\epsilon^2}{4}$.

Proof. It is a direct result of the parallelogram law of norm in Hilbert spaces.

Theorem 2.24. *If* $\Delta(X,H) < \ln\alpha$, *where* H *is a Hilbert space and* α *is the larger root of quadratic equation* $(2+\delta)^2x^2 - 6(1+\delta)x + 2 = 0$ *for any* $0 \le \delta \le 2$, *then* X *is uniformly nonsquare and has uniform normal structure.*

Proof. From Theorem 2.22 and Proposition 2.23,

$$\frac{1}{4}((2+\delta)e^{\Delta(X,Y)} - 2)^2 < \frac{1}{2}(\delta-1)e^{\Delta(X,Y)} + \frac{1}{2} \quad \text{for any} \quad 0 \le \delta \le 2$$

implies X is uniformly nonsquare and has uniform normal structure. The conclusion is derived by solving the above quadratic equation for $e^{\Delta(X,H)}$.

References

[1] M.S. Brodskii and D.P. Milman, On the Center of a Convex Set, Dokl. Akad. Nauk. SSSR(N.S.) 59(1948), 837–840.

[2] J. Diestel, The Geometry of Banach Spaces – Selected Topics. Lecture Notes in Math., Vol. 485, Springer, Berlin and New York (1975).

[3] J. Diestel and J.J. Uhl, Jr., Vector Measures, Math. Surveys, No. 15, Amer. Math. Soc., Providence (1977).

[4] J. Gao, The Parameters $W(\epsilon)$ and Normal Structure Under Norm and Weak Topologies In Banach Spaces, Nonlinear Anal. Vol. 47(8), August 2001, 5709–5722.

[5] J. Gao and K.S. Lau, On The Geometry of Spheres in Normed Linear Spaces, J. Austral. Math. Soc. Ser. A, 48, No.1, 101–112 (1990).

[6] J. Gao and K.S. Lau, On Two Classes of Banach Spaces with Normal Structure, Studia Math., 99, No.1, 41–56 (1991).

[7] R.C. James, Uniformly Nonsquare Banach Spaces, Ann. Math. (2) 80, 542–550 (1964).

[8] J.J. Schäffer, Geometry of Spheres in Normed Spaces, Lecture Notes in Pure and Appl. Math., Vol. 20, Marcel Dekker, New York (1976).

[9] C. Stegall, The Radon-Nikodym property in Conjugate Banach Spaces, II, Trans. Amer. Math. Soc. 264(1981), 507–519.

Ji Gao
Department of Mathematics
Community College of Philadelphia
Philadelphia, PA 19130-3991, USA
E-mail address: jgao@ccp.cc.pa.us

D. Haroske, T. Runst, H.-J. Schmeisser (eds.): Function Spaces, Differential Operators and Nonlinear Analysis. The Hans Triebel Anniversary Volume.

Some Inequalities for Integral Operators, Associated with the Bessel Differential Operator

Vagif S. Guliev

Dedicated to Prof. Hans Triebel on the occasion of his 65th birthday

Abstract. In this paper we consider maximal functions, fractional maximal functions and fractional integrals which are generated by a generalized shift operator, associated with the Bessel differential operator $B = (B_1, \dots, B_n)$, $B_i = \frac{\partial^2}{\partial x_i^2} + \frac{\gamma_i}{x_i}\frac{\partial}{\partial x_i}$, $i = 1, \dots, n$. We present inequalities for these operators in corresponding weighted L_p-spaces. In a special case we have found necessary and sufficient conditions for pairs of weights ensuring the validity of strong type inequalities for fractional integrals.

Introduction

Suppose that R^n is the n-dimensional Euclidean space, $x = (x_1, \dots, x_n)$, $\xi = (\xi_1, \dots, \xi_n)$ are vectors in R^n, $x \cdot \xi = x_1\xi_1 + \dots + x_n\xi_n$, $|x| = (x \cdot x)^{1/2}$, $R_+^n = \{x = (x_1, \dots x_n); x_1 > 0, \dots, x_n > 0\}$, $E_+(x, r) = \{y \in R_+^n : |x - y| < r\}$.

The Bessel differential operator $B = (B_1, \dots, B_n)$ is defined by

$$B_i = \frac{\partial^2}{\partial x_i^2} + \frac{\gamma_i}{x_i}\frac{\partial}{\partial x_i}, \quad i = 1, \dots, n,$$

where $\gamma = (\gamma_1, \dots, \gamma_n)$, $\gamma_1 > 0$, …, $\gamma_n > 0$. We put $|\gamma| = \gamma_1 + \dots + \gamma_n$, $x^\gamma = x_1^{\gamma_1} \dots x_1^{\gamma_n}$, and

$$|E_+(0, r)|_\gamma = \int_{E_+(0,r)} x^\gamma dx = Cr^{n+|\gamma|}.$$

For $1 \le p \le \infty$ let $L_{p,\gamma}(R_+^n) \equiv L_p(R_+^n, x^\gamma dx)$ be the space of functions measurable on R_+^n with the finite norm

$$\|f\|_{L_{p,\gamma}(R_+^n)} = \left(\int_{R_+^n} |f(x)|^p x^\gamma dx \right)^{1/p},$$

$$\|f\|_{L_{\infty,\gamma}(R_+^n)} \equiv \|f\|_{L_\infty(R_+^n)} = \operatorname*{ess\,sup}_{x \in R_+^n} |f(x)|.$$

Denote by T^y the generalized shift operator (B-shift operator) defined by

$$T^y f(x) = \pi^{-\frac{n}{2}} \prod_{i=1}^{n} \Gamma\left(\gamma_i + \frac{1}{2}\right) \Gamma^{-1}(\gamma_i) \int_0^{\pi} \cdots \int_0^{\pi} \prod_{i=1}^{n} \sin^{\gamma_i - 1} \alpha_i$$

$$\times\ f\left(\sqrt{x_1^2 - 2x_1 y_1 \cos\alpha_1 + y_1^2}, \ldots, \sqrt{x_n^2 - 2x_n y_n \cos\alpha_n + y_n^2}\right) d\alpha_1 \ldots d\alpha_n.$$

Let f be in $L_{p,\gamma}(R_+^n)$, $1 \le p \le \infty$. Then for all $y \in R_+^n$, the function $T^y f$ belongs to $L_{p,\gamma}(R_+^n)$, and

$$\|T^y f\|_{L_{p,\gamma}(R_+^n)} \le \|f\|_{L_{p,\gamma}(R_+^n)}. \tag{0.1}$$

Note that T^y is closely connected with the Bessel differential operator $B = (B_1, \ldots, B_n)$ (see [1] for details in the one-dimensional case). The B-shift T^y generates the corresponding B-convolution

$$(f * g)_B(x) = \int_{R_+^n} T^y f(x) g(y) y^\gamma \, dy.$$

The following properties of the B-convolution can be shown:

$$(f * g)_B = (g * f)_B,$$

$$\|(f * g)_B\|_{L_{p,\gamma}(R_+^n)} \le \|f\|_{L_{1,\gamma}(R_+^n)} \|g\|_{L_{p,\gamma}(R_+^n)}.$$

The Bessel (Hankel) transformation can be defined by

$$\overset{\wedge}{\varphi}(\lambda) = (F_B \varphi)(\lambda) = \int_{R_+^n} \varphi(x) \prod_{i=1}^{n} j_{\frac{\gamma_i - 1}{2}}(x_i \lambda_i) x^\gamma \, dx,$$

and its inverse transformation can be given by

$$\overset{\vee}{\varphi}(x) = \left(F_B^{-1} \varphi\right)(x) = C_{n,\gamma} \int_{R_+^n} \varphi(\lambda) \prod_{i=1}^{n} j_{\frac{\gamma_i - 1}{2}}(x_i \lambda_i) \lambda^\gamma \, d\lambda,$$

where $j_\nu(t) = t^{-\nu} J_\nu(t)$, J_ν being the Bessel function of the first kind.

For $f \in L_{p,\gamma}(R_+^n)$, $p = 1$ or 2, we have $F_B(f * g)_B = F_B f F_B g$.

We shall define the Morrey and BMO function spaces, generated by the Bessel differential operator $B = (B_1, \ldots, B_n)$.

Definition 0.1. [1] *Let* $1 \le p < \infty$, $\quad 0 \le \lambda \le n + |\gamma|$, $[t]_1 = \min\{1, t\}$. *We denote by* $L_{p,\gamma,\lambda}(R_+^n)$ *the Morrey-Bessel spaces (B-Morrey spaces) and by* $\widetilde{L}_{p,\gamma,\lambda}(R_+^n)$ *the modified B-Morrey spaces which are the sets of functions* f *being locally integrable* *on* R_+^n, *with finite norms*

$$\|f\|_{L_{p,\gamma,\lambda}(R_+^n)} = \sup_{R_+^n\times(0,\infty)} \left(t^{-\lambda}\int_{E_+(0,t)} T^y|f(x)|^p y^\gamma\, dy\right)^{1/p},$$

$$\|f\|_{\widetilde{L}_{p,\gamma,\lambda}(R_+^n)} = \sup_{R_+^n\times(0,\infty)} \left([t]_1^{-\lambda}\int_{E_+(0,t)} T^y|f(x)|^p y^\gamma\, dy\right)^{1/p}.$$

Definition 0.2. *Let us now introduce, as in* [1], *the BMO-Bessel spaces (B-BMO space)* $BMO_\gamma(R_+^n)$ *as the set of functions being locally integrable on* R_+^n, *with finite norm*

$$\|f\|_{*,\gamma} = \sup_{x,r} |E_+(0,r)|_\gamma^{-1}\int_{E_+(0,r)} |T^y f(x) - f_{E_+(0,r)}(x)| y^\gamma\, dy,$$

where

$$f_{E_+(0,r)}(x) = |E_+(0,r)|_\gamma^{-1}\int_{E_+(0,r)} T^y f(x) y^\gamma\, dy.$$

Note that

$$\widetilde{L}_{p,\gamma,0}(R_+^n) = L_{p,\gamma,0}(R_+^n) = L_{p,\gamma}(R_+^n),$$
$$L_{p,\gamma,n+|\gamma|}(R_+^n) = L_\infty(R_+^n),$$

$$\widetilde{L}_{p,\gamma,\lambda}(R_+^n) \subset_{\succ} L_{p,\gamma}(R_+^n) \quad \text{and} \quad \|f\|_{L_{p,\gamma}(R_+^n)} \le \|f\|_{\widetilde{L}_{p,\gamma,\lambda}(R_+^n)}.$$

Moreover, we have the following embeddings.

Lemma 0.3. *Let* $1 \le p < \infty$, $0 \le \lambda \le n + |\gamma|$. *Then for* $\alpha p = n + |\gamma| - \lambda$

$$L_{p,\gamma,\lambda}(R_+^n) \subset L_{1,\gamma,n+|\gamma|-\alpha}(R_+^n) \quad \textit{and} \quad \|f\|_{L_{1,\gamma,n+|\gamma|-\alpha}(R_+^n)} \le C\,\|f\|_{L_{p,\gamma,\lambda}(R_+^n)}.$$

1. B-maximal functions and fractional B-maximal functions

Let us define B-maximal functions and fractional B-maximal functions (see [1], [2]) by:

$$M_B f(x) = \sup_{r>0} |E_+(0,r)|_\gamma^{-1}\int_{E_+(0,r)} T^y|f(x)| y^\gamma\, dy,$$

and

$$M_B^\beta f(x) = \sup_{r>0} |E_+(0,r)|_\gamma^{\frac{\beta}{n+|\gamma|}-1}\int_{E_+(0,r)} T^y|f(x)| y^\gamma\, dy, \qquad 0 < \beta < n + |\gamma|,$$

respectively. Note that, in the case $\beta = 0$, $M_B^0 f(x) = M_B f(x)$. The aim of this section is to present weak and strong type inequalities for these operators. For illustration we shall give a proof in the most simple case described in the following theorem. It makes clear that the desired inequalities can be reduced to inequalities for maximal functions define on spaces of homogeneous type.

Theorem 1.1.

(i) *If* $f \in L_{1,\gamma}(R_+^n)$, *then for every* $\alpha > 0$

$$|\{x : M_B f(x) > \alpha\}|_\gamma \leq \frac{C}{\alpha} \int_{R_+^n} |f(x)| x^\gamma dx,$$

where $C > 0$ *is independent of* f.

(ii) *If* $f \in L_{p,\gamma}(R_+^n)$, $1 < p \leq \infty$, *then* $M_B f \in L_{p,\gamma}(R_+^n)$ *and*

$$\|M_B f\|_{L_{p,\gamma}(R_+^n)} \leq C_p \|f\|_{L_{p,\gamma}(R_+^n)},$$

where $C_p > 0$ *is independent of* f.

Proof. We need to introduce a maximal function defined on a space of homogeneous type. By this we mean a topological space X equipped with a continuous pseudometric ρ and a positive measure μ satisfying the doubling condition

$$\mu(E(x, 2r)) \leq C\mu(E(x, r)), \tag{1.2}$$

where C is independent of x and $r > 0$. Here $E(x,r) = \{y \in X : \rho(x,y) < r\}$. Let (X, ρ, μ) be a space of homogeneous type. Define

$$M_\mu f(x) = \sup_{r>0} \mu(E(x,r))^{-1} \int_{E(x,r)} |f(y)| d\mu(y).$$

It is well known that the maximal operator M_μ is of weak type (1, 1) and is bounded on $L_p(X, d\mu)$ for $1 < p < \infty$ (see [3]). We shall use this result in the case in which $X = R_+^n$, $\rho(x,y) = |x - y|$, $d\mu(x) = x^\gamma dx$. It is clear that this measure satisfies the doubling condition (1.2).

Also

$$\mu E(x,r) = |E_+(x,r)|_\gamma \leq C r^{n+|\gamma|} \prod_{i=1}^n \max\{1, (x_i/r)^{\gamma_i}\}.$$

We shall show that

$$M_B f(x) \leq C M_\mu f(x).$$

By the definition of the B-shift operator it follows that $T^y \chi_{E_+(0,r)}$ is supported in $E_+(x,r)$ and there exists $C > 0$ such that for all $x \in R_+^n$, $r > 0$ and $y \in E_+(x,r)$

$$T^y \chi_{E_+(0,r)}(x) \leq C \prod_{i=1}^n \min\{1, (r/x_i)^{\gamma_i}\}.$$

Thus

$$\begin{aligned} M_B f(x) &\leq \sum_{k=0}^n M_{B,k} f(x) \\ &= \sum_{k=0}^n \sup_{\substack{r > x_{i_j},\, j=\overline{1,k} \\ r \leq x_{i_j},\, j=\overline{k+1,n} \\ i_j \neq i_p,\, j \neq p}} |E_+(0,r)|_\gamma^{-1} \int_{E_+(x,r)} |f(y)| T^y \chi_{E_+(0,r)}(x) y^\gamma dy, \end{aligned}$$

where

$$M_{B,0}f(x) = \sup_{r\le x_j,\, j=\overline{1,n}} |E_+(0,r)|_\gamma^{-1} \int_{E_+(x,r)} |f(y)| T^y \chi_{E_+(0,r)}(x) y^\gamma dy,$$

$$M_{B,n}f(x) = \sup_{r> x_j,\, j=\overline{1,n}} |E_+(0,r)|_\gamma^{-1} \int_{E_+(x,r)} |f(y)| T^y \chi_{E_+(0,r)}(x) y^\gamma dy.$$

Without loss of generality we assume that $i_j \equiv j$, $j = 1, \dots, n$. Then

$$M_{B,k}f(x) = \sup_{\substack{r> x_j,\, j=\overline{1,k} \\ r\le x_j,\, j=\overline{k+1,n}}} |E_+(0,r)|_\gamma^{-1} \int_{E_+(x,r)} |f(y)| T^y \chi_{E_+(0,r)}(x) y^\gamma dy.$$

Taking into account that $\mu E_+(x,r) \le C r^{n+|\gamma|}$, $|E_+(0,r)|_\gamma = r^{n+|\gamma|}$ and $T^y \chi_{E_+(0,r)}(x) \le 1$, we have in the case $k = 0$,

$$M_{B,0}f(x) \le \sup_{r\le x_j,\, j=\overline{1,n}} \frac{1}{\mu E_+(x,r)} \int_{E_+(x,r)} |f(y)| y^\gamma dy \le C M_\mu f(x).$$

In the case $1 \le k \le n$, we use the estimates

$$\mu E_+(x,r) \le C r^{n+|\gamma|} \prod_{i=1}^{n} \max\{1, (x_i/r)^{\gamma_i}\} = C r^{n+|\gamma|} \prod_{i=1}^{k} (x_i/r)^{\gamma_i},$$

$$T^y \chi_{E_+(0,r)}(x) \le C \prod_{i=1}^{n} \min\{1, (r/x_i)^{\gamma_i}\} = \prod_{i=1}^{k} (r/x_i)^{\gamma_i},$$

to find that

$$M_{B,k}f(x) \le C \sup_{\substack{r> x_j,\, j=\overline{1,k} \\ r\le x_j,\, j=\overline{k+1,n}}} |E_+(0,r)|_\gamma^{-1} r^{n+|\gamma|} \prod_{i=1}^{k} (x_i/r)^{\gamma_i} \prod_{i=1}^{k} (r/x_i)^{\gamma_i}$$
$$\times \frac{1}{\mu E_+(x,r)} \int_{E_+(x,r)} |f(y)| y^\gamma dy \le C M_\mu f(x).$$

Finally we get

$$M_B f(x) \le C M_\mu f(x).$$

This completes the proof.

Corollary 1.2. *If $f \in L_{p,\gamma}(R_+^n)$, $1 \le p \le \infty$, then*

$$\lim_{r\to 0} |E_+(0,r)|_\gamma^{-1} \int\limits_{E_+(0,r)} T^y f(x) y^\gamma dy = f(x)$$

for a.e. $x \in R_+^n$.

Remark 1.3. *In the one-dimensional case Theorem 1.1 was proved earlier by K. Stempak* [4].

Theorem 1.1 can be extended to the case of fractional B-maximal functions in the following way.

Theorem 1.4. *Let $0 \le \beta < n+|\gamma|$, $\frac{1}{p} - \frac{1}{q} = \frac{\beta}{n+|\gamma|}$, $1 \le p \le q < \infty$.*

(i) *If $p = 1$, $f \in L_{1,\gamma}(R_+^n)$, then for all $\lambda > 0$*

$$\int\limits_{\{x\in R_+^n\,:\,M_B^\beta f(x)>\lambda\}} x^\gamma dx \le \left(\frac{C}{\lambda} \int\limits_{R_+^n} |f(x)| x^\gamma dx \right)^q,$$

where C does not depend on f.

(ii) *If $1 < p < \frac{n+|\gamma|}{\beta}$, $f \in L_{p,\gamma}(R_+^n)$, then $M_B^\beta f \in L_{q,\gamma}(R_+^n)$ and*

$$\left(\int_{R_+^n} \left(M_B^\beta f(x) \right)^q x^\gamma dx \right)^{1/q} \le C \left(\int_{R_+^n} |f(x)|^p x^\gamma dx \right)^{1/p},$$

where C-depends only on p, γ.

Next we shall give a survey on weighted inequalities for B- (fractional) maximal functions.

Let w be a positive measurable function on R_+^n. Denote by $L_{p,\gamma,w}(R_+^n)$ the set of measurable functions $f(x)$, $x \in R_+^n$, with finite norm

$$\|f\|_{L_{p,\gamma,w}(R_+^n)} = \left(\int_{R_+^n} |f(x)|^p w(x) x^\gamma dx \right)^{\frac{1}{p}}, \quad 1 \le p < \infty.$$

Definition 1.5. *The weight function w belongs to the class $A_{p,\gamma}(R_+^n)$ for $1 < p < \infty$, if*

$$\sup_{x,r\in R_+^n} |E_+(x,r)|_\gamma^{-1} \int\limits_{E_+(x,r)} w(y) y^\gamma dy \left(|E_+(x,r)|^{-1} \int\limits_{E_+(x,r)} w^{-\frac{1}{p-1}}(y) y^\gamma dy \right)^{p-1} < \infty$$

and w belongs to $A_{1,\gamma}(R_+^n)$, if there exists a positive constant C such that for any $x \in R_+^n$ and $r > 0$

$$|E_+(x,r)|^{-1} \int\limits_{E_+(x,r)} w^{-\frac{1}{p-1}}(y) y^\gamma dy \le C \operatorname*{ess\,sup}_{y\in E_+(x,r)} w(y).$$

The properties of the class $A_{p,\gamma}(R_+^n)$ are analogous to those of the B. Muckenhoupt classes. In particular, if $w \in A_{p,\gamma}(R_+^n)$, then $w \in A_{p-\varepsilon,\gamma}(R_+^n)$ for a certain sufficiently small $\varepsilon > 0$ and $w \in A_{p_1,\gamma}(R_+^n)$ for any $p_1 > p$.

Note that, $|x|^\alpha \in A_{p,\gamma}(R_+^n)$, $1 < p < \infty$, if and only if $-(n+|\gamma|) < \alpha < (n+|\gamma|)(p-1)$ and $|x|^\alpha \in A_{1,\gamma}(R_+^n)$, if and only if $-(n+|\gamma|) < \alpha \le 0$.

Theorem 1.6. *Let $1 < p < \infty$. Then the following two conditions are equivalent:*

(i) *There is a constant $C > 0$ such that for any $f \in L_{p,\gamma,w}(R_+^n)$ the inequality*

$$\left(\int_{R_+^n} (M_B f(x))^p\, w(x) x^\gamma dx\right)^{\frac{1}{p}} \le C \left(\int_{R_+^n} |f(x)|^p w(x) x^\gamma dx\right)^{\frac{1}{p}}$$

holds.

(ii) $w \in A_{p,\gamma}(R_+^n)$.

Theorem 1.7. *The following two conditions are equivalent:*

(i) $$\int\limits_{\{x\in R_+^n\,:\,M_B f(x)>\lambda\}} w(x) x^\gamma dx \le C\lambda^{-1} \left(\int\limits_{R_+^n} |f(x)| w(x) x^\gamma dx\right)$$

with a constant C independent of f and $\lambda > 0$.

(ii) $w \in A_{1,\gamma}(R_+^n)$.

Theorem 1.8. *Let $1 < p < \frac{n+|\gamma|}{\beta}$, $\frac{1}{q} = \frac{1}{p} - \frac{\beta}{n+|\gamma|}$. Then the following two conditions are equivalent:*

(i) *There is a constant $C > 0$ such that for any $f \in L_{p,\gamma,w}(R_+^n)$ the inequality*

$$\left(\int_{R_+^n} \left(M_B^\beta(f w^\beta)(x)\right)^q w(x) x^\gamma dx\right)^{\frac{1}{q}} \le C \left(\int_{R_+^n} |f(x)|^p w(x) x^\gamma dx\right)^{\frac{1}{p}}$$

holds.

(ii) $w \in A_{1+\frac{q}{p'},\gamma}(R_+^n)$, $p' = \frac{p}{p-1}$.

Theorem 1.9. *Let $q = \frac{n+|\gamma|}{n+|\gamma|-\beta}$. Then the following two conditions are equivalent:*

(i) $$\int\limits_{\{x\in R_+^n\,:\,M_B^\beta(f w^\beta)(x)>\lambda\}} w(x) x^\gamma dx \le C\lambda^{-q} \left(\int\limits_{R_+^n} |f(x)| w(x) x^\gamma dx\right)^q$$

with a constant C independent of f and $\lambda > 0$.

(ii) $w \in A_{1,\gamma}(R_+^n)$.

2. Sobolev type theorem for B-fractional integrals

Based on the results of the previous section we are able to establish inequalities for B-Riesz potentials

$$I_B^\alpha f(x) = \int_{R_+^n} T^y |x|^{\alpha-n-|\gamma|} f(y) y^\gamma dy, \quad 0 < \alpha < n + |\gamma|,$$

and modified B-Riesz potentials

$$\left[\widetilde{I}_B^\alpha f\right](x) = \int_{R_+^n} \left(T^y|x|^{\alpha-n-|\gamma|} - |y|^{\alpha-n-|\gamma|}\chi_{E_+^*(0,1)}(y)\right) f(y)y^\gamma dy,$$

where $E_+^*(0,r) = R_+^n \backslash E_+(0,r)$, $r > 0$.

The examples considered below show that if $p \geq \frac{n+|\gamma|}{\alpha}$, then the B-potentials I_B^α are not defined for all functions $f \in L_{p,\gamma}(R_+^n)$.

Example 1. Let $x \in R_+^n$, $0 < \alpha < n + |\gamma|$, $f(x) = \frac{1}{|x|^\alpha \ln |x|}\chi_{E_+^*(0,2)}(x)$. For $p = \frac{n+|\gamma|}{\alpha}$, $f \in L_{p,\gamma}(R_+^n)$ and $I_B^\alpha f(x) = +\infty$.

Example 2. Let $x \in R_+^n$, $0 < \alpha < n+|\gamma|$, $f(x) = |x|^{-\alpha}\chi_{E_+^*(0,2)}(x)$. For $p > \frac{n+|\gamma|}{\alpha}$, $f \in L_{p,\gamma}(R_+^n)$ and $I_B^\alpha f(x) = +\infty$.

For these B-Riesz potentials the following analogue of the Hardy-Littlewood-Sobolev theorem is valid.

Theorem 2.1. *Let* $0 < \alpha < n + |\gamma|$, $1 \leq p < \frac{n+|\gamma|}{\alpha}$, *and* $\frac{1}{p} - \frac{1}{q} = \frac{\alpha}{n+|\gamma|}$.

(i) *If* $f \in L_{p,\gamma}(R_+^n)$, *then the integral* $I_B^\alpha f$ *is absolutely convergent for almost every* $x \in R_+^n$.

(ii) *If* $1 < p < \frac{n+|\gamma|}{\alpha}$, $f \in L_{p,\gamma}(R_+^n)$, *then* $I_{B_n}^\alpha f \in L_{q,\gamma}(R_+^n)$ *and*

$$\|I_B^\alpha f\|_{L_{q,\gamma}(R_+^n)} \leq C_p \|f\|_{L_{p,\gamma}(R_+^n)}, \tag{2.3}$$

where $C_p > 0$ *is independent of* f.

(iii) *If* $f \in L_{1,\gamma}(R_+^n)$, $\frac{1}{q} = 1 - \frac{\alpha}{n+|\gamma|}$, *then*

$$\left|\{x \in R_+^n : \; I_B^\alpha f(x) > \beta\}\right|_\gamma \leq \left(\frac{C}{\beta}\|f\|_{L_{1,\gamma}(R_+^n)}\right)^q, \qquad \beta > 0,$$

where $C > 0$ *is independent of* f.

Proof. We restrict ourselves to the proof of b). We have

$$I_B^\alpha f(x) = \left(f * |y|^{\alpha-n-|\gamma|}\chi_{E_+(0,t)}\right)_B(x) + \left(f * |y|^{\alpha-n-|\gamma|}\chi_{E_+^*(0,t)}\right)_B(x).$$

It is easy to see that the first integral is bounded by $t^\alpha M_B f(x)$. By Hölder's inequality, the second integral can be estimated by

$$\|f\|_{L_{p,\gamma}(R_+^n)} \cdot \||y|^{\alpha-n-|\gamma|}\chi_{E_+^*(0,t)}\|_{L_{p',\gamma}(R_+^n)} = t^{-(n+|\gamma|)/q}\|f\|_{L_{p,\gamma}(R_+^n)}.$$

Summation of the two integrals leads to

$$|I_B^\alpha f(x)| \leq C\left(t^\alpha M_B f(x) + t^{-(n+|\gamma|)/q}\|f\|_{L_{p,\gamma}(R_+^n)}\right). \tag{2.4}$$

If we choose $t = \left[(M_B f(x))^{-1}\|f\|_{L_{p,\gamma}(R_+^n)}\right]^{p/(n+|\gamma|)}$, then (2.4) yields

$$|I_B^\alpha f(x)| \leq C\,(M_B f(x))^{p/q}\|f\|_{L_{p,\gamma}(R_+^n)}^{1-p/q}.$$

Hence, by the Theorem 1.1, we have (2.3).

In addition to (2.3) we can show the following necessary condition.

Theorem 2.2. *Let* $0 < \alpha < n + |\gamma|, \quad 1 < p < \frac{n+|\gamma|}{\alpha}$, *then the condition*

$$\frac{1}{p} - \frac{1}{q} = \frac{\alpha}{n + |\gamma|}$$

is necessary for inequality (2.3) to be valid.

Now, we deal with the modified B-Riesz potential in the limiting case $p = \frac{n+|\gamma|}{\alpha}$.

Theorem 2.3. *Let* $0 < \alpha < n + |\gamma|$, $p = \frac{n+|\gamma|}{\alpha}$, $f \in L_{p,\gamma}(R_+^n)$. *Then* $\widetilde{I}_B^\alpha f \in BMO_\gamma(R_+^n)$ *and*

$$\left\|\widetilde{I}_B^\alpha f\right\|_{BMO_\gamma(R_+^n)} \leq C_p \|f\|_{L_{p,\gamma}(R_+^n)}.$$

Proof. Let $f \in L_p(R_+^n, B)$. Given $t > 0$ we denote $f_1(x) = f(x)\chi_{E_+(0,2t)}(x)$, $f_2(x) = f(x) - f_1(x)$. Then $\widetilde{I}_B^\alpha f(x) = \widetilde{I}_B^\alpha f_1(x) + \widetilde{I}_B^\alpha f_2(x)$.

Denote

$$a_1 = - \int\limits_{E_+(0,2t)\setminus E_+(0,\min\{1,2t\})} |y|^{\alpha-n-|\gamma|} f(y) y^\gamma dy,$$

and

$$a_2 = \int\limits_{E_+(0,\max\{1,2t\})\setminus E_+(0,2t)} |y|^{\alpha-n-|\gamma|} f(y) y^\gamma dy.$$

Note that $\widetilde{I}_B^\alpha f_1(x) - a_1 = I_B^\alpha f_1(x)$. Therefore

$$\left|\widetilde{I}_B^\alpha f_1(x) - a_1\right| \leq \int\limits_{E_+(0,3t)} |y|^{\alpha-n-|\gamma|} \left|T^y f(x)\right| y^\gamma dy, \tag{2.5}$$

if $x \in E_+(0,t)$.

By Theorem 1.1, (0.1) and (2.5) for $\alpha p = n + |\gamma|$

$$|E_+(0,t)|_\gamma^{-1} \int\limits_{E_+(0,t)} \left|T^z \widetilde{I}_B^\alpha f_1(x) - a_1\right| z^\gamma dz \leq C_p \|T^x f\|_{L_{p,\gamma}(R_+^n)} \leq C_p \|f\|_{L_{p,\gamma}(R_+^n)}. \tag{2.6}$$

Applying Hölder's inequality we have $\left|\widetilde{I}_B^\alpha f_2(x) - a_2\right| \leq C|x|t^{-1}\|f\|_{L_{p,\gamma}(R_+^n)}$. Note that if $|x| \leq t$, $|z| \leq 2t$, then $T^z|x| \leq |x| + |z| \leq 3t$. Thus for $\alpha p = n + |\gamma|$ we obtain

$$\left|T^z \widetilde{I}_B^\alpha f_2(x) - a_2\right| \leq C T^z |x| t^{-1} \|f\|_{L_{p,\gamma}(R_+^n)} \leq C \|f\|_{L_{p,\gamma}(R_+^n)}. \tag{2.7}$$

Denote

$$a_f = a_1 + a_2 = \int\limits_{E_+(0,\max\{1,2t\})} |y|^{\alpha-n-|\gamma|} f(y) y^\gamma dy.$$

Finally, by (2.6) and (2.7) we have

$$\left\|\widetilde{I_B^\alpha} f\right\|_{BMO(R_+^n,B)} \le 2\sup_{x,t} |E_+(0,t)|_\gamma^{-1} \int\limits_{E_+(0,t)} \left|T^y \widetilde{I_B^\alpha} f(x) - a_f\right| y^\gamma dy \le C\|f\|_{L_{p,\gamma}(R_+^n)}.$$

and the statement of Theorem 2.3 follows.

Corollary 2.4. *Let* $p = \frac{n+|\gamma|}{\alpha}$, $f \in L_{p,\gamma}(R_+^n)$.

If the integral $I_B^\alpha f$ *exists everywhere, then* $I_B^\alpha f \in BMO_\gamma(R_+^n)$ *and the inequality*

$$\|I_B^\alpha f\|_{BMO_\gamma(R_+^n)} \le C_p \|f\|_{L_{p,\gamma}(R_+^n)}$$

is valid.

We conclude this section stating pointwise estimates for the function $I_B^\alpha f(x)$. Estimates of this type in the case of classical Riesz potentials were obtained in [5], [6].

Lemma 2.5. *Let* $0 < \alpha < n + |\gamma|$, $1 \le p < \frac{\lambda}{\alpha}$. *Then there exists a positive number* C *such that for every* $r > 0$ *and* $x \in R_+^n$

$$|I_B^\alpha f(x)| \le C\left(r^\alpha M_B f(x)) + r^{\alpha - \frac{\lambda}{p}} M_B^{\frac{\lambda}{p}} f(x)\right).$$

Theorem 2.6. *Let* $0 < \lambda \le 1$, $1 < p < \frac{\lambda}{\alpha}$, $1 \le r \le \infty$ *and* $\frac{1}{q} = \frac{1}{p} - \frac{\alpha}{\lambda} + \frac{\alpha p}{\lambda r}$. *Then for every function* $f \in L_{p,\gamma}(R_+^n)$ *such that* $M_B^{\frac{\lambda}{p}} f \in L_{r,\gamma}(R_+^n)$ *the following estimate holds:*

$$\|I_B^\alpha f\|_{L_{q,\gamma}(R_+^n)} \le C\|M_B^{\frac{\lambda}{p}} f\|_{L_{r,\gamma}(R_+^n)}^{\frac{\alpha p}{\lambda}} \|f\|_{L_{p,\gamma}(R_+^n)}^{1-\frac{\alpha p}{\lambda}}.$$

Lemma 2.7. *For any* ε, $0 < \varepsilon < \alpha$, *there exists a constant* $C_\varepsilon > 0$ *such that for any nonnegative function* $\phi : R_+^n \to R$ *and for any point* $x \in R_+^n$ *the following inequality holds:*

$$I_B^\alpha \phi(x) \le C_\varepsilon \sqrt{M_B^{\alpha-\varepsilon}\phi(x) M_B^{\alpha+\varepsilon}\phi(x)}.$$

3. Two-weight inequalities for B-fractional integrals

Finally, we present a collection of weighted inequalities for the B-fractional integrals defined in the previous section. For detailed proofs we refer to the papers [7]–[14].

Theorem 3.1. *Suppose that* $1 < p < \frac{n+|\gamma|}{\alpha}$, $\frac{1}{q} = \frac{1}{p} - \frac{\alpha}{n+|\gamma|}$. *Then the inequality*

$$\left(\int_{R_+^n} |I_B^\alpha(f w^\alpha)(x)|^q\, w(x) x^\gamma dx\right)^{\frac{1}{q}} \le C\left(\int_{R_+^n} |f(x)|^p w(x) x^\gamma dx\right)^{\frac{1}{p}}$$

holds for any $L_{p,\gamma,w}(R_+^n)$ *with a constant* $C > 0$ *independent of* f *if and only if*

$$w \in A_{\beta,\gamma}(R_+^n), \qquad \beta = 1 + \frac{q}{p'}.$$

Theorem 3.2. *Let $0 < \alpha < n + |\gamma|$, $1 < p < \frac{n+|\gamma|}{\alpha}$, $\frac{1}{p} - \frac{1}{q} = \frac{\alpha}{n+\gamma}$ and let σ and u be positive, increasing functions on $(0,\infty)$. Let $v \in A_{1+\frac{q}{p'},\gamma}(R^n_+)$ be a radial function, $w_1 = \sigma v$ and $w = uv$. If for (w_1, w) the condition*

$$\sup_{t>0}\left(\int_t^\infty w_1(\tau)\tau^{-1-\frac{(n+|\gamma|)q}{p'}}d\tau\right)^{\frac{p}{q}}\left(\int_0^{t/2}\left(v^{-\frac{\alpha p}{n+|\gamma|}}(\tau)w(\tau)\right)^{1-p'}\tau^{n+|\gamma|-1}d\tau\right)^{p-1} < \infty. \tag{3.8}$$

is fulfilled, then there exits a constant $c > 0$ such that for all $f \in L_{p,\gamma,w}(R^n_+)$ the inequality

$$\left(\int_{R^n_+}\left|I^\alpha_B(f\cdot v^{\frac{\alpha}{n+|\gamma|}})(x)\right|^q w_1(|x|)x^\gamma dx\right)^{\frac{1}{q}} \le c\left(\int_{R^n_+}|f(x)|^p w(|x|)x^\gamma dx\right)^{\frac{1}{p}} \tag{3.9}$$

holds.

Theorem 3.3. *Let $0 < \alpha < n + |\gamma|$, $1 < p < \frac{n+|\gamma|}{\alpha}$, $\frac{1}{p} - \frac{1}{q} = \frac{\alpha}{n+|\gamma|}$. Let u and σ be positive, decreasing functions on $(0,\infty)$. Let $v \in A_{1+\frac{q}{p'},\gamma}(R^n_+)$ be a radial function. Assume further that $w = uv$ and $w_1 = \sigma v$. If the condition*

$$\sup_{t>0}\left(\int_0^{t/2} w_1(\tau)\tau^{n+|\gamma|-1}d\tau\right)^{\frac{p}{q}}\left(\int_t^\infty\left(v^{-\frac{\alpha p}{n+|\gamma|}}(\tau)w(\tau)\right)^{1-p'}\tau^{-1-\frac{(n+|\gamma|)p'}{q}}d\tau\right)^{p-1} < \infty, \tag{3.10}$$

is fulfilled, then inequality (3.9) holds.

Theorem 3.4. *Let $0 < \alpha < n + |\gamma|$, $1 < p < \frac{n+|\gamma|}{\alpha}$, $\frac{1}{p} - \frac{1}{q} = \frac{\alpha}{n+|\gamma|}$. If the weight functions v, w, w_1 satisfy inequality (3.9), then conditions (3.8) and (3.10) are satisfied as well.*

Theorem 3.5. *Let $0 < \alpha < n + |\gamma|$, $1 < p < \frac{n+|\gamma|}{\alpha}$, $\frac{1}{p} - \frac{1}{q} = \frac{\alpha}{n+|\gamma|}$, let w and w_1 be positive functions on $(0,\infty)$ satisfying the following conditions:*

(i) *there exists a constant $c_1 > 0$ such that for an arbitrary $t > 0$ the inequality*

$$\left(\sup_{t<\tau<8t} w_1(\tau)\right)^{\frac{p}{q}} \le c_1 \inf_{t<\tau<8t} w(\tau)$$

holds.

(ii) $$\sup_{t>0}\left(\int_t^\infty w_1(\tau)\tau^{-1-\frac{(n+|\gamma|)q}{p'}}d\tau\right)^{\frac{p}{q}}\left(\int_0^t w^{1-p'}(\tau)\tau^{n+|\gamma|-1}d\tau\right)^{p-1} < \infty$$

(iii) $$\sup_{t>0}\left(\int_0^t w_1(\tau)\tau^{n+|\gamma|-1}d\tau\right)^{\frac{p}{q}}\left(\int_t^\infty w^{1-p'}(\tau)\tau^{-1-\frac{(n+|\gamma|)p'}{q}}d\tau\right)^{p-1} < \infty.$$

Then the inequality

$$\left(\int_{R^n_+} |I^\alpha_B(f(x)|^q\, w_1(|x|)x^\gamma dx\right)^{\frac{1}{q}} \le c\left(\int_{R^n_+} |f(x)|^p w(|x|)x^\gamma dx\right)^{\frac{1}{p}}$$

is valid.

Acknowledgement. The author would like to thank Prof. H.-J. Schmeißer who has read the manuscript and made a number of valuable corrections and comments.

References

[1] V.S. Guliev, *Sobolev theorems for B-Riesz potentials.* Dokl. RAN, **358 (4)**, 1998, p. 450–451.

[2] V.S. Guliev, *Sobolev theorems for anisotropic Riesz-Bessel potentials on Morrey-Bessel spaces.* Dokl. RAN, **367 (2)**, 1999, p. 155–156.

[3] R.R. Coifman, G. Weiss, *Analyse harmonique non commutative sur certains espaces homogènes.* Lecture Notes in Math., v.242, Springer-Verlag. Berlin, 1971.

[4] K. Stempak, *Almost everywhere summability of Laguerre series*, Studia Math. **100 (2)**(1991), p. 129–147.

[5] L. Hedberg, *On certain convolution inequalities.* Proc. Amer. Math. Soc. **36** (1972), p. 505–510.

[6] D. Adams, *A note on Riesz potentials.* Duke Math. J. 42, 4 (1975), p. 765–777.

[7] E.G. Gusseinov, *Singular integrals in the spaces of functions summable with monotone weight (Russian).* Mat. Sb. 132(174) (1987), 1, p. 28–44.

[8] S.K. Abdullaev, *On some classes of integral operators and weight embedding theorems (Russian).* Dokl. Akad. Nauk SSSR 304 (1989), 6, p. 1289–1293.

[9] I.A. Aliev, A.D. Gadjiev, *Weighted estimates of multidimensional singular integrals generated by the generalized shift operator (English, translated into Russian).* Acad. Sci. Sb. Math. 1994, Vol. 77, No. 1, p. 37–55.

[10] V.S. Guliev, *Two-weighted L_p inequalities for singular integral operators on Heisenberg groups.* Georgian Math. J.1(1994), 4, p. 367–376.

[11] D.E. Edmunds, V.M. Kokilashvili, *Two-weighted inequalities for singular integrals.* Canad. Math. Bull. 38(1995), 3, p. 295–303.

[12] V.M. Kokilashvili, A. Meskhi, *Two-weighted inequalities for singular integrals defined on homogeneous groups.* Proc. Razmadze Mathematical Institute, 1997, Vol. 112, p. 57–90.

[13] A. Meskhi, *Two-weighted inequalities for potentials defined on homogeneous groups.* Proc. Razmadze Mathematical Institute, 1997, Vol. 112, p. 91–111.

[14] V.S. Guliev, *Function spaces, integral operators and two-weighted inequalities on homogeneous groups. Some applications.* Baku. 1996. p. 1–332.

V.S. Guliev
Department of Mathematical Analysis
Institute of Mathematics and Mechanics
Azerbaijan Academy of Sciences, F. Agayev st. 10
370148 Baku, Azerbaijan
E-mail address: guliev@azdata.net

D. Haroske, T. Runst, H.-J. Schmeisser (eds.): Function Spaces, Differential Operators and Nonlinear Analysis. The Hans Triebel Anniversary Volume.

On Determining Individual Behaviour from Population Data

Mats Gyllenberg[1], Andrei Osipov, and Lassi Päivärinta[1]

Dedicated to Prof. Hans Triebel on the occasion of his 65th birthday

Keywords: structured population dynamics, renewal equation, inverse problems.
AMS Subject Classification (2000): Primary: 35R30, Secondary: 92D25.

1. Introduction

Mathematical models of *physiologically structured populations* (Metz and Diekmann 1986; Diekmann *et al.* 1998, 2001) relates mechanisms at the individual level and behaviour at the level of the population. In a typical *direct problem* one prescribes model ingredients that describe mechanisms such at aging, growth and survival at the individual level, lifts the model to the population level and finally studies phenomena at the population level. In the *inverse problem* the situation is reversed. Using knowledge about behaviour at the population level one wants to deduce the underlying mechanisms at the individual level.

The direct problem of structured populations has been extensively studied for many kinds of models using a variety of mathematical techniques (e.g. pde theory, semigroup theory, renewal theory, branching processes). We mention only the book by Webb (1985) on age-structured populations, the book by Metz and Diekmann (1986) on general physiologically structured populations and the book on branching processes by Jagers (1975). On the other hand, results on the inverse problem seem to be rare (van Straalen 1986).

In a series of papers Rundell and coworkers (Rundell 1989, 1993; Pilant and Rundell 1991b; Engl *et al.* 1994) have treated certain inverse problems of age-structured population dynamics. In these papers it is assumed that census data in the form of the age distribution of the population are available. However, for many real populations (e.g. bacterial populations) it is difficult if not impossible to measure the age distribution, whereas other quantities at the population level such as the total population size are easily obtained. Pilant and Rundell (1991a) considered in a pde setting the question of determining the initial age-distribution from data on the total population size, when the birth and death rates are known.

[1]The research of Mats Gyllenberg and Lassi Päivärinta has been supported by the Academy of Finland.

Berndtsson and Jagers (1979) considered the same question in a branching processes framework.

In this paper we discuss the inverse problem of the simplest structured population model: the linear age-structured model (Sharpe and Lotka 1911, McKendrick 1926). We shall, in fact, use the cumulative formulation (Diekmann *et al.* 1993b, 1998; Gyllenberg *et al.* 1997) of the problem. This not only increases the generality but facilitates the analysis considerably. The specific problem we consider is: Under what conditions does knowledge of the total population size and the cumulative number of births uniquely determine the survival and reproduction functions describing individual behaviour? A typical example we have in mind is a population of cells reproducing by fission. Here the inverse problem is indeed relevant, because the age of a cell cannot usually be measured, whereas the total number of cells is easily observed as is the fraction of dividing cells.

2. The direct problem

The classical linear age-structured population model is usually formulated as a hyperbolic pde supplemented by a nonlocal boundary condition (McKendrick 1926):

$$\frac{\partial}{\partial t}n(t,a)+\frac{\partial}{\partial a}n(t,a)=-\mu(a)n(t,a), \tag{2.1}$$

$$n(t,0)=\int_0^\infty \beta(a)n(t,a)da, \tag{2.2}$$

$$n(0,a)=n_0(a). \tag{2.3}$$

Here the solution $n(t,\cdot)$ is the age-density of the population, $\mu(a)$ is the age-specific per capita death rate and $\beta(a)$ is the age-specific per capita fecundity.

Integrating the McKendrick equation (2.1) along characteristics one obtains

$$n(t,a)=\begin{cases} b(t-a)e^{-\int_0^a \mu(\alpha)d\alpha}, & t>a,\\ n_0(a-t)e^{-\int_{a-t}^a \mu(\alpha)d\alpha}, & t<a, \end{cases} \tag{2.4}$$

where

$$b(t):=n(t,0)$$

is the birth rate. Substituting (2.4) into the birth law (2.2) one obtains the renewal equation

$$b(t)=\int_0^t b(t-a)\beta(a)e^{-\int_0^a \mu(\alpha)d\alpha}da+\int_t^\infty n_0(a-t)\beta(a)e^{-\int_{a-t}^a \mu(\alpha)d\alpha}da \tag{2.5}$$

for the birth rate $b(t)$. The direct problem thus reduces to solving the renewal equation (2.5) because once this has been done, (2.4) becomes an explicit formula for the solution $n(t,a)$.

Rates and densities are mathematical abstractions involving limits (derivatives) and they can never be measured. In structured population dynamics it is not even helpful to formulate models in terms of partial differential equations (Diekmann *et al.* 1993ab, 1995, 1998, 2001; Gyllenberg *et al.* 1997). We shall therefore proceed along a slightly different line, the advantages of which will be clear in a while.

From μ and β we obtain two new functions F and L by defining

$$F(a) = e^{-\int_0^a \mu(\alpha)d\alpha} \tag{2.6}$$

and

$$L(a) = \int_0^a \beta(\alpha)e^{-\int_0^\alpha \mu(\tau)d\tau}d\alpha. \tag{2.7}$$

F is called the *survival function* and L the *reproduction function*. They have the following interpretations:

$F(a)$ is the probability that an individual is still alive at age a.
$L(a)$ is the expected number of offspring born to an individual before reaching (dead or alive) age a.

Finally we introduce the cumulative number

$$B(t) := \int_{[0,t)} b(\tau)d\tau$$

of births up to time and replace the initial age density n_0 by a a measure m_0. If m_0 is absolutely continuous with respect to the Lebesgue measure, then of course

$$m_0([0,a)) = \int_{[0,a)} n_0(\alpha)d\alpha$$

for some L^1-function n_0.

We now forget about the rates μ and β and take F and L as the basic ingredients of the population model. We take the measure m_0 as the initial population state. This increases the generality, because we can now allow discontinuities in F and L and handle initial populations described by not necessarily absolutely continuous measures. It also considerably facilitates the analysis.

The integrated versions of (2.5) and (2.4) read as

$$B(t) = \int_{[0,t)} L(t-\tau)B(d\tau) + \int_{\mathbf{R}_+} \frac{L(a+t)-L(a)}{F(a)} m_0(da) \tag{2.8}$$

and

$$m(t,[0,a)) = \int_{[0,\min\{t,a\})} F(t-\tau)B(d\tau) + \int_{[0,\max\{0,a-t\})} \frac{F(\tau+t)}{F(\tau)} m_0(d\tau), \tag{2.9}$$

respectively. The direct problem can now be formulated as follows:

(DP) Given the initial population state m_0, the survival function F and the reproduction function L, solve the integral equation (2.8) for B. The population state $m(t,\cdot)$ at time t is then given by the explicit formula (2.9).

The interpretation of the model ingredients requires that m_0, F and L satisfy certain conditions that we now formulate.

Assumption 2.1. The initial population state m_0 is a finite positive measure defined on $\mathbf{R}_+$ and the functions F and L are defined on $(0,\infty)$ and have the following properties:

(i) F is nonnegative and nonincreasing.
(ii) $\lim_{a\downarrow 0} F(a) = 1$.
(iii) $F(\infty) = \lim_{a\to\infty} F(a) = 0$.
(iv) L is nonnegative, nondecreasing, and nonlattice.
(v) $\lim_{a\downarrow 0} L(a) = 0$.
(vi) $R_0 = L(\infty) = \lim_{a\to\infty} L(a) < \infty$.
(vii) $F(a) = 0 \Rightarrow L(a) = L(\infty)$,
(viii) $\frac{1}{F} \in L^1(m_0)$.

Most of the above conditions are clear from biological point of view. However, we make the following comments: The assumption in (iv) that L is nonlattice means that L is not a step-function with discontinuities in a subset of an additive subgroup of $\mathbf{R}$. We thus rule out the possibility that individuals reproduce only upon exactly reaching a prescribed age a_0 (and possibly upon reaching $2a_0, 3a_0, \ldots$).

If $F(A) = 0$ for some finite A, then no-one can survive beyond A and the initial measure m_0 must be concentrated on $[0, A]$. We impose the slightly stronger condition (viii), which is needed to make the latter integral in (2.9) finite. Moreover, because dead individuals do not give birth, $L(a)$ must be equal to $L(\infty)$ for ages a larger than A. This is formulated in (vii).

The number R_0 is called the *basic reproduction ratio* and it gives the expected life-time production of offspring of an individual.

The direct problem has been studied by a number of authors ever since the appearance of the paper by Sharpe and Lotka (1911). Feller (1941, 1971) finally settled all remaining open problems. Next we give a short account of those results that are important for our investigation of the inverse problem.

Assumption 2.1 ensures that the renewal equation (2.8) has a unique solution. Denote the latter term on the right-hand side of (2.8) by $H(t)$. Then the Laplace transform of b is given by

$$\widehat{b}(s) = \frac{s\widehat{H}(s)}{1 - s\widehat{L}(s)}. \tag{2.10}$$

Assume that the maximum life-time is finite, say 1. Then the support of F, $\operatorname{supp} F = [0,1]$ and $L(a) = L(1)$ for all $a \geq 1$ by Assumption 2.1. It follows that $s\widehat{H}(s)$ and $s\widehat{L}(s)$ are entire functions and hence that $\widehat{b}$ is meromorphic in the whole plane. The zeros of $1 - s\widehat{L}(s)$ (the poles of $\widehat{b}$) are the roots of the *Euler-Lotka equation*

$$\int_{\mathbf{R}_+} e^{-\lambda a} L(da) = 1. \tag{2.11}$$

An easy application of Hadamard's factorization theorem (Titchmarsh 1939, p. 250) shows that the Euler-Lotka equation has infinitely many complex roots λ_k (for details, see Gyllenberg 1985). If $\widehat{b}(s)$ admits an expansion

$$\widehat{b}(s) = \sum \frac{b_k}{s - \lambda_k} \tag{2.12}$$

with $\sum |b_k| < \infty$, then, as proven by Feller (1941), the solution of (2.8) (or rather its derivative b) is representable as the series

$$b(t) = \sum b_k e^{\lambda_k t}, \tag{2.13}$$

where the series converges absolutely for all $t \geq 0$. The coefficients b_k are complex. Because b is positive the characteristic roots λ_k appear as pairs of complex conjugates. We have assumed that all roots are simple. The result above is easily generalized to the case of multiple roots (Feller 1941). Note, however, that due to positivity, the unique real root $\lambda_0 = r$ is necessarily simple and has real part larger than the real parts of all other roots. The root r is called the *Malthusian parameter* and it is positive if and only if $R_0 > 1$. For details we refer to Feller (1941, 1971).

3. The inverse problem

In many practical situations the population state, that is, the age distribution in our case, cannot be directly observed. When this is the case, it is usually also impossible to experimentally measure the model ingredients L and F. What can be observed is only certain linear functionals of the population state, called *population outputs.* The inverse problem consists of determining F and L (or μ and β) in terms of the outputs.

In this paper we shall be concerned only with two outputs, namely the *total population size*

$$N(t) = \int_{\mathbf{R}_+} m(t, da) \tag{3.1}$$

and the *population birth rate*

$$b(t) = \int_{\mathbf{R}_+} \beta(a) m(t, da). \tag{3.2}$$

Of course, a rate cannot be directly measured; the measured quantity is the *cumulative number* $B(t) = \int_{[0,t)} b(\tau) d\tau$ of births up to time t. The use of B instead of b also has the advantage that B makes sense in cases where the model is formulated in terms of the reproduction function L and not in terms of the per capita birth rate β.

It follows from (2.9) that the total population size $N(t)$ satisfies the following equation:

$$N(t) = \int_{[0,t)} F(t - \tau) B(d\tau) + \int_{\mathbf{R}_+} \frac{F(a + t)}{F(a)} m_0(da), \tag{3.3}$$

for $t > 0$. We assume that the outputs $N(t)$ and $B(t)$ are produced by admissible ingredients (that is, by functions F and L satisfying Assumption 2.1) and that they are known on a time-interval of length $T \leq \infty$. We can now give a precise formulation of the inverse problem of linear age-structured population dynamics.

(IP) Given the measure m_0 and the functions B and N defined on $[0,T)$ $(T \leq \infty)$, determine the functions L and F such that the equations (3.3) and (2.8) are satisfied on $[0,T)$.

In particular, we shall be concerned with the question under which conditions the data m_0, B and N *uniquely* determine the ingredients F and L. On the other hand, we shall not attempt to characterize data that guarantee that F and L are admissible in the sense that they satisfy Assumption 2.1.

Once F and L have been determined the death rate μ and the fecundity β are obtained from (2.6) and (2.7) as

$$\mu(a) = -\frac{F'(a)}{F(a)}, \tag{3.4}$$

$$\beta(a) = \frac{L'(a)}{F(a)}. \tag{3.5}$$

Observe that by the monotonicity assumptions (i) and (iv) in Assumption 2.1, F and L are indeed differentiable almost everywhere, so the formulas (3.4) and (3.5) make sense.

Before we formulate our results on the inverse problem, we make a few remarks.

Remark 3.1. A naïve approach to the inverse problem would be to extend F and $b = B'$ as zero to the negative real axis and study the full line convolution equation corresponding to (3.3) with for instance Fourier transform techniques. With this convention N would be defined on the whole real axis and its support $\operatorname{supp} N$ would be $(-\operatorname{ess\,sup} \operatorname{supp} m_0, \infty)$. On the other hand, the measured data contain the values of N on the positive real axis only. The failure of the naïve approach and the difficulty of the inverse problem stem from this fact.

Remark 3.2. Because the age distribution cannot in general be measured, the initial population state m_0 is in many applications unknown. It would therefore be desirable to determine not only the ingredients F and L, but also the initial age distribution m_0 from the outputs N and B. But, as shown by a simple counter example by Gyllenberg *et al.* (2002), this can never be achieved. However, our main motivation comes from the growth dynamics of cell populations and in laboratory experiments the initial population is often well synchronized and it can therefore be accurately approximated by a Dirac measure or a density with a narrow support.

Next we give a simple example, which shows that $N(t)$ and $B(t)$ need not determine the survival function $L(a)$ uniquely.

Example 3.3. Assume that $b(t)$ and $N(t)$ grow not only asymptotically, but *exactly* exponentially, that is, assume that

$$B(t) = \frac{b_0}{r}\left(e^{rt} - 1\right), \tag{3.6}$$

$$N(t) = N_0 e^{rt}, \tag{3.7}$$

and that m_0 is absolutely continuous with distributional derivative n_0. Substituting (3.6) and (3.7) into (3.3), multiplying both sides with e^{-rt} and letting $t \to \infty$, one finds that

$$N_0 = b_0 \int_0^\infty e^{-ra} F(a) da. \tag{3.8}$$

Equation (3.3) is now an equation with F as the only unknown, and it is satisfied by

$$F(a) = \frac{1}{b_0} e^{ra} n_0(a), \quad a \geq 0. \tag{3.9}$$

With N, B and F given by (3.6), (3.7), (3.8) and (3.9) one easily checks that equation (2.8) holds for any function L satisfying

$$\int_{[0,\infty)} e^{-ra} L(da) = 1. \tag{3.10}$$

We thus conclude that the outputs (3.6) and (3.7) do *not* determine the model ingredients uniquely. Note that (3.10) is the same as the Euler–Lotka equation (2.11), but now the Malthusian parameter $\lambda = r$ is given and the survival function L has to be found.

The situation described in Example 3.3 occurs when the initial population is at demographic equilibrium. The (normalized) age-distribution remains the same for all times and it is intuitively clear that the model ingredients cannot be determined from such stable data.

At the end of Section 2 we saw that in many situations the birth rate $b(t)$ has an expansion of the form

$$b(t) = \sum b_k e^{\lambda_k t}. \tag{3.11}$$

We next show that if this expansion has only finitely many terms, then the inverse problem does not have a unique solution

Example 3.4. Assume that the birth rate $b(t)$ is given by a sum of the form (3.11), where the index k ranges over the finite set $\{0, 1, 2, \ldots, q\}$ and that the total population is given by

$$N(t) = \sum_{k=0}^{q} N_k e^{\lambda_k t}. \tag{3.12}$$

Inserting the sums (3.11) and (3.12) into the equation (3.3) on obtains

$$\sum_{k=0}^{q} N_k e^{\lambda_k t} = \sum_{k=0}^{q} b_k e^{\lambda_k t} \int_0^\infty e^{-\lambda_k a} F(a) da, \tag{3.13}$$

which immediately implies

$$N_k = b_k \int_0^\infty e^{-\lambda_k a} F(a) da \quad \text{for all} \quad k \in \{0, 1, 2, \ldots, q\}. \tag{3.14}$$

Substituting this back into Equation (3.3) one obtains

$$\sum_{k=0}^{q} b_k e^{\lambda_k t} \int\limits_0^\infty e^{-\lambda_k a} F(a) da = \sum_{k=0}^{q} b_k e^{\lambda_k t} \int\limits_0^t e^{-\lambda_k a} F(a) da + \int\limits_0^\infty \frac{n_0(a)}{F(a)} F(a+t) da \tag{3.15}$$

or

$$\int_0^\infty \left(\sum_{k=0}^{q} b_k e^{-\lambda_k a} - \frac{n_0(a)}{F(a)} \right) F(a+t) da = 0, \quad t \geq 0, \tag{3.16}$$

where n_0 is the distributional derivative of m_0. Equation (3.16) is an equation in F only and it is obviously satisfied by

$$F(a) = \frac{n_0(a)}{\sum_{k=0}^{q} b_k e^{\lambda_k a}}. \tag{3.17}$$

As a matter of fact, (3.17) is the *unique* solution of equation (3.16), but this is irrelevant for this example.

Substituting (3.17) and (3.11) into (2.8) one finds that every function L satisfying

$$\int_{[0,\infty)} e^{-\lambda_k a} L(da) = 1 \quad \text{for all} \quad k \in \{0, 1, 2, \ldots, q\} \tag{3.18}$$

is a solution of equation (2.8). We shall show that if there exists one such function L satisfying Assumption 2.1, then there are in fact infinitely many such functions L_ε satisfying Assumption 2.1.

View L as a positive measure on $[0, 1]$ and let K be the support of L. Because the functions $e^{-\lambda_k a}$ belong to $L^1(L)$, the Hahn-Banach theorem guarantees the existence of a nonzero measurable and essentially bounded (with respect to the measure L) function g on K such that

$$\int_K e^{-\lambda_k a} g(a) L(da) = 0 \quad \text{for all} \quad k \in \{0, 1, 2, \ldots, q\}. \tag{3.19}$$

Extend g to all of $[0, 1]$ by defining

$$g(a) = 0 \quad \text{for} \quad a \in [0, 1] \backslash K. \tag{3.20}$$

Then the measure

$$L_\varepsilon(da) = (\varepsilon g(a) + 1) L(da) \tag{3.21}$$

is a positive measure on $[0, 1]$ for all $0 < \varepsilon < 1/\|g\|_\infty$ and it satisfies (3.18) because g satisfies (3.19) and L satisfies (3.18).

The situation changes drastically when the sum in (3.11) involves infinitely many terms. Our main theorem (Theorem 3.5) says that the inverse problem has a unique solution whenever the exponential functions $e^{\lambda_k t}$ are complete in the continuous functions.

Theorem 3.5. *Let* supp $m_0 \subset [0,1]$ *and assume that* m_0 *does not have an atom at* 1. *Let* $b(t)$ *and* $N(t)$ *be defined on* $[0, 1+\varepsilon)$ *and representable as series*

$$b(t) = \sum b_k e^{\lambda_k t} \tag{3.22}$$

and

$$N(t) = \sum N_k e^{\mu_k t} \tag{3.23}$$

which converge absolutely for $t \in [0, 1+\varepsilon)$, *where* $\varepsilon > 0$. *Then Equation* (3.3) *has a solution* F *with support* $[0,1]$ *if and only if*

$$\lambda_k = \mu_k \quad \text{for all} \quad k, \tag{3.24}$$

and

$$N_k = b_k \int_0^1 e^{-\lambda_k a} F(a) da \quad \text{for all} \quad k. \tag{3.25}$$

If this is the case, then the function $b(t)$ *can be real-analytically continued to* $(-1, 1+\varepsilon)$. *Moreover, the solution* F *of* (3.3) *is unique and given by*

$$F(a) = \frac{n_0(a)}{b(-a)}, \tag{3.26}$$

where n_0 *is the distributional derivative of* m_0. *Moreover, every function* L *for which*

$$\int_{[0,1]} e^{-\lambda_k a} L(da) = 1 \quad \text{for all} \quad k \tag{3.27}$$

satisfies equation (2.8). *There is a unique such* L *if and only if*

$$\sum \left(1 - \left|\frac{\lambda_k + 2\pi i}{\lambda_k - 2\pi i}\right|\right) = \infty. \tag{3.28}$$

For a complete proof of this theorem we refer to Gyllenberg *et al.* (2002)

References

BERNDTSSON, B. and JAGERS, P. (1979) Exponential growth of a branching process usually implies stable age distribution. *J. Appl. Prob.* **16** 651–656.

DIEKMANN, O., GYLLENBERG, M., and THIEME, H.R. (1993a) Perturbing semigroups by solving Stieltjes renewal equations. *Differential and Integral Equations* **6** 155–181.

DIEKMANN, O., GYLLENBERG, M., METZ, J.A.J., and THIEME, H.R. (1993b) The "cumulative" formulation of (physiologically) structured population models, In "Evolution Equations, Control Theory and Biomathematics", (Ph. Clément and G. Lumer, Eds.), pp. 145–154, Marcel Dekker, New York.

DIEKMANN, O., GYLLENBERG, M., and THIEME, H.R. (1995) Perturbing evolutionary systems by step responses and cumulative outputs. *Differential and Integral Equations* **8** 1205–1244.

DIEKMANN, O., GYLLENBERG, M., METZ, J.A.J., and THIEME, H.R. (1998) On the formulation and analysis of general deterministic structured population models. I. Linear theory. *Journal of Mathematical Biology* **36** 349–388.

DIEKMANN, O., GYLLENBERG, M., HUANG, H., KIRKILIONIS, M., METZ, J.A.J., and THIEME, H.R. (2001) On the formulation and analysis of general deterministic structured population models. II. Nonlinear theory. *Journal of Mathematical Biology* **43** 157–189.

ENGL, H.W., RUNDELL, W., SCHERZER, O. (1994) A regularization scheme for an inverse problem in age-structured populations. *J. Math. Anal. Appl.* **182** 658–679.

FELLER, W. (1941) On the integral equation of renewal theory, *Ann. Math. Statist.* **12** 243–267.

FELLER, W. (1971) *An introduction to probability theory and its applications* Vol.II, Second Edition, Wiley, New York.

GYLLENBERG, M. (1985) The age structure of populations of cells reproducing by asymmetric division, In *Mathematics in Biology and Medicine*, V. Capasso, E. Grosso, and S.L. Paveri-Fontana (Eds.), Springer, Berlin, pp. 320–327.

GYLLENBERG, M., HANSKI, I., and HASTINGS, A. (1997) Structured metapopulation models. In: "Metapopulation biology: ecology, genetics and evolution" (I.A. Hanski and M.E. Gilpin, Eds.) pp. 93–122, Academic Press, San Diego.

GYLLENBERG, M., OSIPOV, A. and PÄIVÄRINTA, L. (2002) An inverse problem of age-structured population dynamics, *Journal of Evolution Equations*, in press.

JAGERS, P. (1975) *Branching Processes with Biological Applications,* Wiley, London.

MCKENDRICK, A.G. (1926) Applications of Mathematics to Medical Problems, *Proc. Edinb. Math. Soc.* **44** 98–130.

METZ, J.A.J. and DIEKMANN, O. (1986) *The Dynamics of Physiologically Structured Populations,* Springer, Berlin.

PILANT, M. AND RUNDELL, W. (1991a) Determining the initial age distribution for an age structured population, *Math. Population Stud.* **3** 3–20.

PILANT, M. AND RUNDELL, W. (1991b) Determining a coefficient in a first-order hyperbolic equation, *SIAM J. Appl. Math.* **51** 294–506.

RUNDELL, W. (1989) Determining the birth function for an age structured population. *Math. Population Stud.* **1** 377–395, 397.

RUNDELL, W. (1993) Determining the death rate for an age-structured population from census data. *SIAM J. Appl. Math.* **53** 1731–1746.

SHARPE, F.R. and LOTKA, A.J. (1911) A problem in age-distribution, *Philosophical Magazine* **21** 435–438.

van Straalen, N.M. (1986) The "inverse problem" in demographic analysis of stage-structured populations, In *The Dynamics of Physiologically Structured Populations,* J.A.J Metz and O. Diekmann (Eds.) pp. 393–408, Springer, Berlin.

Titchmarsh, E.C. (1939) *The Theory of Functions.* Second Edition. Oxford University Press, Glasgow.

Webb, G.F. (1985) *Nonlinear Age-Dependent Population Dynamics*, Marcel Dekker, New York.

Mats Gyllenberg
Dept. of Mathematics
University of Turku
20014 Turku, Finland

Andrei Osipov
Geophysical Observatory
99600 Sodankylä, Finland

Lassi Päivärinta
Dept. of Math. Sciences
University of Oulu
90570 Oulu, Finland
E-mail address: lassi@rieska.oulu.fi

D. Haroske, T. Runst, H.-J. Schmeisser (eds.): Function Spaces, Differential Operators and Nonlinear Analysis. The Hans Triebel Anniversary Volume.

Nonlocal Investigations of Inhomogeneous Indefinite Elliptic Equations via Variational Methods

Yavdat Il'yasov and Thomas Runst

Dedicated to Prof. Hans Triebel on the occasion of his 65th birthday

Abstract. We consider a class of inhomogeneous Neumann boundary value problems with indefinite nonlinearities. We introduce a new and, in some sense, more general variational approach to these problems. Using this idea we prove new results on the existence and multiplicity of positive solutions.

1. Introduction

Let W be a Banach space and I_λ a real functional of class $C^1(W \setminus \{0\})$ which depends on the real parameter λ. It is considered the problem of the existence of critical points of I_λ on W. To solve the problem, firstly it has to be determined a suitable interval $(\lambda_j, \lambda_{j+1}) \subseteq \mathbb{R}$ (a sufficient interval) in such a way that in the future it enables us to find the critical points of I_λ on W for $\lambda \in (\lambda_j, \lambda_{j+1})$. In linear cases, the calculation of the points λ_j is a subject of spectral theory. However in nonlinear cases, it is absent such a general theory; usually, the sufficient interval $(\lambda_j, \lambda_{j+1}) \subseteq \mathbb{R}$ are possibly guessed by using a priori observations.

The main purpose of our paper is to present a method for constructive calculation of the sufficient intervals for the given functional I_λ on Banach space with respect to the considered parameter λ. Our approach is based on development of the fibering method introduced by Pohozaev in [15, 16].

We present the method through the following example of a class of inhomogeneous Neumann boundary value problems with indefinite nonlinearities given by

$$-\Delta_p u - \lambda k(x)|u|^{p-2}u = D(x)|u|^{q-2}u + K(x)|u|^{\gamma-2}u \text{ in } \Omega, \tag{1.1}$$

$$|\nabla u|^{p-2}\frac{\partial u}{\partial n} + d(x)|u|^{p-2}u = 0 \text{ on } \partial\Omega. \tag{1.2}$$

Here Ω is a bounded domain in $\mathbb{R}^n$, $n \geq 2$, with smooth boundary $\partial\Omega$, Δ_p denotes the p-Laplacian, ∇ is the gradient, and $\dfrac{\partial}{\partial n}$ is the normal derivative with respect to the outward normal n on $\partial\Omega$. In what follows we assume that

$$p < q < \gamma \leq p^*, \text{ where } p^* = \begin{cases} \frac{pn}{n-p} & \text{if } p < n, \\ +\infty & \text{if } p \geq n, \end{cases} \tag{1.3}$$

and

$$k(\cdot),\ D(\cdot),\ K(\cdot) \in L_\infty(M),\ d(\cdot) \in L_\infty(\partial M). \tag{1.4}$$

Here p^* is the critical Sobolev exponent of the embedding $W_p^1(M) \subset L_{p^*}(M)$. When both non-linear terms occur in the right-hand side of the differential equation (1.1), i.e., when $K \neq 0$ and $D \neq 0$ in M, one has a inhomogeneous problem. The nonlinearity $K(x)|u|^{\gamma-2}u$ ($D(x)|u|^{q-2}u$) is called indefinite if the function K (D) changes the sign on M (cf. [7]). We shall also say that the problem (1.1)–(1.2) has an indefinite nonlinearity if the functions K and D do not change the sign but they have an opposite one on M, e.g. $K(x) > 0$ and $D(x) < 0$ on M (see [2]). We deal with the existence of positive solutions of (1.1)–(1.2). Note that our results are also new in the case $p = 2$ when the problem corresponds with the classical Laplacian.

The case which is best known in the literature is the homogeneous problem (1.1)–(1.2) (i.e. $D \equiv 0$) when the nonlinearity has definite sign. In this case, the set of bifurcation values of λ, belongs, in general, to the spectral set of the corresponding linearized problem. The indefiniteness and/or the inhomogeneity of nonlinearity change(s) essentially the structure of the solution set. In this case, the dependence of the problem on the parameter λ is more complicated, and the problem of finding the bifurcation values of λ and/or the sufficient intervals $(\lambda_j, \lambda_{j+1}) \subseteq \mathbb{R}$ is not simple (cf. [1, 2, 5, 8, 11, 18]).

The homogeneous case for $p = 2$ with indefinite nonlinearity has been treated in several recent papers [2, 8, 18, 20] where necessary and sufficient conditions for the existence of a positive solution, as well as some results on the existence of a sufficient interval with respect to the parameter λ have been obtained. For the problem with p-Laplacian some local results with respect to λ on the existence of positive solutions have been proved by using the fibering method in [10]. The exact values of the boundaries of the sufficient and necessary interval with respect to the parameter λ for the existence and non-existence of positive solution of the homogeneous problem (1.1) (including the cases $p \neq 2$) with Dirichlet boundary condition have been found in [13].

It is only little known about the inhomogeneous problem (1.1)–(1.2) with indefinite nonlinearities. To our knowledge, only the papers [2, 14, 17] deal with such a problem in the large. In [2], the authors considered (1.1) with Dirichlet boundary condition for $p = 2$ under the assumptions that the functions K and D on M do not change the sign but they have the opposite one. They have shown the existence of a number $\lambda_* < \lambda_1$ (λ_1 is the first eigenvalue of $-\Delta$ in $W_0^{1,2}$) according to certain integrability properties of the ratio $K^{\gamma-2}/D^{q-2}$ such that if $\lambda < \lambda_*$, then the problem has no positive solution. If $\lambda > \lambda_*$ there were proved some existence and multiplicity results. The papers [14] and [17] treated with the inhomogeneous problem (1.1)–(1.2) (in [14] for $p = 2$ and in [17] the case $p \neq 2$

is included) when D has a definite sign whereas K may change the sign. In these papers, existence and multiplicity results on some sufficient intervals with respect to λ are proved by using the fibering method.

There are also related results in the papers of [1, 4, 5, 6, 11]. We remark that [1, 4, 5, 6] treated the inhomogeneous equation (1.1) with Dirichlet boundary condition and on the whole space $\mathbb{R}^n$, respectively, when $0 < q < 1 < p < \gamma$ and $k \equiv 0$ hold and the nonlinearity is not indefinite. In [11] problem (1.1)–(1.2) is considered, when $p = 2$, $q = 1$, $\gamma = p^*$ and $K \equiv const$ hold.

However, the problem of the existence of positive solutions is open in the general formulation of (1.1)–(1.2), where all nonlinear terms occur and the nonlinearities may be indefinite.

2. The fibering scheme

The constrained minimization method is one of the most powerful tools in the study of critical points of a functional I given on a Banach space W (see [19] and the references therein). In general, the method can be described as follows.
We impose that the constraint $M \subseteq W$ satisfies the following conditions:

1.) The constraint M is suitable in the following sense: The critical point $v \in M$ of the restriction of I on M given by $J := I|_M$ defines also a critical point $u(v) \in W$ of I on W.
2.) The restriction J of the functional I on M is bounded from below or/and from above on M.

In some cases, for example when (1.1)–(1.2) is homogeneous, the suitable constraint $M \subseteq W$ can be directly obtained. In the papers [15, 16], Pohozaev suggested the fibering method, which makes it possible to find constructively a constraint which is given by an equality (*constraints of equality type*), i.e., we have

$$M = \{w \in W | H(w) = const\}$$

and M is suitable in the sense of 1.)

However, the application of Pohozaev's method based on constraints of equality type is not always sufficient.

Therefore, we introduce a fibering scheme which allows us to find constructively the constrained minimization problem with mixed constraints which being of equality and inequality type and satisfying both above mentioned conditions 1.) and 2.).

We assume that $(W, ||\cdot||)$ is a real reflexive Banach space. Furthermore, we suppose that the norm $||\cdot||$ defines a C^1-functional $u \to ||u||$ on $W \setminus \{0\}$. In this case, the sphere $S^1 = \{v \in W| \; ||v|| = 1\}$ is a closed submanifold of class C^1 in W and $\mathbb{R}^+ \times S^1$ is C^1-diffeomorphic to $W \setminus \{0\}$. Thus we have the trivial principal fibre bundle $P(S^1, \mathbb{R}^+)$ over S^1 with the structure group $\mathbb{R}^+$ and the bundle space $W \setminus \{0\}$ is C^1-diffeomorphic to $\mathbb{R}^+ \times S^1$.

Let I be a real functional of class $C^1(W \setminus \{0\})$. Corresponding to I we define a functional $\tilde{I} : \mathbb{R}^+ \times S^1 \to \mathbb{R}$ by

$$\tilde{I}(t,v) = I(tv), \ (t,v) \in \mathbb{R}^+ \times S^1. \tag{2.1}$$

Since $\mathbb{R}^+ \times S^1$ is C^1-diffeomorphic to $W \setminus \{0\}$ it follows that $\tilde{I}(t,v)$ is also a C^1-functional on $\mathbb{R}^+ \times S^1$ and the set of critical points of the functional $\tilde{I}(t,v)$ on $\mathbb{R}^+ \times S^1$ as well as the set of the critical points of the functional $I(u)$ on $W \setminus \{0\}$ are one-to-one. Moreover the following holds.

Proposition 2.1. (Pohozaev [15, 16]) *Let $(t_0, v_0) \in \mathbb{R}^+ \times S^1$ be a critical point of $\tilde{I}$. Then $u_0 = t_0 v_0 \in W \setminus \{0\}$ is a critical point of I.*

We make an additional condition on the functional I given by

(RD) *The first derivative $\frac{\partial}{\partial t}\tilde{I}(t,v)$ is a C^1-functional on $\mathbb{R}^+ \times S^1$.*

Now we define

$$Q(t,v) = \frac{\partial}{\partial t}\tilde{I}(t,v), \quad L(t,v) = \frac{\partial^2}{\partial t^2}\tilde{I}(t,v), \ (t,v) \in \mathbb{R}^+ \times S^1. \tag{2.2}$$

We introduce the following subsets of $\mathbb{R}^+ \times S^1$ for the classification of the critical points

$$\Sigma^1 = \{(t,v) \in \mathbb{R}^+ \times S^1 | \, Q(t,v) = 0, \ L(t,v) > 0\}, \tag{2.3}$$

$$\Sigma^2 = \{(t,v) \in \mathbb{R}^+ \times S^1 | \, Q(t,v) = 0, \ L(t,v) < 0\}. \tag{2.4}$$

Proposition 2.2. *Let $(t_0, v_0) \in \Sigma^j$, $j = 1, 2$. Then there exist a neighbourhood $\Lambda(v_0) \subset S^1$ of $v_0 \in S^1$ and a unique C^1-map*

$$t^j : \Lambda(v_0) \to \mathbb{R}$$

such that

$$t^j(v_0) = t_0, \ (t^j(v), v) \in \Sigma^j, \ v \in \Lambda(v_0), \ j = 1, 2. \tag{2.5}$$

Proof. Let $j = 1, 2$. We us assume that $(t_0, v_0) \in \Sigma^j$ is satisfied. Then

$$\frac{\partial Q(t_0, v_0)}{\partial t} = L(t_0, v_0) \neq 0.$$

It follows from assumption **(RD)** that we have $Q(\cdot) \in C^1(\mathbb{R}^+ \times S^1)$. Hence, by application of the implicit function theorem, we can finish the proof. □

By means of Proposition 2.2 we get the following assertion.

Lemma 2.1. *Assume that* **(RD)** *holds. Let $j = 1, 2$. Then the set Σ^j is a C^1-submanifold in $\mathbb{R}^+ \times S^1$ which is locally C^1-diffeomorphic to S^1.*

Using the trivial fibering scheme we are able to study the existence of critical points of the functional I defined on W.

Let $j = 1, 2$. Then $\tilde{J}^j$ denotes the restriction of the functional $\tilde{I}$ to the submanifolds Σ^j, i.e., we have

$$\tilde{J}^j(t,v) = \tilde{I}(t,v), \ (t,v) \in \Sigma^j, \ j = 1, 2.$$

Let $d\tilde{I}(t,v)$ be the differential of $\tilde{I} : \mathbb{R}^+ \times S^1 \to \mathbb{R}$ at the point $(t,v) \in \mathbb{R}^+ \times S^1$, and $T_{(t,v)}(\mathbb{R}^+ \times S^1)$ denotes the tangent space to $\mathbb{R}^+ \times S^1$ at (t,v). Furthermore, let $d\tilde{J}^j(t,v)$ be the differential of $\tilde{J}^i : \Sigma^j \to \mathbb{R}$ at the point $(t,v) \in \Sigma^j$, and $T_{(t,v)}(\Sigma^j)$ denotes the tangent space to Σ^j at (t,v).

Using these definitions we are able to prove the following result which is important for our further considerations in this paper.

Lemma 2.2. *Let $j = 1,2$. We suppose that* **(RD)** *holds. If (t_0, v_0) is a critical point of the functional $\tilde{J}^j$ on the submanifolds Σ^j, i.e., it holds*

$$d\tilde{J}^j(t_0,v_0)(h) = 0, \ \forall h \in T_{(t_0,v_0)}(\Sigma^j), \tag{2.6}$$

then the point (t_0, v_0) is also a critical point for $\tilde{I}$ on $\mathbb{R}^+ \times S^1$. Hence we have

$$d\tilde{I}(t_0,v_0)(l) = 0, \ \forall l \in T_{(t_0,v_0)}(\mathbb{R}^+ \times S^1). \tag{2.7}$$

Proof. We consider the case $j = 1$. The other one can be handled in the same way. Let (t_0, v_0) be a critical point of $\tilde{J}^1$ on Σ^1. At first it holds

$$d\tilde{I}(t_0,v_0)(\tau,\phi) = \frac{\partial}{\partial t}\tilde{I}(t_0,v_0)(\tau) + \frac{\delta}{\delta v}\tilde{I}(t_0,v_0)(\phi) \tag{2.8}$$

for every $\tau \in T_t(\mathbb{R}^+)$ and $\phi \in T_v(S^1)$. Using (2.3) we have $\frac{\partial}{\partial t}\tilde{I}(t_0,v_0)(\tau) = 0$. In order to prove (2.7) it is therefore sufficient to show that

$$\frac{\delta}{\delta v}\tilde{I}(t_0,v_0)(\phi) = 0, \ \forall \phi \in T_{v_0}(S^1) \tag{2.9}$$

holds.

By Proposition 2.2 there exist a neighbourhood $\Lambda(v_0) \subset S^1$ of $v_0 \in S^1$ and a unique C^1-map $t^1 : \Lambda(v_0) \to \mathbb{R}$ such that (2.5) holds. Hence we can introduce the map

$$J^1(v) =: \tilde{I}(t^1(v), v), \ v \in \Lambda(v_0).$$

Then we have

$$J^1(v) \equiv \tilde{J}^1(t^1(v), v), \ \ v \in \Lambda(v_0). \tag{2.10}$$

By Lemma 2.1 we know that Σ^j is locally C^1-diffeomorphic to S^1. Hence condition (2.6) implies that v_0 is a critical point of $J^1(v)$ on $\Lambda(v_0)$, i.e., it holds

$$dJ^1(v_0)(\phi) = 0$$

for all $\phi \in T_{v_0}(S^1)$. By our definition we have $J^1(v) = \tilde{I}(t^1(v), v)$ for all $v \in \Lambda(v_0)$. Hence we get consequently by (2.8) that

$$0 = dJ^1(v_0)(h) = \frac{\partial}{\partial t}\bar{I}(t^1(v_0), v_0)(dt^1(v_0))(h) + \frac{\delta}{\delta v}\tilde{I}(t^1(v_0), v_0)(h) \tag{2.11}$$

holds for all $h \in T_{v_0}(S^1)$. By virtue of (2.3), (2.4) the first term on the right-hand side of (2.11) is equal zero. Thus we have the desired result (2.9), i.e., it holds

$$\frac{\partial}{\partial v}\tilde{I}(t^1(v_0), v_0)(\phi) = 0$$

for all $\phi \in T_{v_0}(S^1)$. In analogy, one can prove the case $j=2$. The proof is finished. □

In the next step, we introduce the so-called ground constrained minimization problems associated with the above given functional I of class $C^1(W \setminus \{0\})$, where $(W, ||\cdot||)$ is a real Banach space and the norm $||\cdot||$ is of class $C^1(W \setminus \{0\})$. Furthermore, we assume that **(RD)** holds.

In what follows, we call the following problems defined by

$$\hat{I}^j = \inf\{\tilde{I}(t,v) |\, (t,v) \in \Sigma^j\},\ j = 1, 2, \tag{2.12}$$

the *ground constrained minimization problems* with respect to the trivial fibering scheme. Hereby we put by definition

$$\hat{I}^j = +\infty, \ \text{ if } \ \Sigma^j = \emptyset,\ j = 1, 2. \tag{2.13}$$

Definition 2.1. *Let* $j = 1, 2$. *The point* $(t_0, v_0) \in \Sigma^j$ *is said to be a solution of* (2.12) *if*

$$-\infty < \hat{I}^j = \tilde{I}(t_0, v_0) < \infty.$$

Remark 2.1 It is also meaningful to consider the corresponding maximization problems given by (2.12)

$$\check{I}^j = \sup\{\tilde{I}(t,v) |\, (t,v) \in \Sigma^j\},\ j = 1, 2. \tag{2.14}$$

Here we replace (2.13) by

$$\check{I}^j = -\infty, \ \text{ if } \ \Sigma^j = \emptyset,\ j = 1, 2. \tag{2.15}$$

However, the substitution $I' = -I$ reduces any maximization problem to a minimization problem again.

Applying Lemma 2.2 and Proposition 2.1 we obtain the following main result.

Theorem 2.1. *Assume that* $I(u) \in C^1(W \setminus \{0\})$ *and* **(RD)** *hold. Let* $j = 1, 2$. *If there exists a solution* (t_0^j, v_0^j) *of* (2.12), *then*

$$u_0^j = t_0^j v_0^j \in W \setminus \{0\} \tag{2.16}$$

is a critical point of I.

Now we show that the constrained minimization problems (2.12) possesses the property of ground states.

Therefore, we denote by Z the set of all (nontrivial) critical points of the functional I on the space $W \setminus \{0\}$. With respect to the trivial fibering scheme we get the following decomposition of Z: $Z = Z_- \cup Z_+ \cup Z_0$, where

$$Z_+ = \{u \in Z |\, (||u||, \frac{u}{||u||}) \in \Sigma_1\},\ Z_- = \{u \in Z |\, (||u||, \frac{u}{||u||}) \in \Sigma_2\},$$

and

$$Z_0 = \{u \in Z |\, (||u||, \frac{u}{||u||}) \in \dot{\delta}\},$$

with

$$\dot{\delta} = \{(t,v) \in \mathbb{R}^+ \times S^1 |\, Q(t,v) = 0,\ L(t,v) = 0\}. \tag{2.17}$$

For later applications, the investigation of the ground states of the functional I is important, see [9]. By definition, the nonzero critical point $u_g \in W \setminus \{0\}$ is said to be the *ground state* if it is a point with the least level of I among all nonzero critical points Z, i.e., it holds

$$\inf\{I(u) | \, u \in Z\} = I(u_g). \tag{2.18}$$

We introduce, in addition, the following term.

Definition 2.2. *The nonzero critical point $u_g^- \in W$ ($u_g^+ \in W$) is said to be a ground state of type (-1) $((0))$ for I if*

$$\min\{I(u) | \, u \in Z_-\} = I(u_g^-), \ (\min\{I(u) | \, u \in Z_+\} = I(u_g^+)) \tag{2.19}$$

holds.

A consequence of the construction of the constrained minimization problems (2.12) is the next result.

Lemma 2.3. *We assume that $I(u) \in C^1(W \setminus \{0\})$ and* **(RD)** *hold.*
Let $j = 1, 2$. If (t_0^j, v_0^j) is a solution of the variational problem (2.12), then

$$u^+ = t_0^1 v_0^1 \in W \setminus \{0\}$$

is a ground state of type (0) for I, and

$$u^- = t_0^2 v_0^2 \in W \setminus \{0\}$$

is a ground state of type (-1) for I.

Furthermore, if $Z_0 = \emptyset$ holds, then one of these solutions u^- or u^+ is a ground state for I, i.e., we have

$$\min\{I(u) | \, u \in Z\} = \min\{I(u_g^-), I(u_g^+)\}. \tag{2.20}$$

In the following, let pr_2 be the canonical projection from $\mathbb{R}^+ \times S^1$ to S^1 and let $\Theta^j = pr_2(\Sigma^j)$, $j = 1, 2$.

As usual, let $j = 1, 2$. We recall that by Proposition 2.2 there exist for every point $v_0^j \in \Theta^j$ a neighbourhood $\Lambda(v_0^j) \subset \Theta^j$ and a unique C^1-map $t^j : \Lambda(v_0^j) \to \mathbb{R}$ such that $(t^j(v), v) \in \Sigma^j$ is satisfied.

Now we give the following definition.

Definition 2.3. *Let $j = 1, 2$. The fibering scheme for I on W is said to be solvable with respect to Σ^j if for every $v \in \Theta^j$ there exists a unique number $t^j(v) \in \mathbb{R}^+$ such that $(t^j(v), v) \in \Sigma^j$ holds. In the case when the fibering scheme for I on W is solvable with respect to both Σ^1 and Σ^2 then we call it a solvable scheme.*

If, in addition, the functional $t^j(v)$ can be found in an exact form, then the fibering scheme is called an exactly solvable one.

We remark that in the papers [8], [10], [19] the constrained minimization method were used for the investigation of homogeneous problems similar to (1.1)–(1.2). These problems can be solved by using the exactly solvable trivial fibering scheme. We point out that in the present paper we concern with applications where the trivial fibering scheme is solvable but not exactly solvable.

3. Characteristic sets

Theorem 2.1 shows that the ground constrained minimization problems (2.12) – constructed with respect to the trivial fibering scheme – are suitable in the sense defined above in condition 1.). In the case of an one-parametric family of functionals we can improve this result and show that condition 1.) and 2.) are satisfied simultaneously.

For a real parameter λ we consider the family of functionals I_λ. We assume that $I_\lambda \in C^1(W \setminus \{0\})$ and **(RD)** hold for all λ. Now we define for $\lambda \in \mathbb{R}$ the functional

$$\tilde{I}_\lambda(t,v) = I_\lambda(tv), \ (t,v) \in \mathbb{R}^+ \times S^1,$$

and the functional $Q_\lambda(t,v)$ and $L_\lambda(t,v)$ are given as above by the fibering scheme. In the same way as before we introduce the subset $\Sigma^j_\lambda \subseteq \mathbb{R}^+ \times S^1$, see (2.3) and (2.4), respectively. In the following,

$$\overline{\mathbb{R}^+} := \mathbb{R}^+ \cup \{0\} \cup \{\infty\}$$

denotes the compactification of $\mathbb{R}^+$ and $\overline{\Sigma}^j$ denotes the closure of Σ^j in $\overline{\mathbb{R}^+} \times S^1$ (with respect to strong topology).

Let $j = 1, 2$ and $\lambda \in \mathbb{R}$. The we denote by $\partial\overline{\Sigma}^j_\lambda$ the boundary of $\overline{\Sigma}^j_\lambda \neq \emptyset$. The condition

$$\delta^j_\lambda = \delta^j_\lambda(\Sigma^j_\lambda) := \partial\overline{\Sigma}^j_\lambda \neq \emptyset$$

we shall call *a system of defining equations of* Σ^j_λ *with respect to the trivial principal fibre bundle* $P(S^1, \mathbb{R}^+)$.

Using (2.17)) it is clear that

$$\dot{\delta}_\lambda \cap \Sigma^j_\lambda \subseteq \delta^j_\lambda, \ j = 1, 2,$$

holds. Now we assume that there exists the following decomposition

$$\delta^j_\lambda = \delta^{j,u}_\lambda \cup \delta^{j,d}_\lambda,$$

where

$$\delta^{j,u}_\lambda = \{(t_0, v_0) \in \delta^j_\lambda | \, \tilde{I}(t_0, v_0) > \hat{I}^j\} \ \text{ and } \ \delta^{j,d}_\lambda = \delta^j_\lambda \setminus \delta^{j,u}_\lambda. \tag{3.1}$$

Furthermore, the set

$$\sigma^j := \{\lambda \in \mathbb{R} : \delta^j_\lambda \neq \emptyset\} \equiv \{\lambda \in \mathbb{R} : \Sigma^j_\lambda \neq \emptyset\}, \ j = 1, 2,$$

is called a *characteristic set* in a wider sense, and the set

$$\sigma^{j,u} := \{\lambda \in \mathbb{R} : \delta^{j,u}_\lambda \neq \emptyset\}, \ \ j = 1, 2, \tag{3.2}$$

$$\sigma^{j,d} := \{\lambda \in \mathbb{R} : \delta^{j,d}_\lambda \neq \emptyset\}, \ j = 1, 2, \tag{3.3}$$

is denoted to be a *upper characteristic* and *residual characteristic set*, respectively, in a wider sense.

Now we consider the following boundaries (a priori bifurcation points) of the characteristic sets for $j = 1, 2$

$$\lambda^j_{\text{inf}} = \inf\{\sigma^j\}, \qquad \lambda^j_{\text{sup}} = \sup\{\sigma^j\}, \tag{3.4}$$

$$\lambda^j_{d,\text{inf}} = \inf\{\sigma^{j,d}\}, \quad \lambda^j_{d,\text{sup}} = \sup\{\sigma^{j,d}\}. \tag{3.5}$$

Here we put

$$\lambda^j_{d,\text{inf}} = +\infty,\ \lambda^j_{d,\text{sup}} = -\infty \text{ if } \sigma^{j,d} = \emptyset. \tag{3.6}$$

It is evident that we have $\lambda^j_{\text{inf}} \le \lambda^j_{d,\text{inf}} \le \lambda^j_{d,\text{sup}} \le \lambda^j_{\text{sup}}$ if $\sigma^j \neq \emptyset$ is satisfied.

It holds the following main theorem on the existence of the sufficient intervals for a given family of the functionals on a Banach space W.

Theorem 3.1. *Assume that $I_\lambda(u) \in C^1(W \setminus \{0\})$ for $\lambda \in \mathbb{R}$ and* **(RD)** *hold. Let $j = 1, 2$. Suppose that $\lambda^j_{\text{inf}} < \lambda^j_{d,\text{inf}}$ ($\lambda^j_{d,\text{sup}} < \lambda^j_{\text{sup}}$) holds, and for every $\lambda \in (\lambda^j_{\text{inf}}, \lambda^j_{d,\text{inf}})$ ($\lambda \in (\lambda^j_{d,\text{sup}}, \lambda^j_{\text{sup}})$) the following conditions are fulfilled:*

(1°) $|\tilde{I}_\lambda(t, v)| \to \infty$ for $(t, v) \in \Sigma^j_\lambda$, if and only if $t \to +\infty$.

(2°) $\overline{\Sigma}^j_\lambda$ is a weakly closed subset with respect to the weak topology of $\mathbb{R}^+ \times S^1$.

Then $\tilde{I}_\lambda$ is bounded from below on Σ^j_λ and attains its infimum in Σ^j_λ.

4. Main results on the existence and multiplicity of positive solutions of (1.1)–(1.2)

Let us state our main results on the existence and multiplicity of positive solutions of (1.1)–(1.2). We consider this problem in the framework of the Sobolev space $W = W^1_p(\Omega)$ equipped with the norm

$$||u|| = \left(\int_\Omega |u|^p dx + \int_\Omega |\nabla u|^p dx\right)^{1/p}. \tag{4.1}$$

Assume that $d(x) \ge 0$ on $\partial\Omega$ holds. Then there exists a first eigenvalue $\lambda_1 \ge 0$ of $-\Delta_p$ under the boundary condition (1.2) in $W^{1,p}$ (cf. [21, 22]). By $\phi_1 \in W^1_p(\Omega)$ we denote the corresponding positive eigenfunction.

Using the fibering scheme and Theorem 3.1 the following boundaries of the sufficient intervals

$$\lambda^*_D = \inf\{\frac{\int_\Omega |\nabla u|^p dx + \int_{\partial\Omega} d(x)|u|^p ds}{\int_\Omega k(x)|u|^p dx} \mid B(u) \ge 0,\ u \in W\}, \tag{4.2}$$

$$\lambda^*_{D,K} = \inf\{\frac{\int_\Omega |\nabla u|^p dx + \int_{\partial\Omega} d(x)|u|^p ds}{\int_\Omega k(x)|u|^p dx} \mid B(u) \ge 0,\ F = 0,\ u \in W\} \tag{4.3}$$

and

$$\lambda^*_{D/K} = \inf\{[(\int_\Omega |\nabla u|^p dx + \int_{\partial\Omega} d(x)|u|^p ds)|F(u)|^{\frac{(q-p)}{(\gamma-p)}} + \tag{4.4}$$

$$+(\frac{q-p}{\gamma-p})^{\frac{(q-p)}{(\gamma-p)}}(\frac{\gamma-q}{\gamma-2})|B(u)|^{\frac{(\gamma-p)}{(\gamma-q)}}]/[\int_\Omega k(x)|u|^p dx|F(u)|^{\frac{(q-p)}{(\gamma-p)}}] \mid$$

$$B(u) \le 0,\ F(u) > 0,\ u \in W\}$$

can be calculated.

Here we put $B(u) = \int_\Omega D(x)|u|^q dx$ and $F(u) = \int_\Omega K(x)|u|^\gamma dx$ for $u \in W$. In the case when $\{u \in W^1_p(\Omega) | \ B(u) \ge 0\}$, $\{u \in W^1_p(\Omega) | \ B(u) \ge 0,\ F = 0\}$ and $\{u \in W^1_p(\Omega) | \ B(u) \le 0,\ F(u) > 0\}$, respectively, is empty we define $\lambda^*_D = +\infty$, $\lambda^*_{D,K} = +\infty$ and $\lambda^*_{D/K} = +\infty$, respectively.

Our first main result on the existence of positive solution for (1.1)–(1.2) is the following.

Theorem 4.1. *Suppose that* (1.4), $p < \gamma \le p^*$, $k(x) \ge 0$ *in* Ω, *and* $d(x) \ge 0$ *on* $\partial\Omega$ *are satisfied.*

I).a) *If* $B(\phi_1) < 0$ *and/or* $F(\phi_1) \ne 0$, *then* $\lambda^*_{D,K} > \lambda_1$.

b) *If* $B(\phi_1) \ne 0$ *and/or* $F(\phi_1) < 0$, *then* $\lambda^*_{D/K} > \lambda_1$.

II). *Assume* $B(\phi_1) < 0$ *and/or* $F(\phi_1) < 0$. *Let* $p < q < \gamma < p^*$. *Then for every* $\lambda \in (\lambda_1, \min\{\lambda^*_{D/K}, \lambda^*_{D,K}\})$ *there exists a weak positive solution* $u^1_\lambda \in W^1_p(\Omega)$ *of* (1.1)–(1.2).

Now we suppose that $p < q < \gamma < p^*$, (1.4), $k(x) \ge 0$ in Ω and $d(x) \ge 0$ on $\partial\Omega$ are satisfied. Furthermore, we assume the set $\{x \in \partial M \mid D(x) > 0\}$ is not empty and $B(\phi_1) < 0$ holds. Then we denote by $\phi_{\lambda,D} \in W^1_p(\Omega)$, $\lambda < \lambda^*_D$, the solution of the following minimization problem

$$\min\{\int_\Omega |\nabla u|^p dx + \int_{\partial\Omega} d(x)|u|^p ds \quad - \quad \lambda\int_\Omega k(x)|u|^p dx \mid$$

$$B(u) = 1,\ u \in W\}. \tag{4.5}$$

The existence of the solution $\phi_{\lambda,D}$ for $\lambda < \lambda^*_D$ and some further properties of problem (4.5) follows from [12].

Our last main result is the following.

Theorem 4.2. *Suppose that* (1.4), $p < q < \gamma < p^*$, $k(x) \ge 0$ *in* Ω *and* $d(x) \ge 0$ *on* $\partial\Omega$ *are satisfied.*

I). *If* $B(\phi_1) < 0$, *then* $\lambda^*_D > \lambda_1$ *holds.*

II). *Let the following assumption be satisfied.*

(i) $B(\phi_1) < 0$,

(ii) *the set* $\{x \in \Omega \mid D(x) > 0\}$ *is not empty and*

(iii) *the set* $\{x \in \Omega \mid K(x) > 0\}$ *is not empty.*

Let $\lambda < \min\{\lambda^*_D, \lambda^*_{D/K}\}$ *and assume that* $F(\phi_{\lambda,D}) \geq 0$. *Then there exists a weak positive solution* $u^2_\lambda \in W^1_p(\Omega)$ *of* (1.1)–(1.2).
Furthermore, if $\lambda_1 < \lambda < \min\{\lambda^*_D, \lambda^*_{D,K}, \lambda^*_{D/K}\})$ *and* $F(\phi_{\lambda,D}) \geq 0$, *then there exist at least two weak positive solutions* $u^1_\lambda, u^2_\lambda \in W^1_p(\Omega)$ *of* (1.1)–(1.2).

Remark. The condition $F(\phi_{\lambda,D}) \geq 0$ holds, for example, if $K(x) \geq 0$ is satisfied for all $x \in \Omega$.

The proof of Theorem 4.1 and 4.2 is based on the fibering method. Applying the fibering scheme we can introduce two different minimization problems with mixed constraints of equality and inequality type for (1.1)–(1.2). Then by studying the corresponding *characteristic sets* we find the boundaries λ^*_D, $\lambda^*_{D,K} and \lambda^*_{D/K}$. Finally, we are able to prove the existence of positive solutions of (1.1)–(1.2).

References

[1] Ambrosetti, A., Brezis, H., Cerami, G., *Combined effect of concave and convex nonlinearities in some elliptic problems.* J. Funct. Anal., **122** (1994), 519–543.

[2] Alama, S., Tarantello, G., *On semilinear elliptic problems with indefinite nonlinearities.* Calculus Var. and Partial Differential Equations **1** (1993), 439–475.

[3] Alama, S., Tarantello, G., *Elliptic problems with nonlinearity in sign.* J. Funct. Anal., **141** (1996), 159–215.

[4] Azorero, J.G., Alonso, I.P., *Multiplicity of solutions for elliptic problems with critical exponent or with a nonsymmetric term.* Trans. Amer. Math. Soc. **323** (1991), 877–895.

[5] Azorero, J.G., Montefusco, E., Peral, I., *Bifurcation for the p-Laplacian in* $\mathbb{R}^N$. Adv. Diff. Eq. **5** (2000), 435–464.

[6] Azorero, J.G., Peral, I., *Some results about the existence of second positive solution in a quasilinear critical problem.* Indiana. Univ. Math. J. **43** (1994), 941–957.

[7] Berestycki, H., Capuzzo–Dolcetta, I., Nirenberg, L., *Superlinear indefinite elliptic problems and nonlinear Liouville theorem.* Topl. Meth. Nonl. Anal. **4** (1994), 59–78.

[8] Berestycki, H., Capuzzo–Dolcetta, I., Nirenberg, L., *Variational methods for indefinite superlinear homogeneous elliptic problems.* NoDEA **2** (1995), 553–572.

[9] Coleman, S., Glazer, V., Martin, A., *Action minima among solution to a class of Euclidean scalar field equations.* Comm. Math. Phys.**58** (1978), 211–221.

[10] Drábek, P., Pohozaev, S.I., *Positive solution for the p-Laplacian: application of the fibering method.* Proc. Roy. Soc. Edinb. Sect. A **127** (1997), 703–726.

[11] Tarantello, G., *Multiplicity results for an inhomogeneous Neumann problems with critical exponent.* Manuscr. Math. **81** (1993), 57–78.

[12] Il'yasov, Y., *Euler's functional for equations involving the p-Laplacian as a functional of the spectral parameter.* Proc. Stekl. Inst. Math. **214** (1996), 175–186.

[13] Il'yasov, Y., *On positive solutions of indefinite elliptic equations.* C. R. Acad. Sci., Paris **333** (2001), 533–538.

[14] Il'yasov, Y., Runst, T., *Existence and multiplicity results for a class of non-linear Neumann boundary value problems.* Jenaer Schriften zur Mathematik und Informatik, Univ. Jena, Math/Inf/99/14, 1–23.

[15] Pohozaev, S.I., *On an approach to Nonlinear equations.* Doklady Acad. Sci. USSR **247** (1979), 1327–1331.

[16] Pohozaev, S.I., *On the method of fibering a solution in nonlinear boundary value problems.* Proc. Stekl. Inst. Math. **192** (1990), 157–173.

[17] Pohozaev, S.I., Véron, L., *Multiple positive solutions of some quasilinear Neumann problems.* Applicable Analysis **74** (2000), 363–391.

[18] Ouyang, T.C., *On the positive solutions of semilinear equations $\Delta u + \lambda u + hu^p = 0$ on the compact manifolds.* Indiana Univ. Math. J. **40** (1991), 1083–1141.

[19] Struwe, M., *Variational Methods, Application to Nonlinear Partial Differential Equations and Hamiltonian Systems.* Springer-Verlag, Berlin, Heidelberg, New-York, 1996.

[20] Tehrani, H.T., *On indefinite superlinear elliptic equations.* Calculus Var. and Partial Differential Equations **4** (1996), 139–153.

[21] Vazquez, J.L., Véron, L., *Solutions positives d'éuations elliptiques semi-linéaires sur des variétés riemanniennes compactes*, C. R. Acad. Sci., Paris **312** (1991), 811–815.

[22] Véron, L., *Première valeur propre non nulle du p-laplacien et équations quasilinéaires elliptiques sur une variété riemannienne compacte*, C. R. Acad. Sci., Paris **314** (1992), 271–276.

Yavdat Il'yasov
Bashkir State University
32 Frunze, Ufa, Russia, 450074
E-mail address: `IlyasovYS@ic.bashedu.ru`

Thomas Runst
Mathematisches Institut
Friedrich-Schiller-Universität Jena
D-07740 Jena, Germany
E-mail address: `runst@minet.uni-jena.de`

D. Haroske, T. Runst, H.-J. Schmeisser (eds.): Function Spaces, Differential Operators and Nonlinear Analysis. The Hans Triebel Anniversary Volume.

Regularity Results and Parametrices of Semi-linear Boundary Problems of Product Type

Jon Johnsen

Dedicated to Prof. Hans Triebel on the occasion of his 65th birthday

1. Introduction

This study focuses on semi-linear problems of the form

$$\begin{aligned} Au + N(u) &= f \quad \text{in} \quad \Omega \\ Tu &= \varphi \quad \text{on} \quad \Gamma := \partial\Omega. \end{aligned} \tag{1}$$

Here (f, φ) are the given data, and u the unknown. Problem (1) should be elliptic in some bounded, C^∞-smooth region $\Omega \subset \mathbb{R}^n$; that is A should be a linear differential operator in Ω while T should be a trace operator such that the system $\{A, T\}$ is elliptic in Ω. More generally, A could be suitably "pseudo-differential" as long as $\{A, T\}$ is injectively elliptic in the Boutet de Monvel calculus of boundary problems.

$N(u)$ stands for a non-linearity which combines $u(x)$ and its derivatives $D^\alpha u$ in a polynomial way, roughly speaking.

The main point is the following frequently asked question: given a solution u, does the presence of $N(u)$ influence the regularity of u?

This problem can of course be phrased in various frameworks: to measure regularity, the Besov and Triebel-Lizorkin spaces $B^s_{p,q}$ and $F^s_{p,q}$ could be adopted for $s \in \mathbb{R}$ and p, $q \in]0, \infty]$ (though with $p < \infty$ for $F^s_{p,q}$). But to simplify matters – and indeed to fix ideas – this survey deals with the Sobolev, or Bessel potential, spaces $H^s_p(\overline{\Omega})$, where $s \in \mathbb{R}$ and $1 < p < \infty$. Now the solution may be known to exist in some a priori space, denoted $H^{s_0}_{p_0}$ throughout, while data are given in other spaces having some integral exponent $r \in]1, \infty[$. The case with $r \neq p_0$ requires extra efforts, and the present paper deals with a flexible way of handling this.

The word "semi-linear" is often taken as an indication that solutions of problems like (1) will have practically the same regularity as in the linear case, i.e., as when $N \equiv 0$. However, when $r \neq p_0$ is allowed, it is more demanding to describe for which a priori spaces and data spaces this property of semi-linearity holds.

A classical way to obtain such conclusions is to improve the knowledge of u in finitely many steps (i.e., a boot-strap method). But one faces rather pains-taking difficulties when this method is applied to cases in which the a priori space for u is not "close enough" to the solution space associated to the data f and φ in the linear theory. (Such phenomena have been described in [Joh93, Joh95b] and in a joint work with T. Runst [JR97].)

However, in a recent article [Joh01] a different technique was worked out – it requires rather weaker assumptions than boot-strap methods do, it has cleaner proofs and in particular it also avoids the technicalities mentioned above. In short this approach is a much more flexible tool. It was exemplified for elliptic problems in full generality in [Joh01], where the crucial point was a specific parametrix formula for the non-linear problem (1); this formula is useful because one can read off a given solution's regularity directly.

The purpose of the present paper is to give a concise account of the resulting technique and to present how the parametrices straightforwardly give regularity improvements.

To give a very brief account of the outcome of the study (with examples to follow further below), it is useful to introduce three *parameter domains*:

$$\mathbb{D}(\mathcal{A}), \quad \mathbb{D}(N), \quad \mathbb{D}(L_u). \tag{2}$$

Here $\mathbb{D}(\mathcal{A})$ consists of all the (pairs of) parameters (s,p) for which the matrix-formed operator $\mathcal{A} := \binom{A}{T}$ is defined on the space H^s_p. This takes into account the class of T (and of A in the pseudo-differential case).

Similarly $\mathbb{D}(N)$ contains all the (s,p) for which N is defined on H^s_p and has order *less* than that of A on this space. Finally, and most importantly, for any $(s_0,p_0) \in \mathbb{D}(N)$ and any given u in $H^{s_0}_{p_0}$, there should exist some linear but possibly u-dependent operator L_u such that

$$N(u) = -L_u(u). \tag{3}$$

When the operator L_u is studied in its own right on the scale H^s_p, then $\mathbb{D}(L_u)$ contains the (s,p) for which L_u is defined and has lower order than A. In addition it is necessary to require of N and L_u that $\mathbb{D}(N) \subset \mathbb{D}(L_u)$.

In practice $\mathbb{D}(L_u)$ is much larger than $\mathbb{D}(N)$, and the more regular u is known a priori to be, the larger $\mathbb{D}(L_u)$ will be. (While $\mathbb{D}(N)$ is the same independently of any given solution u.) This leads to a main feature:

On the one hand, using a boot-strap method it turns out that one can work inside the domain $D(A) \cap \mathbb{D}(N)$; which is logical because N would lose "more derivatives" on spaces outside $\mathbb{D}(N)$. On the other hand, the present parametrix methods work well on the larger set

$$\mathbb{D}_u := \mathbb{D}(\mathcal{A}) \cap \mathbb{D}(L_u). \tag{4}$$

For this reason a given solution may be treated under much weaker initial assumptions on the data (f,φ). Indeed, the regularity of u is read off from the following *parametrix formula* (derived in [Joh01])

$$u = P^{(N)}_u(Rf + K\varphi) + \mathcal{R}u + (RL_u)^N u. \tag{5}$$

Here $(\, R \; K \,)$ denotes a left-parametrix of $(\frac{A}{T})$, with associated smoothing operator $\mathcal{R}$; that is $RA + KT = I - \mathcal{R}$. The parametrix $P_u^{(N)}$ is a finite Neumann series with the linear operator RL_u as 'quotient'. Consequently, with known mapping properties of R, K and L_u, the above formula (5) shows directly how the regularity of u is determined by the data together with the a priori regularity of u itself (the latter enters the term $(RL_u)^N$).

Below this is explained in detail by means of an example.

2. The framework

As a another simplification we may consider the following model problem, which is rich enough to illustrate the points. Here and below γ_0 denotes the standard trace (restriction) on Γ:

$$\begin{aligned} -\Delta u + u\partial_{x_1} u &= f \quad \text{in} \quad \Omega \\ \gamma_0 u &= \varphi \quad \text{on} \quad \Gamma. \end{aligned} \tag{6}$$

The discussion of (6) will be carried out under the hypothesis that a solution u is given for some specific data (f, φ) fulfilling

$$u \in H^{s_0}_{p_0}(\overline{\Omega}) \tag{7}$$

$$f \in H^{t-2}_r(\overline{\Omega}), \qquad \varphi \in B^{t-1/r}_{r,r}(\Gamma). \tag{8}$$

In general, the space $H^s_p(\overline{\Omega})$ is defined by restriction to Ω and $B^s_{p,p}(\Gamma)$ is defined via local coordinates on Γ.

With this set-up, the theme is whether u belongs to the space $H^t_r(\overline{\Omega})$ too. (There are of course necessary conditions for this to be true, eg $s_0 > 1/p_0$ must hold for the boundary condition to make sense. It is tempting to require in analogy that $t > 1/r$, but it is a point that weaker assumptions will suffice; hence this discussion is postponed a little.) Using boot-strap arguments in treating this, difficulties occur as mentioned in the introduction. Indeed, for cases with, say p_0 and r close to 1 and ∞ respectively, and small values of $s_0 > t$, boot-strapping is possible, but careful arguments based on special estimates of $u\partial_1 u$ are needed to avoid auxiliary spaces on which $\gamma_0 u$ is undefined; cf. [Joh95b].

With a more direct approach, the aim is to "invert" (6) by means of the formula

$$u = P_u^{(N)}(R_D f + K_D \varphi) + \mathcal{R}u + (R_D L_u)^N u. \tag{9}$$

To explain the various quantities in (9), it is first noted that the linear problem corresponding to (6) is considered as an equation for the elliptic Green operator (when $s > 1/p$, $1 < p < \infty$),

$$\mathcal{A} = \begin{pmatrix} -\Delta \\ \gamma_0 \end{pmatrix} : H^s_p(\overline{\Omega}) \to \begin{matrix} H^{s-2}_p(\overline{\Omega}) \\ \oplus \\ B^{s-1/p}_{p,p}(\Gamma). \end{matrix} \tag{10}$$

Then $(R_D \ K_D)$ is taken as a parametrix (belonging to the Boutet de Monvel calculus), i.e., it is continuous in the opposite direction in (10) and

$$(R_D \quad K_D)\begin{pmatrix} -\Delta \\ \gamma_0 \end{pmatrix} = I - \mathcal{R}; \tag{11}$$

here the range $\mathcal{R}(H_p^s) \subset C^\infty(\overline{\Omega})$ for all $s > 1/p$. (In fact $\mathcal{R} = 0$ is possible for this Dirichlet problem, but it is retained here to make it clear that also in general its presence is harmless.)

The second ingredient in (9) is a decomposition of the non-linear term as

$$u\partial_{x_1} u = -L_u(u), \qquad L_u \text{ linear.} \tag{12}$$

More precisely, it is necessary to ensure that L_u has certain mapping properties, hence it is defined by means of a universal extension operator from Ω to $\mathbb{R}^n$, say ℓ_Ω, to be

$$-L_u(v) = \pi_1(\ell_\Omega u, \ell_\Omega \partial_1 v) + \pi_2(\ell_\Omega u, \ell_\Omega \partial_1 v) + \pi_3(\ell_\Omega v, \ell_\Omega \partial_1 u). \tag{13}$$

Here the $\pi_j(\cdot,\cdot)$ are para-multiplication operators defined on $\mathbb{R}^n$ (so that restriction to Ω of each term on the right-hand side of (13) is understood). These are introduced using a Littlewood-Paley partition $1 = \sum_{j=0}^\infty \Phi_j(\xi)$ with smooth functions Φ_j supported at $\{ 2^{j-1} \le |\xi| \le 2^{j+1} \}$ for $j > 0$; then

$$\pi_1(g,h) = \sum_{j=2}^{\infty} (\Phi_0(D) + \cdots + \Phi_{j-2}(D))g \cdot \Phi_j(D)h, \tag{14}$$

and $\pi_3(g,h) := \pi_1(h,g)$ whilst $\pi_2(g,h)$ gives the remainder in the formal decomposition of $g \cdot h$.

Using this, the parametrices $P_u^{(N)}$ of the non-linear problem (6) are now finally introduced as

$$P_u^{(N)} = I + R_D L_u + \cdots + (R_D L_u)^{(N-1)}, \qquad N \in \mathbb{N}. \tag{15}$$

They clearly depend on the given solution u, and since both R_D and L_u are linear, so are the $P_u^{(N)}$ on every $H_p^s(\overline{\Omega})$ with $(s,p) \in \mathbb{D}(\mathcal{A}) \cap \mathbb{D}(L_u)$; cf. (16) ff.

For the above model problem, the parameter domains from the introduction are, when $N(u) = u\partial_1 u$ and $t_+ = \max(0,t)$ denotes the positive part,

$$\mathbb{D}(\mathcal{A}) = \{ (s,p) \mid s > 1/p \} \tag{16}$$

$$\mathbb{D}(N) = \{ (s,p) \mid s > \tfrac{1}{2} + (\tfrac{n}{p} - \tfrac{n}{2})_+, \quad s > \tfrac{n}{p} - 1 \} \tag{17}$$

$$\mathbb{D}(L_u) = \{ (s,p) \mid s + s_0 > 1 + (\tfrac{n}{p_0} + \tfrac{n}{p} - n)_+ \}. \tag{18}$$

The two first restrictions make the trace $\gamma_0 u$ and the product $u \cdot \partial_1 u$ well defined on $H_p^s(\overline{\Omega})$, whilst the second condition in (17) implies that for some $s' > s - 2$ it holds that $u\partial_1 u \in H_p^{s'}(\overline{\Omega})$ for every $u \in H_p^s(\overline{\Omega})$.

More noteworthy is it that the requirement in (18) is effectively weaker the better the a priori knowledge of u is: for higher values of s_0 or larger values of p_0, more pairs (s,p) fulfil the inequality.

A closer analysis shows that L_u has order ω in the sense that

$$L_u \colon H^s_p(\overline{\Omega}) \to H^{s-\omega}_p(\overline{\Omega}) \quad \text{for all } (s,p) \in \mathbb{D}(L_u), \tag{19}$$

$$\omega = 1 + (\tfrac{n}{p_0} - s_0)_+ + \varepsilon, \qquad \varepsilon \geq 0. \tag{20}$$

Here $\varepsilon > 0$ is only necessary for $s_0 = \frac{n}{p_0}$. If one removes ℓ_Ω and the restriction to Ω from L_u, then the resulting operator is in the Hörmander class $\mathrm{OPS}^{\omega}_{1,1}(\mathbb{R}^n \times \mathbb{R}^n)$, leading to an analogous continuity property.

It is important to observe that with $u \in H^{s_0}_{p_0}(\overline{\Omega})$ for some (s_0, p_0) in $\mathbb{D}(N)$, it follows from (17) that the order of L_u satisfies $\omega < 2$ in (19). In other words, L_u loses fewer derivatives than Δ. It is a peculiar fact that L_u, once u is chosen, actually has constant order on all spaces regardless of whether they have parameter inside or outside $\mathbb{D}(N)$ (by comparison, the boundary of $\mathbb{D}(N)$ contains points where N attains the order 2).

Using the above continuity results for R_D, K_D and L_u, one can now show that the parametrix formula holds and derive the regularity results.

Remark 2.1. *The operator L_u in* (13) *differs from the linearisations in J.M. Bony's work* [Bon81] *because the π_2-term is a part of the operator instead of being treated as a negligible error term. In the present context this has to be so, for it does occur that this term has a non-negligible regularity and in addition it would not be natural to violate the identity $L_u u = u\partial_1 u$. For this reason it is suggested that one could call L_u the* full *paralinearisation of $u\partial_1 u$.*

Moreover, the perhaps more natural linearisations $v \mapsto u\partial_1 v$ and the differential $u\partial_1 v + v\partial_1 u$ do not work in this context, because they are not moderate in the terminology of [Joh01]. *Indeed, on H^s_p they have order equal to $s - 1 - s_0$ for large s, and this has no upper limits for $s \to \infty$; unlike L_u that has constant order with respect to s as observed above.*

Remark 2.2. *About the above results it should be mentioned that the properties of linear elliptic problems were deduced for the H^s_p-scale in full generality by G. Grubb* [Gru90], *who extended the Boutet de Monvel calculus to these spaces (and to the classical Besov spaces). In particular this implies* (10) *and the statements following it. (For the $B^s_{p,q}$ and $F^s_{p,q}$ scales there is a similar extension of the calculus in* [Joh96], *which applies to the present problems in the same way.) For introductions to the calculus the reader may consult* [Gru97, Gru91].

In the definition of L_u, the universal extension operator was constructed by V. Rychkov [Ryc99b, Ryc99a]. *He showed that ℓ_Ω can be taken such that for all s and p it is continuous*

$$\ell_\Omega \colon H^s_p(\overline{\Omega}) \to H^s_p(\mathbb{R}^n) \tag{21}$$

and that $r_\Omega \ell_\Omega = I$ holds on $H^s_p(\overline{\Omega})$ (in fact it was carried out for the Besov and Triebel-Lizorkin scales).

The para-multiplication operators $\pi_j(\cdot,\cdot)$ in (13) *follow M. Yamazaki* [Yam86] *in the notation and the definition. To prove* (19)–(20) *it suffices to combine* (21) *with standard estimates of the $\pi_j(\cdot,\cdot)$; these are essentially found in* [Yam86], *but for a proof the reader may consult* [Joh01], *which presents a general study of non-linear operators of product type (encompassing sums of terms* $D^\alpha u \cdot D^\beta u$ *and more general expressions). For a full set of estimates of para-multiplication operators proved directly (without the somewhat heavier paradifferential techniques in* [Yam86]*), the reader may eg consult* [Joh95a, Th 5.1].

3. Results involving parametrices

We shall now proceed to state the results for the model problem, that one obtains from the parametrices. Recall that $\mathcal{A} = \binom{-\Delta}{\gamma_0}$ and $N(u) = u\partial_1 u$. Furthermore $\mathbb{D}(\mathcal{A})$, $\mathbb{D}(N)$ and $\mathbb{D}(L_u)$ are given as in (16) ff, so that they fulfil the conditions described after (2).

The first point is to establish that the parametrix formula (9) really holds, and to give basic properties of the entering operators.

Theorem 3.1. *Assume that* (6) *and* (7)–(8) *hold for parameters fulfilling*

$$\begin{aligned} (s_0, p_0) &\in \mathbb{D}(\mathcal{A}) \cap \mathbb{D}(N) \\ (t, r) &\in \mathbb{D}(\mathcal{A}) \cap \mathbb{D}(L_u) =: \mathbb{D}_u. \end{aligned} \tag{22}$$

Then (9) *holds, and for* $P_u^{(N)}$ *as in* (15),

$$\forall N \in \mathbb{N}_0,\ \forall (s,p) \in \mathbb{D}_u: \qquad P_u^{(N)} \colon H_p^s(\overline{\Omega}) \to H_p^s(\overline{\Omega}), \tag{23}$$

$$\exists N \in \mathbb{N}_0: \quad (R_D L_u)^N \colon H_{p_0}^{s_0}(\overline{\Omega}) \to H_r^t(\overline{\Omega}). \tag{24}$$

In these lines the arrows stand for continuous, linear maps.

Actually (24) holds for all sufficiently large values of N, but usually it is enough have a single such N.

The above Theorem 3.1 is a special case of an abstract result proved in [Joh01, Th 2.2]. The proof is not difficult in itself; it is formulated for a general situation specified by some lengthy, but essentially rather mild conditions labelled (I)–(V) in [Joh01, Sect. 2], and that these are fulfilled for the model problem considered in this paper is a consequence of the above Section 2.

Now one immediately gets

Corollary 3.2. *If* (6), (7)–(8) *and* (22) *all hold, then* u *is also an element of* $H_r^t(\overline{\Omega})$.

It is a major point of the paper that, to prove this, one may take N as in (24); then the properties in (24), (23), (19) together with formula (9) show that $u \in H_r^t(\overline{\Omega})$.

It deserves to be emphasised that (t, r) is assumed to lie in $\mathbb{D}(L_u)$ but not necessarily in the smaller parameter domain for the non-linear term, $\mathbb{D}(N)$. For this reason it is possible to conclude that given solutions may belong to spaces that are beyond the reach of the boot-strap method.

4. Final remarks

This paper focuses on semi-linear elliptic boundary problems, and even specialises to a simple model problem in order not to burden the exposition; the possible extensions are many, but the reader should get a good impression of the possibilities from the above. Within the framework of elliptic problems some of the generalisations are indicated after (1), but is is also possible to include semi-linear elliptic systems like the stationary Navier-Stokes equation or von Karman's equations. This requires an extended notion of product type non-linearities, defined on sections of vector bundles. The reader may consult [Joh01] for this.

The general study in [Joh01, Sect. 2] also allow some applications to parabolic initial-boundary problems with non-linearities of product type. However, when such problems are non-homogeneous the compatibility conditions on the data set severe restrictions to how much the regularity can be improved; but even so it should be possible to work out some results in this area.

Concerning the tools, it is on the one hand clear that it is a fine theory of linear elliptic problems that enter, namely that of the Boutet de Monvel calculus. On the other hand, the treatment of the non-linear terms is based on para-multiplication operators on $\mathbb{R}^n$. This technique was essentially introduced (independently) by J. Peetre and H. Triebel around 1976–77 [Pee76, Tri77, Tri78] in order to analyse the pointwise product. One way to sum up the present paper could be to say that para-multiplication also may enter in a crucial way in treatments of certain non-linear perturbations of elliptic boundary problems.

References

[Bon81] J.-M. Bony, *Calcul symbolique et propagations des singularités pour les équations aux dérivées partielles non linéaires*, Ann. scient. Éc. Norm. Sup. **14** (1981), 209–246.

[Gru90] G. Grubb, *Pseudo-differential boundary problems in L_p-spaces*, Comm. Part. Diff. Equations **15** (1990), 289–340.

[Gru91] G. Grubb, *Parabolic pseudo-differential boundary problems and applications*, Microlocal analysis and applications, Montecatini Terme, Italy, July 3–11, 1989 (Berlin) (L. Cattabriga and L. Rodino, eds.), Lecture Notes in Mathematics, vol. 1495, Springer, 1991.

[Gru97] G. Grubb, *Pseudodifferential boundary problems and applications*, Jahresber. Deutsch. Math.-Verein. **99** (1997), no. 3, 110–121.

[Joh93] J. Johnsen, *The stationary Navier-Stokes equations in L_p-related spaces*, Ph.D. thesis, University of Copenhagen, Denmark, 1993, Ph.D.-series **1**.

[Joh95a] J. Johnsen, *Pointwise multiplication of Besov and Triebel-Lizorkin spaces*, Math. Nachr. **175** (1995), 85–133.

[Joh95b] J. Johnsen, *Regularity properties of semi-linear boundary problems in Besov and Triebel-Lizorkin spaces*, Journées "équations derivées partielles", St. Jean de Monts, 1995, Grp. de Recherche CNRS no. 1151, 1995, pp. XIV1–XIV10.

[Joh96] J. Johnsen, *Elliptic boundary problems and the Boutet de Monvel calculus in Besov and Triebel-Lizorkin spaces*, Math. Scand. **79** (1996), 25–85.

[Joh01] J. Johnsen, *Parametrices and L_p-theory of semi-linear boundary problems*, Tech. Report R-01-2006, Department of Mathematical Sciences, Aalborg University, 2001, (preprint).

[JR97] J. Johnsen and T. Runst, *Semilinear boundary problems of composition type in L_p-related spaces*, Comm. P. D. E. **22** (1997), no. 7–8, 1283–1324.

[Pee76] J. Peetre, *New thoughts on Besov spaces*, Duke Univ. Math. Series, Durham, 1976.

[Ryc99a] V.S. Rychkov, *On a theorem of Bui, Paluszyński, and Taibleson*, (Russian, English) Proc. Steklov Inst. Math. **227** (1999), 280–292; translation from Tr. Mat. Inst. Steklova **227** (1999), 286–298.

[Ryc99b] V.S. Rychkov, *On restrictions and extensions of the Besov and Triebel-Lizorkin spaces with respect to Lipschitz domains*, J. London Math. Soc. (2) **60** (1999), no. 1, 237–257.

[Tri77] H. Triebel, *Multiplication properties of the spaces $B^s_{p,q}$ and $F^s_{p,q}$*, Ann. Mat. Pura Appl. **113** (1977), 33–42.

[Tri78] H. Triebel, *Spaces of Besov-Hardy-Sobolev type*, Teubner-Texte zur Mathematik, vol. 15, Teubner Verlagsgesellschaft, Leipzig, 1978.

[Yam86] M. Yamazaki, *A quasi-homogeneous version of paradifferential operators, I. Boundedness on spaces of Besov type*, J. Fac. Sci. Univ. Tokyo Sect. IA, Math. **33** (1986), 131–174.

Jon Johnsen
Department of Mathematical Sciences
Aalborg University
Fredrik Bajers Vej 7G
DK–9220 Aalborg East, Denmark
E-mail address: `jjohnsen@math.auc.dk`

D. Haroske, T. Runst, H.-J. Schmeisser (eds.): Function Spaces, Differential Operators and Nonlinear Analysis. The Hans Triebel Anniversary Volume.

Potential Estimates for Large Solutions of Semilinear Elliptic Equations

Denis A. Labutin

Dedicated to Prof. Hans Triebel on the occasion of his 65th birthday

1. Introduction

In this report we describe our recent existence results [26], [27] for two not quite standard problems for semilinear elliptic equations in general domains.

The first problem concerns *large solutions* to nonlinear elliptic equations in arbitrary bounded domains $\Omega \subset \mathbf{R}^n$, $n \geq 3$. These are solutions $u \in C^2_{\mathrm{loc}}(\Omega)$ to the nonlinear problem

$$\begin{cases} \Delta u - |u|^{q-1}u = 0 & \text{in} \quad \Omega \\ \qquad u(x) \to +\infty & \text{when} \quad x \to \partial\Omega. \end{cases} \tag{1.1}$$

For the parameter q we always assume

$$q > 1.$$

Note that the strong maximum principle for elliptic equations implies that u from (1.1) satisfies

$$u > 0, \quad \Delta u - u^q = 0 \quad \text{in} \quad \Omega. \tag{1.2}$$

Hence without loss of generality we can consider only positive solutions of (1.1).

Let now $\mathbf{S}^n$ be the unit n-dimensional sphere with the standard metric g_0 induced by the embedding $\mathbf{S}^n \hookrightarrow \mathbf{R}^{n+1}$. Our second problem concerns finding *geometrically complete solutions* in arbitrary domains $\Omega \subset \mathbf{S}^n$, $n \geq 3$. These are solutions $u \in C^2_{\mathrm{loc}}(\Omega)$ to the problem

$$\begin{cases} \Delta_{g_0} u - R(g_0)\dfrac{(n-2)}{4(n-1)}u - 1\,\dfrac{(n-2)}{4(n-1)}u^{\frac{n+2}{n-2}} = 0 & \text{in} \quad \Omega \\ \qquad u > 0 & \text{in} \quad \Omega \\ \qquad u^{\frac{4}{n-2}} g_0 \text{ is a complete metric} & \text{in} \quad \Omega. \end{cases} \tag{1.3}$$

Here Δ_{g_0} is the Laplace-Beltrami operator on $(\mathbf{S}^n, g_0)$, and

$$R(g_0) = n(n-1)$$

is the scalar curvature of g_0. Geometrically (1.3) means that in the domain Ω we conformally deform the standard metric g_0 to the new metric

$$g = u^{\frac{4}{n-2}} g_0,$$

which is complete in Ω, and has the scalar curvature

$$R(g) \equiv -1.$$

By definition, the metric g in (1.3) is complete in Ω if for any semi-open curve $\gamma\colon [0,1) \to \Omega$ such that

$$\operatorname*{dist}_{g_0}(\gamma(t), \partial\Omega) \to 0 \quad \text{when} \quad t \to 1-0,$$

we have

$$\operatorname*{length}_{g}(\gamma) = \int_0^1 u^{\frac{2}{n-2}}(\gamma) \left(g_0(\gamma)_{ij}\dot{\gamma}^i\dot{\gamma}^j\right)^{1/2} dt = +\infty.$$

As compared to (1.1), (1.2), we have the particular case

$$q = (n+2)/(n-2)$$

in (1.3).

Thus instead of more common boundary value problems (Dirichlet, Neumann, and their nonlinear analogies) we impose in (1.1), (1.3) the pointwise blow-up condition, or the condition of the completeness of the corresponding metric. The crucial fact about the equation

$$\Delta u - |u|^{q-1}u = 0, \quad q > 1, \tag{1.4}$$

is that the elliptic comparison principle holds for its sub- and supersolutions. Equation (1.4) is the Euler–Lagrange equation for the functional

$$F(u) = \frac{1}{2}\int_\Omega |Du|^2\,dx + \frac{1}{q+1}\int_\Omega |u|^{q+1}\,dx.$$

Another nice property of (1.4) is that the functional F is bounded from below, obviously $F \geq 0$. All these facts make the treatment of the traditional boundary value problems for (1.4) rather straightforward. In particular such effects for unbounded functionals as Pohozaev identities, quantisation of energy for the strong convergence, and so on, do not arise for the "minus sign" nonlinearity in (1.1)–(1.4).

The main result of our work is the necessary and sufficient condition on the domains Ω for the solubility of (1.1), (1.3). Our criteria for solvability of (1.1), (1.3) use the language of potential theory. Let us recall a fundamental result from the classical potential theory for the Laplace equation. This is the Wiener test for solvability of the classical Dirichlet problem for harmonic functions. Wiener's theorem states that in a bounded domain $D \subset \mathbf{R}^n$, $n \geq 3$, the Dirichlet problem

$$\begin{cases} \Delta u = 0 & \text{in} \quad D \\ u = f & \text{on} \quad \partial D \end{cases}$$

is solvable for all boundary data $f \in C(\partial D)$, if and only if $\mathbf{R}^n \setminus D$ is not thin. Formally the latter means that

$$\int_0^1 \frac{cap(B(x,\rho)\setminus D)}{cap(B(x,\rho))}\frac{d\rho}{\rho} = +\infty \quad \text{for any } x \in \partial D.$$

Here $cap(\cdot)$ is the classical (electrostatic) capacity, and the upper integration limit 1 can be replaced by any $\delta > 0$. Our main theorems state that (1.1), (1.3) admit a solution if and only if the corresponding Wiener-type tests with certain capacities hold. We describe the results for problem (1.1) in Section 2, and for problem (1.3) in Section 3.

I wish to thank members of Professor Triebel's analysis group at Friedrich Schiller Universität Jena for their interest in my work, help, and encouragement since I was a student. I am very grateful to Winfried Sickel, Hans-Jürgen Schmeisser, Thomas Runst, and Dorothee Haroske for the opportunity to attend the conference in Teistungen in June-July 2001, for their hospitality during my visit to Jena in December 2001, and for their friendly patience during the preparation of this note. Although with a big delay, I wish Professor Triebel a happy birthday.

This report was completed during my stay at Centre de Recerca Matematica in March 2002, and I wish to thank colleagues in Barcelona for their hospitality.

2. Solutions with pointwise blowup

Of the two basic questions on problem (1.1) in arbitrary domains Ω namely existence and uniqueness, our main result completely resolves the first. Theorem 2.1 states that the solubility of (1.1) is equivalent to the Wiener-type test with respect to a certain capacity. On the second question it is well known that the uniqueness for (1.1) fails in general domains [25], [19].

After the groundbreaking papers by Perkins Dynkin and LeGall solutions of (1.1), (1.2) attracted a lot of attention of probabilists. Currently this is a very active area of research on the interface between the theory of random processes, nonlinear partial differential equations, and analysis. We refer to the ICM reports by Perkins [45] and LeGall [30] for a survey of the progress in the field and bibliography, cf. also [29]. Recent monographs [11] [12], [14], and [31] are dedicated to different aspects of the theory. At the moment the probabilistic methods are limited to the case

$$1 < q \leq 2$$

in (1.1), (1.2). Our research was inspired by a recent result of Dhersin and LeGall [10]. They proved that the existence for the problem (1.1) for $q = 2$ is equivalent to a Wiener type criterion for $\Omega^c = \mathbf{R}^n \setminus \Omega$. This result is one of the milestones of the theory, cf. [30], [31]. The crucial idea of Dhersin and LeGall was to combine the classical potential theory with the sharp bounds on the hitting probability for the super-Brownian motion related to positive solutions of

$$\Delta u - u^2 = 0.$$

An open problem in this area was to extend the result to the full range $q > 1$, cf., e.g., [31]. Relying upon entirely analytic ideas we prove the Wiener test for solubility of (1.1) for all $q > 1$.

Large solutions (1.1) were initially studied by Loewner and Nirenberg [34], as well as in the earlier papers of Keller [23] and Osserman [44]. Loewner and Nirenberg considered the case $q = (n+2)/(n-2)$ arising in conformal differential geometry. They proved that in smooth domains Ω there exists the unique solution of (1.1). Later the questions of existence, uniqueness, and the rate of the boundary blow-up were investigated by many authors to which the bibliography is very extensive [54]. For example, Brézis and Veron [8] proved that singletons are regular boundary points for (1.1) if and only if

$$1 < q < \frac{n}{n-2}.$$

Aviles, Bandle, Essen, Finn, Marcus, McOwen, Veron, and others investigated the questions for domains bounded by non-smooth hypersurfaces or manifolds of lower dimensions, as well as for more general semilinear equations. Kondratiev and Nikishkin [25] discovered the non-uniqueness for (1.1), cf. also [19]. We refer to survey [43] and monograph [54], for further description and references. Additionally papers [18] and [53] contain the very recent results. However, up to this point, the analytic approach has not obtained the necessary and sufficient condition for the existence in (1.1).

The capacity suitable for problem (1.1) is defined as follows. Fix $x_0 \in \mathbf{R}^n$, $n \geq 3$. Let $K \subset \mathbf{R}^n$ be a compact subset of the ball $B(x_0, 3/2)$. For $1 < p < \infty$ define

$$\mathcal{C}_p(K) = \inf\left\{ \int_{B(x_0,2)} |D^2\varphi|^p \;:\; \varphi \in C_0^\infty(B(x_0,2)),\; \varphi|_K \geq 1 \right\}. \tag{2.1}$$

Following the axiomatic potential theory we extend $\mathcal{C}_p$ as an outer capacity to any set E, $\overline{E} \subset B(x_0, 3/2)$. Capacities defined with different x_0 are equivalent. Capacity $\mathcal{C}_p$ is essentially the Bessel capacity associated with the Sobolev space $W^{2,p}(\mathbf{R}^n)$. Such capacities were carefully investigated in the theory of nonlinear potentials. The theory originates in early works by Maz'ya and Serrin in the 1960s, later developed in the 1970–1980s in papers by Adams, Fuglede, Havin, Hedberg, Maz'ya, Meyers, and many others. Good references are monographs [4], [39], and [58] wherein the reader can also find a rich bibliography along with historical notes. Now our main result [26] on (1.1) is stated below.

Theorem 2.1. *Let $\Omega \subset \mathbf{R}^n$, $n \geq 3$, be a bounded domain, and let $q > 1$. The following statements are equivalent:*

(i) *Problem (1.1) has a solution $u \in C^2_{\mathrm{loc}}(\Omega)$.*

(ii) *The set $\Omega^c = \mathbf{R}^n \setminus \Omega$ is not thin, that is*

$$\int_0^1 \frac{\mathcal{C}_{q'}\left(\Omega^c \cap B(x,r)\right)}{r^{n-2}} \frac{dr}{r} = +\infty \quad \text{for all} \quad x \in \Omega^c, \tag{2.2}$$

where $\frac{1}{q} + \frac{1}{q'} = 1$.

For $q = 2$, Theorem 2.1 was proved in [10] using probabilistic methods. It is very likely that the proof from [10] can be generalised for $1 < q \leq 2$ using ideas from [13]. Condition (2.2) and well-known properties of the capacity easily imply the solubility of (1.1) in specific classes of domains Ω.

Wiener proved his criterion for solvability of the Dirichlet problem for harmonic functions in the fundamental papers [55], [56]. Later Wiener tests for the Dirichlet problem for more general *linear* second-order (degenerate) elliptic and parabolic equations were proven in [33], [17], [7], [9], [15], [16]. Recently the first complete results were obtained for linear elliptic equations of higher order [41]. Seminal papers [38] and [20] started the research on Wiener regularity of the Dirichlet problem for *quasilinear* equations of the second order by proving the sufficiency of a Wiener type criterion. A recent paper [24] completed the investigation of the basic question by proving the necessity, cf. also earlier contribution [32]. Monographs [21] and [35] give a comprehensive exposition of these results. Trudinger and Wang [49] presented an alternative more general and concise approach to quasilinear equations of the second order. In [28], a Wiener criterion was proved for Hessian equations. Hessian equations [50], [51], [52] are *fully nonlinear* (nonlinear on the second derivatives) elliptic equations. We refer to surveys [2], [40] for further description of this area and the bibliography. In connection with Theorem 2.1 we mention the following result. Consider the classical (finite data) Dirichlet problem

$$\begin{cases} \Delta u - |u|^{q-1}u = 0 & \text{in} \quad \Omega \\ u = f & \text{on} \quad \partial\Omega, \end{cases}$$

$q > 1$. Adams and Heard [3], [1] proved that it is solvable for all $f \in C(\partial\Omega)$ if and only if the classical Wiener test from [55], [56] holds for Ω.

Capacity (2.1) has been used in the previous works on potential theory for semilinear equations. Baras and Pierre [6] used it to characterise removable singularities for solutions of (1.2). In [5], [22] capacity (2.1) was used to investigate a different class of semilinear equations. We also mention the continuing series of papers by Marcus and Veron [36], [37] on Riesz-Herglotz type effects for equation (1.2) and its parabolic counterpart, questions which are also under current active study from the probabilistic point of view by Dynkin, Kuznetsov, LeGall, Delmas, Dhersin, and others.

3. Geometrically complete solutions

The problem of characterisation of domains on the sphere, which admit a complete metric with constant scalar curvature conformal to the standard metric on the sphere, originates in the works of Loewner and Nirenberg [34], and Schoen and Yau [47]. We refer to problem #36 from Yau's problem list [57], [48], and to the survey by McOwen [43] for more detailed description. Analytically for a given domain $\Omega \subset \mathbf{S}^n$ and a given constant

$$R \in \{-1, 0, 1\}$$

one seeks a solution $u \in C^2_{\text{loc}}(\Omega)$ to the problem

$$\begin{cases} \Delta_{g_0} u - \dfrac{n(n-2)}{4} u + R \dfrac{(n-2)}{4(n-1)} u^{\frac{n+2}{n-2}} = 0 & \text{in} \quad \Omega \\ u > 0 & \text{in} \quad \Omega \\ u^{\frac{4}{n-2}} g_0 \text{ is complete metric} & \text{in} \quad \Omega. \end{cases} \tag{3.1}$$

In our paper [27] we resolve problem (3.1) in the case of negative constant scalar curvature

$$R = -1.$$

In other words we find the necessary and sufficient condition on the domain Ω for the existence of a solution to (1.3). As it was mentioned in the introduction, our condition is inspired by the notion of thinness from potential theory. Let us remark that our work concerns the basic problem of existence for (1.3), although other questions about solutions of (1.3), (3.1) can be asked as well, e.g., [46], [42], [43].

Now we briefly describe the previous results on solvability of (1.3) and more generally (3.1). For more detailed description and references cf. McOwen's survey [43]. We will formulate the results only for $\mathbf{S}^n$ although some of them have been proved for arbitrary compact manifolds. The case $R = 1$ is regarded as the hardest among the three.

Set

$$K = \mathbf{S}^n \setminus \Omega.$$

General *necessary* conditions for solvability of (3.1) in terms of the Hausdorff measure of K were given by Schoen and Yau [47]. They are

$$\Lambda_{\frac{n-2}{2}+\varepsilon}(K) = 0 \ \forall \varepsilon > 0 \quad \text{for} \quad R = 1,$$
$$\Lambda_{\frac{n}{2}+\varepsilon}(K) = 0 \ \forall \varepsilon > 0 \quad \text{for} \quad R = 0.$$

Earlier Loewner and Nirenberg [34] found that a necessary condition for solvability of (1.3) is

$$\Lambda_{\frac{n-2}{2}}(K) = +\infty \quad \text{for} \quad R = -1.$$

In the converse direction up to the present moment the existence was established under much more restrictive *sufficient* conditions on K than those in terms of Hausdorff measure. In papers by Schoen, Mazzeo, and Pacard existence for (3.1) with $R = 1$ was proved if K is a disjoint union of points and smooth submanifolds of dimensions less or equal $(n-2)/2$. It is easy to show, that the statement holds in the case $R = 0$ as well. For (3.1) with $R = -1$ Veron, Aviles, and McOwen proved that the problem is solvable if K is a disjoint union of smooth submanifolds of dimensions strictly greater than $(n-2)/2$. There are further results by Finn for submanifolds with boundary, conical edges, and certain types of their unions. We refer to Finn [18] for the latest results and references.

Clearly there was still a substantial gap between such sufficient and necessary conditions.

The capacity suitable for problem (1.3) is defined for subsets $E \subset \mathbf{S}^n$, $n \geq 3$, satisfying

$$\operatorname{diam}_{g_0}(\overline{E}) < \pi/2.$$

Any such set lies in a hemisphere. After a rotation we can assume that it lies in the south hemisphere. Set

$$[g_0^{ij}] = g_0^{-1}.$$

Then for any compactum E we define

$$\begin{aligned} \mathcal{C}(E) &= \inf\left\{\int_{\mathbf{S}^n} |\nabla_{g_0}^2 \varphi|_{g_0}^{\frac{n+2}{4}} \, dV_{g_0}\right\} \\ &= \inf\left\{\int_{\mathbf{S}^n} \left(g_0^{ij} g_0^{kl} (\nabla_{g_0}^2 \varphi)_{ij} (\nabla_{g_0}^2 \varphi)_{kl}\right)^{\frac{n+2}{8}} \, dV_{g_0}\right\}. \end{aligned} \tag{3.2}$$

Note that

$$\frac{1}{(n+2)/(n-2)} + \frac{1}{(n+2)/4} = 1.$$

In (3.2) symbols ∇_{g_0}, $|\cdot|_{g_0}$, dV_{g_0} stand respectively for the Riemannian connection, the norm with respect to the inner product of tensors, the Riemannian volume element of the metric g_0. The infimum in (3.2) is taken over the set

$$\Big\{\varphi \in C_0^\infty(\mathbf{S}^n) : \quad \varphi \equiv 1 \text{ in a neighbourhood of } E \\ \text{and } \varphi|_{B(N,\pi/4)} \equiv 0\Big\},$$

where $B(N, \pi/4)$ is the ball in the metric g_0 of radius $\pi/4$ centered at the north pole N. Essentially, $\mathcal{C}(\cdot)$ is the Bessel capacity $\mathcal{C}_{(n+2)/4}(\cdot)$ for the Sobolev space $W^{2,(n+2)/4}(\mathbf{R}^n)$ defined in the previous section. Now we state the main result [27] for problem (1.3).

Theorem 3.1. *Let $\Omega \subset \mathbf{S}^n$, $n \geq 3$, be an open set, $K = \mathbf{S}^n \setminus \Omega$. The following statements are equivalent:*

(i) *There exists a complete metric in Ω with constant negative scalar curvature conformal to g_0.*

(ii) *The compactum K is not thin, that is*

$$\int_0^1 \left(\frac{\mathcal{C}(B(x,\rho) \cap K)}{\mathcal{C}(B(x,\rho))}\right)^{\frac{2}{n-2}} \frac{d\rho}{\rho} = +\infty \quad \text{for all } x \in K.$$

In view of Theorem 3.1 it would now be interesting to clarify how the condition

$$\mathcal{C}(\mathbf{S}^n \setminus \Omega) = 0$$

relates to the solvability of (3.1) with $R = 0, 1$.

References

[1] D.R. Adams, L^p potential theory techniques and nonlinear PDE, in *Potential theory (Nagoya, 1990)*, 1–15, de Gruyter, Berlin, 1992.

[2] D.R. Adams, Potential and capacity before and after Wiener. Proceedings of the Norbert Wiener Centenary Congress, 1994 (East Lansing, MI, 1994), 63–83. *Proc. Sympos. Appl. Math.*, **52**, Amer. Math. Soc., Providence, RI, 1997.

[3] D.R. Adams, A. Heard, The necessity of the Wiener test for some semi-linear elliptic equations. *Indiana Univ. Math. J.*, **41** (1992), 109–124.

[4] D.R. Adams, L.I. Hedberg, *Function spaces and potential theory*, Springer-Verlag, Berlin, Heidelberg, 1996.

[5] D.R. Adams, M. Pierre, Capacitary strong type estimates in semilinear problems. *Ann. Inst. Fourier (Grenoble)*, **41** (1991), 117–135.

[6] P. Baras, M. Pierre, Removable singularities for semilinear equations. *Ann. Inst. Fourier (Grenoble)*, **34** (1984), 185–206.

[7] P. Bauman, A Wiener test for nondivergence structure, second-order elliptic equations. *Indiana Univ. Math. J.*, **34** (1985), 825–844.

[8] H. Brézis, L. Veron, Removable singularities for some nonlinear elliptic equations. *Arch. Rational Mech. Anal.*, 75 (1980/81), 1–6.

[9] G. Dal Maso, U. Mosco, Wiener criteria and energy decay for relaxed Dirichlet problems. *Arch. Rational Mech. Anal.*, **95** (1986), 345–387.

[10] J.-S. Dhersin, J.-F. Le Gall, Wiener's test for super-Brownian motion and the Brownian snake. *Probab. Theory Related Fields*, **108** (1997), 103–129.

[11] E.B. Dynkin, *Diffusions, superdiffusions and partial differential equations*, American Mathematical Society Colloquium Publications, American Mathematical Society, Providence, RI, *to appear.*

[12] E.B. Dynkin, *An introduction to branching measure-valued processes*, American Mathematical Society, Providence, RI, 1994.

[13] E.B. Dynkin, S.E. Kuznetsov, Superdiffusions and removable singularities for quasilinear partial differential equations. *Comm. Pure Appl. Math.*, 49 (1996), 125–176.

[14] A.M. Etheridge, *An introduction to superprocesses*, American Mathematical Society, Providence, RI, 2000.

[15] L.C. Evans, R.F. Gariepy, Wiener's criterion for the heat equation. *Arch. Rational Mech. Anal.*, **78** (1982), 293–314.

[16] E.B. Fabes, N. Garofalo, E. Lanconelli, Wiener's criterion for divergence form parabolic operators with C^1-Dini continuous coefficients. *Duke Math. J.*, **59** (1989), 191–232.

[17] E. Fabes, D. Jerison, C. Kenig, The Wiener test for degenerate elliptic equations. *Ann. Inst. Fourier (Grenoble)*, **32** (1982), 151–182.

[18] D.L. Finn, Behavior of positive solutions to $\Delta_g u = u^q + Su$ with prescribed singularities. *Indiana Univ. Math. J.*, 49 (2000), 177–219.

[19] D.L. Finn, R.C. McOwen, Singularities and asymptotics for the equation $\Delta_g u - u^q = Su$. *Indiana Univ. Math. J.*, 42 (1993), 1487–1523.

[20] R. Gariepy, W.P. Ziemer, A regularity condition at the boundary for solutions of quasilinear elliptic equations. *Arch. Rational Mech. Anal.*, **67** (1977), 25–39.

[21] J. Heinonen, T. Kilpeläinen, O. Martio, *Nonlinear potential theory of degenerate elliptic equations.* Oxford University Press, New York, 1993.

[22] N.J. Kalton, I.E. Verbitsky, Nonlinear equations and weighted norm inequalities. *Trans. Amer. Math. Soc.*, **351** (1999), 3441–3497.

[23] J.B. Keller, On solutions of $\Delta u = f(u)$. *Comm. Pure Appl. Math.*, **10** (1957), 503–510.

[24] T. Kilpeläinen, J. Malý, The Wiener test and potential estimates for quasilinear elliptic equations. *Acta Math.*, **172** (1994), 137–161.

[25] V.A. Kondratiev, V.A. Nikishkin, On positive solutions of singular boundary value problems for the equation $\Delta u = u^k$. *Russian J. Math. Phys.*, **1** (1993), 131–135.

[26] D.A. Labutin, Wiener regularity for large solution of nonlinear equations. *Preprint.*

[27] D.A. Labutin, Thinness for scalar-negative singular Yamabe metrics. *in preparation.*

[28] D.A. Labutin, Potential estimates for a class of fully nonlinear elliptic equations. *Duke Math. J.*, to appear.

[29] J.-F. LeGall, A path-valued Markov process and its connections with partial differential equations. *First European Congress of Mathematics (Paris, 1992)*, Vol. II 185–212, Birkhäuser, Basel, 1994.

[30] J.-F. LeGall, Branching processes, random trees and superprocesses. *Proceedings of the International Congress of Mathematicians (Berlin, 1998), Doc. Math.*, Extra Vol. III (1998), 279–289.

[31] J.-F. LeGall, *Spatial branching processes, random snakes and partial differential equations.* Lectures in Mathematics ETH Zürich. Birkhäuser Verlag, Basel, 1999.

[32] P. Lindqvist, O. Martio, Two theorems of N. Wiener for solutions of quasilinear elliptic equations. *Acta Math.*, **155** (1985), 153–171.

[33] W. Littman, G. Stampacchia, H.F. Weinberger, Regular points for elliptic equations with discontinuous coefficients. *Ann. Scuola Norm. Sup. Pisa (3)*, **17** (1963), 43–77.

[34] C. Loewner, L. Nirenberg, Partial differential equations invariant under conformal or projective transformations, Contributions to analysis (a collection of papers dedicated to Lipman Bers), Academic Press, New York, 1974, 245–272.

[35] J. Malý, W.P. Ziemer, *Fine regularity of solutions of elliptic partial differential equations.* American Mathematical Society, Providence, RI, 1997.

[36] M. Marcus, L. Veron, The boundary trace of positive solutions of semilinear elliptic equations: the subcritical case. *Arch. Rational Mech. Anal.*, 144 (1998), 201–231.

[37] M. Marcus, L. Veron, The boundary trace of positive solutions of semilinear elliptic equations: the supercritical case. *J. Math. Pures Appl.*, 77 (1998), 481–524.

[38] V.G. Maz'ya, On the continuity at a boundary point of solutions of quasilinear equations. *Vestnik Leningrad. Univ.*, **25** (1970), 42–55.

[39] V.G. Maz'ya, *Sobolev spaces.* Springer-Verlag, Berlin-New York, 1985.

[40] V.G. Maz'ya Unsolved problems connected with the Wiener criterion. The Legacy of Norbert Wiener: A Centennial Symposium (Cambridge, MA, 1994), 199–208. *Proc. Sympos. Pure Math.*, **60**, Amer. Math. Soc., Providence, RI, 1997.

[41] V.G. Maz'ya, The Wiener test for higher order elliptic equations. *Preprint Institut Mittag-Leffler*, report No. 38, 1999/2000.

[42] R. Mazzeo, D. Pollack, K. Uhlenbeck, Moduli spaces of singular Yamabe metrics. *J. Amer. Math. Soc.*, **9** (1996), 303–344.

[43] R.C. McOwen, Results and open questions on the singular Yamabe problem, Dynamical systems and differential equations, Vol. II (Springfield, MO, 1996). *Discrete Contin. Dynam. Systems*, Added Volume II (1998), 123–132.

[44] R. Osserman, On the inequality $\Delta u \geq f(u)$. *Pacific J. Math.*, **7** (1957), 1641–1647.

[45] E.A. Perkins, Measure-valued branching diffusions and interactions. *Proceedings of the International Congress of Mathematicians (Zürich, 1994)*, Vol. 2 1036–1046, Birkhäuser, Basel, 1995.

[46] N. Korevaar, R. Mazzeo, F. Pacard, R. Schoen, Refined asymptotics for constant scalar curvature metrics with isolated singularities. *Invent. Math.*, **135** (1999), 233–272.

[47] R. Schoen, S.T. Yau, Conformally flat manifolds, Kleinian groups and scalar curvature, Invent. Math. **92** (1988), 47–71.

[48] R. Schoen, S.T. Yau, Lectures on differential geometry. International Press, Cambridge, MA, 1994.

[49] N.S. Trudinger, X.-J. Wang, On the weak continuity of elliptic operators and applications to potential theory. *Preprint Australian National University*, MRR00-018, 2000, *Amer. J. Math.*, to appear.

[50] N.S. Trudinger, X.-J. Wang, Hessian Measures III. *Preprint Australian National University*, MRR00-016, 2000, *J. Funct. Anal.*, to appear.

[51] N.S. Trudinger, X.-J. Wang, Hessian Measures II. *Ann. of Math. (2)*, **150** (1999), 579–604.

[52] N.S. Trudinger, X.-J. Wang, Hessian Measures I. *Topol. Methods Nonlinear Anal.*, **10** (1997), 225–239.

[53] L. Véron, Generalized boundary value problems for nonlinear elliptic equations. *Electron. J. Diff. Eqns.*, **6** (2001), 313–342.

[54] L. Véron, *Singularities of solutions of second order quasilinear equations.* Longman, Harlow, 1996.

[55] N. Wiener, The Dirichlet problem. *J. Math. and Phys.*, **3** (1924), 127–146.

[56] N. Wiener, Certain notions in potential theory. *J. Math. and Phys.*, **3** (1924), 24–51.

[57] Open problems in geometry. *Differential geometry: partial differential equations on manifolds (Los Angeles, CA, 1990)*, 1–28, Proc. Sympos. Pure Math., **54**, Amer. Math. Soc., Providence, RI, 1993.

[58] W.P. Ziemer, *Weakly differentiable functions. Sobolev spaces and functions of bounded variation.* Springer-Verlag, New York, 1989.

Denis A. Labutin
Department of Mathematics
ETH Zentrum
Zürich CH-8092, Switzerland
E-mail address: denis@math.ethz.ch

D. Haroske, T. Runst, H.-J. Schmeisser (eds.): Function Spaces, Differential Operators and Nonlinear Analysis. The Hans Triebel Anniversary Volume.

Coarea Properties of Sobolev Functions

Jan Malý[1]

Dedicated to Prof. Hans Triebel on the occasion of his 65th birthday

Abstract. The Lusin N-property is well known as a criterion for validity of theorems on change of variables in integral. Here we consider related properties motivated by the coarea formula. They also imply a generalization of Eilenberg's inequality. We prove them for functions with gradient in the Lorentz space $L_{m,1}$. This relies on estimates of Hausdorff content of level sets for Sobolev functions and analysis of their Lebesgue points. A significant part of the presented results has its origin in a joint work with David Swanson and William P. Ziemer.

1. Introduction

The formula on change of variables in integral has its "hard version" first proved by Federer, see [3, 3.2.3, 3.2.12, 3.2.22].

Theorem 1.1. *Let $\Omega \subset \mathbb{R}^n$ be an open set, $E \subset \Omega$ be a measurable set and $f : \Omega \to \mathbb{R}^d$ be a Lipschitz function. Let $u : \Omega \to \mathbb{R}$ be a measurable function. Suppose that $m \in \{1, \dots, n\}$. Suppose that one of the following situations occurs:*

(a) $m = d$

(b) $n = d$

(c) $m \in \{1, \dots, d\}$ *and* $f(\Omega)$ *is* $\mathcal{H}^m$*-rectifiable.*

Then

$$\int_E u(x)|J_m f(x)|\,dx = \int_{\mathbb{R}^d} \Big(\int_{E\cap f^{-1}(y)} u(x)\,d\mathcal{H}^{n-m}(x) \Big)\,d\mathcal{H}^m(y), \tag{1.1}$$

provided the integral on the left makes sense.

The formula (1.1) is termed *area formula* in case (a) and *coarea formula* (in narrow sense) in case (b). In wide sense, we call "coarea formula" all cases of (1.1). We will investigate the possibilities of its generalization to non-Lipschitz functions f.

---

[1]The research is supported in part by the Research Project MSM 113200007 from the Czech Ministry of Education, Grant No. 201/00/0767 from the Grant Agency of the Czech republic (GA ČR) and Grant No. 165/99 from the Grant Agency of Charles University (GA UK)

Throughout the paper Ω will always denote an open subset of $\mathbb{R}^n$. We denote by $\mathcal{H}^s$ the s-dimensional Hausdorff measure and by $\mathcal{H}^s_\delta$ the s-dimensional Hausdorff content built from coverings by sets with diameter less than δ. The Lebesgue measure of a set E will be denoted by $|E|$. A function f on Ω is said to be a Sobolev function if its partial derivatives in the sense of distribution can be identified with locally integrable functions. Given a Sobolev function $f : \Omega \to \mathbb{R}^d$, $J_m f$ stands for its m-dimensional Jacobian.

It is well known that the problem can be reduced to analysis of Lebesgue null sets E. Indeed, as shown by Federer, if $f : \Omega \to \mathbb{R}^d$ is a Sobolev function, we can find Lipschitz functions $f_j : \mathbb{R}^n \to \mathbb{R}^d$ and disjoint subsets E_j of Ω such that $f = f_j$ on E_j and $|\Omega \setminus \bigcup E_j| = 0$.

This follows from the a.e. approximate differentiability of f [3, 3.1.4], [1] and a general property of a.e. approximately differentiable functions [3, 3.1.8].

Now, using the version for Lipschitz functions, we observe that (1.1) holds for $E \subset \bigcup_j E_j$ and it remains to investigate its validity for the Lebesgue null set $E \setminus \bigcup_j E_j$.

For $m = n$ this leads to the so-called Lusin N-property: if $|E| = 0$, then $\mathcal{H}^n(f(E)) = 0$. We will consider a more general version, which fits also to $m < n$.

Let $1 \le m \le n$ be a real number. We say that a Sobolev mapping $f : \Omega \to \mathbb{R}^d$ satisfies the *m-coarea property* in a measurable set $\Omega' \subset \Omega$ if for each Lebesgue null set $E \subset \Omega'$ and $\mathcal{H}^m$-almost each $y \in \mathbb{R}^d$ we have $\mathcal{H}^{n-m}(E \cap f^{-1}(y)) = 0$.

Now, we observe the following consequence of Federer's results

Theorem 1.2. *Let $\Omega \subset \mathbb{R}^n$ be an open set, $E \subset \Omega$ be a measurable set and $f : \Omega \to \mathbb{R}^d$ be a Sobolev function. Let $u : \Omega \to \mathbb{R}$ be a measurable function. Suppose that $m \in \{1, \dots, n\}$ and f has the m-coarea property. Suppose that one of the cases (a), (b), (c) of Theorem* 1.1 *occurs. Then* (1.1) *is valid provided the integral on the left makes sense.*

The next theorem is a version of Eilenberg's inequality for Sobolev functions. The classical Eilenberg's inequality assumes Lipschitz functions [2], [3, 2.10.25, 2.10.26]. Our statement is an easy consequence of known results, standard methods and definitions, but it illustrates the importance of the m-coarea property. We write

$$\boldsymbol{\alpha}(s) = \frac{\pi^{s/2}}{\Gamma(\frac{s}{2} + 1)} .$$

Theorem 1.3. *Let $\Omega \subset \mathbb{R}^n$ be an open set, $E \subset \Omega$ be a measurable set and $f : \Omega \to \mathbb{R}^d$ be a Sobolev function. Let $u : \Omega \to \mathbb{R}$ be a nonnegative measurable function. Suppose that $1 \le m \le n$ is a real number and f has the m-coarea property. Then*

$$\int_{\mathbb{R}^d} \Big(\int_{E \cap f^{-1}(y)} u(x)\, d\mathcal{H}^{n-m}(x) \Big)\, d\mathcal{H}^m(y) \le \frac{\boldsymbol{\alpha}(n-m)\boldsymbol{\alpha}(m)}{\boldsymbol{\alpha}(n)} \int_E u(x)\, |\nabla f(x)|^m \, dx.$$

Now we state our main results.

Theorem 1.4. *Let $f : \Omega \to \mathbb{R}^d$ be a Sobolev function and $|\nabla f| \in L_{m,1}(\Omega)$. Let $\Omega' \subset \Omega$ be a measurable set. Suppose that $\mathcal{H}^{n-m}$-a.e.point $z \in \Omega'$ is a Lebesgue point for f $\mathcal{H}^{n-m}$-a.e.in Ω'. Then f has the m-coarea property in Ω'.*

In Section 3 we will see that any Sobolev function with $\nabla u \in L_{m,1}$ has a representative with Lebesgue points $\mathcal{H}^{n-m}$-a.e. Taking such a good representative we can consider $\Omega' = \Omega$ in the previous theorem.

A counterexample due to Cesari, see [11], shows that there is a continuous function $f \in W^{1,n}(\Omega; \mathbb{R}^n)$ such that f maps a set of measure zero onto a solid cube. It follows that we cannot replace the Lorentz space by L_m in Theorem 1.4.

The coarea property however holds for Hölder continuous functions in $W^{1,m}(\Omega; \mathbb{R}^d)$. More generally, we say that $z \in \Omega$ is a Hölder-Lebesgue point for a measurable function f if there exists $\beta > 0$ such that

$$\lim_{r \to 0} r^{-\beta} \fint_{B(z,r)} |f(x) - f(z)|\, dx = 0. \tag{1.2}$$

Then we obtain the following

Theorem 1.5. *Let $f : \Omega \to \mathbb{R}^d$ be a Sobolev function and $|\nabla f| \in L_m(\Omega)$. Let $\Omega' \subset \Omega$ be a measurable set. Suppose that $\mathcal{H}^{n-m}$-a.e.point $z \in \Omega'$ is a Hölder-Lebesgue point for f. Then f has the m-coarea property in Ω'.*

The coarea formula for scalar $W^{1,1}$-function is due to Federer [3, 4.5.9(14)]. The area case $m = n$ of Theorem 1.4 has be shown by Kauhanen, Koskela and Malý [7]. The area case of Theorem 1.5 is due to Martio and Malý [11], [9]. The coarea case $m = d$ of both theorems has been established by Malý, Swanson and Ziemer [12].

The area case Theorem 1.4 implies that the m-coarea holds if f belongs to the Sobolev space $W^{1,p}(\Omega)$ with $p > m$. The area formula for $p > n$ has been proved by Marcus and Mizel [15] and the coarea formula for $p > n$ by Van der Putten [18]. The result in [12] improves the range for the coarea formula to $p > d$. Related results are in the paper by Hajłasz [6].

This paper is organized as follows. In Section 2 we recall the relation between Lorentz and Orlicz spaces. In Section 3 we follow the development of estimates of Hausdorff content of level sets established in [10], [13]. Here also we discuss Lebesgue points of Sobolev mappings. Section 4 is devoted to proofs of the main results.

2. Lorentz and Orlicz spaces

In this section we discuss the relation between Lorentz and Orlicz spaces, This has been observed by Hauhanen, Koskela and Malý [7] and adapted to our purposes in [13].

We define the distribution function μ_g of a measurable function g as

$$\mu_g(t) := |\{|g| > t\}|, \qquad t \geq 0.$$

Recall that the norm in the *Lorentz space* $L_{m,1}(\Omega)$ can be expressed as

$$\|g\|_{L_{m,1}} := \int_0^\infty \mu_g(t)^{1/m}\, dt.$$

Let $\mathbf{F}\colon [0,+\infty) \to \mathbb{R}$ be a Young function. The Luxemburg norm of the *Orlicz space* $L^{\mathbf{F}}(\Omega)$ is the expression

$$\|g\|_{\mathbf{F}} = \inf\Big\{\lambda > 0 : \int_\Omega \mathbf{F}(|g|/\lambda)\, dx \le 1\Big\}.$$

We can express the Lorentz space $L_{m,1}$ as a union of a family of Orlicz spaces [7]. It can be seen from the following theorem. Its simple proof is due to A. Cianchi.

Theorem 2.1. *Suppose that $m > 1$. Let g be a measurable function on Ω. Then*

$$\|g\|^m_{L_{m,1}} = \inf\Big\{\int_\Omega \mathbf{F}(|g|) : \mathbf{F} \text{ is a Young function, } \int_0^\infty \mathbf{F}'(t)^{\frac{1}{1-m}}\, dt \le 1\Big\}\,.$$

Proof. Recall the well-known formula

$$\int_\Omega \mathbf{F}(|g|)\, dx = \int_0^\infty \mathbf{F}'(t)\mu_g(t)\, dt \tag{2.1}$$

which holds for any Young function. Using Hölder's inequality we obtain

$$\int_0^\infty \mu_g(t)^{1/m}\, dt \le \Big(\int_0^\infty \mathbf{F}'(t)\mu_g(t)\, dt\Big)^{1/m} \Big(\int_0^\infty \mathbf{F}'(t)^{1/(1-m)}\, dt\Big)^{(m-1)/m}$$

which gives one inequality. The infimum is attained at the function

$$\mathbf{F}_g(t) = \|f\|^{m-1}_{L_{m,1}} \int_0^t \mu_g(s)^{\frac{1}{m}-1}\, ds. \tag{2.2}$$

□

A regularization of the function $\mathbf{F}_g$ which minimizes in the previous theorem yields a function $\mathbf{F}$ with the global ∇_2 & Δ_2 property. This is expressed in the following proposition. For the proof we refer to [13].

Proposition 2.2. *Let $g \in L_{m,1}(\Omega)$ and $1 < q < m < p$. Then there is a C^1 Young function $\mathbf{F}$ and a constant $C = C(m,p,q) > 0$ such that*

$$\int_\Omega \mathbf{F}(|g|)\, dx \le C\|g\|^m_{L_{m,1}(\Omega)}\,, \tag{2.3}$$

$$\int_0^\infty \mathbf{F}'(t)^{-\frac{1}{m-1}}\, dt \le C \tag{2.4}$$

and

$$q \le \frac{t\mathbf{F}'(t)}{\mathbf{F}(t)} \le p, \quad t > 0. \tag{2.5}$$

3. Estimates of level sets

In this section we estimate the $(n-m)$-dimensional Hausdorff content of level sets of Sobolev functions. For example, the inequality

$$\mathcal{H}^{n-m}_\infty(\{u>1\}) \le C \int_{\mathbb{R}^n} \left(|u|^p + |\nabla u|^p\right) dx \tag{3.1}$$

holds if $p > m$ and u is a suitably represented Sobolev function on $\mathbb{R}^n$. This can be read as the inequality between the $(n-m)$-dimensional Hausdorff content and the $W^{1,p}$-capacity. These estimates go back to Frostman ($p = 2$) and the case of general p is due to Reshetnyak [17] and Maz'ya and Havin [16]. In fact, the Hausdorff content $\mathcal{H}^h_\infty$ with respect to a general growth function h in place of t^{n-m} can be estimated. Now, we are interested to leave the $(n-m)$-dimensional content on the left of (3.1), but use a more refined growth on the right.

Let $\mathbf{F}$ be a Young function. We denote by $W^{1,\mathbf{F}}(\Omega)$ the Sobolev-Orlicz space of all Sobolev functions on Ω which together with their first order partial derivatives belong to $L^{\mathbf{F}}(\Omega)$. Then $W^{1,\mathbf{F}}_c(\Omega)$ is the set of all functions from $W^{1,\mathbf{F}}(\Omega)$ with a compact support in Ω and $W^{1,\mathbf{F}}_0(\Omega)$ is the closure of $W^{1,\mathbf{F}}_c(\Omega)$ in $W^{1,\mathbf{F}}(\Omega)$.

We denote by $\mathcal{D}(\Omega)$ the set of all infinitely smooth functions on Ω with compact support in Ω and by $\mathcal{D}^+(\Omega)$ the set of all nonnegative functions from $\mathcal{D}(\Omega)$.

Our aim is to estimate the $(n-m)$-dimensional Hausdorff content of level sets of functions from $W^{1,\mathbf{F}}$.

We first consider the case of an open set $G \subset B(0,1)$. We minimize the integral

$$\mathbf{I}(u) = \int_{\mathbb{R}^n} \mathbf{F}(|\nabla u|)\, dx$$

in the class

$$\mathcal{K}(G) = \left\{u \in W^{1,\mathbf{F}}_0(B):\ u = 0 \text{ outside } B,\ u \ge 1 \text{ on } G\right\},$$

where $B = B(0,2)$. The elements of $\mathcal{K}(G)$ are called *competitors* for the capacitary problem. The function $u_G \in W^{1,\mathbf{F}}(\mathbb{R}^n)$ which minimizes $\mathbf{I}$ in $\mathcal{K}(G)$ (it is unique if $\mathbf{F}$ is strictly convex, which is true in our applications) is said to be the $\mathbf{F}$-*capacitary extremal* for G. Similarly to the standard growth case, there exists a unique measure μ_G, called the *capacitary distribution* for G, such that

$$\int_\Omega \mathbf{A}(\nabla u_G)\cdot \nabla\varphi\, dx = \int_\Omega \varphi\, d\mu_G \quad \forall \varphi \in \mathcal{D}(\Omega), \tag{3.2}$$

where

$$\mathbf{A}(\xi) = \mathbf{F}'(|\xi|)\,\frac{\xi}{|\xi|}, \qquad \xi \in \mathbb{R}^n,$$

and the following is true.

Proposition 3.1. *Suppose that* $\mathbf{F}$ *satisfies* (2.5). *Then for all* $v \in \mathcal{K}(G)$ *we have*

$$\mu_G(G) \le p\,\mathbf{I}(v).$$

The key to the capacitary estimates is the following theorem established in [10]. For this we assume that $\mathbf{F}$ satisfies (2.5). The expression on the right is a generalization of the Wolff potential. The case of the standard growth $\mathbf{F}(t) = t^p$ is due to Kilpeläinen and Malý [8].

Theorem 3.2. *Let $\Omega \subset \mathbb{R}^n$ be an open set and $R = \operatorname{diam}(\Omega)$. Let $\mu \geq 0$ be finite Radon measure in Ω. Let u be a solution of*

$$\int_\Omega \mathbf{A}(\nabla u) \cdot \nabla\varphi \, dx \leq \int_\Omega \varphi \, d\mu \quad \text{for all } \varphi \in \mathcal{D}^+(\Omega). \tag{3.3}$$

Then

$$\text{app-}\limsup_{x \to z} u(x) \leq C \int_0^{2R} [\mathbf{F}']^{-1}\Big(\frac{\mu(B(z,r))}{r^{n-1}}\Big) \, dr \;,$$

where $C = C(n,p,q)$.

We define the k-th order (fractional) maximal function of a measure $\mu \geq 0$ as

$$M_k\mu(x) = \sup_{r>0} \frac{r^k \mu(B(x,r))}{|B(x,r)|}.$$

We skip the subscript k if $k = 0$. The maximal function of a function f is defined as the maximal function of the measure with density $|f|$,

The following generalization of the Hardy-Littlewood maximal theorem is due to Bagby and Ziemer [1].

Proposition 3.3. *Let $\mu \geq 0$ be a Radon measure on $\mathbb{R}^n$ and $k \geq 0$. Then*

$$\mathcal{H}^{n-k}_\infty(\{M_k\mu > \lambda\}) \leq \frac{C\mu(\mathbb{R}^n)}{\lambda} \;,$$

where

$$C = C(n,k).$$

Now we are ready to estimate the level sets. We follow the development in [13].

Lemma 3.4. *Let $\mathbf{F}$ be a Young function satisfying (2.5) and (2.4). Let $B = B(0,2)$ and $G \subset B(0,1)$ be an open set. Let $u \in W^{1,\mathbf{F}}(\mathbb{R}^n) \cap \mathcal{K}(G)$. Then*

$$\mathcal{H}^{n-m}_\infty(G) \leq C \int_B \mathbf{F}(|\nabla u|) \, dx \;,$$

where $C = C(n,p,q,m)$.

Proof. We may assume that $u = u_G$. We write $\mu = \mu_G$. For each $z \in G$ we have by Theorem 3.2

$$1 \leq \text{app-}\limsup_{x \to z} u(x) \leq \int_0^8 \mathbf{a}^{-1}\Big(\frac{\mu(B(z,r))}{r^{n-1}}\Big) \, dr \leq \int_0^8 \mathbf{a}^{-1}\big(r^{1-m} M_m\mu(z)\big) \, dr.$$

We use the change of variables

$$r^{1-m} M_m\mu(z) = \mathbf{a}(t).$$

Then

$$1 \le -M_m\mu(z)^{1/(m-1)} \int_0^\infty t\Big(\mathbf{a}(t)^{1/(1-m)}\Big)' \, dt$$
$$= M_m\mu(z)^{1/(m-1)} \int_0^\infty \mathbf{a}(t)^{1/(1-m)} \, dt \le C M_m\mu(z)^{1/(m-1)}.$$

We have shown that there is $\lambda > 0$ such that $M_m\mu > \lambda$ on G. By Proposition 3.3 and Proposition 3.1

$$\mathcal{H}^{n-m}_\infty(G) \le C\mu(B) \le C \int_B \mathbf{F}(|\nabla u|)\, dx.$$

□

Theorem 3.5. *Let* $\mathbf{F}$ *be a Young function satisfying* (2.5) *and* (2.4). *Let* $G \subset \mathbb{R}^n$ *be a measurable set. Let* $u \in W^{1,\mathbf{F}}(\mathbb{R}^n)$, $u \ge 1$ *on* G. *Suppose that* $\mathcal{H}^{n-m}$*-a.e. point of* G *is a Lebesgue point for* u. *Then*

$$\mathcal{H}^{n-m}_\infty(G) \le C \int_{\mathbb{R}^n} \Big(\mathbf{F}(|\nabla u|) + \mathbf{F}(|u|)\Big)\, dx \ ,$$

where $C = C(n, p, q, m)$.

Sketch of the proof. By standard approximation techniques we may reduce to the case when G is open. We find a set $Z \subset \mathbb{R}^n$ such that the balls $B(z,1)$, $z \in Z$, cover $\mathbb{R}^n$ and the balls $B(z,2)$, $z \in Z$, have bounded overlap multiplicity by a constant depending only on n. Then on the balls $B(z,2)$ we multiply u with suitable cut-off functions and use Lemma 3.4 together with a Poincaré type inequality. Finally we sum over $z \in Z$. □

The case $m = 1$ is due to Fleming [5], see also the elementary proof in [12].

Theorem 3.6. *Let* $G \subset \mathbb{R}^n$ *be a measurable set. Let* $u \in W^{1,1}(\mathbb{R}^n)$, $u \ge 1$ *on* G. *Suppose that* $\mathcal{H}^{n-1}$*-a.e. point of* G *is a Lebesgue point for* u. *Then*

$$\mathcal{H}^{n-1}_\infty(G) \le C \int_{\mathbb{R}^n} \Big(|\nabla u|) + |u|\Big)\, dx \ ,$$

where $C = C(n)$.

Federer and Ziemer [4] proved that each function from $W^{1,p}(\Omega)$ has a representative which has Lebesgue points γ_p-quasi everywhere, where γ_p is the $W^{1,p}$-capacity. It follows that such a representative has Lebesgue points $\mathcal{H}^{n-m}$-a.e. for any $m < p$ (by (3.1)) or with $m = p = 1$ (by Fleming's Theorem 3.6).

We can apply Theorem 3.5 to the analysis of Lebesgue points of Sobolev functions. The result is obtained by Malý, Swanson and Ziemer [13]. Its proof uses also Proposition 2.2 and the method developed in [4].

Theorem 3.7. *Suppose that* u *is a Sobolev function with gradient in* $L_{m,1}$. *Then there exists a representative of* u *which has Lebesgue points* $\mathcal{H}^{n-m}$*-a.e.*

4. Coarea properties

Besides the ordinary d-dimensional Hausdorff content $\mathcal{H}^m_\delta$ we consider the set functions

$$\lambda^m_\delta : A \mapsto \inf\Big\{\sum_j \alpha_j(\operatorname{diam} A_j)^m : \ \alpha_j \ge 0, \ \operatorname{diam} A_j \le \delta, \\ \chi_{_A} \le \sum_j \alpha_j \chi_{_{A_j}}\Big\}, \qquad \delta > 0.$$

By [3, 2.10.24],

$$\mathcal{H}^m(A) = \lim_{\delta\to 0} \lambda^m_\delta(A) \tag{4.1}$$

for any set $A \subset \mathbb{R}^d$.

Lemma 4.1. *Let* $\mathbf{F}$ *be a Young function satisfying* (2.5) *and* (2.4). *Let* $f \in W^{1,\mathbf{F}}(\Omega, \mathbb{R}^m)$. *Suppose that* f *has Lebesgue points a.e. in a set* $E \subset \Omega$. *Let* $(x_0, y_0) \in E \times \mathbb{R}^m$ *and* $r > 0$. *Then*

$$r^m \mathcal{H}^{n-m}_\infty\Big(E \cap B(x_0,r) \cap f^{-1}(B(y_0,r))\Big) \le C \int_{B(x_0,2r)\cap f^{-1}(B(y_0,2r))} \Big(1 + \mathbf{F}(|\nabla f|)\Big)\,dx.$$

Sketch of the proof. We use Theorem 3.5 with

$$u(x) = \begin{cases} 1, & |f(x_0+rx) - y_0| \le r, \\ 2 - \left|\frac{f(x_0+rx)-y_0)}{r}\right|, & r < |f(x_0+rx) - y_0| < 2r, \\ 0, & |f(x_0+rx) - y_0| \ge 2r. \end{cases}$$

Then there is a set E' of Lebesgue points for u such that we can compute

$$\begin{aligned} r^m \mathcal{H}^{n-m}_\infty\Big(E \cap B(x_0,r) \cap f^{-1}(B(y_0,r))\Big) &= \mathcal{H}^{n-m}_\infty\Big(E' \cap B(0,1) \cap \{u \ge 1\}\Big) \\ &\le C \int_{B(0,2)} \Big(\mathbf{F}(|\nabla u|) + \mathbf{F}(u)\Big)\,dx \\ &\le C \int_{B(x_0,2r)\cap f^{-1}(B(y_0,2r))} \Big(\mathbf{F}(|\nabla f|) + 1\Big)\,dx. \end{aligned}$$

□

Proof of Theorem 1.4. In the cartesian product $\mathbb{R}^n \times \mathbb{R}^d$ we will use "balls" $B([x,y],r) = B(x,r) \times B(y,r)$. Let $E \subset \Omega'$ be a Lebesgue null set. We may assume that E consists only of Lebesgue points for f. Given $\varepsilon > 0$, we find an open set $G \subset \Omega$ such that

$$E \subset G \quad \text{and} \quad \int_G (1 + \mathbf{F}(|\nabla f|))\,dx < \varepsilon.$$

Choose $\delta > 0$. Let $x \in E$. We find $r_{x,0} < \delta$ such that $B(x, r_{x,0}) \subset G$ and denote $r_{x,i} = 10^{-i} r_{x,0}$. If $\{a_i\}$ is a bounded sequence of positive real numbers, then there exists i such that $a_{i+1} \le 2a_i$. Applying this trick to

$$a_i = \left(1 + r_{x,i}^{-n-d} \int_{B(x,r_{x,i})\cap f^{-1}(B(f(x),r_{x,i}))} (1 + \mathbf{F}(|\nabla f|))\,dx\right)^{-1}$$

we find $r_x \in (0, \delta/10)$ such that

$$\begin{aligned}(10r_x)^{n+d} &+ \int_{B(x,10r_{x,i})\cap f^{-1}(B(f(x),10r_{x,i}))} (1 + \mathbf{F}(|\nabla f|))\,dx \\ &\le Cr_x^{n+d} + \int_{B(x,r_{x,i})\cap f^{-1}(B(f(x),r_{x,i}))} (1 + \mathbf{F}(|\nabla f|))\,dx.\end{aligned} \tag{4.2}$$

The system $\{B(x,r_x) \times B(f(x),r_x)\}$ forms a covering of the graph of f over E and by a Vitali-type covering theorem we find a pairwise disjoint sequence $\{B_j\}$ of "balls" $B_j = B(x_j,r_j) \times B(f(x_j),r_j)$ such that $x_j \in E$, $r_j = r_{x_j}$, and

$$\{[x, f(x)] : x \in E\} \subset \bigcup_j B(x_j, 5r_j) \times B(f(x_j), 5r_j). \tag{4.3}$$

Given $c > 0$, we consider the set

$$A = \{y \in \mathbb{R}^d : \ |y| < c, \ \mathcal{H}^m_\infty(E \cap f^{-1}(y)) > 1/c\}.$$

We denote

$$A_j = A \cap B(f(x_j), 5r_j), \qquad \alpha_j = c\,\mathcal{H}^{n-m}_\infty\Big(E \cap B(x_j,5r_j) \cap f^{-1}(A_j)\Big).$$

Then for each $y \in A$ we have by (4.3)

$$\begin{aligned}\chi_A(y) = 1 &\le c\sum_j \mathcal{H}^{n-m}_\infty\Big(E \cap B(x_j,5r_j) \cap f^{-1}(A_j)\Big)\chi_{A_j}(y) \\ &= \sum_j \alpha_j \chi_{A_j}\end{aligned}$$

so that for the set function introduced at the beginning of this section we have

$$\begin{aligned}\lambda^m_\delta(A) &\le \sum_j \alpha_j \operatorname{diam}(A_j)^m \\ &\le Cc\sum_j \mathcal{H}^{n-m}_\infty\Big(E \cap B(x_j,5r_j) \cap f^{-1}(A_j)\Big)\operatorname{diam}(A_j)^m.\end{aligned} \tag{4.4}$$

By Lemma 4.1 and (4.2)

$$\begin{aligned}&\mathcal{H}^{n-m}_\infty\Big(E \cap B(x_j,5r_j) \cap f^{-1}(A_j)\Big)\operatorname{diam}(A_j)^m \\ &\le Cr_j^m \mathcal{H}^{n-m}_\infty\Big(E \cap B(x_j,5r_j) \cap f^{-1}(B(f(x_j),5r_j))\Big) \\ &\le C\int_{B(x_j,10r_j)\cap f^{-1}(B(f(x_j,10r_j)))} (1 + \mathbf{F}(|\nabla f|))\,dx \\ &\le C\Big(r_j^{n+d} + \int_{B(x_j,r_j)\cap f^{-1}(B(f(x_j,r_j)))} (1 + \mathbf{F}(|\nabla f|))\,dx\Big).\end{aligned} \tag{4.5}$$

Since the "balls" B_j are disjoint and contained in G, by (4.5) and (4.4), summing only over those j with $A_j \neq \emptyset$ and observing that for these j we have $B_j \subset G \cap B(0, 2c)$, we obtain

$$\lambda_\delta^m(A) \leq Cc\Big(c^d|G| + \int_G (1 + \mathbf{F}(|\nabla f|))\,dx\Big) \leq Cc(c^d + 1)\varepsilon.$$

Letting ε and $\delta \to 0$ we obtain that $\mathcal{H}^m(A) = 0$. This concludes the proof. □

Proof of Theorem 1.5. This can be obtained following the lines of the proof of Theorem 1.4 in [12]. □

References

[1] T. Bagby and W. P. Ziemer, Pointwise differentiability and absolute continuity, *Trans. Amer. Math. Soc.* **191** (1974), 129–148.

[2] S. Eilenberg, On φ measures, *Ann. Soc. Pol. de Math.* **17** (1938), 251–252.

[3] H. Federer, *Geometric Measure Theory*, Springer-Verlag, New York, Heidelberg, 1969.

[4] H. Federer and W. P. Ziemer, The Lebesgue set of a function whose partial derivatives are p-th power summable, *Indiana Univ. Math. J.* **22** (1972), 139–158.

[5] W. H. Fleming, Functions whose partial derivatives are measures, *Illinois J. Math.* **4** (1960), 452–478.

[6] P. Hajłasz, *Sobolev mappings, co-area formula and related topics*, Proceedings on Analysis and Geometry, Sobolev Institute Press, Novosibirsk, 2000, 227–254.

[7] J. Kauhanen, P. Koskela and J. Malý, On functions with derivatives in a Lorentz space, *Manuscripta Math.* **100,1** (1999), 87–101.

[8] T. Kilpeläinen and J. Malý, The Wiener test and potential estimates for quasilinear elliptic equations, *Acta Math.* **172** (1994), 137–161.

[9] J. Malý, *Sufficient Conditions for Change of Variables in Integral*, Proceedings on Analysis and Geometry, Sobolev Institute Press, Novosibirsk, 2000, 370–386.

[10] J. Malý, *Wolff potential estimates of superminimizers of Orlicz type Dirichlet integrals*, Preprint MATH-KMA, Charles University, Praha 2002.

[11] J. Malý and O. Martio, Lusin's condition (N) and mappings of the class $W^{1,n}$, *J. Reine Angew. Math.* **458** (1995), 19–36.

[12] J. Malý, D. Swanson, and W. P. Ziemer, *Coarea formula for Sobolev mappings*, Preprint MATH-KMA-2001/68, Charles University, Praha.

[13] J. Malý, D. Swanson, and W. P. Ziemer, *Fine behavior of functions with gradients in a Lorentz space*, in preparation.

[14] J. Malý and W. P. Ziemer, *Fine regularity of solutions of elliptic partial differential equations*, AMS Mathematical Surveys and Monographs Vol. 51, Amer. Math. Soc., Providence, 1997.

[15] M. Marcus and V. J. Mizel, Transformations by functions in Sobolev spaces and lower semicontinuity for parametric variational problems, *Bull. Amer. Math. Soc.* **79** no. 4 (1973), 790–795.

[16] V. G. Maz'ya and V. P. Havin, Nonlinear potential theory, *Uspekhi Mat. Nauk* **27** (1972), 67–138. English translation: *Russian Math. Surveys* **27** (1972), 71–148.

[17] Yu. G. Reshetnyak, On the concept of capacity in the theory of functions with generalized derivatives (Russian), *Sibirsk. Mat. Zh.* **10** (1969), 1109–1138. English translation: *Siberian Math. J.* **10** (1969), 818–842.

[18] R. Van der Putten, On the critical-values lemma and the coarea formula (Italian), *Boll. Un. Mat. Ital. B.* **6,3** (1992), 561–578.

Jan Malý
School of Mathematics, KMA
Charles University in Prague
Sokolovská 83
CZ – 18675 Praha 8, Czech Republic
E-mail address: maly@karlin.mff.cuni.cz

D. Haroske, T. Runst, H.-J. Schmeisser (eds.): Function Spaces, Differential Operators and
Nonlinear Analysis. The Hans Triebel Anniversary Volume.

Banach Envelopes of the Besov and Triebel-Lizorkin Spaces and Applications to PDE's

Osvaldo Méndez and Marius Mitrea

Dedicated to Prof. Hans Triebel on the occasion of his 65th birthday

Abstract. If the "hat" denotes the Banach envelope, we show that for $s \in \mathbb{R}$, then

(i) For $0 < p, q < 1$, one has

$$\widehat{B_p^{s,q}(\mathbb{R}^n)} = \widehat{F_p^{s,q}(\mathbb{R}^n)} = B_1^{s-n(\frac{1}{p}-1),1}(\mathbb{R}^n),$$

(ii) for $0 < p < 1$, $1 \leq q < \infty$,

$$\widehat{B_p^{s,q}(\mathbb{R}^n)} = B_1^{s-n(\frac{1}{p}-1),q}(\mathbb{R}^n)\ , \ \widehat{F_p^{s,q}(\mathbb{R}^n)} = B_1^{s-n(\frac{1}{p}-1),1}(\mathbb{R}^n),$$

(iii) and for $1 \leq p < \infty$, $0 < q < 1$,

$$\widehat{B_p^{s,q}(\mathbb{R}^n)} = B_p^{s,1}(\mathbb{R}^n)\ , \ \widehat{F_p^{s,q}(\mathbb{R}^n)} = F_p^{s,1}(\mathbb{R}^n)$$

We present applications of these results to global regularity of PDE's in non-smooth domains, emphasizing the three-dimensional Lamé system of elastostatics (Theorem 10).

Introduction

The present paper concerns the computation of the Banach envelopes of the Besov and Triebel-Lizorkin spaces. Our primary motivation is the application of these concepts to some problems related to the regularity of systems of partial differential equations in non-smooth domains. Our functional-analytic approach yields Sobolev regularity of any system for which "endpoint" L^2- and Hardy-space-type estimates are known (see Theorem 10). Of particular interest is the three-dimensional Lamé system of elastostatics.

1. Quasi-Banach spaces and Banach envelopes

We start by recalling that a quasi-norm on a vector space X is a non-negative, real-valued function $\|\cdot\|$ on X such that for $x \in X$, $\|x\| = 0 \Leftrightarrow x = 0$, for $x, y \in X$ and $\lambda \in \mathbb{C}$,

$$\|\lambda x\| = |\lambda| \|x\|$$

and, for some constant $C \geq 1$,

$$\|x + y\| \leq C(\|x\| + \|y\|).$$

A quasi-Banach space is a complete topological vector space, whose topology is induced by a quasi-norm. Given a quasi-Banach space X whose dual separates points, we define the Banach envelope of X (denoted by $\hat{X}$) as the completion of X under the norm

$$\|x\| = \inf\{r : r > 0 \text{ and } x \in r\text{co}(B)\} \tag{1.1}$$

where B is an arbitrary bounded neighborhood of 0 in X, and, for $Y \subset X$, $\text{co}(Y)$ denotes the convex hull of Y. Notice that the norms generated by different neighborhoods B via (1.1) are all equivalent. For a more exhaustive study of quasi-Banach spaces we refer the interested reader to [10]. It is not hard to see that the inclusion

$$X \hookrightarrow \hat{X}$$

is bounded and dense, and that $X^* = (\hat{X})^*$. Given two quasi-Banach spaces X_1 and X_2, the space of all bounded linear maps from X_1 to X_2 (denoted by $\mathcal{B}(X_1, X_2)$) is easily seen to be quasi-Banach space when furnished with any of the equivalent quasi-norms

$$\|T\| = \sup_{x \in B} \|T(x)\|$$

for $T \in \mathcal{B}(X_1, X_2)$, obtained for different choices of the bounded neighborhood of the origin B. Also, for T as above there is a natural bounded, linear extension $\hat{T} \in \mathcal{B}(\hat{X}_1, \hat{X}_2)$, in such a way that the "hat" operation defines a linear bounded and injective operator from $\mathcal{B}(X_1, X_2)$ into $\mathcal{B}(\hat{X}_1, \hat{X}_2)$. We point out that several properties of T are automatically transferred to its extension $\hat{T}$ (see [11] and [6]). For our present purposes, it is sufficient to mention the following theorem, which plays a central role in our discussion:

Theorem 1. *In the above setting, T is an isomorphism if and only if $\hat{T}$ is.*

We now turn to the main theorem of this section:

Theorem 2. *Let X and Y be a quasi-Banach and a Banach space respectively, such that X is densely and boundedly embedded in Y and $X^* = Y^*$. Then (up to an isomorphism) $\hat{X} = Y$.*

Proof. See [11]. □

2. Besov and Triebel-Lizorkin spaces

In this section we collect some relevant facts about the Besov and Triebel-Lizorkin spaces and apply Theorem 2 to compute their Banach envelopes. Let $\mathcal{F}$ stand for the Fourier transform on $\mathcal{S}'$ and let Φ, $\varphi \in \mathcal{S}$ be such that

1. $\mathcal{F}(\Phi) \geq c > 0$ uniformly on $\{\xi \in \mathbb{R}^n : |\xi| \leq \frac{5}{3}\}$ and $\operatorname{supp} \mathcal{F}(\Phi) \subseteq \{\xi \in \mathbb{R}^n : |\xi| \leq 2\}$
2. $\mathcal{F}(\varphi) \geq c > 0$ uniformly on $\{\xi \in \mathbb{R}^n : \frac{3}{5} \leq |\xi| \leq \frac{5}{3}\}$ and $\operatorname{supp} \mathcal{F}(\varphi) \subseteq \{\xi \in \mathbb{R}^n : \frac{1}{2} \leq |\xi| \leq 2\}$
3. For $\xi \neq 0$, $\sum_{i \in \mathbb{Z}} |\mathcal{F}(\varphi)(2^i \xi)|^2 = 1$.

For $0 < p, q \leq \infty$ ($p < \infty$ in case of $F_p^{s,q}$) and $s \in \mathbb{R}$, the non-homogeneous Besov and Triebel-Lizorkin spaces are defined respectively as

$$B_p^{s,q}(\mathbb{R}^n) = \{f \in \mathcal{S}' : \|f\|_{B_p^{s,q}(\mathbb{R}^n)} = \|\Phi * f\|_{L^p} + (\sum_{i=0}^{\infty} (2^{is} \|\varphi_i * f\|_{L^p})^q)^{\frac{1}{q}} < \infty\}$$

and

$$F_p^{s,q}(\mathbb{R}^n) = \{f \in \mathcal{S}' : \|f\|_{F_p^{s,q}(\mathbb{R}^n)} = \|\Phi * f\|_{L^p} + \|(\sum_{i=0}^{\infty} (2^{is} |\varphi_i * f|)^q)^{\frac{1}{q}}\|_{L^p} < \infty\},$$

with the usual modifications if either $q = \infty$ or $p = \infty$.
In order to simplify the following statements, we put, for $0 < p < \infty$,

$$p \wedge 1 := \min\{p, 1\} \quad , \quad p \vee 1 := \max\{p, 1\} \quad , \quad p' := (\max\{0, 1 - \frac{1}{p}\})^{-1}.$$

We recall the following duality and embedding theorems (see [11], [13], [14] and [15]):

Theorem 3. *For $s \in \mathbb{R}, 0 < p \leq 1$ and $0 < q < \infty$, it holds that:*

$$(B_p^{s,q}(\mathbb{R}^n))^* = B_\infty^{-s+n(\frac{1}{p}-1),q'}(\mathbb{R}^n)$$

and

$$(F_p^{s,q}(\mathbb{R}^n))^* = F_\infty^{-s+n(\frac{1}{p}-1),\infty}(\mathbb{R}^n) = B_\infty^{-s+n(\frac{1}{p}-1),\infty}(\mathbb{R}^n),$$

whereas for $s \in \mathbb{R}$, $1 \leq p < \infty$ and $0 < q < \infty$

$$(B_p^{s,q}(\mathbb{R}^n))^* = B_{p'}^{-s,q'}(\mathbb{R}^n)$$

and

$$(F_p^{s,q}(\mathbb{R}^n))^* = F_{p'}^{-s,q'}(\mathbb{R}^n).$$

Theorem 4. *For $s_i \in \mathbb{R}, i = 1, 2$, $0 < p_1 \leq p_2 \leq \infty$, $0 < q_1 \leq q_2 \leq \infty$ and $s_1 - \frac{n}{p_1} = s_2 - \frac{n}{p_2}$, the embedding*

$$B_{p_1}^{s_1,q_1}(\mathbb{R}^n) \hookrightarrow B_{p_2}^{s_2,q_2}(\mathbb{R}^n)$$

is bounded and dense. Moreover, if $s_i \in \mathbb{R}, i = 1, 2$, $0 < p_1 < p_2 < \infty$, $0 < q_1, q_2 \leq \infty$ and $s_1 - \frac{n}{p_1} = s_2 - \frac{n}{p_2}$, the same holds for the embedding

$$F_{p_1}^{s_1,q_1}(\mathbb{R}^n) \hookrightarrow F_{p_2}^{s_2,q_2}(\mathbb{R}^n).$$

Theorems 3 and 4 coupled with Theorem 2, yield the following result:

Theorem 5. *For $s \in \mathbb{R}$ one has for $0 < p, q < 1$*

$$\widehat{B_p^{s,q}(\mathbb{R}^n)} = \widehat{F_p^{s,q}(\mathbb{R}^n)} = B_1^{s-n(\frac{1}{p}-1),1}(\mathbb{R}^n),$$

whereas for $0 < p < 1$ and $1 \leq q < \infty$,

$$\widehat{B_p^{s,q}(\mathbb{R}^n)} = B_1^{s-n(\frac{1}{p}-1),q}(\mathbb{R}^n)\ ,\ \widehat{F_p^{s,q}(\mathbb{R}^n)} = B_1^{s-n(\frac{1}{p}-1),1}(\mathbb{R}^n)$$

and for $1 \leq p < \infty$, $0 < q < 1$,

$$\widehat{B_p^{s,q}(\mathbb{R}^n)} = B_p^{s,1}(\mathbb{R}^n)\ ,\ \widehat{F_p^{s,q}(\mathbb{R}^n)} = F_p^{s,1}(\mathbb{R}^n).$$

We also recall that $(B_p^{s,q}(\mathbb{R}^n))_{s,p,q}$ and $(F_p^{s,q}(\mathbb{R}^n))_{s,p,q}$ are complex interpolation scales (see [13], [15] and [11]).
Theorem 1 and the paragraph preceding it, Theorem 5 and the stability results in [9] yield:

Theorem 6. *Let T be a bounded linear operator on $F_p^{s,q}(\mathbb{R}^n)$ for (s,p,q) in some neighborhood $\mathcal{U} \subset \mathbb{R} \times (0,\infty) \times (0,\infty)$ of (s_0, p_0, q_0), which is an isomorphism on $F_{p_0}^{s_0,q_0}(\mathbb{R}^n)$. Then there is a neighborhood $\mathcal{V}$ of (s_0, p_0, q_0) such that T extends to an isomorphism on $F_{p\vee 1}^{s-n(\frac{1}{p\wedge 1}-1),r}(\mathbb{R}^n)$ and its transpose T^* extends to an isomorphism on $F_{p\vee 1}^{-s+n(\frac{1}{p\wedge 1}-1),r'}(\mathbb{R}^n)$ for $(s,p,q) \in \mathcal{V}$. Here $r = q$ if both, $p, q \geq 1$ or $r = 1$ otherwise.*

Theorem 7. *Let T be a bounded linear operator on $B_p^{s,q}(\mathbb{R}^n)$ for (s,p,q) in some neighborhood $\mathcal{U} \subset \mathbb{R} \times (0,\infty) \times (0,\infty)$ of (s_0, p_0, q_0), which is an isomorphism on $B_{p_0}^{s_0,q_0}(\mathbb{R}^n)$. Then there is a neighborhood $\mathcal{V}$ of (s_0, p_0, q_0) such that T extends to an isomorphism on $B_{p\vee 1}^{s-n(\frac{1}{p\wedge 1}-1),q\vee 1}(\mathbb{R}^n)$ and its transpose T^* extends to an isomorphism on $B_{p'}^{-s+n(\frac{1}{p\wedge 1}-1),q'}(\mathbb{R}^n)$ for $(s,p,q) \in \mathcal{V}$, with $s \neq s_0$.*

To finish this section we point out that for $s \in \mathbb{R}$, $1 < p < \infty$,

$$F_p^{s,2}(\mathbb{R}^n) = L_s^p(\mathbb{R}^n),$$

where $L_s^p(\mathbb{R}^n)$ stands for the potential space defined via the Fourier transform as

$$\left\{f \in \mathcal{S}' : \|f\|_{L_s^p} = \|(I-\Delta)^{\frac{s}{2}}(f)\|_{L^p} < \infty\right\}.$$

Furthermore, for $\frac{n}{n+1} < p \leq 1$, $F_p^{0,2}(\mathbb{R}^n) = H^p(\mathbb{R}^n)$, the atomic Hardy space ([15]). In view of the above identification and for matter of tradition, we will rather use the L_s^p notation instead of $F_p^{s,2}$, specially in the framework of boundary value problems.

3. Applications to PDE's

The functional analytic machinery developed in the previous section provides a valuable tool for the treatment of certain problems arising in the interface between Harmonic Analysis and PDE's, more specifically, in the study of global regularity of (systems of) PDE's in non-smooth domains. Recall that a domain Ω is said to be Lipschitz if its boundary (denoted by $\partial\Omega$) is locally the graph of a Lipschitz function (see [16]). We start by observing that the spaces $B_p^{s,q}(\Omega)$ $(F_p^{s,q}(\Omega))$ can be defined via restrictions to Ω of distributions in $B_p^{s,q}(\mathbb{R}^n)$ $(F_p^{s,q}(\mathbb{R}^n))$ (see [8], [15]). We relax the notation by making no distinction between the space of vector-valued functions with components in $A_p^{s,q}(\Omega)$ and $A_p^{s,q}(\Omega)$ itself, A standing for either B or F. For $0 < s < 1$, the spaces on the boundary, $B_p^{s,q}(\partial\Omega)$ $(F_p^{s,q}(\partial\Omega))$ can be defined via the pull-back induced by local coordinates (see [6], [11] and the references therein). Theorems 5, 6 and 7, remain valid for the corresponding scales of spaces defined on $\partial\Omega$, which also inherits the interpolation properties of the scales in $\mathbb{R}^n$ ([6], [11]). For the rest of this work, we fix a Lipschitz domain $\Omega \subset \mathbb{R}^3$ with $\partial\Omega$ connected, (almost everywhere defined) exterior unit normal N and two real numbers λ, μ, with $\lambda > 0$ and $\lambda > -\frac{2}{3}\mu$. We consider the three-dimensional Lamé operator,

$$\mathcal{L}u := \mu\Delta u + (\lambda + \mu)\nabla(\mathrm{div} u),$$

with $u = (u_1, u_2, u_3) \in \mathbb{R}^3$, the Poisson problem for the Dirichlet boundary condition,

$$\begin{cases} \mathcal{L}u = f \in L^p_{s+\frac{1}{p}-2}(\Omega) \\ Tru = g \in B_p^{s,p}(\partial\Omega) \\ u \in L^p_{s+\frac{1}{p}}(\Omega) \end{cases} \tag{3.1}$$

and for the Neumann type boundary condition

$$\begin{cases} \mathcal{L}u = f \in L^q_{-s+\frac{1}{q}-1}(\Omega) \\ \frac{\partial u}{\partial \nu} = g \in B_q^{-s,q}(\partial\Omega) \\ u \in L^q_{1-s+\frac{1}{q}}(\Omega). \end{cases} \tag{3.2}$$

In problem (3.2),

$$\frac{\partial u}{\partial \nu} := (\lambda(\mathrm{div} u)N + \mu[\nabla u + \nabla u^T]N)|_{\partial\Omega} \tag{3.3}$$

(where the superscript denotes transposition), is the traction of u on $\partial\Omega$. In the present setting, (3.3) is interpreted as a distribution in $B_q^{-s,q}(\partial\Omega)$, and the (vector-valued) function $u \in L^p_{s+\frac{1}{p}}(\Omega)$ is said to be a solution of problem (3.2) iff for every $\phi \in B_p^{s,p}(\partial\Omega)$ extended to $\tilde{\phi} \in L^p_{s+\frac{1}{p}}(\Omega)$ (in the trace sense), the equality

$$\langle \frac{\partial u}{\partial \nu}, \phi\rangle = \langle f, \tilde{\phi}\rangle + \mu\langle \nabla u + \nabla u^T, \nabla\phi + \nabla\phi^T\rangle + \lambda\langle \mathrm{div} u, \mathrm{div}\tilde{\phi}\rangle \tag{3.4}$$

holds. The above pairings are well defined via duality of the intervening spaces (see [6] for details).

The above problems can be approached via integral equations: to that end, we recall the definition of the Kelvin matrix (here δ_{ij} stands for the Kronecker delta)

,

$$\Gamma_{ij}(X) := \frac{3}{8\pi}\left(\frac{1}{\mu}+\frac{1}{2\mu+\lambda}\right)\frac{\delta_{ij}}{|X|}+\frac{3}{8\pi}\left(\frac{1}{\mu}-\frac{1}{2\mu+\lambda}\right)\frac{X_iX_j}{|X|^3},\ X\in\mathbb{R}^3\setminus\{0\}.$$

and the single and double-layer potentials,

$$\mathcal{S}f(Q)(X) = \int_{\partial\Omega}\Gamma(X-Q)f(Q)d\sigma(Q)$$

and

$$\mathcal{D}f(Q)(X) = \int_{\partial\Omega}\left(\frac{\partial}{\partial\nu}\Gamma(X-Q)\right)^T f(Q)d\sigma(Q),$$

respectively, defined for $X\in\mathbb{R}^3\setminus\partial\Omega$, where $d\sigma$ stands for the surface measure on $\partial\Omega$. We also recall the boundary behavior of these potentials for $f\in L^p(\partial\Omega)$

$$\lim_{X\to P}\mathcal{D}f(X) = \mathrm{pv}\int_{\partial\Omega}\left(\frac{\partial}{\partial\nu}\Gamma(P-Q)\right)^T f(Q)d\sigma(Q) = (\frac{1}{2}I+K)f(P) \tag{3.5}$$

and

$$\lim_{X\to P}\frac{\partial}{\partial\nu}\mathcal{S}f(Q)(X) = (-\frac{1}{2}+K^*)(f)(P), \tag{3.6}$$

where the convergence is understood in a suitable non-tangential manner (see [16] and [5]). We will also write S for the boundary operator

$$Sf(P) = \int_{\partial\Omega}\Gamma(P-Q)\,f(Q)\,d\sigma(Q).$$

We remark that the boundary singular-integral operators K and its formal transpose K^* are bounded on L^p for $1<p<\infty$ ([2]). Finally, we define the elastic Newtonian potential $\mathcal{N}$ on Ω as

$$\mathcal{N}(f): L^p_{s-2}(\Omega)\to L^p_s(\Omega)\ ,\ \mathcal{N}(f) = \int_\Omega\Gamma(X-Y)f(y)\,dy \tag{3.7}$$

for $1<p<\infty$, $0\le s\le 2$ and a vector-valued, compactly supported function f, and extend $\mathcal{N}$ by density in a standard manner. Notice that for f as in problem (3.2), the traction $\frac{\partial}{\partial\nu}\mathcal{N}(f)$ is well defined as a distribution in $B_q^{-s,q}(\partial(\Omega)$ (see [6] for the proofs of the assertions in this paragraph).

3.1. Mapping properties of the layer potentials

The proof of the following theorem can be found in [6]

Theorem 8. *Let $\Omega\subset\mathbb{R}^n$ be a Lipschitz domain. Let $1<p<\infty$ and $0<s<1$. Then there exists a positive constant C depending on Ω such that for all $f\in B_p^{-s,p}(\partial\Omega)$,*

$$\max\left\{\|\mathcal{S}f\|_{B_p^{1-s+\frac{1}{p},p}(\Omega)},\|\mathcal{S}f\|_{L^p_{1-s+\frac{1}{p}}(\Omega)}\right\}\le C\,\|f\|_{B_p^{-s,p}(\partial\Omega)}$$

and for all $f \in B_p^{s,p}(\partial\Omega)$

$$\max\left\{\|\mathcal{D}f\|_{B_p^{s+\frac{1}{p},p}(\Omega)}, \|\mathcal{D}f\|_{L^p_{s+\frac{1}{p}}(\Omega)}\right\} \leq C\,\|f\|_{B_p^{s,p}(\partial\Omega)}.$$

Let Ψ denote the six-dimensional space of vector valued functions ψ satisfying

$$\partial_i\psi^j + \partial_j\psi^i = 0,$$

and, for any space of distributions $S(\partial\Omega)$ on $\partial\Omega$, let

$$S(\partial\Omega)_\Psi = \{\theta : \theta \in S(\partial\Omega), \langle\theta, \varphi|_{\partial\Omega}\rangle = 0 \text{ for all } \varphi \in \Psi\}.$$

For $\epsilon > 0$ let $\mathcal{R}_\epsilon$ be the region in the $(s, \frac{1}{p})$ plane inside the hexagon with vertices $A = (1-\epsilon, 1)$, $B = (1,1)$ $C = (1, \frac{1}{2}-\epsilon)$, $D = (\epsilon, 0)$, $E = (0,0)$ and $F = (0, \frac{1}{2}+\epsilon)$. We now present the main results in this section:

Theorem 9. *Let* $\Omega \subset \mathbb{R}^3$ *be a bounded Lipschitz domain. Then there exists* $\epsilon = \epsilon(\Omega) \in (0,1]$, *such that for* $(s, \frac{1}{p}) \in \mathcal{R}_\epsilon$, $\frac{1}{p} + \frac{1}{q} = 1$, *the operators listed below (which are well defined and bounded) are isomorphisms:*

1. $\frac{1}{2}I + K : B_p^{s,p}(\partial\Omega) \to B_p^{s,p}(\partial\Omega)$
2. $\frac{1}{2}I + K^* : B_q^{-s,q}(\partial\Omega) \to B_q^{-s,q}(\partial\Omega)$
3. $S : B_q^{-s,q}(\partial\Omega) \to B_q^{1-s,q}(\partial\Omega)$
4. $\pm\frac{1}{2}I + K : B_p^{s,p}(\partial\Omega)/\Psi \to B_p^{s,p}(\partial\Omega)/\Psi$
5. $\pm\frac{1}{2}I + K^* : B_q^{-s,q}(\partial\Omega)_\Psi \to B_q^{-s,q}(\partial\Omega)_\Psi$
6. $S : B_q^{-s,q}(\partial\Omega)_\Psi \to B_q^{1-s,q}(\partial\Omega)/\Psi$.

Proof. As it follows from [4], [3], [11] and [16], there is a positive number $\delta = \delta(\Omega)$ such that the operators (which are well defined)

1. $\mathcal{S} : F_p^{0,2}(\partial\Omega) \to F_p^{1,2}(\partial\Omega)$
2. $\frac{1}{2}I + K : F_p^{1,2}(\partial\Omega) \to F_p^{1,2}(\partial\Omega)$
3. $\frac{1}{2}I + K^* : F_p^{0,2}(\partial\Omega) \to F_p^{0,2}(\partial\Omega)$

are bounded for $1-\delta < p \leq 1$ and are isomorphisms for $p = 1$. From Theorems 1, 6 and 7, one concludes that there exists a positive number $\epsilon = \epsilon(\Omega)$ such that for $0 < s < \epsilon$, the operators

1. $\mathcal{S} : B_1^{-s,1}(\partial\Omega) \to B_1^{1-s,1}(\partial\Omega)$
2. $\frac{1}{2}I + K : B_1^{1-s,1}(\partial\Omega) \to B_1^{1-s,1}(\partial\Omega)$
3. $\frac{1}{2}I + K^* : B_1^{-s,1}(\partial\Omega) \to B_1^{-s,1}(\partial\Omega)$

are isomorphisms. Notice that these results cover the invertibility of $\frac{1}{2}I + K$ on the upper horizontal segment of the boundary of $\mathcal{R}_\epsilon$. Theorem 9 now follows from the L^2-results in [5], [16], duality and complex interpolation. We omit the details, which the interested reader will find in [11] and the references cited there.

We are now ready to prove:

Theorem 10. *Let $\Omega \subset \mathbb{R}^3$ be a bounded, Lipschitz domain and ϵ be as in Theorem 9. Then, for $(s, \frac{1}{p}) \in \mathcal{R}_\epsilon$, the problems (3.1) and (3.2) are solvable, (in the case of (3.2), we assume the necessary compatibility condition $\langle f, \psi \rangle = \langle g, \psi|_{\partial\Omega} \rangle$ for all $\psi \in \Psi$), i.e. a solution u exists, it is unique (modulo Ψ for (3.2)) and there exists a positive constant $C = C(\Omega)$ such that*

$$\|u\|_{L^p_{s+\frac{1}{p}}(\Omega)} \leq C \left(\|f\|_{L^p_{s+\frac{1}{p}-2}(\Omega)} + \|g\|_{B^{s,p}_p(\partial\Omega)} \right)$$

for the solution u of (3.1) and

$$\|u\|_{L^p_{1-s+\frac{1}{q}}(\Omega)} \leq C \left(\|f\|_{L^p_{-s-1+\frac{1}{q}}(\Omega)} + \|g\|_{B^{-s,q}_q(\partial\Omega)} \right)$$

for the solution u of (3.2). Moreover, the solutions are given by

$$u(x) = \mathcal{N}(f) + \mathcal{D}\left((\frac{1}{2}I + K)^{-1}(g - \mathcal{N}(f)|_\Omega) \right)(X)$$

and

$$v(x) = \mathcal{N}(f) + \mathcal{S}\left((-\frac{1}{2}I + K^*)^{-1}(g - \frac{\partial}{\nu}\mathcal{N}(f)) \right)(X)$$

respectively. These results are optimal, in the sense that for $(s, \frac{1}{p}) \notin \mathcal{R}_\epsilon$ there exist a Lipschitz domain $\Omega^D_{s,p}$ on which (3.1) it is not solvable and a Lipschitz domain $\Omega^N_{s,p}$, in which (3.2) is not solvable in the sense defined above.

Proof. The existence proof follows directly from (3.5), (3.6), (3.7) and the remark following it, Section 3.1 and Theorem 9. See [6] and [11] for the proof of uniqueness and the optimality of the range $\mathcal{R}_\epsilon$. Notice that $\mathcal{R}_0$ is a region of solvability common to all Lipschitz domains. □

We would like to underline the fact that similar results follow in the same fashion for any (system of) partial differential equation(s) for which endpoint estimates as in [3] and [16] are known. The dimensional limitation in the case of the Lamé system stems precisely from the fact that the corresponding Hardy-space estimates are known only for $n = 3$ ([4]).

References

[1] J. Bergh, J. Löfström *Interpolation Spaces. An Introduction.* Springer Verlag, Berlin, 1976.

[2] R. Coifman, A. Mc Intosh, Y. Meyer *L'Intégrale de Cauchy définit un opérateur borné sur L^2 pour les courbes Lipschitziennes.* Ann. Math. (2) 116, 1982, 361–387.

[3] B. Dahlberg, C. Kenig. *Hardy Spaces and the Neumann Problem in L^p for Laplace's equation in Lipschitz Domains.* Ann. Math. (2) 125, 1987, 437–465.

[4] B. Dahlberg, C. Kenig. *L^q estimates for the three-dimensional system of elastostatics in Lipschitz domains.* Lecture Notes in Pure and Applied Math., Vol. 122, 1990, 621–634.

[5] B. Dahlberg, C. Kenig, G. Verchota. *Boundary Value problems for system of elastostatics on Lipschitz domains.* Duke Math. J., 57, No.3, 1988, 795–818.

[6] E. Fabes, O. Méndez, M. Mitrea. *Boundary Layers of Sobolev-Besov Spaces and Poisson's equation for the Laplacian on Lipschitz domains.* J. Func. Anal., 159, 1998, 323–368.

[7] M. Frazier, B. Jawerth *A Discrete Transform and Decomposition of Function Spaces.* J. Func. Anal., 93(1), 1990, 35–160.

[8] D. Jerison, C. Kenig. *The Inhomogeneous Dirichlet Problem in Lipschitz Domains.* J. Func. Anal., 130, 1995, 161–219.

[9] N. Kalton, M. Mitrea. *Stability results on interpolation scales of quasi-Banach spaces and applications.* Trans. Am. Math. Soc., 350(10), 1998, 3903–3922.

[10] N. Kalton, N. Peck, J. Roberts *An F space Sampler.* London Math. Soc., LNS, n. 89, 1984.

[11] O. Méndez, M. Mitrea. *The Banach Envelopes of Besov and Triebel-Lizorkin Spaces and Applications to Partial Differential Equations* J. Four. Anal. and Appl., 6(5), 2000, 503–531.

[12] T. Runst, W. Sickel *Sobolev Spaces of Fractional Order, Nemytskij Operators and Nonlinear Partial differential Operators.* De Gruyter, Berlin, New York, 1996.

[13] H. Triebel. *Interpolation Theory, Function Spaces, Differential Operators.* North-Holland, Amsterdam, 1978.

[14] H. Triebel. *Theory of Function Spaces.* Birkhäuser, Basel, 1983.

[15] H. Triebel. *Theory of Function Spaces II.* Birkhäuser, Basel, 1992.

[16] G. Verchota. *Layer Potentials and regularity for the Dirichlet Problem for Laplace's equation in Lipschitz Domains.* J. Funct. Anal., 59, 1984, 572–611.

Osvaldo Méndez
Department of Mathematics
University of Texas at El Paso
500W University Ave.
El Paso, TX 79968, USA
E-mail address: `mendez@math.utep.edu`

Marius Mitrea
Department of Mathematics
University of Missouri at Columbia
201 Math. Sciences Bldg.
Columbia, MO 65211, USA
E-mail address: `marius@math.missouri.edu`

D. Haroske, T. Runst, H.-J. Schmeisser (eds.): Function Spaces, Differential Operators and Nonlinear Analysis. The Hans Triebel Anniversary Volume.

On the Flow Map for a Class of Parabolic Equations

Luc Molinet, Francis Ribaud, and Abdellah Youssfi

Dedicated to Prof. Hans Triebel on the occasion of his 65th birthday

Abstract. We consider the Cauchy problem for the one-dimensional parabolic equations $\partial_t u - \partial_{xx} u \pm \partial_x^d u^k = 0$, $k \in \mathbb{N}^*$, $d \in \{0,1\}$, with initial data in $H^s(\mathbb{R})$. We study the flow map corresponding to the integral equation. Our results complete the known results on *ill-posedness* in $H^s(\mathbb{R})$ and show the particularity of the case $(k,d) = (2,0)$ for which we prove that the critical space $H^{s_c}(\mathbb{R}) = H^{-3/2}(\mathbb{R})$ suggesting by standard scaling arguments cannot be reached. Our results hold also in the periodic setting.

1. Introduction and main results

We study the Cauchy problem for the following class of one-dimensional parabolic equations (on $\mathbb{R}$ or on $\mathbb{T}$)

$$\begin{cases} \partial_t u - \partial_{xx} u = \pm\, \partial_x^d u^k \\ u(0,x) = \varphi(x) \end{cases}, \tag{1.1}$$

where $k \in \mathbb{N}^*$, $d \in \{0,1\}$, $\partial_x^0 = Id$ and $\partial_x^1 = \partial_x$. These both kinds of equations arise in a lot of physical contexts (semilinear heat equation for $d = 0$ and generalized Burgers equation for $d = 1$).

Here we will only consider the solutions of the integral equation corresponding to (1)

$$u(t) = S(t)\varphi \pm \int_0^t S(t-t')\partial_x^d u^k(t')\, dt' , \tag{1.2}$$

where $S(t) = exp(t\,\partial_{xx})$ denotes the generator of the free linear heat equation.

It is well known that (2) is locally well posed in $H^s(\mathbb{R})$ for $s > -1$ if $(k,d) = (2,0)$ and for $s \geq s_c(k,d) = 1/2 - (2-d)/(k-1)$ otherwise ($H^{s_c(k,d)}(\mathbb{R})$ is the critical space suggested by scaling arguments). Furthermore it has also been proved that some of those results are sharp. When $(k,d) = (2,1)$ a nonuniqueness result in $H^s(\mathbb{R})$, $s < s_c(2,1) = -1/2$, has been obtained in [6], and when $(k,d) = (3,0)$ a nonexistence result in $H^s(\mathbb{R})$, $s < s_c(3,0) = -1/2$, has been derived in [5].

In this paper, our aim is to prove in some sense that equation (2) cannot be solved in $H^s(\mathbb{R})$ by iterative methods whenever $s < -1$ if $(k,d) = (2,0)$ and whenever $s < s_c(k,d) = 1/2 - (2-d)/(k-1)$ otherwise. This emphasizes the particularity of the case $(k,d) = (2,0)$ for which the Cauchy problem cannot be solved in $H^s(\mathbb{R})$ for $s < -1$ while the critical exponent suggested by scaling arguments is $s_c(2,0) = -3/2$. In our knowledge this is the first example of semilinear parabolic equation for which the scaling exponent cannot be reached. Similar results hold for some dispersive equations [3], [4], [8], [12], [16], and for some semilinear wave equations, see [10].

Now it is convenient to introduce the "*well-posedness*" and "*ill-posedness*" of (2).

Definition.

1) *We will say that* (2) *is locally well posed in* $H^s(\mathbb{R})$ *if:*
   - $\mathbf{P_1}$*: For all initial data* $\varphi \in H^s(\mathbb{R})$ *there exist* $T = T(\|\varphi\|_{H^s}) > 0$*, a subspace* $X_T \subset C([0,T]; H^s(\mathbb{R}))$ *and a unique solution* $u \in X_T$ *of* (2).
   - $\mathbf{P_2}$*: For all initial data* $\varphi_0 \in H^s(\mathbb{R})$ *there exist* V_{φ_0}*, a neighbourhood of* φ_0 *in* $H^s(\mathbb{R})$*, and* $T > 0$ *such that the map* $\varphi \mapsto u(\varphi)$ *is of class* C^∞ *from* V_{φ_0} *to* $C([0,T]; H^s(\mathbb{R}))$.
2) *We say that* (2) *is ill posed in* $H^s(\mathbb{R})$ *if* $\mathbf{P_1}$ *or* $\mathbf{P_2}$ *is false.*

The known results obtained by iterative methods are the following.

Theorem 1. *Let* $s \geq s_c(k,d)$ *where*

$$s_c(k,d) = \frac{1}{2} - \frac{2-d}{k-1}.$$

Then

1) *If* $(k,d) \neq (2,0)$*, then* $\mathbf{P_1}$ *holds. Furthermore if* $s > s_c(k,d)$*, then* (2) *is locally well posed in* $H^s(\mathbb{R})$.
2) *Let* $(k,d) = (2,0)$*. Then* (2) *is locally well posed for* $s > -1$.

We refer the reader to [1] and [6] for a proof of Theorem 1 when $d = 1$ and to [2], [14] and [15] when $d = 0$. The reader who is interested in local well-posedness results for (2) in other functional spaces (like $L^p(\mathbb{R})$, $H^s_p(\mathbb{R})$, ...) could also consult [7], [9] and [17], where also further references can be found.

Our method to derive "ill-posedness" results is different from those used in [5], [6] and [17]. This method rather relies on an idea first introduced in [4] (see also [16]) to prove the "ill-posedness" of the KdV equation below $H^{-3/4}(\mathbb{R})$ in the continuous case and below $H^{-1/2}(\mathbb{T})$ in the periodic case. It was also recently used in [12], [13] to show the bad behaviour of a class of semilinear dispersive wave equations with respect to the iterative methods as well as the nonsmoothness of the flow associated with these equations.

Our first result is the following.

Theorem 2.

1) *The equation* (2) *is ill posed if* $s < -1$ *and* $(k,d) = (2,0)$*, and if* $s < s_c(k,d)$ *otherwise.*
2) *For the periodic setting the result* 1) *is valid.*

Our second result deals with the fixed point method for which we have the following.

Theorem 3. *Let $s < -1$ if $(k,d) = (2,0)$ or $s < s_c(k,d)$ otherwise, and let $T > 0$. Then for all subspaces $X_T \hookrightarrow C([0,T]; H^s(\mathbb{R}))$ the following two estimates*

$$\|S(t)\varphi\|_X \le C\|\varphi\|_{H^s}, \quad \forall \varphi \in H^s(\mathbb{R}), \tag{1.3}$$

and

$$\left\| \int_0^t S(t-t')\partial_x^d u^k(t')\, dt' \right\|_X \le C\|u\|_X^k, \quad \forall u \in X_T, \tag{1.4}$$

never hold together.

Since (1.3)–(1.4) would be needed to implement a Picard iterative scheme on the integral equation (2), Theorem 3 shows clearly the impossibility of performing such a method to solve (2) for these values of (s,k,d).

2. Proofs

The proofs of Theorem 2 and Theorem 3 are essentially based on the following key lemma.

Lemma. *Let $s < -1$ if $(k,d) = (2,0)$, and $s < s_c(k,d)$ otherwise. Then there exists a sequence of functions $\{h_N\} \subset H^s(\mathbb{R})$ such that for all $T > 0$*

$$\|h_N\|_{H^s} \simeq 1\,, \tag{2.1}$$

$$\lim_{N\to+\infty} \sup_{[0,T]} \left\| \int_0^t S(t-t')\partial_x^d \left(S(t')h_N\right)^k dt' \right\|_{H^s} = +\infty\,. \tag{2.2}$$

A) Proof of Theorem 2

Assume that (2) is well posed in H^s. Then the flow map $\varphi \mapsto u(\varphi)$ is in particular of class C^k at $\varphi = 0$. Property $\mathbf{P_1}$ implies that $u(t,x,0) = u(0)(t,x) = 0$. It follows by straightforward calculation that

$$u_k(t,x) = \frac{\partial^k u}{\partial \varphi^k}(t,x,0)[h_N,\dots,h_N]$$

$$= k! \int_0^t S(t-t')\partial_x^d \left(S(t')h_N\right)^k dt'\,.$$

But the flow map is C^k differentiable. Hence one must have

$$\sup_{[0,T]} \|u_k(t)\|_{H^s} \le C_T\, \|h_N\|_{H^s}^k, \tag{2.3}$$

which contradicts the lemma.

B) Proof of Theorem 3

Assume that there exist $T>0$ and $X_T \hookrightarrow C([0,T[;H^s(\mathbb{R}))$ such that (1.3)–(1.4) hold. Then choosing $u(t,x)=(S(t)h_N)(x)$ in (1.4) we obtain that

$$\left\| \int_0^t S(t-t')\partial_x^d \left(S(t')h_N\right)^k \, dt' \right\|_X \leq C\|S(t')h_N\|_X^k$$

and since $X \hookrightarrow C([0,T[;H^s)$ it follows that

$$\sup_{[0,T]} \left\| \int_0^t S(t-t')\partial_x^d \left(S(t')h_N\right)^k \, dt' \right\|_{H^s} \leq C\|S(t')h_N\|_X^k .$$

Then (1.3) yields

$$\sup_{[0,T]} \left\| \int_0^t S(t-t')\partial_x^d \left(S(t')h_N\right)^k \, dt' \right\|_{H^s} \leq C\,\|h_N\|_{H^s}^k$$

which again contradicts the lemma.

C) Proof of the key lemma

We prove the lemma only on $\mathbb{R}$. The periodic case is similar, see ([11]). For the proof we are led to consider the particular case $(k,d)=(2,0)$, and next we deal with the case $(k,d)\neq(2,0)$.

2.1. The case $(k,d)=(2,0)$

We define h_N with the help of its Fourier transform by

$$\hat{h}_N(\xi) = N^{-s}\gamma^{-\frac{1}{2}}\left(\chi_{[-N-\gamma,-N]}(\xi) + \chi_{[N,N+\gamma]}(\xi)\right), \tag{2.4}$$

where $\gamma = N^{-\theta}$, $\theta \geq 0$ to be chosen later.

Note that the $(h_N)_N$ are real-valued functions and

$$\|h_N\|_{H^s} \simeq 1 . \tag{2.5}$$

On the other hand we have

$$\begin{aligned}
\hat{u}_2(t,\xi) &= \int_0^t e^{-(t-t')\xi^2}\mathcal{F}(S(t')h_N) * \mathcal{F}(S(t')h_N)(\xi)\,dt' \\
&= \int_0^t e^{-t\xi^2}e^{t'\xi^2}\int_{-\infty}^{+\infty} e^{-t'\xi_1^2}\hat{h}_N(\xi_1)e^{-t'(\xi-\xi_1)^2}\hat{h}_N(\xi-\xi_1)\,d\xi_1\,dt' \\
&= \int_{-\infty}^{+\infty} \frac{\hat{h}_N(\xi_1)}{\xi_1}\frac{\hat{h}_N(\xi-\xi_1)}{\xi-\xi_1}\left[e^{-t((\xi-\xi_1)^2+\xi_1^2)} - e^{-t\xi^2}\right] d\xi_1 .
\end{aligned}$$

It follows that

$$\|u_2(t)\|_{H^s}^2 \geq C\int_{\gamma/2}^{\gamma}\left|\int_{-\infty}^{+\infty} \frac{\hat{h}_N(\xi_1)}{\xi_1}\frac{\hat{h}_N(\xi-\xi_1)}{\xi-\xi_1}\left[e^{-t((\xi-\xi_1)^2+\xi_1^2)} - e^{-t\xi^2}\right] d\xi_1\right|^2 d\xi.$$

Observe now that we can rewrite

$$\int_{-\infty}^{+\infty} \frac{\hat{h}_N(\xi_1)}{\xi_1}\frac{\hat{h}_N(\xi-\xi_1)}{\xi-\xi_1}\left[e^{-t\left((\xi-\xi_1)^2+\xi_1^2\right)}-e^{-t\xi^2}\right]d\xi_1 = f_1+f_2+f_3,$$

where f_1, f_2, and f_3 are some functions with compact supports K_1, K_2 and K_3 such that

$$K_1 \subset [-\gamma,\gamma]\,,\ K_2 \subset [2N, 2N+2\gamma]\,,\ K_3 \subset [-2N-2\gamma,-2N]\,.$$

Hence we have

$$\|u_2(t)\|_{H^s}^2 \geq C\int_{\gamma/2}^{\gamma}\left|\int_{K_1(\xi)} \frac{\hat{h}_N(\xi_1)}{\xi_1}\frac{\hat{h}_N(\xi-\xi_1)}{\xi-\xi_1}\left[e^{-t\left((\xi-\xi_1)^2+\xi_1^2\right)}-e^{-t\xi^2}\right]d\xi_1\right|^2 d\xi,$$

where

$$\begin{aligned}K_1(\xi) &= \{\xi_1\,:\,\xi_1\in[-N-\gamma,-N]\,,\ \xi-\xi_1\in[N,N+\gamma]\}\\ &\cup\{\xi_1\,:\,\xi-\xi_1\in[-N-\gamma,-N]\,,\ \xi_1\in[N,N+\gamma]\}\end{aligned}$$

and so,

$$\|u_2(t)\|_{H^s}^2 \geq C\int_{\gamma/2}^{\gamma}\left|e^{-2tN^2}-e^{-t\xi^2}\right|\left|\int_{K_1(\xi)} \frac{\hat{h}_N(\xi_1)}{N}\frac{\hat{h}_N(\xi-\xi_1)}{N}\,d\xi_1\right|^2 d\xi.$$

This proves that

$$\|u_2(t)\|_{H^s} \geq C\sqrt{\left|e^{-2tN^2}-e^{-t\gamma^2}\right|}\,N^{-2s}N^{-2}\gamma^{\frac{1}{2}}$$

and so for $T>0$,

$$\sup_{t\in[0,T]}\|u_2(t)\|_{H^s} \geq CN^{-2s}N^{-2}\gamma^{\frac{1}{2}}\,.$$

Since $s<-1$, we can always choose $\gamma = N^{-\theta}$ with $0\leq\theta<-4s-4$ which implies that

$$\lim_{N\to+\infty}\sup_{t\in[0,T]}\|u_{2,N}\|_{H^s} = +\infty\,.$$

This finishes the proof of the lemma in this case.

2.2. The case $(k,d)\neq(2,0)$.

In this case, we define the Fourier transform of h_N by

$$\hat{h}_N(\xi) = N^{-\frac{2s+1}{2}}\left[\psi_+\left(\frac{\xi}{N}\right)+\psi_-\left(\frac{\xi}{N}\right)\right] \tag{2.6}$$

where $\psi_+\geq 0$ is C^∞ and supported by $\{\xi\ :\ 0<A\leq\xi\leq B\}$ (A and B to be chosen later) such that

$$\psi_+(\xi)=1,\quad \forall\xi\in[A+(B-A)/4, B-(B-A)/4], \tag{2.7}$$

and we define ψ_- by $\psi_-(\xi)=\psi_+(-\xi)$. First remark that h_N is real-valued and that for all $s\in\mathbb{R}$ and all $N\in\mathbb{N}$,

$$\|h_N\|_{H^s}\simeq 1\,. \tag{2.8}$$

Now we compute the Fourier transform of u_k which is defined by

$$u_k = \int_0^t S(t-t')\partial_x^d \left(S(t')h_N\right)^k \, dt'. \tag{2.9}$$

By Fubini's theorem

$$\begin{aligned}
&\hat{u}_k(t,\xi_0) \\
&= \int_0^t \xi_0^d \, e^{-(t-t')\xi_0^2} \mathcal{F}([S(t')h_N]^k)(\xi_0)\, dt' \\
&= N^{-k\frac{2s+1}{2}} \xi_0^d e^{-t\xi_0^2} \sum_{p=0}^{k} \left(\begin{array}{c} k \\ p \end{array} \right) \int_0^t e^{t'\xi_0^2} \mathcal{F}\left((S(t')\psi_+)^{k-p}(S(t')\psi_-)^p\right)(\xi_0)\, dt'.
\end{aligned}$$

Since for all t' in $[0,t]$, the Fourier transform of $(S(t')\psi_+)^{k-p}(S(t')\psi_-)^p$ is supported in

$$\{\xi \;:\; pN(A+B) - kBN \leq \xi \leq pN(A+B) - kAN\,\}$$

it follows that

$$\hat{u}_k(t,\xi_0) = N^{-k\frac{2s+1}{2}} \xi_0^d e^{-t\xi_0^2} \int_0^t e^{t'\xi_0^2} \mathcal{F}\left((S(t')\psi_+)^k\right)(\xi_0)\, dt' + g(t,\xi_0), \tag{2.10}$$

where the first term on the right-hand side of equation (2.10) is supported in the set

$$\{\xi \;:\; kNA \leq \xi \leq kNB\,\}$$

while the function g is supported in

$$\{\xi \,/\, -kNB \leq \xi \leq (k-1)NB - NA\,\}.$$

Hence for $A > B(k-1)/(k+1)$ we have

$$\begin{aligned}
&\hat{u}_k(t,\xi_0)\chi_{[kAN\,,\,kBN]}(\xi_0) \\
&= N^{-k\frac{2s+1}{2}} \xi_0^d e^{-t\xi_0^2} \int_0^t e^{t'\xi_0^2} \mathcal{F}\left((S(t')\psi_+)^k\right)(\xi_0)\, dt' \\
&= N^{-k\frac{2s+1}{2}} \xi_0^d e^{-t\xi_0^2} \int_0^t \int_{\mathbb{R}^{k-1}} e^{t'\xi_0^2} e^{-(t-t')\xi_0^2} e^{-t'(\xi_0-\xi_1)^2} \cdots e^{-t'(\xi_{k-2}-\xi_{k-1})^2} e^{-t'\xi_{k-1}^2} \\
&\quad \times \psi_+\left(\frac{\xi_0-\xi_1}{N}\right) \cdots \psi_+\left(\frac{\xi_{k-2}-\xi_{k-1}}{N}\right) \psi_+\left(\frac{\xi_{k-1}}{N}\right) d\xi_1 \cdots d\xi_{k-1} dt' \\
&= N^{-k\frac{2s+1}{2}} \xi_0^d \, e^{-t\xi_0^2} \int_0^t \int_{\mathbb{R}^{k-1}} e^{2t' \sum_{i=0}^{k-1} \xi_i(\xi_{i-1}-\xi_i)} \\
&\quad \times \psi_+\left(\frac{\xi_0-\xi_1}{N}\right) \cdots \psi_+\left(\frac{\xi_{k-2}-\xi_{k-1}}{N}\right) \psi_+\left(\frac{\xi_{k-1}}{N}\right) d\xi_1 \cdots d\xi_{k-1} dt'
\end{aligned}$$

By Fubini's theorem and integrating with respect to t', this leads to

$$\begin{aligned}&\hat{u}_k(t,\xi_0)\chi_{[kAN\,,\,kBN]}\\&= N^{-k\frac{2s+1}{2}}\xi_0^d\, e^{-t\xi_0^2}\int_{\mathbb{R}^{k-1}}\frac{e^{2t\sum_{i=0}^{k-1}\xi_i(\xi_{i-1}-\xi_i)}-1}{2\sum_{i=0}^{k-1}\xi_i(\xi_{i-1}-\xi_i)}\\&\times\psi_+\left(\frac{\xi_0-\xi_1}{N}\right)\cdots\psi_+\left(\frac{\xi_{k-2}-\xi_{k-1}}{N}\right)\psi_+\left(\frac{\xi_{k-1}}{N}\right)d\xi_1\cdots d\xi_{k-1}.\end{aligned}$$

Now, since $\psi_+(\xi/N)$ is supported by $[AN,BN]$ it follows that

$$\sum_{i=0}^{k-1}\xi_i(\xi_{i-1}-\xi_i)\geq 0$$

and so by convexity of the exponential function,

$$\begin{aligned}&\hat{u}_k(t,\xi_0)\chi_{[kAN\,,\,kBN]}\\&\geq N^{-k\frac{2s+1}{2}}t\,\xi_0^d\, e^{-t\xi_0^2}\times\\&\times\int_{\mathbb{R}^{k-1}}\psi_+\left(\frac{\xi_0-\xi_1}{N}\right)\cdots\psi_+\left(\frac{\xi_{k-2}-\xi_{k-1}}{N}\right)\psi_+\left(\frac{\xi_{k-1}}{N}\right)d\xi_1\cdots d\xi_{k-1}\end{aligned}$$

which proves that

$$\hat{u}_k(t,\xi_0)\chi_{[kAN\,,\,kBN]}\geq N^{-k\frac{2s+1}{2}}t\,\xi_0^d\, e^{-t\xi_0^2}\left[\psi_+\left(\frac{\cdot}{N}\right)*\cdots*\psi_+\left(\frac{\cdot}{N}\right)\right](\xi_0).$$

Now from (2.7) it follows that for $\xi_0\in[kAN\,,\,kBN]$

$$\hat{u}_k(t,\xi_0)\geq N^{-k\frac{2s+1}{2}}t\,\xi_0^d\, e^{-t\xi_0^2}\left(\chi_{[cN,dN]}*\cdots*\ \chi_{[cN,dN]}\right)(\xi_0),\tag{2.11}$$

where

$$c=A+\frac{B-A}{4}\,,\ d=B-\frac{B-A}{4}\,.$$

Recalling that

$$\hat{\chi}_{[cN,dN]}(\xi)=2\exp\left(-i\frac{c+d}{2}N\xi\right)\frac{\sin\left(\frac{d-c}{2}N\xi\right)}{\xi}$$

which yields to

$$\mathcal{F}\left(\chi_{[cN,dN]}*\cdots*\chi_{[dN,dN]}\right)(\xi)=2^k\exp\left(-ik\frac{c+d}{2}N\xi\right)\left(\frac{\sin\left(\frac{d-c}{2}N\xi\right)}{\xi}\right)^k,$$

we obtain that

$$\chi_{c[N,dN]}*\cdots*\chi_{[cN,dN]}(x)=2^k\int_{\mathbb{R}}e^{ix\xi}e^{-ik\frac{c+d}{2}N\xi}\left(\frac{\sin\left(\frac{d-c}{2}N\xi\right)}{\xi}\right)^k d\xi,$$

and hence we have

$$\chi_{[cN,dN]}*\cdots*\chi_{[cN,dN]}\left(k\frac{A+B}{2}N+\alpha N\right)=2^kN^{k-1}\int_{\mathbb{R}}\cos(\alpha u)\left(\frac{\sin(u)}{u}\right)^k du.$$

This proves (by continuity) that for $\alpha > 0$ small enough there exists $C > 0$ which does not depend on N such that

$$\hat{u}_k(t,\xi)\chi_{[kN(A+B)/2-\alpha N, kN(A+B)/2+\alpha N]} \geq CN^{-k\frac{2s+1}{2}} t\,\xi^d\,\xi^{k-1}\,e^{-t\xi^2}\chi_{[kN(A+B)/2-\alpha N, kN(A+B)/2+\alpha N]}.$$

This gives the following lower bound for $\|u_k(t)\|_{H^s}$,

$$\|u_k(t)\|_{H^s}^2 \geq N^{-k(2s+1)} \int_{kN(A+B)/2-\alpha N}^{kN(A+B)/2+\alpha N} (1+|\xi|^2)^s t^2 e^{-2t\xi^2} |\xi|^{2(d+k-1)}\, d\xi$$

$$= N^{-k(2s+1)} \int_{kN(A+B)/2-\alpha N}^{kN(A+B)/2+\alpha N} (1+|\xi|^2)^s (t^2\xi^4) e^{-2t\xi^2} |\xi|^{2(d+k-1)} |\xi|^{-4}\, d\xi\,.$$

Hence it follows that

$$\sup_{[0,T]} \|u_k(t)\|_{H^s}^2 \geq N^{-k(2s+1)} \int_{kN(A+B)/2-\alpha N}^{kN(A+B)/2+\alpha N} (1+|\xi|^2)^s |\xi|^{2(d+k-1)} |\xi|^{-4}\, d\xi$$

which leads to

$$\sup_{[0,T]} \|u_k(t)\|_{H^s}^2 \geq CN^\theta$$

where

$$\theta = -k(2s+1) + 1 + 2s + 2(d+k-1) - 4. \tag{2.12}$$

Finally, we remark that $\theta > 0$ holds if and only if $s < s_c$ is satisfied.

References

[1] D. Bekiranov, *The initial-value problem for the generalized Burgers' equation*, Diff. Int. Eq., 9 (6) (1996), 1253–1265.

[2] H. Biagioni, L. Cadeddu and T. Gramchev, *Semilinear parabolic equations with singular initial data in anisotropic weighted spaces*, Diff. Int. Eq., 12 (5) (1999), 613–636.

[3] H.A. Biagioni and F. Linares, *Ill-posedness for the derivative Schrödinger and generalized Benjamin-Ono equations*, Trans. A.M.S., 9 (253) (2001), 3649–3659.

[4] J. Bourgain, *Periodic Korteweg de Vries equation with measures as initial data*, Sel. Math. New. Ser. 3 (1993), 115–159.

[5] H. Brezis and A. Friedman, *Nonlinear parabolic equations involving measures as initial conditions*, J. Math. pures et appl. 62 (1983), 73–97.

[6] D.B. Dix, *Nonuniqueness and uniqueness in the initial-value problem for Burger's equation*, SIAM J. Math. Anal., 27 (3) (1996), 708-724.

[7] Y. Giga, *Solutions for semilinear parabolic equations in L^p and the regularity of weak solutions of the Navier-Stokes system*, J. Diff. Eq. 61 (1986), 186–212.

[8] C. Kenig, G. Ponce and L. Vega, *On the ill-posedness of some canonical dispersive equations*, to appear in Duke Math. J.

[9] H. Kozono and M. Yamazaki, *Semilinear heat equations and the Navier-Stokes equation with distributions in new function spaces as initial data*, Comm. P. D.E., 19 (5/6) (1994), 959–1014.

[10] H. Linblad and C. Sogge, *On existence and scattering with minimal regularity for semilinear wave equations*, J.F.A. 130 (1995), 357–426.

[11] L. Molinet, F. Ribaud and A. Youssfi, *Ill-posedness issues for a class of parabolic equations*, Preprint.

[12] L. Molinet, J.C. Saut and N. Tzvetkov, *Ill-posedness issues for the Benjamin-Ono equations and related equations*, to appear in SIAM J. Math. Anal.

[13] L. Molinet, J.C. Saut and N. Tzvetkov, *Well-posedness and ill-posedness results for the Kadomtsev-Petviashvili I equation*, to appear in Proc. Roy. Soc. Edinburgh, Sect. A.

[14] F. Ribaud, *Cauchy problem for semilinear parabolic equations with initial data in $H^s_p(\mathbb{R}^n)$ spaces*, Rev. Math. Iberoamericana 14 (1) (1998), 1–54.

[15] F. Ribaud, *Semilinear parabolic equations with distributions as initial data*, Disc. Cont. Dyn. Syst. 3 (1997), 305–316.

[16] N. Tzvetkov, *Remark on the ill-posedness for KdV equation*, C. R. Acad. Sci. Paris, 329 (I) (1999), 1043–1047.

[17] F.B. Weissler, *Local existence and nonexistence for semilinear parabolic equation in L^p*, Indiana Univ. Math. J. 29 (1980), 79–102.

Luc Molinet
L.A.G.A., Institut Galilée
Université Paris-Nord
93430 Villetaneuse, France
E-mail address: molinet@math.univ-paris13.fr

Francis Ribaud
Equipe d'Analyse et de Mathématiques Appliquées
Université de Marne-La-Vallée
5 bd Descartes, Cité Descartes
Champs-sur-Marne
77454 Marne-La-Vallée Cedex 2, France
E-mail address: ribaud@math.univ-mlv.fr

Abdellah Youssfi
Equipe d'Analyse et de Mathématiques Appliquées
Université de Marne-La-Vallée
5 bd Descartes, Cité Descartes
Champs-sur-Marne
77454 Marne-La-Vallée Cedex 2, France
E-mail address: youssfi@math.univ-mlv.fr

D. Haroske, T. Runst, H.-J. Schmeisser (eds.): Function Spaces, Differential Operators and Nonlinear Analysis. The Hans Triebel Anniversary Volume.

Spaces of Functions with Bounded and Vanishing Mean Oscillation

David Opěla

Dedicated to Prof. Hans Triebel on the occasion of his 65th birthday

Abstract. We study generalized Campanato spaces and its vanishing subspaces. Our main interest is the connection between the geometry of the domain and the relation of the Campanato spaces to convenient Hölder spaces. We define the vanishing subspace, an analogue of VMO, and study its properties. In particular, we characterize compact subsets of VMO.

Introduction

The present paper deals with with some spaces of real functions defined on an open subset of $\mathbb{R}^N$. Their definition involves the mean oscillation over certain cubes. In the most common example, BMO (bounded mean oscillation), the oscillation is bounded, in the general case it is estimated by a function of volume of the cube. We study the relation of those spaces with other closely related spaces, namely, the (generalized) Hölder spaces. This relation depends heavily on the geometry of the domain, which is one of our main concerns. We define a "vanishing" subspace whose role is analogous to that of VMO and we study its properties. Our last, but not least goal is to study the last mentioned spaces from the topological point of view.

1. Preliminaries

The letter N always denotes the dimension of the real space we are working in, for $x \in \mathbb{R}^N$ we denote by $|x|_2$, $|x|_\infty$ the Euclidean and ℓ_∞ norm of x, respectively. The symbol Ω always denotes an open bounded connected set (bounded domain), $|\Omega|$ stands for its N-dimensional Lebesgue measure. For a set of positive measure A and a function $f \in L^1(\Omega)$, $x \in \mathbb{R}^N$ and $\delta > 0$ we set

$$f_A := \frac{1}{|A \cap \Omega|} \int_{A\cap\Omega} f(y)\, dy, \qquad Q_{x,\delta} = \{y \in \mathbb{R}^N;\ |y - x|_\infty < \frac{\delta}{2}\}.$$

If X, Y are Banach spaces, the symbols $X \hookrightarrow Y$, $X \hookrightarrow\hookrightarrow Y$ stand for (continuous) embedding and compact embedding, respectively. We say that two functions f, g

are *equivalent*, if there is a constant $c > 0$ such that $c^{-1}f(x) \le g(x) \le cf(x)$, for all x. A function f satisfies the *doubling condition*, if it is equivalent to g given by $g(x) := f(2x)$.

Let $\varphi : (0;\infty) \to (0;\infty)$ be a function. Our main focus is on the so-called *generalized Campanato space* given by

$$L^C_\varphi(\Omega) = \{f \in L^1(\Omega);\ \|f\|_{L^C_\varphi(\Omega)} < \infty\},$$

where

$$\|f\|_{L^C_\varphi(\Omega)} := \sup_{x\in\Omega,\ 0<\delta\le 2\operatorname{diam}\Omega}\ \inf_{c\in\mathbb{R}} \frac{1}{|Q_{x,\delta}|\cdot\varphi(|Q_{x,\delta}|^{\frac{1}{N}})}\int_{Q_{x,\delta}\cap\Omega} |f(y)-c|\,dy.$$

We will often refer to the function φ as to the *modulus*. Some authors (usually harmonic analysts) study essentially the same spaces and call them BMO_φ.

Similarly we can define the closely related *generalized Morrey space*, which is the following space

$$L^M_\varphi(\Omega) = \{f \in L^1(\Omega);\ \|f\|_{L^M_\varphi(\Omega)} < \infty\},$$

where

$$\|f\|_{L^M_\varphi(\Omega)} := \inf_{c\in\mathbb{R}}\ \sup_{x\in\Omega,\ 0<\delta\le 2\operatorname{diam}\Omega} \frac{1}{|Q_{x,\delta}|\cdot\varphi(|Q_{x,\delta}|^{\frac{1}{N}})}\int_{Q_{x,\delta}\cap\Omega} |f(y)-c|\,dy.$$

Finally, we recall the definition of the *generalized Hölder space*, i.e.,

$$\mathcal{C}^{0,\sigma}(\overline{\Omega}) = \{f \in L^1(\Omega);\ \|f\|_{\mathcal{C}^{0,\sigma}(\overline{\Omega})} < \infty\},$$

where

$$\|f\|_{\mathcal{C}^{0,\sigma}(\overline{\Omega})} := \operatorname*{ess\,sup}_{x\ne y\in\Omega} \frac{|f(x)-f(y)|}{\sigma(|x-y|_\infty)}.$$

If the function σ is a power, $\sigma(t) = t^\alpha$, we will abbreviate $\mathcal{C}^{0,t^\alpha}(\overline{\Omega})$ by $\mathcal{C}^{0,\alpha}(\overline{\Omega})$.

All three introduced functionals are not norms, but pseudonorms, more precisely, they are equal to zero if and only if the function is constant. To overcome this problem we identify all function differing by a constant and work with the corresponding factor space. In this way we obtain Banach spaces in all three cases.

The generalized Campanato spaces have been studied by many authors. We cite the paper of J. Kovats [K-99], who employed (a more general version of) the embedding into Hölder spaces to establish some regularity results for fully nonlinear elliptic PDEs.

We briefly recall well-known facts about the relation of those spaces (we state them in the case of a (bounded) domain, although the connectedness is not always necessary).

Definition 1.1. *An open bounded set $\Omega \subset \mathbb{R}^N$ is said to be of type A if*

$$\inf_{x\in\Omega,\ 0<\delta\le 2\operatorname{diam}\Omega} \frac{|\Omega\cap Q_{x,\delta}|}{|Q_{x,\delta}|} > 0.$$

Note that, in the infimum above, it is enough to consider only δ's less that some $\varepsilon > 0$. It is not difficult to show that any Lipschitz domain is of type A, hence the notion is fairly general. An easy example of a domain which is not of type A is a subgraph of a quadratic function.

Morrey and Campanato spaces first appeared in the 60's, invented in order to study the Hölder regularity of elliptic partial differential equations. Those authors considered as the function φ the power function and essentially proved what is contained in the following theorem.

Theorem 1.2. *Let $\Omega \subset \mathbb{R}^N$ be a bounded domain. Then the following relations hold true:*

(i) *If $\alpha \leq -N$, then $L^C_{t\alpha}(\Omega) = L^M_{t\alpha}(\Omega) = L^1(\Omega)/C$,*

(ii) *if $-N < \alpha < 0$, then $L^C_{t\alpha}(\Omega) = L^M_{t\alpha}(\Omega) \subsetneq L^1(\Omega)/C$,*

(iii) *if $\alpha = 0$, then $L^M_{t\alpha}(\Omega) = L^\infty(\Omega)/C \subsetneq L^C_{t\alpha}(\Omega) = \mathrm{BMO}(\Omega)$,*

(iv) *if $0 < \alpha$, then $L^M_{t\alpha}(\Omega) = \{0\}$,*

(v) *if $0 < \alpha \leq 1$ and Ω is of type A, then $L^C_{t\alpha}(\Omega) = \mathcal{C}^{0,\alpha}(\overline{\Omega})$,*

(vi) *if $\alpha > 1$, then $L^C_{t\alpha}(\Omega) = \{0\}$.*

Remark 1.3. For $\alpha > 1$ is useful to replace the infimum over all constants in the definition of $L^C_{t\alpha}(\Omega)$ by the infimum over all polynomials of degree less (or less or equal) than α. Very thorough treatment of those spaces can be found in [DS-84]. See also Kovats' paper [K-99].

The embedding

$$\mathcal{C}^{0,\varphi}(\overline{\Omega}) \hookrightarrow L^C_\varphi(\Omega)$$

is quite easy and holds on any domain and for more general moduli. The inverse embedding is also true for more general moduli than powers (but not arbitrary), as was proved by S. Spanne who was the first to study generalized Campanato spaces. In [Sp-65] he proved the following theorem.

Theorem 1.4. *Let $\varphi : (0;\infty) \to (0;\infty)$ be an increasing function, $Q \subset \mathbb{R}^N$ an open cube. If the function φ satisfies the Dini condition, i.e., there is a positive $\varepsilon > 0$ such that*

$$\int_0^\varepsilon \frac{\varphi(t)}{t}\,dt < \infty,$$

then

$$L^C_\varphi(Q) \hookrightarrow \mathcal{C}^{0,\omega}(\overline{Q}), \quad \omega(s) = \int_0^s \frac{\varphi(t)}{t}\,dt. \tag{1.1}$$

On the other hand, if φ does not satisfy the Dini condition, then there exists a function f satisfying

$$f \in L^C_\varphi(Q) \setminus L^\infty(Q).$$

Though the theorem is stated only for cubes and increasing moduli, the first part holds for more general moduli (e.g. satisfying the doubling condition) and domains of type A. The second part holds for increasing moduli and any Ω.

It follows from the results of A. Cianchi and L. Pick (see [CP-01]) that whenever φ and ω are such that $\frac{\varphi(t)}{t}$ is nonincreasing, $t^{N-1}\varphi(t)$ is nondecreasing and $\limsup_{t\to\infty} \frac{\omega(t)}{\varphi(t)} = \infty$, then the embedding (1.1) is strict.

2. Basic and auxiliary results

For the development of the forthcoming results it was very important to find conditions on the modulus, that can be assumed without loss of generality. In the following lemma, the second condition is equivalent to the fact that the corresponding Campanato space is different from $L^1(\Omega)/C$ while the third one assures its nontriviality (i.e., it is equivalent to the assertion $L^C_\varphi(\Omega) \neq \{0\}$) and implies that any Lipschitz function belongs to $L^C_\varphi(\Omega)$. The phrase "without loss of generality" below means that given a domain Ω and a modulus φ satisfying (ii) and (iii) — we assume those two conditions for the reasons described above — we can find a modulus $\tilde{\varphi}$ satisfying condition (i)–(iv) such that $L^C_\varphi(\Omega) = L^C_{\tilde{\varphi}}(\Omega)$ with equivalent norms.

Lemma 2.1. *Let $\Omega \subset \mathbb{R}^N$ be a bounded domain, then the following conditions on the modulus φ can be assumed without loss of generality.*

(i) $\varphi(t)t^N$ *is nondecreasing,*
(ii) $\lim_{t\to 0+} \varphi(t)t^N = 0$,
(iii) $\liminf_{t\to 0+} \frac{\varphi(t)}{t} > 0$,
(iv) φ *is continuous.*

From now on we will assume that all moduli φ satisfy the four conditions of the previous lemma.

The following lemma is used throughout almost all the proofs. Its first part is well known, while the second follows easily from some lemmas in Spanne's paper, but only for special domains (cubes). One of the reasons why cubes are very convenient as domains of Campanato spaces is the third (easy and known) assertion of the lemma.

Lemma 2.2. *Let $\Omega \subset \mathbb{R}^N$ be a bounded domain, then the following expression is an equivalent norm on $L^C_\varphi(\Omega)$*

$$\|f\| := \sup_{x\in\Omega,\ 0<\delta\leq 2\operatorname{diam}\Omega} \frac{1}{|Q_{x,\delta}| \cdot \varphi(|Q_{x,\delta}|^{\frac{1}{N}})} \int_{Q_{x,\delta}\cap\Omega} |f(y) - f_{Q_{x,\delta}}|\, dy.$$

For any $\varepsilon > 0$, the following expression is also an equivalent norm

$$\|f\|_{L^C_\varphi(\Omega),\varepsilon} := \sup_{x\in\Omega,\ 0<\delta\leq\varepsilon} \inf_{c\in\mathbb{R}} \frac{1}{|Q_{x,\delta}| \cdot \varphi(|Q_{x,\delta}|^{\frac{1}{N}})} \int_{Q_{x,\delta}\cap\Omega} |f(y) - c|\, dy.$$

If $\Omega = Q$ is a cube, then also the following norm is equivalent on $L^C_\varphi(\Omega)$

$$\|f\|^\square := \sup_{Q' \subset Q} \inf_{c \in \mathbb{R}} \frac{1}{|Q'| \cdot \varphi(|Q'|^{\frac{1}{N}})} \int_{Q'} |f(y) - c| \, dy.$$

In the classical case (i.e., when the modulus is a power), we can replace the cubes in the definition of Campanato spaces by balls. It is easy to check that the same is true, whenever the modulus satisfies the doubling condition. On the other hand, there is an example, that this is not true in general, but the domain is bad too (not of type A).

3. Geometry of the domain

Our first aim concerns the shape of the domain involved in the relation to the Hölder spaces. Note that the notion of the geometry of the domain does not make sense if $N = 1$, hence throughout this section we will assume that $N \geq 2$. It turns out that the validity of the reverse implication in Spanne's theorem in the classical case is different for $\alpha < 1$ and $\alpha = 1$. The following theorems clarify the situation.

Theorem 3.1. *Let $\Omega \subset \mathbb{R}^N$ be a bounded domain, $0 < \alpha < 1$. Then*

$$L^C_{t^\alpha}(\Omega) = C^{0,\alpha}(\overline{\Omega}),$$

if and only if the domain is of type A.

For the embedding in the Lipschitz case (i.e., $\varphi(t) = t$), it suffices for Ω to satisfy the following condition (defined in e.g. [KJF-77], p. 24). We use the ℓ_∞-norm, but any norm yields the same condition.

Definition 3.2. *A bounded domain $\Omega \subset \mathbb{R}^N$ is said to satisfy the condition* (S), *if there exists a constant $M > 0$ such that for every $x, y \in \Omega$ there is a chain of points $x = z_0, z_1, \ldots, z_n = y$ such that the segments $[z_i z_{i+1}]$ are contained in Ω and*

$$\sum_{i=0}^{n-1} |z_i z_{i+1}|_\infty \leq M |x - y|_\infty.$$

Theorem 3.3. *Let $\Omega \subset \mathbb{R}^N$ be a bounded domain satisfying the condition* (S). *Then*

$$L^C_t(\Omega) = C^{0,1}(\overline{\Omega}). \tag{3.1}$$

Nevertheless, there is a domain $\Omega \subset \mathbb{R}^N$ ($N \geq 2$) such that (3.1) *does not hold. Clearly, this domain is neither of type A nor satisfies the condition* (S).

We have seen that functions in $L^C_\varphi(\Omega)$ for Ω which is not of type A are not so nice as those in $L^C_\varphi(\Omega)$ with Ω of type A. However, the following result shows that some continuity still may be deduced, if the domain is not too bad. To state the result precisely we have to generalize the notion of domain of type A.

Definition 3.4. *Let $\Omega \subset \mathbb{R}^N$ be a bounded domain and δ be a positive number. We set*

$$A_\Omega(\delta) := \inf_{x\in\Omega,\ 0<r\le\delta} \frac{|Q_{x,r}\cap\Omega|}{|Q_{x,r}|}.$$

We will say that the domain is of type $A_\Omega(\delta)$ and sometimes abbreviate the notation to $A(\delta)$, if there is no risk of confusion.

It is not difficult to show that for any Ω, the function $A_\Omega(\delta)$ is continuous and that Ω is of type A if and only if $A_\Omega(\delta) \approx A$ (a positive constant).

Theorem 3.5. *Let $\Omega \subset \mathbb{R}^N$ be a bounded domain and let the modulus φ satisfy the doubling condition. Suppose that there is a constant $C > 0$ such that the following inequality holds for any $t > 0$*

$$\frac{\varphi(t)}{A_\Omega(t)} \le C \inf_{\tau\in[\frac{t}{2};t]} \frac{\varphi(\tau)}{A_\Omega(\tau)}.$$

Then the following embedding holds

$$L^C_\varphi(\Omega) \hookrightarrow \mathcal{C}^{0,\omega_\Omega}(\overline{\Omega}),$$

if the function

$$\omega_\Omega(s) := \int_0^s \frac{\varphi(t)}{A_\Omega(t)t}\, dt$$

is well defined.

4. The vanishing subspace

The concern of the present section is the following generalization of VMO. Since BMO is a special case of the (generalized) Campanato space, we can extend the concept of "vanishing norm" to those spaces in an obvious way. It seems to the author that the study of these more general vanishing spaces is new, although in a special case of a power they appeared in the literature (see [Ch-94]).

Definition 4.1. *The generalized vanishing Campanato space is the following set*

$$V^C_\varphi(\Omega) := \{f \in L^C_\varphi(\Omega);\ \lim_{\varepsilon\to 0+} \|f\|_{L^C_\varphi(\Omega),\varepsilon} = 0\},$$

endowed with the norm $\|\cdot\|_{L^C_\varphi(\Omega)}$.

Proposition 4.2. *Suppose that $\Omega \subset \mathbb{R}^N$ is a bounded domain. Then $V^C_\varphi(\Omega)$ is a closed subspace of $L^C_\varphi(\Omega)$.*

Remark 4.3. In fact, more or less the same proof as that of Theorem 1.2(v), provides the following result. For any bounded domain Ω which is of type A

$$V^C_{t^\alpha}(\Omega) = c^{0,\alpha}(\overline{\Omega}),$$

where

$$c^{0,\alpha}(\overline{\Omega}) = \left\{ f \in \mathcal{C}^{0,\alpha}(\overline{\Omega});\ \lim_{\delta \to 0+} \operatorname*{ess\,sup}_{x,y\in\Omega,\ 0<|x-y|_\infty<\delta} \frac{|f(x)-f(y)|}{|x-y|_\infty^\alpha} = 0 \right\}$$

is the so-called *little Hölder space.*

Our next result is the identification of the vanishing space with the union of certain Campanato spaces. The special case of VMO is an exercise in [T-86], p. 220.

Proposition 4.4. *Let $\Omega \subset \mathbb{R}^N$ be a bounded domain. Then we have*

$$V_\varphi^C(\Omega) = \bigcup_{\tilde{\varphi} \ll \varphi} L_{\tilde{\varphi}}^C(\Omega),$$

with the following notation

$$\tilde{\varphi} \ll \varphi \quad \text{means} \quad \lim_{t \to 0+} \frac{\tilde{\varphi}(t)}{\varphi(t)} = 0.$$

Note that the preceding proposition together with the fact that Campanato spaces which are more "smooth" than $L_t^C(\Omega)$ are trivial, implies that the vanishing subspace of $L_t^C(\Omega)$ is also trivial.

It is a well-known fact that if $f(x) = \log|x|$ then f belongs to $\mathrm{BMO}((-1;1))$ but not to $\mathrm{VMO}((-1;1))$. Therefore VMO is properly contained in BMO. It turns out that this phenomenon is true in general.

Proposition 4.5. *Let $\Omega \subset \mathbb{R}^N$ be a bounded domain and let φ be a modulus. Then*

$$V_\varphi^C(\Omega) \subsetneq L_\varphi^C(\Omega).$$

It was D. Sarason (see [Sa-75]) who invented VMO and showed that bounded uniformly continuous functions are dense in it. This certainly cannot be true for smoother Campanato spaces, since not every $\mathcal{BUC}$-function is Hölder continuous. However, one can show (under certain assumptions) that $\mathcal{C}^\infty$-functions are dense in $V_\varphi^C(\Omega)$. The proof involves quite standard convolution techniques and some extension theorems.

Remark 4.6. If $\varphi(t) = t^\alpha$, with $0 < \alpha < 1$, then a function $f \in L_\varphi^C(\Omega_0)$, $\Omega_0 \subset \Omega$, can be extended to a function $\tilde{f} \in L_\varphi^C(\Omega)$, where Ω is of type A, if and only if Ω_0 is of type A. Indeed, if Ω_0 is of type A, we use Theorem 1.2(v) and the fact that a Hölder function can be extended from any set (see [DDK-74]). If it is not the case, we can find a function $f \in L_{t^\alpha}^C(\Omega_0) \setminus \mathcal{C}^{0,\alpha}(\overline{\Omega_0})$. This function certainly cannot be extended to a function $\tilde{f} \in L_\varphi^C(\Omega)$, because its extension $\tilde{f}$ would belong to $\mathcal{C}^{0,\alpha}(\overline{\Omega})$.

Before proceeding with the extension theorems, we first state a lemma which enables us to use the technique of partition of unity. It characterizes the so-called *pointwise multipliers* of $L_\varphi^C(\Omega)$. Its special case (for Ω being a torus) can be found in [T-86]. However, since functions in $L_\varphi^C(\Omega)$ are members of equivalence classes modulo constants, we should first clarify what we mean by fg for $f \in L^C_\varphi(\Omega)$. It is the equivalence class modulo constants which contains $\tilde{f}g$, where $\tilde{f}$ is the representative of f with $\tilde{f}_\Omega = 0$.

Lemma 4.7. *Let φ be a modulus satisfying the doubling condition and let Ω be a domain of type A. We define ψ by*

$$\psi(s) = \frac{\varphi(s)}{\int_t^{2\operatorname{diam}\Omega} \frac{\varphi(t)}{t}\, dt}. \tag{4.1}$$

Then for any $g \in L^C_\psi(\Omega) \cap L^\infty(\Omega)$ there exists a constant $c > 0$ such that

$$\forall f \in L^C_\varphi(\Omega): \quad \|gf\|_{L^C_\varphi(\Omega)} \le c\|f\|_{L^C_\varphi(\Omega)}.$$

Now we can state the theorem. Note that if Ω is a cube, the extension is much easier and we do not have to require anything from the modulus φ.

Theorem 4.8. *Let one of the following two conditions be satisfied*

(i) $\Omega \subset \mathbb{R}^N$ *is a cube* Q,

(ii) $\Omega \subset \mathbb{R}^N$ *is a bounded domain with a Lipschitz boundary and φ is a modulus satisfying the doubling condition, and let the function ψ given by (4.1) satisfy* $\liminf_{t\to 0+} \frac{\psi(t)}{t} > 0$.

Let $\tilde{\Omega} \subset \mathbb{R}^N$ be a bounded domain containing Ω. Then there exists a bounded extension operator

$$E : L^C_\varphi(\Omega) \to L^C_\varphi(\tilde{\Omega}).$$

We remark, that the condition on ψ in (ii) is automatically satisfied if $\frac{\varphi(t)}{t}$ is nonincreasing, which holds in most interesting examples.

Now we state the theorem concerning density of C^∞-functions. A similar result is known for Hölder spaces (see [DDK-74]), hence for Campanato spaces with moduli which are powers on domains of type A.

Theorem 4.9. *Let φ satisfy $\lim_{t\to 0+} \frac{\varphi(t)}{t} = \infty$ and let the assumptions of the previous theorem hold. Then the following is true*

$$\overline{C^\infty(\overline{\Omega})}^{\|\cdot\|_{L^C_\varphi(\Omega)}} = V^C_\varphi(\Omega).$$

5. Topological properties

It is known that any Hölder space is a nonseparable space, while the little Hölder space is a separable subspace. This suggests together with Theorem 1.2(v) and Remark 4.3 that the following theorem might be true. Indeed, it is.

Theorem 5.1. *Let Ω be a bounded domain, then $L^C_\varphi(\Omega)$ is nonseparable and $V^C_\varphi(\Omega)$ is a separable subspace.*

Our next results characterize compact subsets of $V^C_\varphi(\Omega)$.

Theorem 5.2. *Let the modulus φ be almost increasing, i.e., there is a positive number B, such that*

$$\frac{\varphi(t_1)}{\varphi(t_2)} < B, \text{ for every } 0 < t_1 < t_2 < 2 \operatorname{diam} \Omega.$$

Then for a set $K \subset V_\varphi^C(\Omega)$ the two following conditions are equivalent.

(i) *K is relatively compact in $L_\varphi^C(\Omega)$,*

(ii) *functions from K are equiuniformly vanishing, i.e.,*

$$\lim_{\delta \to 0+} \sup_{f \in K} \|f\|_{L_\varphi^C(\Omega),\delta} = 0$$

holds.

Note that the previous theorem in particular applies to VMO.

Remark 5.3. A similar condition, namely,

$$\lim_{\delta \to 0+} \sup_{f \in K} \operatorname*{ess\,sup}_{0<|x-y|_\infty<\delta,\ x,y\in\Omega} \frac{|f(x) - f(y)|}{|x - y|_\infty^\alpha} = 0, \tag{5.1}$$

characterizes (relatively) compact subsets in little Hölder spaces, see [DDK-74]. In [BFT-69] (see also [W-99], pp. 86–91, 99 for some corrections), the authors showed that $C^{0,\alpha}(\overline{\Omega})$ is isometrically isomorphic to ℓ^∞ and the isomorphism maps $c^{0,\alpha}(\overline{\Omega})$ onto c_0. Moreover, a short inspection of the construction shows that sets in $c^{0,\alpha}(\overline{\Omega})$ satisfying (5.1) are mapped onto a set satisfying

$$\exists \{y_n\}_{n=1}^\infty \in c_0 : \ \forall \{x_n\}_{n=1}^\infty \in K : \ |x_n| \le y_n, \text{ for all } n \in \mathbb{N}. \tag{5.2}$$

The condition (5.2) characterizes (relatively) compact subsets of c_0, as can be shown either directly, or by applying the Arzelà-Ascoli theorem on $\mathcal{C}(\mathbb{N}^*)/C = c_0$, where $\mathbb{N}^*$ is the one-point compactification of $\mathbb{N}$. This gives another proof of the special (Hölder) case.

The previous theorem does not hold if φ is not almost increasing, which is illustrated by the following example.

Example 5.4. Consider the function $f_n \in L^\infty((0;1))$, $n \in \mathbb{N}$, given by

$$f_n(x) = \sum_{k=0}^{2^n-1} \chi_{(k2^{-n};k2^{-n}+2^{-(n+1)})}(x), \quad x \in (0;1).$$

Since f_n are (essentially) bounded by 1, we see that $\{f_n\}_{n=1}^\infty \subset V_{t^{-1/2}}^C((0;1))$ and, moreover, they are equiuniformly vanishing in this Campanato space. On the other hand, for $m \neq n$ we have

$$\|f_n - f_m\|_{L_{t^{-1/2}}^C((0;1))} \ge \|f_n - f_m\|_1 = 1,$$

hence the sequence $\{f_n\}_{n=1}^\infty$ has no convergent subsequences and so it is not relatively compact.

However, the following theorem shows that to characterize compact subsets of any vanishing space it is enough to add another condition, which resembles one of the conditions for compact subsets in $L^1(\Omega)$.

Theorem 5.5. *A set $K \subset V_\varphi^C(\Omega)$ is compact in $L_\varphi^C(\Omega)$ if and only if the two following conditions are satisfied.*

(i) *Functions from K are equiuniformly vanishing,*

(ii) *functions from K satisfy the following condition*

$$\lim_{h\to 0} \sup_{f\in K} \int_\Omega \left| \tilde{f}(y+h) - \tilde{f}(y) \right| \, dy = 0,$$

where

$$\tilde{f}(x) = \begin{cases} f(x), & x \in \Omega, \\ 0, & x \notin \Omega, \end{cases}$$

f being the representative satisfying $f_\Omega = 0$.

Our last result is an application of the characterization of compact subsets of VMO to some optimal Sobolev embedding. In [CP-98], A. Cianchi and L. Pick proved the equivalence of the first two conditions in the following theorem. Note, that we still have to work a little bit to show that the third condition implies one of the first two.

Theorem 5.6. *Let X be a rearrangement invariant Banach function space, let $Q \subset \mathbb{R}^N$ be a cube. We set*

$$V^1X(Q) = \{f \in L^1(Q);\ |Du|_2 \in X(Q)\},$$

where Du denotes the weak gradient of u. Then the following conditions are equivalent

(i) *$V^1X(Q) \subset \mathrm{VMO}(Q)$ uniformly, i.e.,*

$$\lim_{s\to 0+} \sup_{\|Du\|_{X(Q)}\leq 1} \|u\|_{L_1^C(Q),s} = 0.$$

(ii) $\lim_{s\to 0+} \frac{1}{s} \|r^{\frac{1}{N}} \chi_{(0;s)}(r)\|_{\overline{X'}} = 0,$

(iii) $V^1X(Q) \hookrightarrow\hookrightarrow \mathrm{BMO}(Q)$.

Acknowledgement. The article contains the main results of my Diploma Thesis. Some more results, examples, proofs and more references will appear in an article I am preparing. I would like to thank my advisor Luboš Pick for his encouragements and mathematical advice.

References

[BFT-69] R. Bonic, J. Framton, A. Tromba; *Λ-Manifolds*; J. Funct. Anal. **3** (1969), 310–320

[Ch-94] F. Chiarenza; *L^p-Regularity for Systems of PDE's with Coefficients in* VMO; Nonlinear Analysis, Function Spaces and Applications **5** (1994), 1–32

[CP-98] A. Cianchi, L. Pick; *Sobolev Embedding into* BMO, VMO *and* L^∞; Ark. Mat. **36** (1998), 317–340

[CP-01] A. Cianchi, L. Pick; *Sobolev Embedding into Spaces of Campanato, Morrey and Hölder Type*; submitted to J. Math. Anal. Appl.

[DDK-74] A. Doktor, M. Kučera, A. Kufner; *Function Spaces II – Smooth Functions* (in Czech); Státní pedagogické nakladatelství, Praha, 1974

[DS-84] R. A. DeVore, R. C. Sharpley; *Maximal Functions Measuring Smoothness*; Mem. Amer. Math. Soc. **293** (1984), 1–115,

[K-99] J. Kovats; *Dini-Campanato Spaces and Applications to Nonlinear Elliptic Equations*; Electron. J. Differential Equations, **1999** (1999), No. 37, 1–20

[KJF-77] A. Kufner, O. John, S. Fučík; *Function Spaces*; Publishing House of the Czechoslovak Academy of Sciences, Academia, Prague, 1977

[Sa-75] D. Sarason; *Functions of Vanishing Mean Oscillation*; Trans. Amer. Math. Soc. **207** (1975), 391–405

[Sp-65] S. Spanne; *Some Function Spaces Defined Using the Mean Oscillation Over Cubes*; Ann. Scuola Norm. Sup. Pisa **19** (1965), 593–608

[T-86] A. Torchinsky; *Real-Variable Methods in Harmonic Analysis*; Academic Press, Orlando, Florida, 1986

[W-99] N. Weaver; *Lipschitz Algebras*; World Scientific, Singapore, 1999

David Opěla
Department of Mathematics
Campus Box 1146
Washington University in Saint Louis
Saint Louis, MO 63130, USA
E-mail address: opela@math.wustl.edu, opela@karlin.mff.cuni.cz

D. Haroske, T. Runst, H.-J. Schmeisser (eds.): Function Spaces, Differential Operators and Nonlinear Analysis. The Hans Triebel Anniversary Volume.

On Equivalent Quasi-norms on Lorentz Spaces

Bohumír Opic

Dedicated to Prof. Hans Triebel on the occasion of his 65th birthday

Abstract. We present a survey of new formulas which provide equivalent quasi-norms on Lorentz spaces.

1. Introduction and notation

Lorentz spaces are significant in many branches of mathematical analysis (cf. [BS], [DVL], [Z], etc.). The aim of this paper is to present new characterization of Lorentz spaces by means of certain quasi-norms which are equivalent to the classical ones. Our results are motivated by mapping properties of fractional maximal operators, Riesz potentials, the Hilbert transform and the Calderón operator associated with an operator of joint weak type. The proofs are based on properties of certain averaging operators on the cone of non-negative and non-increasing functions in convenient weighted Lebesgue spaces.

Given two quasi-Banach spaces X and Y, we write $X = Y$ if X and Y are equal in the algebraic and the topological sense (their quasi-norms are equivalent). The symbol $X \hookrightarrow Y$ means that $X \subset Y$ and the natural embedding of X in Y is continuous.

We write $A \lesssim B$ (or $A \gtrsim B$) if $A \le cB$ (or $cA \ge B$) for some positive constant c independent of appropriate quantities involved in the expressions A and B, and $A \approx B$ if $A \lesssim B$ and $B \lesssim A$. Throughout the paper we use the abbreviation LHS($*$) (RHS($*$)) for the left- (right-) hand side of the relation ($*$). Moreover, we adopt the convention that $1/\infty = 0$.

Let $\Omega \subset \mathbb{R}^n$ be a measurable subset (with respect to n-dimensional Lebesgue measure) and $|\Omega|$ its measure. The symbol $\mathcal{M}(\Omega)$ is used to denote the family of all scalar-valued (real or complex) measurable functions on the set Ω. By $\mathcal{M}^+(\Omega)$ we mean the subset of $\mathcal{M}(\Omega)$ consisting of all non-negative functions on Ω. If $\Omega = (a,b) \subseteq \mathbb{R}$, we write simply $\mathcal{M}(a,b)$ and $\mathcal{M}^+(a,b)$ instead of $\mathcal{M}((a,b))$ and $\mathcal{M}^+((a,b))$. Finally, $\mathcal{M}^+(a,b;\downarrow)$ stands for the collection of all $f \in \mathcal{M}^+(a,b)$ which are non-increasing on (a,b).

Given $p, r \in (0,\infty]$, the Lorentz space $L^{p,r}(\Omega)$ is defined by (cf. [L1], [L2], [BS])

$$L^{p,r}(\Omega) = \{f \in \mathcal{M}(\Omega);\ \|f\|_{p,r} = \|f\|_{p,r;\Omega} < \infty\}, \tag{1.1}$$

where

$$\|f\|_{p,r} := \|t^{\frac{1}{p}-\frac{1}{r}} f^*(t)\|_{r,(0,|\Omega|)}. \tag{1.2}$$

Here f^* stands for the non-increasing rearrangement of f given by

$$f^*(t) := \inf\{\lambda > 0;\ |\{x \in \Omega;\ |f(x)| > \lambda\}| \le t\}, \quad t \in (0, \infty),$$

and $\|\cdot\|_{r,(a,b)}$, $-\infty \le a < b \le \infty$, is the usual quasi-norm in the Lebesgue space $L^r(a,b)$.

The functional (1.2) is not always a norm, even when $p, r \ge 1$. Following A.P. Calderón [C], we replace f^* by its maximal function

$$f^{**}(t) := t^{-1} \int_0^t f^*(s)\, ds, \qquad t \in (0, \infty),$$

in (1.2) and define the Lorentz space $L^{(p,r)}(\Omega)$, $p, r \in (0, \infty]$, by

$$L^{(p,r)}(\Omega) = \{f \in \mathcal{M}(\Omega);\ \|f\|_{(p,r)} = \|f\|_{(p,r);\Omega} < \infty\}, \tag{1.3}$$

where

$$\|f\|_{(p,r)} := \|t^{\frac{1}{p}-\frac{1}{r}} f^{**}(t)\|_{r,(0,|\Omega|)}. \tag{1.4}$$

One can see that the functional (1.4) is a norm if $r \ge 1$. Moreover, (cf. e.g. [OP, Theorem 3.8 (i)])

$$L^{(p,r)}(\Omega) = L^{p,r}(\Omega) \qquad \text{if} \quad 1 < p \le \infty \text{ and } 0 < r \le \infty. \tag{1.5}$$

In general, $L^{(p,r)}(\Omega) \hookrightarrow L^{p,r}(\Omega)$.

2. Equivalent quasi-norms on spaces $L^{(p,r)}(\Omega)$

The equality (1.5) is a consequence of the fact that $f^* \le f^{**}$ and that the Hardy-Littlewood maximal operator is bounded on the space $L^{p,r}(\Omega)$ when $1 < p \le \infty$ and $0 < r \le \infty$. If the role of the Hardy-Littlewood maximal operator is played by a fractional maximal operator, we arrive at the following result.

Theorem 2.1. ([EdO2, Th. 1.1]) *Let $0 < p, r \le \infty$ and $0 < q < \infty$. Then, for all $f \in \mathcal{M}(\Omega)$,*

$$\|f\|_{(p,r);\Omega} \approx \|t^{\frac{1}{q}-\frac{1}{r}} \sup_{\tau \in (t,|\Omega|)} \tau^{\frac{1}{p}-\frac{1}{q}} f^{**}(\tau)\|_{r,(0,|\Omega|)}. \tag{2.1}$$

Corollaries.

(i) Let $0 < p, r < \infty$. Then, for all $f \in \mathcal{M}(\Omega)$,

$$\|f\|_{(p,r);\Omega} \approx \| \sup_{\tau \in (t,|\Omega|)} \tau^{\frac{1}{p}-\frac{1}{r}} f^{**}(\tau)\|_{r,(0,|\Omega|)}.$$

(ii) Let $0 < p \le \infty$ and $0 < q < \infty$. Then, for all $f \in \mathcal{M}(\Omega)$,

$$\|f\|_{(p,p);\Omega} \approx \|t^{\frac{1}{q}-\frac{1}{p}} \sup_{\tau \in (t,|\Omega|)} \tau^{\frac{1}{p}-\frac{1}{q}} f^{**}(\tau)\|_{p,(0,|\Omega|)}.$$

The main part of Theorem 2.1 concerns the case when $0 < p < q < \infty$ and $1/p < 1+1/q$ (cf. the proof of Theorem 1.1 in [EdO2]). To explain the idea which is behind (2.1), assume in addition that $p > 1$ and $\Omega = \mathbb{R}^n$. Putting $\gamma = n(1/p - 1/q)$, we see that $\gamma \in (0, n)$. Let M_γ be the fractional maximal operator given by

$$(M_\gamma f)(x) = \sup_{Q \ni x} |Q|^{\frac{\gamma}{n}-1} \int_Q |f(y)|\, dy, \quad f \in \mathcal{M}(\mathbb{R}^n), \quad x \in \mathbb{R}^n,$$

where the supremum is extended over all the cubes Q in $\mathbb{R}^n$ with sides parallel to the coordinate axes. By [CKOP, Theorem 1.1],

$$(M_\gamma f)^*(t) \lesssim \sup_{\tau \in (t,\infty)} \tau^{\frac{\gamma}{n}} f^{**}(\tau), \quad t \in (0, \infty), \tag{2.2}$$

for all $f \in \mathcal{M}(\mathbb{R}^n)$ and this estimate is sharp (in the sense that for any $f \in \mathcal{M}^+(\mathbb{R}^n)$ which is radially non-increasing – notation $f \in \mathcal{M}_r^+(\mathbb{R}^n; \downarrow)$ – the symbol $\lesssim$ can be replaced by $\gtrsim$ in (2.2)). Now, given the space $L^{q,r}(\mathbb{R}^n) =: Y$ with $r \in (0, \infty]$, put $\overline{Y} = L^{q,r}((0, \infty))$. Then one can show that the space X,

$$X := \{f \in \mathcal{M}(\mathbb{R}^n);\ \|f\|_X < \infty\}, \tag{2.3}$$

where

$$\|f\|_X := \| \sup_{\tau \in (t,\infty)} \tau^{\frac{\gamma}{n}} f^{**}(\tau)\|_{\overline{Y}},$$

is the largest rearrangement-invariant space which is mapped by M_γ into Y. On the other hand, Theorem 2.1 asserts that $X = L^{(p,r)}(\mathbb{R}^n)$.

If the role of fractional maximal operators is played by Riesz potentials, we obtain the following assertion.

Theorem 2.2. ([EdO2, Th. 1.4]) *Let $0 < p, r \leq \infty$ and $0 < q < \infty$. Assume that either $1/p < 1 + 1/q$ or $|\Omega| = \infty$. Then, for all $f \in \mathcal{M}(\Omega)$,*

$$\|f\|_{(p,r);\Omega} \approx \left\| t^{\frac{1}{q}-\frac{1}{r}} \int_t^\infty \sigma^{\frac{1}{p}-\frac{1}{q}-1} f^{**}(\sigma)\, d\sigma \right\|_{r,(0,|\Omega|)}. \tag{2.4}$$

Corollaries.

(i) Let p, q and Ω be as in Theorem 2.2. Then, for all $f \in \mathcal{M}(\Omega)$,

$$\|f\|_{(p,q);\Omega} \approx \left\| \int_t^\infty \sigma^{\frac{1}{p}-\frac{1}{q}-1} f^{**}(\sigma)\, d\sigma \right\|_{q,(0,|\Omega|)}.$$

(ii) Let $0 < p < \infty$ and $0 < r \leq \infty$. Then, for all $f \in \mathcal{M}(\Omega)$,

$$\|f\|_{(p,r);\Omega} \approx \left\| t^{\frac{1}{p}-\frac{1}{r}} \int_t^\infty \sigma^{-1} f^{**}(\sigma)\, d\sigma \right\|_{r,(0,|\Omega|)}. \tag{2.5}$$

In particular,

$$\|f\|_{p,p;\Omega} \approx \left\| \int_t^\infty \sigma^{-1} f^{**}(\sigma)\, d\sigma \right\|_{p,(0,|\Omega|)} \quad \text{if} \quad 1 < p < \infty. \tag{2.6}$$

Just as we have explained the idea behind Theorem 2.1, so we can verify the idea hidden behind Theorem 2.2. To this end suppose that $0 < r \leq \infty$,

$1 < p < q < \infty$, $1/p < 1 + 1/q$ and $\Omega = \mathbb{R}^n$. Put $\gamma = n(1/p - 1/q)$ and define the Riesz potential I_γ by

$$(I_\gamma f)(x) = \int_{\mathbb{R}^n} \frac{f(y)}{|x-y|^{n-\gamma}}\, dy, \quad x \in \mathbb{R}^n.$$

It is well known that, for all $t \in (0, \infty)$,

$$(I_\gamma f)^*(t) \lesssim t^{\frac{1}{p}-\frac{1}{q}-1} \int_0^t f^*(\sigma)\, d\sigma + \int_t^\infty \sigma^{\frac{1}{p}-\frac{1}{q}-1} f^*(\sigma)\, d\sigma \tag{2.7}$$

and that this estimate is sharp (in the same sense as (2.2) – cf. [S,(1.20)] or [EGO, Lemma 3.4]). Rewriting RHS(2.7) by Fubini's theorem, we arrive at

$$(I_\gamma f)^*(t) \lesssim \int_t^\infty \sigma^{\frac{1}{p}-\frac{1}{q}-1} f^{**}(\sigma)\, d\sigma \quad \text{for all } t \in (0, \infty).$$

Now, given the space $L^{q,r}(\mathbb{R}^n) =: Y$ with $r \in (0, \infty]$, put $\overline{Y} = L^{q,r}((0, \infty))$. Then one can show that the space X from (2.3), where

$$\|f\|_X := \left\| \int_t^\infty \sigma^{\frac{1}{p}-\frac{1}{q}-1} f^{**}(\sigma)\, d\sigma \right\|_{\overline{Y}},$$

is the largest rearrangement-invariant space which is mapped by I_γ into Y. On the other hand, Theorem 2.2 asserts that $X = L^{(p,r)}(\mathbb{R}^n)$.

Since (cf. Theorem 4.7 and Proposition 4.10 in Chapter 3 of [BS])

$$\int_t^\infty \sigma^{-1} f^{**}(\sigma)\, d\sigma = t^{-1} \int_0^t f^*(\sigma)\, d\sigma + \int_t^\infty \sigma^{-1} f^*(\sigma)\, d\sigma$$

gives the sharp estimate of $(Hf)^*(t)$, $t \in (0, \infty)$, where H is the Hilbert transform, defined by

$$(Hf)(x) = \text{p.v.} \int_{\mathbb{R}} \frac{f(y)}{x-y}\, dy, \quad x \in \mathbb{R},$$

one can similarly explain the idea behind formula (2.6).

If $1 < p < q < \infty$ and $\gamma := n\ (1/p - 1/q) < n$, then, by the result of B. Muckenhoupt and R.L. Wheeden (cf. [MW] or [AH, Theorem 3.6.1]),

$$\|M_\gamma f\|_{q;\mathbb{R}^n} \approx \|I_\gamma f\|_{q;\mathbb{R}^n} \quad \text{for all non-negative } f \in L^p(\mathbb{R}^n) \tag{2.8}$$

(here $\|\cdot\|_{q;\mathbb{R}^n}$ stands for the usual $L^q(\mathbb{R}^n)$-norm). Together with the fact that estimates (2.2) and (2.7) are sharp for all $f \in \mathcal{M}_r^+(\mathbb{R}^n; \downarrow)$, Theorems 2.1 and 2.2 imply that there is the following variant of (2.8) involving quasi-norms of Lorentz spaces.

Corollary. Let $0 < r \le \infty$, $1 < p < q < \infty$ and $\gamma := n\ (1/p - 1/q) < n$. Then, for all $f \in \mathcal{M}_r^+(\mathbb{R}^n; \downarrow)$,

$$\|M_\gamma f\|_{q,r;\mathbb{R}^n} \approx \|I_\gamma f\|_{q,r;\mathbb{R}^n} \approx \|f\|_{p,r;\mathbb{R}^n}\,. \tag{2.9}$$

Similarly, one can obtain:

Corollary. Let $0 < r \leq \infty$ and $1 < p < \infty$. Then, for all $f \in \mathcal{M}_r^+(\mathbb{R};\downarrow)$,

$$\|Hf\|_{p,r;\mathbb{R}} \approx \|f\|_{p,r;\mathbb{R}}\,.$$

Remark. The estimate (2.9) motivates us to look for an extension of [AH, Theorem 3.6.1] to the scale of Lorentz spaces. It reads as follows:

Let $0 < r \leq \infty$, $1 < p < q < \infty$ and $\gamma := n\,(1/p - 1/q) < n$. Then, for all positive Radon measures μ on $\mathbb{R}^n$,

$$\|M_\gamma \mu\|_{q,r;\mathbb{R}^n} \approx \|I_\gamma \mu\|_{q,r;\mathbb{R}^n}\,. \tag{2.10}$$

Indeed, the pointwise estimate (cf. [AH, p. 72])

$$(M_\gamma \mu)(x) \lesssim (I_\gamma \mu)(x), \quad x \in \mathbb{R}^n,$$

shows that LHS(2.10) $\lesssim$ RHS(2.10). To prove the converse inequality, one applies the fact that, for all $g \in \mathcal{M}(\mathbb{R}^n)$,

$$\|g\|_{q,r;\mathbb{R}^n} = \|t^{\frac{1}{q}-\frac{1}{r}} g^*(t)\|_{r,(0,\infty)} \approx \|\lambda^{1-\frac{1}{r}}\ |\{x \in \mathbb{R}^n;\ |g(x)| > \lambda\}|^{\frac{1}{q}}\ \|_{r,(0,\infty)}$$

and the good λ inequality (3.6.1) from [AH].

There is the following counterpart of Theorems 2.1 and 2.2.

Theorem 2.3. ([EdO2], Theorems 1.2 and 1.3) *Let* $0 < p, r \leq \infty$ *and* $0 > q > -\infty$. *Then, for all* $f \in \mathcal{M}(\Omega)$,

$$\|f\|_{(p,r);\Omega} \approx \|t^{\frac{1}{q}-\frac{1}{r}} \sup_{\tau \in (0,t)} \tau^{\frac{1}{p}-\frac{1}{q}} f^{**}(\tau)\|_{r,(0,|\Omega|)} \tag{2.11}$$

and

$$\|f\|_{(p,r);\Omega} \approx \left\|t^{\frac{1}{q}-\frac{1}{r}} \int_0^t \tau^{\frac{1}{p}-\frac{1}{q}-1} f^{**}(\tau)\right\|_{r,(0,|\Omega|)}. \tag{2.12}$$

If we take $1/q = -1 + 1/p$ and $p > 1$ in (2.12), we arrive at the next result.

Corollary. Let $0 < r \leq \infty$ and $1 < p \leq \infty$. Then, for all $f \in \mathcal{M}(\Omega)$,

$$\|f\|_{(p,r);\Omega} \approx \left\|t^{\frac{1}{p}-\frac{1}{r}} \left(\frac{1}{t}\int_0^t f^{**}(\sigma)\,d\sigma\right)\right\|_{r,(0,|\Omega|)}$$

(compare with (1.4)).

Remarks.

(i) The constants of equivalence in the relations (2.1), (2.4), (2.11) and (2.12) depend only on p, q and r (see proofs of these results in [EdO2]).

(ii) One can easily extend Theorems 2.1–2.3 to the case when the Lorentz space $L^{(p,r)}(\Omega)$ is replaced by the Lorentz-Zygmund space $L^{(p,r)}(\log L)^\alpha(\Omega)$ ($p, r \in (0,\infty]$, $\alpha \in \mathbb{R}$) defined by (cf. [BR])

$$L^{(p,r)}(\log L)^\alpha(\Omega) = \{f \in \mathcal{M}(\Omega);\ \|f\|_{(p,r),\alpha} = \|f\|_{(p,r),\alpha;\Omega} < \infty\}, \tag{2.13}$$

where

$$\|f\|_{(p,r),\alpha} := \|t^{\frac{1}{p}-\frac{1}{r}} \ell^\alpha(t) f^{**}(t)\|_{r,(0,|\Omega|)} \tag{2.14}$$

and $\ell(t) := 1 + |\log t|$, $t \in (0,\infty)$.

(iii) Theorems 2.1–2.3 remain true if the role of the measure space (Ω, dx) is played by a totally σ-finite measure space (R, μ) with a non-atomic measure μ.

3. Equivalent quasi-norms on spaces $L^{p,r}(\Omega)$

Theorem 2.3 can be rewritten as

$$\|f\|_{(p,r);\Omega} \approx \|t^{\frac{1}{q}-\frac{1}{r}} \|\tau^{\frac{1}{p}-\frac{1}{q}-\frac{1}{s}} f^{**}(\tau)\|_{s,(0,t)}\|_{r,(0,|\Omega|)} \tag{3.1}$$

for all $f \in \mathcal{M}(\Omega)$ provided that $0 < p, r \leq \infty$, $0 > q > -\infty$ and $s \in \{1, \infty\}$. In this connection a natural question arises: Does an analogue of (3.1) hold if f^{**} on the right-hand side of (3.1) is replaced by f^* and $\|f\|_{(p,r);\Omega}$ on the left-hand side of (3.1) by $\|f\|_{p,r;\Omega}$? The positive answer is given by the following result.

Theorem 3.1. ([O, Th. 1.2]) *Let $0 < p, r, s \leq \infty$ and $0 > q > -\infty$. Then, for all $f \in \mathcal{M}(\Omega)$,*

$$\|f\|_{p,r;\Omega} \approx \|t^{\frac{1}{q}-\frac{1}{r}} \|\tau^{\frac{1}{p}-\frac{1}{q}-\frac{1}{s}} f^{*}(\tau)\|_{s,(0,t)}\|_{r,(0,\infty)}. \tag{3.2}$$

Corollaries. Let $0 < p,\ r \leq \infty$ and $0 > q > -\infty$.

(i) Then, for all $f \in \mathcal{M}(\Omega)$,

$$\|f\|_{p,r;\Omega} \approx \|t^{\frac{1}{q}-\frac{1}{r}} \|\tau^{-\frac{1}{q}} f^{*}(\tau)\|_{p,(0,t)}\|_{r,(0,\infty)}. \tag{3.3}$$

In particular, if $p > 1$ and $1/q = -1+1/p$, then (3.3) implies that, for all $f \in \mathcal{M}(\Omega)$,

$$\|f\|_{p,r;\Omega} \approx \left\|t^{\frac{1}{p}-\frac{1}{r}} \left(\frac{1}{t} \|\tau^{1-\frac{1}{p}} f^{*}(\tau)\|_{p,(0,t)}\right)\right\|_{r,(0,\infty)}$$

and

$$\|f\|_{p,p;\Omega} \approx \left\|\frac{1}{t} \|\tau^{1-\frac{1}{p}} f^{*}(\tau)\|_{p,(0,t)}\right\|_{p,(0,\infty)}.$$

(ii) Then, for all $f \in \mathcal{M}(\Omega)$,

$$\|f\|_{p,r;\Omega} \approx \|t^{\frac{1}{q}-\frac{1}{r}} \sup_{\tau\in(0,t)} \tau^{\frac{1}{p}-\frac{1}{q}} f^{*}(\tau)\|_{r,(0,\infty)}. \tag{3.4}$$

In particular, if $p > 1$ and $1/q = -1+1/p$, then (3.4) implies that, for all $f \in \mathcal{M}(\Omega)$,

$$\|f\|_{p,r;\Omega} \approx \left\|t^{\frac{1}{p}-\frac{1}{r}} \left(\frac{1}{t} \sup_{\tau\in(0,t)} \tau f^{*}(\tau)\right)\right\|_{r,(0,\infty)}$$

and

$$\|f\|_{p,p;\Omega} \approx \left\|\frac{1}{t} \sup_{\tau\in(0,t)} \tau f^{*}(\tau)\right\|_{r,(0,\infty)}.$$

(iii) Then, for all $f \in \mathcal{M}(\Omega)$,

$$\|f\|_{p,r;\Omega} \approx \left\|t^{\frac{1}{q}-\frac{1}{r}} \int_0^t \tau^{\frac{1}{p}-\frac{1}{q}-1} f^{*}(\tau)\, d\tau\right\|_{r,(0,\infty)}. \tag{3.5}$$

In particular, if $p > 1$ and $1/q = -1+1/p$, then (3.5) implies that, for all $f \in \mathcal{M}(\Omega)$,

$$\|f\|_{p,r;\Omega} \approx \|t^{\frac{1}{p}-\frac{1}{r}} f^{**}(t)\|_{r,(0,\infty)} \tag{3.6}$$

(which corresponds to the result of A.P. Calderón [C, Theorem 6]).

There is the following counterpart of Theorem 3.1.

Theorem 3.2. ([O, Th. 1.1]) *Let $0 < p, r, s \le \infty$ and $0 < q < \infty$. Then, for all $f \in \mathcal{M}(\Omega)$,*

$$\|f\|_{p,r;\Omega} \approx \|t^{\frac{1}{q}-\frac{1}{r}} \|\tau^{\frac{1}{p}-\frac{1}{q}-\frac{1}{s}} f^*(\tau)\|_{s,(t,\infty)}\|_{r,(0,\infty)}. \tag{3.7}$$

Corollaries. Let $0 < p, r, s \le \infty$ and $0 < q < \infty$.

(i) Then, for all $f \in \mathcal{M}(\Omega)$,

$$\|f\|_{p,r;\Omega} \approx \|t^{\frac{1}{q}-\frac{1}{r}} \|\tau^{-\frac{1}{q}} f^*(\tau)\|_{p,(t,\infty)}\|_{r,(0,\infty)}.$$

In particular, for all $f \in \mathcal{M}(\Omega)$,

$$\|f\|_{p,q;\Omega} \approx \| \; \|\tau^{-\frac{1}{q}} f^*(\tau)\|_{p,(t,\infty)}\|_{q,(0,\infty)}$$

and

$$\|f\|_{q,q;\Omega} \approx \Big(\int_0^\infty \Big(\int_t^\infty \tau^{-1}[f^*(\tau)]^q \, d\tau\Big) \, dt\Big)^{1/q}.$$

(ii) Then, for all $f \in \mathcal{M}(\Omega)$,

$$\|f\|_{p,r;\Omega} \approx \|t^{\frac{1}{q}-\frac{1}{r}} \sup_{\tau\in(t,\infty)} \tau^{\frac{1}{p}-\frac{1}{q}} f^*(\tau)\|_{r,(0,\infty)}.$$

In particular, for all $f \in \mathcal{M}(\Omega)$,

$$\|f\|_{p,q;\Omega} \approx \| \sup_{\tau\in(t,\infty)} \tau^{\frac{1}{p}-\frac{1}{q}} f^*(\tau)\|_{q,(0,\infty)}.$$

(iii) Then, for all $f \in \mathcal{M}(\Omega)$,

$$\|f\|_{p,r;\Omega} \approx \Big\|t^{\frac{1}{q}-\frac{1}{r}} \int_t^\infty \tau^{\frac{1}{p}-\frac{1}{q}-1} f^*(\tau) \, d\tau\Big\|_{r,(0,\infty)}.$$

In particular, for all $f \in \mathcal{M}(\Omega)$,

$$\|f\|_{p,q;\Omega} \approx \Big\|\int_t^\infty \tau^{\frac{1}{p}-\frac{1}{q}-1} f^*(\tau) \, d\tau\Big\|_{q,(0,\infty)}$$

and

$$\|f\|_{q,q;\Omega} \approx \Big(\int_0^\infty \Big(\int_t^\infty \tau^{-1} f^*(\tau) \, d\tau\Big)^q dt\Big)^{1/q}.$$

(iv) Then, for all $f \in \mathcal{M}(\Omega)$,

$$\|f\|_{q,r;\Omega} \approx \|t^{\frac{1}{q}-\frac{1}{r}} \|\tau^{-\frac{1}{s}} f^*(\tau)\|_{s,(t,\infty)}\|_{r,(0,\infty)}.$$

In particular, for all $f \in \mathcal{M}(\Omega)$,

$$\|f\|_{q,r;\Omega} \approx \Big\|t^{\frac{1}{q}-\frac{1}{r}} \int_t^\infty \tau^{-1} f^*(\tau) \, d\tau\Big\|_{r,(0,\infty)}$$

(which is a "dual result" to [C, Theorem 6]; cf. (3.6)).

(v) If $s > p$ and $1/q = 1/p - 1/s$, then (3.7) implies that, for all $f \in \mathcal{M}(\Omega)$,

$$\|f\|_{p,r;\Omega} \approx \Big\|t^{\frac{1}{p}-\frac{1}{r}} \Big(t^{-\frac{1}{s}} \|f^*(\tau)\|_{s,(t,\infty)}\Big)\Big\|_{r,(0,\infty)}.$$

In particular, for all $f \in \mathcal{M}(\Omega)$,

$$\|f\|_{p,r;\Omega} \approx \Big\|t^{\frac{1}{p}-\frac{1}{r}}\Big(\frac{1}{t}\int_t^\infty f^*(\tau)\,d\tau\Big)\Big\|_{r,(0,\infty)} \quad \text{if} \quad p \in (0,1)$$

(which is a counterpart of [C, Theorem 6]; cf. (3.6)).

Remarks.
(i) To explain the idea behind (3.7), let $X = L^r(v)$, where the weight function v is given by $v(t) = t^{1/p-1/r}$, $t \in (0,\infty)$, and $L^r(v)$ is the weighted Lebesgue space defined by

$$L^r(v) = \{g \in \mathcal{M}(0,\infty);\ \|g|\ L^r(v)\| := \|vg\|_{r,(0,\infty)} < \infty\}.$$

Consider the weighted averaging operator T given by

$$(Tg)(t) = \frac{1}{t^{\frac{1}{p}-\frac{1}{q}}}\|\tau^{\frac{1}{p}-\frac{1}{q}-\frac{1}{s}}g(\tau)\|_{s,(t,\infty)}, \quad t \in (0,\infty), \tag{3.8}$$

Then:

(a) T is bounded on X if $r \in [1,\infty]$;
(b) T is bounded on $X \cap \mathcal{M}^+(0,\infty;\downarrow)$ if $r \in (0,1)$;
(c) T has a bounded inverse on $X \cap \mathcal{M}^+(0,\infty;\downarrow)$.

Consequently,

$$\|g\|_X \approx \|Tg\|_X \quad \text{for all} \quad g \in X \cap \mathcal{M}^+(0,\infty;\downarrow). \tag{3.9}$$

Now, if $f \in \mathcal{M}(\Omega)$, the estimate (3.7) follows from (3.9) on putting $g = f^*$ since

$$\|g\|_X = \|t^{\frac{1}{p}-\frac{1}{r}}f^*(t)\|_{r,(0,\infty)} = \|f\|_{p,r;\Omega}$$

and

$$\|Tg\|_X = \|t^{\frac{1}{q}-\frac{1}{r}}\|\tau^{\frac{1}{p}-\frac{1}{q}-\frac{1}{s}}f^*(\tau)\|_{s,(t,\infty)}\|_{r,(0,\infty)}.$$

Similarly, replacing the operator T from (3.8) by

$$(Tg)(t) = \frac{1}{t^{\frac{1}{p}-\frac{1}{q}}}\|\tau^{\frac{1}{p}-\frac{1}{q}-\frac{1}{s}}g(\tau)\|_{s,(0,t)}, \quad t \in (0,\infty),$$

one can explain the idea behind (3.2).

(ii) Note that expressions similar to RHS(3.7) (or RHS(3.2)) with the limiting value $q = \infty$ (or $q = -\infty$) and involving logarithmic terms appeared in [EvO], [CP, (0.1)], [P] and [D] in connection with "limiting" real interpolation to define spaces which, in general, differ from Lorentz-Zygmund ones. Our results show that in the non-limiting case (when q is finite) the situation is quite different.

(iii) It is easy to see that Theorem 3.2 remains true if $p = r = s = q = \infty$.

(iv) It follows from the proofs of Theorems 3.1 and 3.2 (see [O]) that the constants of equivalence in the relations (3.2) and (3.7) depend only on p, q and r.

(v) Analogous results to those of Theorems 3.1 and 3.2 can be proved in the case when the Lorentz space $L^{p,r}(\Omega)$ is replaced by the Lorentz–Zygmund space $L^{p,r}(\log L)^\alpha(\Omega)$ ($p, r \in (0,\infty]$, $\alpha \in \mathbb{R}$) defined by (cf. (2.13), (2.14))

$$L^{p,r}(\log L)^\alpha(\Omega) = \{f \in \mathcal{M}(\Omega);\ \|f\|_{p,r,\alpha} = \|f\|_{p,r,\alpha;\Omega} < \infty\}$$

where

$$\|f\|_{p,r,\alpha} := \|t^{\frac{1}{p}-\frac{1}{r}}\ell^\alpha(t) f^*(t)\|_{r,(0,|\Omega|)}.$$

Note also that Theorems 3.1 and 3.2 continue to hold if the measure space (Ω, dx) is replaced by a general measure space.

We conclude this section with results concerning the Calderón operator appearing in the weak type theory of interpolation. Suppose that

$$0 < p_0 < p_1 \le \infty, \quad 0 < q_0, q_1 \le \infty \quad \text{with} \quad q_0 \ne q_1$$

and

$$m = (1/q_0 - 1/q_1)/(1/p_0 - 1/p_1).$$

Let S be the Calderón operator associated with the interpolation segment

$$[(1/p_0, 1/q_0), (1/p_1, 1/q_1)],$$

that is,

$$(Sf)(t) := t^{-\frac{1}{q_0}} \int_0^{t^m} \tau^{\frac{1}{p_0}} f^*(\tau)\frac{d\tau}{\tau} + t^{-\frac{1}{q_1}} \int_{t^m}^\infty \tau^{\frac{1}{p_1}} f^*(\tau)\frac{d\tau}{\tau}, \tag{3.10}$$

for all $f \in \mathcal{M}(\Omega)$. (Recall that $Sf \in \mathcal{M}^+(0,\infty;\downarrow)$ for any $f \in \mathcal{M}(\Omega)$.) Assume that $0 < \theta < 1$ and define p and q by

$$\frac{1}{p} = \frac{1-\theta}{p_0} + \frac{\theta}{p_1}, \quad \frac{1}{q} = \frac{1-\theta}{q_0} + \frac{\theta}{q_1}. \tag{3.11}$$

Since the operator S is of joint weak type $(p_0, q_0; p_1, q_1)$, cf. [BR] or [BS], the Marcinkiewicz interpolation theorem implies that the operator

$$S : L^{p,r}(\Omega) \to L^{q,r}((0,\infty)), \quad 0 < r \le \infty,$$

is bounded, that is,

$$\|Sf\|_{q,r;(0,\infty)} \lesssim \|f\|_{p,r;\Omega} \quad \text{for all} \quad f \in L^{p,r}(\Omega). \tag{3.12}$$

We shall even show that the reverse inequality in (3.12) holds. Consequently, the quantity $\|Sf\|_{q,r;(0,\infty)}$ is equivalent to the quasi-norm $\|f\|_{p,r;\Omega}$ for all $f \in L^{p,r}(\Omega)$. To prove this assertion, we define the operators S_0 and S_1 on $\mathcal{M}(\Omega)$ by

$$(S_0 f)(t) := t^{-\frac{1}{q_0}} \int_0^{t^m} \tau^{\frac{1}{p_0}} f^*(\tau)\frac{d\tau}{\tau}, \quad t \in (0,\infty), \tag{3.13}$$

and

$$(S_1 f)(t) := t^{-\frac{1}{q_1}} \int_{t^m}^\infty \tau^{\frac{1}{p_1}} f^*(\tau)\frac{d\tau}{\tau}, \quad t \in (0,\infty). \tag{3.14}$$

Then $S = S_0 + S_1$ on $\mathcal{M}(\Omega)$ and the result mentioned above is a corollary of the following theorem.

Theorem 3.3. *Let $0 < p_0 < p_1 \le \infty$, $0 < q_0, q_1 \le \infty$ with $q_0 \ne q_1$, and let $0 < r \le \infty$. Assume that the numbers p and q are given by (3.11), and the operators S_i $(i = 0, 1)$ are defined by (3.13) and (3.14). Then, for all $f \in \mathcal{M}(\Omega)$,*

$$\|f\|_{p,r;\Omega} \approx \|t^{\frac{1}{q}-\frac{1}{r}}(S_i f)(t)\|_{r,(0,\infty)}, \quad i = 0, 1. \tag{3.15}$$

Proof. Since $0 \le (S_i f)(t) \le (Sf)(t)$ for all $t \in (0,\infty)$, every $f \in \mathcal{M}(\Omega)$ and $i = 0, 1$, the inequality RHS(3.15) $\lesssim$ LHS(3.15) follows from a convenient version of the Marcinkiewicz interpolation theorem (cf. e.g. [BS, Chapter 4, Th. 4.13] if $p_0, p_1, q_0, q_1, r \ge 1$; for a general case see e.g. [EOP, Th. 5.12]). Thus, it remains to prove that, for all $f \in \mathcal{M}(\Omega)$ and $i = 0, 1$,

$$\|t^{\frac{1}{q}-\frac{1}{r}}(S_i f)(t)\|_{r,(0,\infty)} \gtrsim \|f\|_{p,r;\Omega}. \tag{3.16}$$

Let $f \in \mathcal{M}(\Omega)$ and $t \in (0,\infty)$. Then

$$(S_0 f)(t) \ge f^*(t^m)\, t^{-\frac{1}{q_0}} \int_0^{t^m} \tau^{\frac{1}{p_0}-1} d\tau \approx f^*(t^m)\, t^{-\frac{1}{q_0}+\frac{m}{p_0}}.$$

Consequently,

$$\begin{aligned} \|t^{\frac{1}{q}-\frac{1}{r}}(S_0 f)(t)\|_{r,(0,\infty)} &\gtrsim \|t^{\frac{1}{q}-\frac{1}{q_0}+\frac{m}{p_0}-\frac{1}{r}} f^*(t^m)\|_{r,(0,\infty)} \\ &\approx \|\tau^{\frac{1}{m}(\frac{1}{q}-\frac{1}{q_0})+\frac{1}{p_0}-\frac{1}{r}} f^*(\tau)\|_{r,(0,\infty)}. \end{aligned} \tag{3.17}$$

Moreover, since $1/q - 1/q_0 = \theta\,(1/q_1 - 1/q_0)$, we obtain that

$$\frac{1}{m}\Big(\frac{1}{q} - \frac{1}{q_0}\Big) + \frac{1}{p_0} = \frac{1}{p}.$$

Therefore,

$$\text{RHS(3.17)} = \|\tau^{\frac{1}{p}-\frac{1}{r}} f^*(\tau)\|_{r,(0,\infty)} = \|f\|_{p,r;\Omega} \tag{3.18}$$

and the inequality (3.16) with $i = 0$ follows from (3.17) and (3.18).

Similarly, we have

$$(S_1 f)(t) \ge t^{-\frac{1}{q_1}} \int_{t^m}^{2t^m} \tau^{\frac{1}{p_1}-1} f^*(\tau) d\tau \ge f^*(2t^m)\, t^{-\frac{1}{q_1}+\frac{m}{p_1}}.$$

Hence,

$$\begin{aligned} \|t^{\frac{1}{q}-\frac{1}{r}}(S_1 f)(t)\|_{r,(0,\infty)} &\gtrsim \|t^{\frac{1}{q}-\frac{1}{q_1}+\frac{m}{p_1}-\frac{1}{r}} f^*(2t^m)\|_{r,(0,\infty)} \\ &\approx \|\tau^{\frac{1}{m}(\frac{1}{q}-\frac{1}{q_1})+\frac{1}{p_1}-\frac{1}{r}} f^*(\tau)\|_{r,(0,\infty)}. \end{aligned} \tag{3.19}$$

Moreover, since $1/q - 1/q_1 = (1-\theta)\,(1/q_0 - 1/q_1)$, we obtain that

$$\frac{1}{m}\Big(\frac{1}{q} - \frac{1}{q_1}\Big) + \frac{1}{p_1} = \frac{1}{p}.$$

Therefore,

$$\text{RHS(3.19)} = \|\tau^{\frac{1}{p}-\frac{1}{r}} f^*(\tau)\|_{r,(0,\infty)} = \|f\|_{p,r;\Omega} \tag{3.20}$$

and the inequality (3.16) with $i = 1$ follows from (3.19) and (3.20). □

Corollary. Let the numbers p_0, q_0, p_1, q, p, q and r be as in Theorem 3.3. Let S be the Calderón operator defined by (3.10). Then, for all $f \in \mathcal{M}(\Omega)$,

$$\|f\|_{p,r;\Omega} \approx \|Sf\|_{q,r;(0,\infty)}.$$

Acknowledgement. The research was supported by the grant no. 201/01/0333 of the Grant Agency of the Czech Republic.

References

[AH] D.R. Adams and L.I. Hedberg. *Function spaces and potential theory.* Springer, Berlin, 1996.

[BR] C. Bennett and K. Rudnick. *On Lorentz-Zygmund spaces.* Dissert. Math. 175 (1980), 1–72.

[BS] C. Bennett and R. Sharpley *Interpolation of operators.* Pure and Appl. Math. 129, Academic Press, New York, 1988.

[C] A.P. Calderón. *Spaces between L^1 and L^∞ and the theorem of Marcinkiewicz.* Studia Math. 26 (1966), 273–299.

[CKOP] A. Cianchi, R. Kerman, B. Opic and L. Pick. *Sharp rearrangement inequality for the fractional maximal operator.* Studia Math. 138 (2000), 277–284.

[CP] M. Cwikel and E. Pustylnik. *Weak type interpolation near "endpoint" spaces.* J. Funct. Anal. 171 (2000), 235–277.

[D] R.Ya. Doktorskii. *Reiteration relations of the real interpolation method.* Soviet Math. Dokl. 44 (1992), 665–669 (Russian).

[DVL] R.A. DeVore and G.G. Lorentz. *Constructive approximation.* Springer, Berlin, 1993.

[EGO] D.E. Edmunds, P. Gurka and B. Opic. *Double exponential integrability of convolution operators in generalized Lorentz–Zygmund spaces.* Indiana Univ. Math. J. 44 (1995), 19–43.

[EdO1] D.E. Edmunds and B. Opic. *Boundedness of fractional maximal operators between classical and weak type Lorentz spaces.* Research Report No. 2000–15, CMAIA, University of Sussex at Brighton, 2000, pp. 40.

[EdO2] D.E. Edmunds and B. Opic. *Equivalent quasi-norms on Lorentz spaces.* Research Report No. 2001–04, CMAIA, University of Sussex at Brighton, 2001, pp. 10.

[EvO] W.D. Evans and B. Opic. *Real interpolation with logarithmic functors and reiterations.* Canad. J. Math. 52 (2000), 920–960.

[EOP] W.D. Evans, B. Opic and L. Pick. *Real interpolation with logarithmic functors.* J. Inequal. Appl. (to appear).

[L1] G.G. Lorentz. *Some new function spaces.* Ann. of Math. 51 (1950), 37–55.

[L2] G.G. Lorentz. *On the theory of spaces* Λ. Pacific J. Math. 1 (1951), 411–429.

[MW] B. Muckenhoupt and R.L. Wheeden. *Weighted norm inequalities for fractional integrals.* Trans. Amer. Math. Soc. 192 (1974), 261–274.

[O] B. Opic. *New characterizations of Lorentz spaces.* Preprint, Prague 2001, pp. 10.

[OP] B. Opic and L. Pick. *On generalized Lorentz-Zygmund spaces.* Math. Inequal. 2 (1999), 391–467.

[P] E. Pustylnik. *Optimal interpolation in spaces of Lorentz-Zygmund type.* J. d'Anal. Math. 79 (1999), 113–157.

[S] E.T. Sawyer. *Boundedness of classical operators on classical Lorentz spaces.* Studia Math. 96 (1990), 145–158.

[Z] W.P. Ziemer. *Weakly Differentiable Functions.* Springer Verlag, New York, 1989.

Bohumír Opic
Mathematical Institute
Academy of Sciences of the Czech Republic
Žitná 25, 115 67 Praha 1, Czech Republic
E-mail address: opic@math.cas.cz

D. Haroske, T. Runst, H.-J. Schmeisser (eds.): Function Spaces, Differential Operators and
Nonlinear Analysis. The Hans Triebel Anniversary Volume.

Concave Functions of Second Order Elliptic Operators, Kernel Estimates and Applications

Evgeniy Pustylnik

Dedicated to Prof. Hans Triebel on the occasion of his 65th birthday

Abstract. This is a survey of new possibilities given by the operator function calculus for obtaining rather sharp estimates in various approximation problems connected with elliptic operators. More detailed applications are given to the Fourier method for the solution of hyperbolic and parabolic equations.

1. One of the main problems of the theory of approximation methods and other problems connected with converging sequences $u_n \to u$ is estimating the rate of convergence, namely, the difference $u - u_n$ with respect to $n = 1, 2, \ldots$. In many cases this estimate has a power character with a non-limiting exponent like

$$\|u - u_n\| = o(n^{-\tau+\epsilon}), \qquad \forall\epsilon > 0.$$

Moreover, in the most of such cases, one has some counterexamples showing that the limiting estimate with $\epsilon = 0$ does not hold. Sometimes, instead of this, one succeeds to replace the power inexactness to the logarithmic one

$$\|u - u_n\| = o\left(n^{-\tau}(\ln n)^{1+\epsilon}\right), \qquad \forall\epsilon > 0$$

and this result usually regards as a "very nice achievement". Of course, it regards as "still more excellent", if somebody succeeds to make the next step and to obtain the estimate of the type

$$\|u - u_n\| = o\left(n^{-\tau}\ln n(\ln\ln n)^{1+\epsilon}\right), \qquad \forall\epsilon > 0.$$

It is clear that any next step of such a kind in any particular problem is obtained by means of some great efforts and intricate technique, thus it is achieved more and more seldom. And extremely seldom the matter is ended by the general estimate

$$\|u - u_n\| = o(n^{-\tau}\psi(n)), \qquad \forall\psi : \int^{\infty} \frac{dt}{t\psi(t)} < \infty,$$

which apparently alone may be regarded here as final one.

To say the truth, using the methods of classical analysis, even the power estimate could be obtained rather not simply. More possibilities are given by the functional analysis where any approximation problem corresponds to some linear operator P and the desired estimate is obtained via studying the fractional powers of this operator P^τ. For example, the limiting exponent in the estimate is often equal to the limiting value of τ such that the operator P^τ is continuous or compact in some special spaces. This is possible because the fractional powers of linear operators are studied very well, starting from the classical scale of L_p-spaces in 60-th and up to very abstract scales of Banach spaces, that could be found in the famous monograph on interpolation theory by Prof. Triebel [Tr].

In order to obtain the logarithmic estimates, following this way, one should define and study the operator $\ln P$. One of the best methods is to define $\ln P$ as a generator of the semigroup P^τ, $\tau > 0$. The situation with arbitrary functions $\psi(t)$ which are needed for optimal estimation is much more complicated. The classical definition of operator functions $\psi(P)$ in a Hilbert space is not suitable for studying them in other spaces; many other operator calculi require analyticity of functions or their power type decreasing at infinity. At the same time the functions we need are not analytical at zero and may be slowly changing at infinity. Generally speaking, we should consider the functions $\alpha(t) : (0, \infty) \mapsto (0, \infty)$ such that $\alpha(t) = t^m \phi(t)$, $m = 0, 1, 2, \ldots$, while $\phi(t)$ is concave and increasing.

In order to show which basic properties of operator functions we need to know, let us consider an equation $Pu = f$ with some linear operator P. If we know that $P^{-1} : F \mapsto E$ for some spaces E, F then we immediately obtain the existence of a solution $u = P^{-1}f \in E$ for each element $f \in F$. Let $P = \phi(P)\psi(P)$ for two increasing functions $\phi(t), \psi(t)$. Were we able to show that $\phi^{-1}(P) : F \mapsto E$, the second factor $\psi(P)$ could be used for obtaining some additional information about the solution u, because in this case $u \in \mathcal{D}(\psi(P)) \subsetneqq E$.

Another variant: P^{-1} does not act from F to E, but there exists an increasing function $\phi(t)$ such that $P^{-1}\phi^{-1}(P) : F \mapsto E$. In this case we obtain the existence of solution u for each $f \in \mathcal{D}(\phi(P))$ in F. Therefore *the basic problems that we should be able to solve are the definition of possibly "smaller" functions $\alpha(P^{-1})$ acting between given spaces and of possibly "bigger" functions $\beta(P)$ such that the given element $f \in \mathcal{D}(\beta(P))$.*

The most powerful methods of these problems solution are given by interpolation theory of linear operators. For instance, the case of fractional powers was studied and solved just so due to the very fruitful complex interpolation method. Recall that this method works only with numerical parameters of interpolation. But passing to arbitrary operator functions, the parameter of interpolation also should be taken as functional one. Unfortunately, all known till now interpolation methods with functional parameter are much weaker than the complex method. That is why the abstract results for arbitrary operator functions are far from optimality.

A question is arisen: to what extent could the results be stronger if to consider particular operators in special spaces? For example, we will be interested in positive elliptic operators of the second order, acting between the spaces $L_2(\Omega)$ and $C(\Omega)$ for bounded sets $\Omega \subset \mathbb{R}^n$. For the negative fractional powers $P^{-\tau}$, we have in this case an integral representation

$$P^{-\tau} f(x) = \int_\Omega G_\tau(x, y) f(y) dy, \qquad x \in \Omega \tag{1}$$

with the kernel $G_\tau(x,y)$ continuous for $x \neq y$ and having near $x = y$ the power singularity:

$$0 \leq G_\tau(x,y) \leq C|x-y|^{2\tau-n}, \qquad n \geq 2, \quad \tau \in (0,1). \tag{2}$$

Notice that, in fact, this estimate is true for all $\tau < n/2$, but we do not need the values $\tau \geq 1$, because $P^{[\tau]}$ is a simply studied iteration of P.

The mapping properties of integral operators (1) with the estimate (2) are well studied in the whole scale of spaces between L_2 and C. This enabled various optimal convergence rate estimates of the power type to be obtained by the methods of functional analysis. Of course, an analogous representation for arbitrary functions of elliptic operators could be useful for studying them and obtaining still better estimates than in a power case. In one of the results below I shall show the difference which is gained just on this way in comparison with results for abstract operators.

Let me note one more peculiarity of the formula (2): for $n \geq 3$, it is valid also for $\tau = 1$, passing continuously from the fractional powers to the operator P^{-1} itself. In the case of $n = 2$ we have a "jump" from the fractional powers to the logarithm:

$$G_1(x,y) \leq C\big|\ln|x-y|\big|, \qquad n = 2. \tag{3}$$

An analogous estimate for arbitrary operator functions is able to show in this case too a smooth passing from parts of operator to the operator itself. For example: what a singularity has the kernel $G_\phi(x,y)$ for the function $\phi(P^{-1})$ when $\phi(t) = t\left(\ln\frac{1}{t}\right)^\epsilon$, $\epsilon > 0$, i.e., when the function is increasing faster than any fractional power but slower than t? Running ahead, I can answer that in this case

$$G_\phi(x,y) \leq C\big|\ln|x-y|\big|^{1+\epsilon},$$

i.e., the estimate remains true for $\epsilon = 0$ as well.

2. So, let P be a linear differential operator of the second order generated by a differential expression

$$Pu(x) = \sum_{i,j=1}^{n} a_{ij}(x)\frac{\partial^2 u}{\partial x_i \partial x_j} + \sum_{i=1}^{n} b_i(x)\frac{\partial u}{\partial x_i} + c(x)u, \tag{4}$$

where $x = (x_1, \ldots, x_n) \in \Omega \subset \mathbb{R}^n$. There are also some admissible boundary conditions on $\partial\Omega$ that we will not indicate explicitly. The expression (4) is assumed to be uniformly elliptic, i.e., the corresponding quadratic form

$$p(\xi) = \sum_{i,j=1}^{n} a_{ij}(x)\xi_i\xi_j$$

is strictly negatively defined for all $x \in \Omega$. As shown by S. Agmon in 1962, for sufficiently large $c(x)$, this operator is positive in each L_p, $p < \infty$, i.e.,

$$\|(P+tI)^{-1}\| \leq \frac{M}{t+1}, \qquad \forall t \geq 0.$$

Moreover, under rather broad assumptions on coefficients and boundary, the operator $(P+tI)^{-1}$ has an integral representation

$$(P+tI)^{-1}u(x) = \int_\Omega G(t,x,y)\,u(y)\,dy, \qquad x \in \Omega. \tag{5}$$

In fact, in all our study of functions of the operator P, we use only some estimates for the kernel $G(t,x,y)$ that I shall indicate below. Various sufficient conditions for such estimates to hold can be found in the rich literature devoted to this topic – for example, in the monograph [Sh]. These conditions proceed to get weaker in the modern literature, enlarging possibilities for applying the results which I intend to present.

The basic estimate which I need is

$$0 \le G(t,x,y) \le Cr^{2-n}e^{-kr\sqrt{t}}, \quad r = |x-y|, \quad n \ge 3, \tag{6}$$

uniformly over all $x, y \in \Omega$, $t \ge 0$. In the case of $n = 2$ the power r^{2-n} must be replaced by $\ln(1/r)$ which is insufficient for us, and we will need then the estimate for derivatives

$$\left|\frac{\partial}{\partial x_i}G(t,x,y)\right| \le Cr^{-1}e^{-kr\sqrt{t}}, \qquad i = 1, 2. \tag{7}$$

Some inverse estimates will be needed when we will prove optimality of our results for kernels of operator functions. Mainly we need a local estimate

$$G(x,y) = G(0,x,y) \ge C_1 r^{2-n}, \quad x, y \in B(x_0, R) \subset \Omega, \tag{8}$$

being valid for some ball $B(x_0, R)$ centered in some point $x_0 \in \Omega$. Sometimes an operator P, like the Laplace operator, satisfies more strong inequality

$$G(t,x,y) \ge C_1(R)r^{2-n}e^{-kr\sqrt{t}}, \quad y \in B(x,R) \subset \Omega \tag{9}$$

for all $t \ge 0$. In this case we shall call P as an operator of *classical type.* For such operators, it is possible to obtain the exact description of kernels $G_\phi(x,y)$ behaviour on the diagonal $x = y$.

There are various ways to define functions of operators. If P is selfadjoint, its continuous functions can be defined in accordance with the classical Riesz calculus. Moreover, the operator P^{-1}, being inverse to elliptic one, is usually compact and has a representation

$$P^{-1}u = \sum_{i=1}^{\infty} \lambda_i (u, e_i) e_i, \tag{10}$$

where $\{e_i\}$ is an orthonormal basis of the operator's P^{-1} eigenfunctions corresponding to eigenvalues $\lambda_i \to 0$. As a rule, we arrange these eigenvalues in the descending order. For an operator $\phi(P^{-1})$, we then obtain a representation

$$\phi(P^{-1})u = \sum_{i=1}^{\infty} \phi(\lambda_i)(u, e_i)e_i, \tag{11}$$

which converges in L_2 for all $u \in L_2$. The functions $\phi(P)$ can be defined as $\phi(P) = [\psi(P^{-1})]^{-1}$ for $\psi(t) = 1/\phi(1/t)$ or again as the series

$$\phi(P)u = \sum_{i=1}^{\infty} \phi(1/\lambda_i)(u, e_i)e_i, \tag{12}$$

which converges only for u from some subset of L_2, forming the domain $\mathcal{D}(\phi(P))$.

The representation (11) is hardly suitable for obtaining all needed properties of the operators $\phi(P^{-1})$. Even for the fractional powers of the Laplace operator, its direct application to proving desired estimates takes tens of pages (see, e.g., [Il]) More convenient is to use an integral relation

$$\phi(P^{-1})u = \int_0^{\infty} (P + sI)^{-1} d\sigma(s), \tag{13}$$

which can be proved in the Riesz theory for any function $\phi(t)$ representable in a form

$$\phi\left(\frac{1}{t}\right) = \int_0^{\infty} \frac{d\sigma(s)}{t+s}, \qquad \int_0^{\infty} \frac{d\sigma(s)}{1+s} < \infty. \tag{14}$$

At the first sight, we have here an essential decrease of the set of possible functions. In fact, this is not so, because for any increasing concave function $\phi(t)$ with $\phi(0) = \phi'(\infty) = 0$, the following relation holds:

$$\phi\left(\frac{1}{t}\right) \sim \int_0^{\infty} \frac{d\phi'(1/s)}{t+s}, \tag{15}$$

where $f \sim g$ is meant as $C_1 f \le g \le C_2 f$ for some $C_1, C_2 > 0$.

The relation (13) can be used as definition of an operator function $\phi(P^{-1})$ even in the case when the operator P is not selfadjoint. Also L_2 may be replaced by any Banach space where the operator P is positive and thus the integral in (13) exists. The corresponding operator calculus is now well studied, having all needed algebraic and other properties [Pu1]. Therefore all kernel estimates presented below hold for the general case as well.

3. Using the representations (5) and (13), we obtain immediately that

$$\phi(P^{-1})u(x) = \int_\Omega G_\phi(x,y)\, u(y)\, dy, \qquad G_\phi(x,y) = \int_0^{\infty} G(s,x,y)\, d\sigma(s), \tag{16}$$

whence for $n \ge 3$

$$0 \le G_\phi(x,y) \le C r^{2-n} \int_0^{\infty} e^{-kr\sqrt{s}} d\sigma(s), \qquad r = |x-y|.$$

Of course, this estimate is not so good for direct applications, since the measure $d\sigma$ is not expressed via the corresponding function $\phi(t)$ analytically, and the equivalence (15) does not help in comparing operators. After some special transformations we can derive here that for each $m = 1, 2, \ldots$

$$G_\phi(x,y) \le C r^{2-n-2m} (-1)^{m-1} \tilde{\phi}^{(m-1)}(r^{-2}), \qquad \tilde{\phi}(s) = \phi(1/s). \tag{17}$$

In a partial case of $\phi(s) = s^\tau,\ 0 < \tau < 1$, the choice of m is unessential, because for any m the estimate (17) gives the same order of growth $r^{2\tau-n}$ as $r \to 0$. For arbitrary $\phi(s)$, the number $m = 1$ gives the estimate

$$G_\phi(x,y) \le Cr^{-n}\phi(r^2),$$

that appears to be very rough. Indeed, the right-hand side can be here unsummable with respect to y for any x if the function ϕ is slowly changing at zero. At the same time, the integral of the left-hand side is

$$\int_\Omega G_\phi(x,y)\,dy = \phi(P^{-1})u_0(x)$$

with $u_0(x) \equiv 1$, which is finite, at least, for almost all $x \in \Omega$.

The next value $m = 2$ gives the estimate

$$G_\phi(x,y) \le Cr^{2-n}\phi'(r^2), \tag{18}$$

where the right-hand side is summable for any $x \in \Omega$ and any ϕ. Despite of this m being only second value in an infinite sequence, the estimate (18) turns out to be sufficient for optimal solution of all problems on operator functions posed above. For example, it allows us to study the mapping properties of $\phi(P^{-1})$ as operators in rearrangement invariant spaces. Recall that the *Lorentz space* $\Lambda_{\alpha(t)}$ with the fundamental function $\alpha(t)$ is the space of all measurable functions having the finite norm

$$\|u(x)\|_{\Lambda_{\alpha(t)}} = \int_0^\infty u^*(t)\,d\alpha(t),$$

and the *Marcinkiewicz space* $M_{\beta(t)}$ with the fundamental function $\beta(t)$ is the space of all measurable functions having the finite norm

$$\|u(x)\|_{M_{\beta(t)}} = \sup_{0<t<\infty} \frac{\beta(t)}{t}\int_0^t u^*(s)\,ds.$$

where u^* means the nonincreasing rearrangement of the function u.

The following assertion was proved in [Pu2]:

Theorem 1. *If the kernel $G_\phi(x,y)$ satisfies the estimate (18) then the operator $\phi(P^{-1})$ is bounded from $\Lambda_{\alpha(t)}$ into $M_{\beta(t)}$ if and only if*

$$\frac{\alpha(t)}{\beta(t)} \ge C\phi(t^{2/n}). \tag{19}$$

The estimate (18) may be regarded as optimal one up to equivalence of functions. This can be seen from the following assertions.

Proposition 2. *The function $\phi(t)$ in the right-hand side of (18) cannot be replaced by any other concave function $\psi(t) \le \phi(t)$ unless $\psi(t)$ is equivalent to $\phi(t)$.*

Proposition 3. *If P is an elliptic operator of classical type then among all concave functions equivalent to $\phi(t)$ there exists a function $\psi(t)$ such that*

$$G_\phi(x,y) \sim r^{2-n}\psi'(r^2) \tag{20}$$

in each inner point $x \in \Omega$.

In the case of a selfadjoint elliptic operator P of classical type, the function $\psi'(r^2)$ in (20) can even be indicated explicitly:

$$\psi'(r^2) = \int_0^\infty e^{-kr\sqrt{s}}d\sigma(s),$$

but this expression is useless for application. More effective relation could be obtained if the function $\phi(t)$ is increasing at zero not too slowly. More exactly, we should use the extension indices

$$p_\phi = \lim_{t\to 0}\frac{\ln m_\phi(t)}{\ln t}, \qquad q_\phi = \lim_{t\to\infty}\frac{\ln m_\phi(t)}{\ln t},$$
$$m_\phi(t) = \sup_s \frac{\phi(ts)}{\phi(s)},$$

and then, in the case of $p_\phi > 0$, for a selfadjoint elliptic operator of classical type, we will have that

$$G_\phi(x,y) \sim r^{-n}\phi(r^2). \tag{21}$$

For the case $n = 2$, the estimate (18) (and also (20) or (21)) remains true if the function $\phi(t)$ is not "too close" to the function t, i.e., if $q_\phi < 1$. Otherwise, using the estimate (7), we can prove that

$$0 \le G_\phi(x,y) \le Cg(r^2), \qquad \text{where} \quad g(r) = \int_r^{a^2}\phi'(s)\frac{ds}{s}. \tag{22}$$

Here a is an arbitrary number greater than $2r$ for all considered x, y, e.g., $a \ge 2\operatorname{diam}\Omega$. Correspondingly Theorem 1 remains true for $n = 2$ if $q_\phi < 1$; otherwise the inequality (19) must be replaced by another necessary and sufficient condition

$$\frac{\alpha(t)}{\beta(t)} \ge C\big(\phi(t) + t\,g(t)\big). \tag{23}$$

4. Now we can pass to solution of the previously posed problems. The first of them is: to find a function $\phi(t)$, as small as possible, such that $\phi(P^{-1}) : L_2 \mapsto C$. Just solving this problem for fractional powers $P^{-\tau}$ enables the functional analysis methods to be used for estimating the rate of convergence of various approximation processes in a power form. The corresponding result said that $P^{-\tau} : L_2 \mapsto C$ for any number $\tau > n/4$; moreover, for $\tau = n/4$ such an action does not hold. With the help of Theorem 1 we are able now to obtain more precise result, using arbitrary concave functions of the operator P^{-1}.

Theorem 4. *Let P be a second order positive elliptic operator, satisfying all conditions needed for Theorem 1. Let $\Phi(t) = t^{n/4}\phi(t)$. Then the condition*

$$\int_0^1 (\phi(t))^{\frac{n}{n-1}} \frac{dt}{t} < \infty, \qquad n \geq 2, \tag{24}$$

ensures that $\Phi(P^{-1}) : L_2 \mapsto C$.

Let us compare this result with the former assertion for fractional powers which allowed us to take $\phi(t) = t^\epsilon$ for any $\epsilon > 0$. Now the condition (24) allows us to take, e.g., any function of the form

$$\phi(t) = \left(\frac{1}{\ln \frac{e}{t}}\right)^{1-\frac{1}{n}+\epsilon}, \qquad \forall \epsilon > 0,$$

which may increase essentially slower than any power.

Let me note a rather unexpected subtraction of $1/n$ in the last exponent. This is just the same gain that I have promised to show, the gain of using the integral representation of operators $\phi(P^{-1})$ and their kernel estimate. Basing only on general properties of operator functions, without such estimates, it is possible to prove an analogous result, but with the exponent 1 instead of $\frac{n}{n-1}$ in (24). For small n, the gain is quite notable.

The next problem is to describe the domain $\mathcal{D}(\phi(P))$ in L_2. A possible form of its solution is to find some sufficient (as broad as possible) conditions for a given element u to belong to this domain. As far as we consider only the concave functions $\phi(t)$, we must previously determine the maximal iteration $v = P^m u$ belonging to L_2, i.e., such that $P^{m+1}u \notin L_2$. I shall give two kinds of results: using only general properties of operator functions or using their integral representation. Unfortunately, I cannot compare them and show the gain as before. The first criterion seems more complicated, since it uses the modulus of continuity

$$w_2^{(2)}(t, v) = \sup_{|h|\leq t} \|v(x+2h) - 2v(x+h) + v(x)\|_{L_2}.$$

Theorem 5. *Let P be a second order positive selfadjoint elliptic operator in L_2. Then the inequality*

$$\int_0^1 \frac{1}{t} w_2^{(2)}(t^{1/2}, v)\, d\phi'\left(\frac{1}{t}\right) < \infty \tag{25}$$

ensures the inclusion $v \in \mathcal{D}(\phi(P))$ and thus $u \in \mathcal{D}(\Psi(P))$, where $\Psi(t) = t^m \phi(t)$.

The second criterion uses the properties of the element v itself or, more precisely, of Pv that can be calculated directly and easily.

Theorem 6. *Let an operator P satisfy all conditions for obtaining the estimate (18) for the kernels $G_\phi(x, y)$ and let $Pv(x)$ be a summable function. Then the inequality*

$$\int_\Omega \left[\int_\Omega \frac{|Pv(y)|\phi'(|x-y|^2)}{|x-y|^{n-2}}\, dy\right]^2 dx < \infty \tag{26}$$

ensures the inclusion $v \in \mathcal{D}(P\phi(P^{-1}))$ *and thus* $u \in \mathcal{D}(\Psi_1(P))$, *where* $\Psi_1(t) = t^{m+1}\phi(1/t)$.

Remark. This theorem is true even for $n = 2$ if the estimate (18) is valid. Otherwise we should replace the function $\phi'(t)$ in (26) by the function $g(t)$ from (22).

5. In conclusion I want to consider some examples of an approximation problem, where we just need all presented results for operator functions. This problem relates to the Fourier method for solution of differential equations of hyperbolic or parabolic type containing a second order elliptic operator (4). Let $e_i = e_i(x)$ be orthonormalized eigenfunctions of this operator and let $\lambda_i = \|P^{-1}e_i\|_{L_2}$ be the corresponding characteristic numbers, arranged in the descending order. The Fourier method consists in constructing the successive approximations

$$u_n(x,t) = \sum_{i=1}^{n}(u, e_i)e_i(x),$$

which obviously converge to the exact solution $u(x,t)$ in L_2. But we will be interested in obtaining stronger results — for example, in proving their absolute and uniform convergence on the domain Ω and, moreover, in estimating the rate of such a convergence. For this purpose, we may use the following general result proved in [Pu3]:

Proposition 7. *Let the eigenvectors* e_i *of a symmetrical positive operator* A *form an orthonormal basis in a Hilbert space* H *and let the positive functions* $\alpha(t)$, $\beta(t)$ *be such that the function* $\gamma(t) = \alpha(t)\beta(1/t)$ *is monotone decreasing, tending to* ∞ *as* $t \to 0$. *Suppose that* $\alpha(A^{-1})$ *acts from* H *to some other Banach space* E *and that a vector* $u \in \mathcal{D}(\beta(A))$. *Then* $u \in E$ *and can be expanded as a Fourier series* $u = \sum(u, e_i)e_i$ *which converges in* E. *The rate of convergence can be characterized by the estimate*

$$\|u - \sum_{i=1}^{n}(u, e_i)e_i\|_E = o\left(\frac{1}{\gamma(\lambda_n)}\right), \tag{27}$$

where λ_n *is the* n*-th eigenvalue of the operator* A^{-1} *after arrangement in the descending order.*

In order to derive from this assertion the absolute and uniform convergence on Ω it suffices to take $H = L_2(\Omega)$ and $E = C(\Omega)$.

Consider now a hyperbolic equation

$$\frac{d^2u}{dt^2}(x,t) + Pu(x,t) = f(x,t), \qquad u(x,0) = u_0(x), \quad u'_t(x,t) = v_0(x), \tag{28}$$

where $x \in \Omega,\ t \in [0,T]$ and P is a second order positive selfadjoint elliptic operator with coefficients and boundary conditions independent of t. The exact solution of this equation can be written in an operator form

$$u(t) = \cos(P^{\frac{1}{2}}t)u_0 + P^{-\frac{1}{2}}\sin(P^{\frac{1}{2}}t)v_0 + P^{-\frac{1}{2}}\int_0^t \sin(P^{\frac{1}{2}}t - P^{\frac{1}{2}}s)f(s)ds, \tag{29}$$

thus the properties of the unknown solution $u(t,x)$ can be expressed via corresponding properties of the given functions $u_0(x), v_0(x)$ and $f(t,x)$.

Theorem 8. *Let* $\Phi(s), \Psi(s) : (0,\infty) \mapsto (0,\infty)$ *and let* $\Theta(s) = \Psi(s)/\sqrt{s}$. *Assume that* $\Phi(P^{-1}) : L_2 \mapsto C$ *and also*

$$u_0 \in \mathcal{D}\big(\Psi(P)\big), \qquad v_0 \in \mathcal{D}\big(\Theta(P)\big), \qquad f(t) \in \mathcal{D}\big(\Theta(P)\big)$$

uniformly in $t \in [0,T]$, *i.e.,* $\sup_t \|\Theta(P)f(t)\|_{L_2} < \infty$. *If*

$$\lim_{s\to 0} \Phi(s)\Psi(1/s) = \infty, \tag{30}$$

then the successive approximations $u_n(x,t)$ *converge to the exact solution* $u(x,t)$ *absolutely and uniformly on* $\Omega \times [0,T]$, *and the rate of this convergence is characterized by the estimate*

$$\sup_{x\in\Omega, 0\le t\le T} |u(x,t) - u_n(x,t)| = O\big(\kappa(\lambda_n)\big), \tag{31}$$

where

$$\kappa(t) = \sup_{s\le t} \big(\Phi(s)\Psi(1/s)\big)^{-1}. \tag{32}$$

Analogous results can be obtained for the parabolic equation

$$\frac{du}{dt}(x,t) + Pu(x,t) = f(x,t), \qquad u(x,0) = u_0(x), \tag{33}$$

with $x \in \Omega$, $t \in [0,T]$ and the same operator P as above in (28). The operator formula for the exact solution looks now as

$$u(x,t) = e^{-Pt}u_0 + \int_0^t e^{-P(t-s)} f(x,s)\, ds \tag{34}$$

and convergence of the Fourier method can be studied like the previous case.

Theorem 9. *Let* $\Phi(s), \Psi(s) : (0,\infty) \mapsto (0,\infty)$. *Assume that* $\Phi(P^{-1}) : L_2 \mapsto C$ *and also*

$$u_0 \in \mathcal{D}\big(\Psi(P)\big), \qquad f(t) \in \mathcal{D}\big(\Psi(P)\big)$$

uniformly in $t \in [0,T]$, *i.e.,* $\sup_t \|\Psi(P)f(t)\|_{L_2} < \infty$. *If the functions* $\Phi(s)$ *and* $\Psi(s)$ *satisfy (30) then the successive approximations* $u_n(x,t)$ *converge to the exact solution* $u(x,t)$ *absolutely and uniformly on* $\Omega \times [0,T]$, *and the rate of this convergence is characterized by the estimate (31) with* $\kappa(t)$ *defined by (32).*

The operator e^{-Pt} may be regarded as "improving" one, since for any $u_0 \in L_2$ and all $t > 0$, the elements $e^{-Pt}u_0$ belong to domains of all positive functions of the operator P. Therefore the last theorem remains true in a slightly weaker form even without the condition $u_0 \in \mathcal{D}\big(\Psi(P)\big)$. Namely, for any $u_0 \in L_2$ the Fourier method converges for $t \in (0,T]$ and this convergence is uniform on any segment $[t_1,T]$ with $t_1 > 0$ (the estimate (31) also remains true).

There are certain problems in which a Fourier series undergoes, term by term, some additional operations (differentiation, special integral transforms etc.) and the rate of its convergence is considered only thereafter. One way to investigate such a problem is to replace the space C by some more complicated space E, including these operations into the definition of its norm (for example, by some Sobolev space if the series should be differentiated). Unfortunately, the space E could get too complicated for searching a function $\Phi(s)$ such that the operator $\Phi(P^{-1})$ acts from L_2 to E.

Another way is to state the subordination of these operations to some functions of the operator P^{-1} in the space L_2. Suppose, e.g., we want to investigate the convergence of some Fourier series after applying to it, term by term, a differential operator D^r of the order $|r|$. The classical result gives that this operator is subordinated in L_2 to the power $P^{|r|/2}$, i.e., the operator $D^r P^{-|r|/2}$ is bounded on L_2. This allowed us to obtain the convergence of successive approximations $u_n(x,t)$ after applying to them the operator D^r.

Theorem 10. *Let, in addition to all conditions of Theorem 8 or 9, the operator* $\Phi(P^{-1})$ *be such that* $\Phi(P^{-1})P^{|r|/2} : L_2 \mapsto C$. *Then the functions* $D^r u_n(x,t)$ *converge to the function* $D^r u(x,t)$ *absolutely and uniformly on* $\Omega \times [0,T]$, *and the rate of this convergence is again characterized by the estimate (31).*

References

[Tr] H. Triebel. *Interpolation Theory, Function Spaces, Differential Operators.* VEB Deutscher Verlag, Berlin, 1978.

[Sh] Norio Shimakura. *Partial Differential Operators of Elliptic Type.* Transl. Math. Monographs, v. 99, AMS, 1992.

[Il] V. A. Ilyin. Kernels of fractional order. Mat. Sbornik, v. 41 (1957), 459–480 (Russian).

[Pu1] E. Pustylnik. On functions of positive operator. Mat. Sbornik, v. 119 (1982), 32–47 (Russian). *English transl.*: Math. USSR, Sbornik, v. 47 (1984), 27–42.

[Pu2] E. Pustylnik. Functions of a second order elliptic operator in rearrangement invariant spaces. Int. Eq. Oper. Th., v. 22 (1995), 476–498.

[Pu3] E. Pustylnik. The rate of convergence of Fourier series with respect to the eigenfunctions of a positive operator. Funct. Diff. Eq., v. 4 (1997), 391–403.

Evgeniy Pustylnik
Dept. of Mathematics
Technion
32000 Haifa, Israel
E-mail address: evg@techunix.technion.ac.il

D. Haroske, T. Runst, H.-J. Schmeisser (eds.): Function Spaces, Differential Operators and Nonlinear Analysis. The Hans Triebel Anniversary Volume.

On Approximation of Solutions of Parabolic Functional Differential Equations in Unbounded Domains

László Simon[1]

Dedicated to Prof. Hans Triebel on the occasion of his 65th birthday

Introduction

We shall consider initial-boundary value problems for the equation

$$D_t u(t,x) - \sum_{j=1}^{n} D_j[f_j(t,x,u(t,x),\nabla u(t,x))] + f_0(t,x,u(t,x),\nabla u(t,x)) +$$
$$h(t,x,[H(u)](t,x)) = F(t,x), \quad (t,x) \in Q_T = (0,T) \times \Omega$$

where $\Omega \subset R^n$ is an unbounded domain with sufficiently smooth boundary, H is a linear continuous operator in $L^p(Q_T)$, the functions f_j, h satisfy the Carathéodory conditions and certain polynomial growth conditions. We shall show that the weak solutions of this problem can be obtained as the limit (as $k \to \infty$) similar problems, considered in $(0,T) \times \Omega_k$ where $\Omega_k \subset \Omega$ are bounded domains with sufficiently smooth boundary, having the property $\Omega_k \supset \Omega \cap B_k$ ($B_k = \{x \in R^n : |x| < k\}$). Similar statements were proved in [13] for more special equations. There will be also proved a uniqueness theorem and the boundedness of the solutions if some additional conditions are satisfied. We shall prove a theorem on the stabilization of the solutions as $t \to \infty$.

The problem was motivated by the climate model considered by J.I. Díaz and G. Hetzer [8] where this type of equation was considered on the unit sphere in R^3 (instead of Ω). Similar qualitative properties were proved in [1] and [7] for the climate model. Functional partial differential equations arise also in population dynamics, plasticity, hysteresis (see, e.g., [2], [5], [12], [19]).

In [15] a similar problem was considered for bounded Ω, where the equation contained a rapidly increasing term with respect to u and also discontinuous terms in u. It is not difficult to extend the results of the present paper to higher order parabolic equations, containing discontinuous terms with respect to the unknown function.

[1]This work was supported by the Hungarian National Foundation for Scientific Research under grant OTKA T 031807

1. Existence theorems

Let $\Omega \subset R^n$ be an unbounded domain with sufficiently smooth boundary, $p \geq 2$. Denote by $W^{1,p}(\Omega)$ the usual Sobolev space with the norm

$$\| u \| = \left[\int_\Omega (|\nabla u|^p + |u|^p) \right]^{1/p} .$$

Let V be a closed linear subspace of $W^{1,p}(\Omega)$ and denote by $L^p(0,T;V)$ the Banach space of the set of measurable functions $u : (0,T) \to V$ such that $\| u \|^p$ is integrable and $X_T = L^p(0,T;V) \cap L^2(Q_T)$ with the norm

$$\| \cdot \|_{X_T} = \| \cdot \|_{L^p(0,T;V)} + \| \cdot \|_{L^2(Q_T)} .$$

The dual space of X_T is $X_T^\star = L^q(0,T;V^\star) + L^2(Q_T)$ where $1/p + 1/q = 1$ and $V^\star$ is the dual space of V (see [9], [11], [18]).

Let $\varphi \in C_0^\infty(R^n)$ be a fixed function having the properties

$$0 \leq \varphi(x) \leq 1, \quad \varphi(x) = 1 \text{ if } |x| \leq 1/2, \quad \varphi(x) = 0 \text{ if } |x| \geq 3/4$$

and define function φ_k by $\varphi_k(x) = \varphi(x/k)$.

Assume that

A V_k is a closed linear subspace of $W^{1,p}(\Omega_k)$ such that

$$\text{for any } w \in V, \quad (\varphi_k w) |_{\Omega_k} \in V_k .$$

Further, there exist linear and continuous (extension) operators $L_k : V_k \to V$ such that for any $w_k \in V_k$, $(L_k w_k) |_{\Omega_k} = w_k$, for any $w \in V$, $(L_k \varphi_k w) |_{\Omega_k} = \varphi_k w$, the sequence $\| L_k \|$ is bounded. Moreover, the operators L_k have linear and continuous extensions $\hat{L}_k : L^2(\Omega_k) \to L^2(\Omega)$ such that the sequence $\| \hat{L}_k \|$ is bounded, too.

Remark 1. It is easy to show that assumption **A** is satisfied e.g. in the following special cases:

a) $V = W_0^{1,p}(\Omega)$, $V_k = W_0^{1,p}(\Omega_k)$;

b) $\partial\Omega$ is bounded, $\Omega_k = \Omega \cap B_k$, $V = W^{1,p}(\Omega)$ and $V_k = W^{1,p}(\Omega_k)$;

c) $\partial\Omega \in C^m$ is bounded, $\Omega_k = \Omega \cap B_k$, $V = W_0^{1,p}(\Omega)$ and $V_k = \{v \in W^{1,p}(\Omega_k) : v|_{\partial\Omega} = 0\}$.

Define the operators M_k by $(M_k v)(t,x) = v(t,\cdot) |_{\Omega_k} (x)$, $v \in X_T$. Then we have $M_k(\varphi_k v) \in X_T^k = L^p(0,T;V_k)$.

Similarly, define the operators N_k by $(N_k v)(t,x) = (L_k v(t,\cdot))(x)$, $v \in X_T^k$. Then $N_k : L^p(0,T;V_k) \to L^p(0,T;V)$ are linear and continuous, their norms are bounded.

On the functions f_j we assume that

B (i) $f_j : Q_T \times R \times R^n \to R$ are measurable in $(t,x) \in Q_T$ and continuous in $\eta \in R, \zeta \in R^n$;

(ii) $|f_j(t,x,\eta,\zeta)| \leq c_1(|\eta|^{p-1} + |\zeta|^{p-1} + |\eta|) + k_1(x)$ with some constant c_1 and a function $k_1 \in L^q(\Omega)$ $(j = 0,1,\ldots,n)$;

(iii) $\sum_{j=1}^{n}[f_j(t,x,\eta,\zeta) - f_j(t,x,\eta,\tilde{\zeta})](\zeta_j - \tilde{\zeta}_j) > 0$ if $\zeta \neq \tilde{\zeta}$;

(iv) $\sum_{j=1}^{n} f_j(t,x,\eta,\zeta)\zeta_j + f_0(t,x,\eta,\zeta)\eta \geq c_2[|\zeta|^p + |\eta|^p] - k_2(x)$ with some constant $c_2 > 0$ and $k_2 \in L^1(\Omega)$.

Remark 2. A simple example for f_j satisfying **B** is

$$\begin{aligned} f_j(t,x,\eta,\zeta) &= a_j(t,x)\zeta_j|\zeta_j|^{p-2}, \quad (j = 1,\dots,n), \\ f_0(t,x,\eta,\zeta) &= a_0(t,x)\eta|\eta|^{p-2} + b_0(t,x)\eta, \end{aligned}$$

where a_j, b_0 are measurable functions, satisfying $0 < c_0 \leq a_j(t,x) \leq c_0'$, $0 \leq b_0(t,x) \leq c_0'$ with some constants c_0, c_0'.

On function h we assume

C (i) $h(t,x,\theta)$ is measurable in (t,x) and continuous in θ.

(ii) $|h(t,x,\theta)| \leq k_3(x)k_4(|\theta|)|\theta|^{p-1} + k_5(x)$, where $k_3 \in L^1(\Omega) \cap L^\infty(\Omega)$, $k_5 \in L^q(\Omega)$ and k_4 is a continuous function, satisfying $\lim_\infty k_4 = 0$.

Finally, assume that

D $H : L^p(Q_T) \cap L^2(Q_T) \to L^p(Q_T) \cap L^2(Q_T)$ is a linear and continuous operator in the norm of $L^p(Q_T)$ and $L^2(Q_T)$ such that for any compact $K \subset \Omega$ there is a compact $\tilde{K} \subset \Omega$ with the following property: the restriction of $H(u)$ to $(0,t) \times K$ depends only on the restriction of u to $(0,t) \times \tilde{K}$ for all $t \in (0,T]$ and it is continuous as an operator $L^p(Q_t) \to L^p(Q_t)$ with the same bounds for all t.

Remark 3. The operator H may have e.g. one of the following forms:

$$[H(u)](t,x) = \int_0^t \beta_0(s,t,x)u(s,x)ds \text{ or } [H(u)](t,x) = u(\tau(t),x)$$

with some $\beta_0 \in L^\infty((0,T) \times Q_T)$ and a continuously differentiable function τ satisfying $\tau' > 0, 0 < \tau(t) \leq t$.

Define operators $A, B : X_T \to X_T^\star$ and $A_k, B_k : X_T^k \to (X_T^k)^\star$ by

$$\begin{aligned} [A(u),v] &= \int_0^T \langle A(u)(t), v(t)\rangle dt \\ &= \int_0^T \left[\sum_{j=1}^{n} \int_\Omega f_j(t,x,u,\nabla u) D_j v dx + \int_\Omega f_0(t,x,u,\nabla u) v dx \right] dt, \end{aligned}$$

$$\begin{aligned} [B(u),v] &= \int_0^T \langle B(u)(t), v(t)\rangle dt \\ &= \int_0^T \left[\int_\Omega h(t,x,H(u)(t,x)) v dx \right] dt, \quad u,v \in X_T; \end{aligned}$$

$$[A_k(u_k),v_k]=\int_0^T \langle A_k(u_k)(t),v_k(t)\rangle dt$$
$$=\int_0^T \left[\sum_{j=1}^n \int_{\Omega_k} f_j(t,x,u_k,\nabla u_k)D_j v_k dx+\int_{\Omega_k} f_0(t,x,u_k,\nabla u_k)v_k dx\right] dt,$$

$$[B_k(u_k),v_k]=\int_0^T \langle B_k(u_k)(t),v_k(t)\rangle dt$$
$$=\int_0^T \left[\int_{\Omega_k} h(t,x,H(N_k u_k)(t,x))v_k dx\right] dt, \quad u_k,v_k\in X_T^k.$$

Finally, define for $F \in X_T^\star$ its "restriction" $F_k \in (X_T^k)^\star$ by

$$[F_k, v_k] = [F, N_k v_k], \quad v_k \in X_T^k.$$

Theorem 1. *Assume* **A**–**D**. *Then for any* $F \in X_T^\star$, $u_0 \in V$ *there exists* $u_k \in X_T^k$ *satisfying*

$$\frac{du_k}{dt} + (A_k + B_k)(u_k) = F_k, \quad \frac{du_k}{dt} \in L^q(0,T;V_k^\star),$$
$$u_k(0) = M_k(\varphi_k u_0), \quad k = 1,2,\dots \tag{1.1}$$

Further, there exist a subsequence (u_{k_l}) *of the sequence* (u_k) *and* $u \in X_T$ *such that*

$$(N_{k_l} u_{k_l}) \to u \text{ weakly in } X_T$$

and u *satisfies*

$$\frac{du}{dt} + (A + B)(u) = F, \quad \frac{du}{dt} \in X_T^\star,$$
$$u(0) = u_0. \tag{1.2}$$

Proof. The existence of solutions u_k of (1.1) follows from the fact that $(A_k + B_k) : X_T^k \to (X_T^k)^\star$ is bounded, demicontinuous, pseudomonotone with respect to

$$D(L) := \{u \in X_T^k : \frac{du}{dt} \in (X_T^k)^\star, u(0) = 0\}$$

and it is coercive (see, e.g., [14], [15]). Thus by a known existence theorem (see, e.g., [3]) there exists a solution of (1.1).

Similarly to [10], [11], [16] we perform the substitution $u = e^{ct}\tilde{u}$, $u_k = e^{ct}\tilde{u}_k$ with c a fixed positive constant. Then (1.1) is equivalent with the problem to find $\tilde{u}_k \in X_T^k$ such that

$$\frac{d\tilde{u}_k}{dt} + (\tilde{A}_k + \tilde{B}_k)(\tilde{u}_k) + c\tilde{u}_k = \tilde{F}_k, \quad \frac{d\tilde{u}_k}{dt} \in L^q(0,T;V_k^\star),$$
$$\tilde{u}_k(0) = M_k(\varphi_k u_0), \quad k = 1,2,\dots \tag{1.3}$$

where

$$[\tilde{A}_k(\tilde{u}_k),v_k]=\int_0^T \left[\sum_{j=1}^n \int_{\Omega_k} \tilde{f}_j(t,x,\tilde{u}_k,\nabla\tilde{u}_k)D_j v_k dx+\int_{\Omega_k} \tilde{f}_0(t,x,\tilde{u}_k,\nabla\tilde{u}_k)v_k dx\right] dt$$

with

$$\tilde{f}_j(t,x,\eta,\zeta) = e^{-ct} f_j(t,x,e^{ct}\eta,e^{ct}\zeta),$$

$$[\tilde{B}_k(\tilde{u}_k), v_k] = \int_0^T \left[\int_{\Omega_k} e^{-ct} h(t,x,H(e^{ct}N_k\tilde{u}_k)(t,x))v_k dx\right] dt, \quad \tilde{F}_k = e^{-ct}F_k.$$

Similarly, (1.2) is equivalent with the problem to find $\tilde{u} \in X_T$ such that

$$\frac{d\tilde{u}}{dt} + (\tilde{A}+\tilde{B})(\tilde{u}) + c\tilde{u} = \tilde{F}, \quad \frac{d\tilde{u}}{dt} \in X_T^\star, \qquad \tilde{u}(0) = u_0, \tag{1.4}$$

where

$$[\tilde{A}(\tilde{u}), v] = \int_0^T \left[\sum_{j=1}^n \int_\Omega \tilde{f}_j(t,x,\tilde{u},\nabla\tilde{u})D_j v dx + \int_\Omega \tilde{f}_0(t,x,\tilde{u},\nabla\tilde{u})v dx\right] dt,$$

$$[\tilde{B}(\tilde{u}), v] = \int_0^T \left[\int_\Omega e^{-ct} h(t,x,H(e^{ct}\tilde{u})(t,x))v dx\right] dt, \quad \tilde{F} = e^{-ct}F.$$

Applying both sides of (1.3) to $\tilde{u}_k$ we find

$$\frac{\|N_k\tilde{u}_k(t)\|^2_{L^2(\Omega)}}{2} - \frac{\| u_0 \|^2_{L^2(\Omega)}}{2} + c_3 \| N_k\tilde{u}_k \|^p_{L^p(0,t;V)} + c_4 \| N_k\tilde{u}_k \|^2_{L^2(Q_t)} \le [\| F \|_{X_T^\star} + c_5] \| N_k\tilde{u}_k \|_{L^p(0,t;V)} + c_6 \tag{1.5}$$

for all $t \in [0,T]$ with some positive constants c_3, c_4, c_5, c_6. This inequality implies that

$$\| N_k\tilde{u}_k \|_{X_T}, \quad \| N_k\tilde{u}_k \|_{L^\infty(0,T;L^2(\Omega))} \text{ are bounded.} \tag{1.6}$$

Hence

$$\tilde{A}_k(\tilde{u}_k), \quad \tilde{B}_k(\tilde{u}_k) \text{ are bounded in } L^q(0,T;V_k^\star). \tag{1.7}$$

Further,

$$(N_k\tilde{u}_k) \to \tilde{u} \text{ weakly in } X_T \text{ for a subsequence} \tag{1.8}$$

with some $\tilde{u} \in X_T$. Define the "extensions" $\hat{A}_k(\tilde{u}_k)$ by

$$[\hat{A}_k(\tilde{u}_k), v] = [\tilde{A}_k(\tilde{u}_k), M_k(\varphi_k v)], \quad v \in X_T,$$

then $\| \hat{A}_k(\tilde{u}_k) \|_{X_T^\star}$ is bounded. Consequently, for a subsequence

$$(\hat{A}_k(\tilde{u}_k)) \to \tilde{w} \text{ weakly in } X_T^\star \tag{1.9}$$

with some $\tilde{w} \in X_T^\star$.

Since by (1.3), (1.6), (1.7) $\| \frac{d\tilde{u}_k}{dt} \|_{(X_T^k)^\star}$ is bounded, by using also (1.6) we can choose a subsequence of $(\tilde{u}_k)$ such that for any bounded $\Omega_0 \subset \Omega$,

$$(N_k \tilde{u}_k) \to \tilde{u} \text{ in } L^p((0,T) \times \Omega_0) \text{ and} \tag{1.10}$$

$$(N_k \tilde{u}_k) \to \tilde{u} \text{ a.e. in } Q_T. \tag{1.11}$$

By (1.10) and assumption **D** for a suitable subsequence

$$H(e^{ct} N_k \tilde{u}_k) \to H(e^{ct}\tilde{u}) \text{ a.e. in } Q_T. \tag{1.12}$$

Since for an arbitrary $v \in X_T$

$$(M_k(\varphi_k v)) \to v \text{ in the norm of } X_T \text{ and}$$

$$[\tilde{F}_k, M_k(\varphi_k v)] = [\tilde{F}, N_k(M_k(\varphi_k v))] = [\tilde{F}, \varphi_k v],$$

applying (1.3) to $M_k(\varphi_k v)$ with an arbitrary fixed $v \in X_T$, we obtain as $k \to \infty$

$$\begin{gathered} \frac{d\tilde{u}}{dt} + \tilde{w} + \tilde{B}(\tilde{u}) = \tilde{F}, \quad \frac{d\tilde{u}}{dt} \in X_T^\star, \\ \tilde{u}(0) = u_0 \end{gathered} \tag{1.13}$$

(see, e.g., [18]).

Now we prove $\tilde{w} = \tilde{A}(\tilde{u})$, i.e. $w = A(u)$ (where $w = e^{ct}\tilde{w}$). Apply (1.1) to $M_k(u_k - u)\zeta$ with arbitrary fixed $\zeta \in C_0^\infty(\Omega)$ having the properties $\zeta \geq 0$, $\zeta(x) = 1$ in a compact subset K of Ω. So we obtain for sufficiently large k

$$\begin{aligned} &[D_t u_k - D_t u, M_k((u_k - u)\zeta)] + [D_t u, M_k((u_k - u)\zeta)] \\ &\quad + [A_k(u_k), M_k((u_k - u)\zeta)] + [B_k(u_k), M_k((u_k - u)\zeta)] \\ &= [F_k, M_k((u_k - u)\zeta)]\,. \end{aligned} \tag{1.14}$$

For the first term (for sufficiently large k) we have

$$\begin{aligned} [D_t u_k - D_t u, M_k((u_k - u)\zeta)] &= \frac{1}{2}\int_0^T \left[\frac{d}{dt}\int_\Omega (u_k(t) - u(t))^2 \zeta dx\right] dt \\ &= \frac{1}{2}\int_\Omega (u_k(T) - u(T))^2 \zeta dx \geq 0. \end{aligned} \tag{1.15}$$

Further, by (1.8)

$$\begin{aligned} &\lim_{k\to\infty} [D_t u, M_k((u_k - u)\zeta)] = 0, \\ &\lim_{k\to\infty} [F_k, M_k((u_k - u)\zeta)] = \lim_{k\to\infty} [F, N_k(M_k((u_k - u)\zeta))] \\ &\qquad = \lim_{k\to\infty} [F, (u_k - u)\zeta] = 0. \end{aligned} \tag{1.16}$$

By **D**, Hölder's inequality and (1.10)

$$\lim_{k\to\infty} [B_k(u_k), M_k((u_k - u)\zeta)] = 0. \tag{1.17}$$

Thus (1.14)–(1.17) imply

$$\limsup_{k\to\infty} [A_k(u_k), M_k((u_k - u)\zeta)] \leq 0. \tag{1.18}$$

Since by **D**, Hölder's inequality and (1.10)

$$\lim_{k\to\infty}\int_{Q_{T,k}} f_0(t,x,u_k,\nabla u_k)(u_k-u)\zeta dtdx = 0 \quad (\text{where } Q_{T,k}=(0,T)\times\Omega_k),$$

(1.18) implies

$$\limsup_{k\to\infty}\sum_{j=1}^{n}\int_{Q_{T,k}} f_j(t,x,u_k,\nabla u_k)D_j[(u_k-u)\zeta]dtdx \le 0. \tag{1.19}$$

By using arguments of [6] (see also [13]) we obtain from (1.19)

$$\nabla u_k \to \nabla u \text{ a.e. in } (0,T)\times K.$$

Since K can be chosen as any compact subset of Ω, we find

$$\nabla(N_k u_k) \to \nabla u \text{ a.e. in } Q_T.$$

Thus Vitali's theorem and Hölder's inequality imply (see, e.g. [6])

$$(\bar{A}_k(u_k)) \to A(u) \text{ weakly in } X_T^\star,$$

where

$$[\bar{A}_k(u_k),v] = [A_k(u_k), M_k(\varphi_k v)],$$

i.e. $w = A(u)$ which completes the proof of our theorem.

Remark 4. It follows from the above proof that if the solution of (1.2) is unique then also for the original sequence (u_k) of solutions to (1.1), $(N_k u_k)$ converges weakly in X_T to the solution u of (1.2).

If some additional conditions are satisfied then one can prove the uniqueness of the solution.

Theorem 2. *Assume* **A–D** *and the following monotonicity condition is satisfied:*

$$\sum_{j=1}^{n}\left[f_j(t,x,\xi)-f_j(t,x,\tilde{\xi})\right]\left(\xi_j-\tilde{\xi}_j\right)+\left[f_0(t,x,\xi)-f_0(t,x,\tilde{\xi})\right]\left(\xi_0-\tilde{\xi}_0\right)$$

$$\ge -c_0\left(\xi_0-\tilde{\xi}_0\right)^2 \tag{1.20}$$

with some constant c_0. Further, there exists a constant k_0 such that

$$|h(t,x,\theta)-h(t,x,\tilde{\theta})| \le k_0|\theta-\tilde{\theta}| \tag{1.21}$$

for any $(t,x)\in Q_T$ and $\theta,\tilde{\theta}\in R$. Finally, H is positive, i.e. $u\ge 0$ implies $H(u)\ge 0$.

Then the solution of (1.2) is unique.

Proof. Perform the substitution $u=e^{ct}\tilde{u}$. Then (1.2) is equivalent with (1.4) and for sufficiently large c we obtain that the solution of (1.4) is unique because then (by (1.20), (1.21)) the operator $\tilde{u}\mapsto(\tilde{A}+\tilde{B})\tilde{u}+c\tilde{u}$ is monotone.

It is not difficult to prove an existence theorem for the interval $[0,\infty)$. Denote by X_∞ and $X_\infty^\star$ the set of functions $u:[0,\infty)\to V$, $w:[0,\infty)\to V^\star$, respectively, such that for any finite T, $u\in X_T$, $w\in X_T^\star$, respectively. Further, define $Q_\infty =$

$(0,\infty)\times\Omega$ and let $L^p_{loc}(Q_\infty)$ be the set of functions $v : Q_\infty \to R$ such that $v \in L^p(Q_T)$ for arbitrary finite T.

Theorem 3. *Assume* **A**, *further assume that* $f_j : Q_\infty \times R^{n+1} \to R$, $h : Q_\infty \times R \to R$ *satisfy* **B** *and* **C** *for any finite* $T > 0$ *and* $H : L^p_{loc}(Q_\infty) \to L^p_{loc}(Q_\infty)$ *satisfies* **D** *for any finite* T. *Then for arbitrary* $F \in X^\star_\infty$ *there exists* $u \in X_\infty$ *such that* u *satisfies (1.2) for any finite* T.

Now we formulate another existence theorem (for $p > 2$) which holds for equations satisfying certain modified conditions.

Let γ be a continuous weight function satisfying

$$\gamma(x) \geq c_1 > 0 \qquad \text{and} \qquad \int_\Omega \frac{dx}{\gamma^{2/(p-2)}} < \infty$$

with some constant c_1. Denote by $W^{1,p}_\gamma(\Omega)$ the space of functions having a finite norm

$$\| w \| = \left[\int_\Omega [|\nabla w|^p + \gamma |w|^p]\right]^{1/p}.$$

By Hölder's inequality it is easy to show that $W^{1,p}_\gamma(\Omega)$ is continuously imbedded into $L^2(\Omega)$. Let V^γ be a closed linear subspace of $W^{1,p}_\gamma(\Omega)$ and $X^\gamma_T = L^p(0,T;V^\gamma)$.

Assume that

$\tilde{\mathbf{A}}$ $\tilde{V}_k$ is a closed linear subspace of $W^{1,p}(\Omega_k)$ such that for any $w \in V^\gamma$, $(\varphi_k w) |_{\Omega_k} \in \tilde{V}_k$. Further, there exist linear and continuous (extension) operators $\tilde{L}_k : \tilde{V}_k \to V^\gamma$ such that for any $w_k \in \tilde{V}_k$, $(\tilde{L}_k w_k) |_{\Omega_k} = w_k$ a.e.; for any $w \in V^\gamma$, $(\tilde{L}_k \varphi_k w) |_{\Omega_k} = \varphi_k w$ a.e., the sequence $\| \tilde{L}_k \|$ is bounded. Define the operators $\tilde{N}_k$ by

$$(\tilde{N}_k v)(t,x) = (\tilde{L}_k v(t,\cdot))(x), \quad v \in L^p(0,T;\tilde{V}_k).$$

On the functions f_j now we assume that

$\tilde{\mathbf{B}}$ (i) $f_j : Q_T \times R \times R^n \to R$ are measurable in $(t,x) \in Q_T$ and continuous in $\eta \in R, \zeta \in R^n$;

(ii) $|f_j(t,x,\eta,\zeta)| \leq c_1[(\gamma(x))^{1/q}|\eta|^{p-1} + |\zeta|^{p-1} + |\eta|] + k_1(x)$, $j = 1,\dots,n$, $|f_0(t,x,\eta,\zeta)| \leq c_1[(\gamma(x))^{p-1}|\eta|^{p-1} + |\zeta|^{p-1} + |\eta|] + k_1(x)$ with some constant c_1 and a function $k_1 \in L^q(\Omega)$;

(iii) $\sum_{j=1}^n [f_j(t,x,\eta,\zeta) - f_j(t,x,\eta,\tilde{\zeta})](\zeta_j - \tilde{\zeta}_j) > 0$ if $\zeta \neq \tilde{\zeta}$;

(iv) $\sum_{j=1}^n f_j(t,x,\eta,\zeta)\zeta_j + f_0(t,x,\eta,\zeta)\eta \geq c_2[|\zeta|^p + (\gamma(x))^{p-1}|\eta|^p] - k_2(x)$ with some constant $c_2 > 0$ and $k_2 \in L^1(\Omega)$.

$\tilde{\mathbf{C}}$ (i) $h(t,x,\theta)$ is measurable in (t,x) and continuous in θ.

(ii) $|h(t,x,\theta)| \leq k_3(x)k_4(|\theta|)(\gamma(x))^{p-1}|\theta|^{p-1} + k_5(x)$, where $k_3 \in L^1(\Omega) \cap L^\infty(\Omega)$, $\int_\Omega |k_5|^q \frac{1}{\gamma^{p-1}} < \infty$ and k_4 is a continuous function, satisfying $\lim_\infty k_4 = 0$.

Finally, assume

$\tilde{\mathbf{D}}$ $H : L^p_\gamma(Q_T) \to L^p_\gamma(Q_T)$ is a linear and continuous operator (in the L^p space with the weight function γ) having the same properties as in **D**.

Theorem 4. *Assume* $\tilde{\mathbf{A}}$–$\tilde{\mathbf{D}}$. *Then for any* $F \in (X^\gamma_T)^\star$, $u_0 \in V^\gamma$ *there exists* $u_k \in L^p(0,T;\tilde{V}_k)$ *satisfying*

$$\frac{du_k}{dt} + (A_k + B_k)(u_k) = F_k, \quad \frac{du_k}{dt} \in L^q(0,T;\tilde{V}^\star_k),$$
$$u_k(0) = M_k(\varphi_k u_0), \quad k = 1, 2, \ldots$$

Further, there exists a subsequence (u_{k_l}) *of the sequence* (u_k) *and* $u \in X^\gamma_T$ *such that*

$$(N_{k_l} u_{k_l}) \to u \text{ weakly in } X^\gamma_T$$

and u *satisfies*

$$\frac{du}{dt} + (A+B)(u) = F, \quad \frac{du}{dt} \in (X^\gamma_T)^\star, \qquad u(0) = u_0. \tag{1.22}$$

The proof is similar to that of Theorem 1.

Analogously to Theorem 2, one can prove

Theorem 5. *Assume* $\tilde{\mathbf{A}}$–$\tilde{\mathbf{D}}$ *and (1.20), (1.21). Then the solution of (1.22) is unique.*

Similarly to Theorem 3, one can formulate and prove an existence theorem on solutions for $t \in (0, \infty)$.

2. Boundedness and stabilization

Theorem 6. *Assume that the conditions of Theorem 3 are satisfied such that (instead of* **A** *(iv)) for all* $t \in [0, \infty)$

$$\sum_{j=1}^n [f_j(t,x,\eta,\zeta)\zeta_j + f_0(t,x,\eta,\zeta)\eta \geq c_2(|\zeta|^p + |\eta|^p + |\eta|^2) - k_2(x) \tag{2.1}$$

with some constant $c_2 > 0$ *and* $k_2 \in L^1(\Omega)$, $\| F(t) \|_{V^\star}$ *is bounded,*

$$|h(t,x,\theta)|^q \leq c_4^\star |\theta|^2 + k_4^\star(x) \tag{2.2}$$

with some constant $c_4^\star$ *and a function* $k_4^\star \in L^1(\Omega)$. *Further, for any* $u \in L^p_{loc}(Q_\infty) \cap L^2_{loc}(Q_\infty)$

$$\int_\Omega |H(u)|^2(t,x)dx \leq c_5^\star \sup_{\tau\in[0,t]} \int_\Omega |u(\tau,x)|^2 dx. \tag{2.3}$$

Finally,

$$c_4^\star c_5^\star < c_2^{1+q/p} p^{q/p} q. \tag{2.4}$$

Then for the solution u of the problem in Q_∞, *the function*

$$y(t) = \int_\Omega |u(t,x)|^2 dx$$

is bounded in $(0,\infty)$ *and there exist positive numbers* c', c'' *such that*

$$\int_{T_1}^{T_2} \| u(t) \|_V^p \, dt \le c'(T_2 - T_1) + c'' \text{ for sufficiently large } T_1 < T_2. \tag{2.5}$$

Remark 5. The examples in Remark 3. satisfy (2.3).

Sketch of the proof. Apply (1.2) to u, then it is not difficult to derive the inequality

$$\begin{aligned}\frac{1}{2}[y(T_2) - y(T_1)] + \tilde{c}_2 \int_{T_1}^{T_2} \| u(t) \|_V^p \, dt + c_2 \int_{T_1}^{T_2} y(t)dt \\ \le c_0 \int_{T_1}^{T_2} \sup_{\tau\in[0,t]} y(\tau)dt + c_6^\star(T_2 - T_1)\end{aligned} \tag{2.6}$$

with some positive numbers $c_2, \tilde{c}_2, c_0, c_6^\star$ satisfying $c_2 > c_0$. This inequality implies the boundedness of y in $(0,\infty)$.

Remark 6. From the proof it follows the inequality $y(t) \le \frac{c_6^\star}{c_2 - c_0}$.

Theorem 7. *Assume that* $\tilde{A} - \tilde{D}$ *are satisfied for any finite* $T > 0$ *such that the constants and functions are independent of* T. *Further, assume (2.2) and (2.3) (for any u with the property* $u \in X_T^\gamma$ *for each finite* T*). Then for the solution u of (1.22) in* $(0,\infty)$,

$$y(t) = \int_\Omega |u(t,x)|^2 dx$$

is bounded in $(0,\infty)$ *and there exist positive numbers* c', c'' *such that*

$$\int_{T_1}^{T_2} \| u(t) \|_{V^\gamma}^p \, dt \le c'(T_2 - T_1) + c''.$$

Idea of the proof. Similarly to (2.6) one gets the inequality

$$\frac{1}{2}[y(T_2) - y(T_1)] + \tilde{c}_2 \int_{T_1}^{T_2} \| u(t) \|_{V^\gamma}^p \, dt \le c_0 \int_{T_1}^{T_2} \sup_{\tau\in[0,t]} y(\tau)dt + c_6^\star(T_2 - T_1).$$

Since

$$y(t) = \int_\Omega |u(t,x)|^2 dx \le const \left[\int_\Omega |u(t,x)|^p \gamma(x) dx\right]^{2/p},$$

we obtain

$$\frac{1}{2}[y(T_2) - y(T_1)] + \tilde{c}_3 \int_{T_1}^{T_2} [y(t)]^{p/2} dt \le c_0 \int_{T_1}^{T_2} \sup_{\tau\in[0,t]} y(\tau)dt + c_6^\star(T_2 - T_1).$$

It is not difficult to show that the last inequality implies the boundedness of y if $p > 2$.

By using some additional assumptions one can prove (by using arguments of [7]) that the solution u of (1.2) in $(0,\infty)$ satisfies $D_t u \in L^2(0,\infty;L^2(\Omega))$ which implies the stabilization result (Theorem 4).

Theorem 8. *Let the functions f_j be defined by*

$$f_j(t,x,\eta,\zeta) = a_j(x)\zeta_j|\zeta_j|^{p-2}, \quad j=1,\ldots,n,$$
$$f_0(t,x,\eta,\zeta) = a_0(x)(\eta|\eta|^{p-2}+\eta)+g(x,\eta),$$

where the measurable functions a_j satisfy $0 < c_0 \le a_j(x) \le c_0'$ with some constants c_0,c_0' and g is a Carathéodory function satisfying

$$|g(x,\eta)| \le k_3(x)|\eta| + k_5(x) \text{ with } k_3 \in L^1(\Omega)\cap L^\infty(\Omega),\ k_5 \in L^q(\Omega \cap L^2(\Omega). \tag{2.7}$$

Assume that the conditions of Theorem 6 are fulfilled such that

$$|h(t,x,\theta)| \le \chi(t)[|\theta| + \tilde{k}(x)] \tag{2.8}$$

with some functions $\chi \in L^2(0,\infty)$, $\tilde{k} \in L^2(\Omega)$. Let u be a weak solution of (1.2) in $(0,\infty)$ with

$$F \in L^\infty(0,\infty;L^2(\Omega)) \cap W^{1,2}_{loc}(0,\infty;L^2(\Omega)),$$
$$\int_0^\infty \left[\int_\Omega |D_t F(t,x)|^2 dx\right]^{1/2} dt < \infty. \tag{2.9}$$

Then

$$D_t u \in L^2(0,\infty;L^2(\Omega)) \text{ and } u \in L^\infty(0,\infty;V). \tag{2.10}$$

Sketch of the proof. Define the functional $\Phi : L^2(\Omega) \to R$ by

$$\Phi(w) = \int_\Omega \left[\sum_{j=1}^n a_j(x)|D_j w|^p + a_0(x)(|w|^p + |w|^2)\right] dx,$$
$$w \in D(\Phi) = V \cap L^2(\Omega), \quad \Phi(w) = +\infty \text{ if } w \notin D(\Phi).$$

Then Φ is a convex nonnegative lower semicontinuous functional (see, e.g., [4], [17]) and set $A_1 = \partial\Phi$, the subdifferential of Φ. One can show that the weak solution u of (1.2) is the (unique) strong solution of

$$\frac{du}{dt} + A_1 u(t) \ni b(t) = F - g(x,u) - h(t,x,H(u)) \in L^2(0,T;L^2(\Omega))$$

for any finite T and for a.e. t

$$\| \frac{du}{dt} \|^2_{L^2(\Omega)} + \frac{d}{dt}[\Phi(u(t))] = \langle b(t), \frac{du}{dt}(t)\rangle_{L^2(\Omega)}$$

(see [4]). Integrating the last inequality over (σ,τ), by the assumptions of the theorem we find

$$\frac{1}{2}\int_\sigma^\tau \left[\int_\Omega |D_t u|^2 dx\right] dt + \Phi(u(\tau)) - \Phi(u(\sigma)) \le const.$$

Since $\Phi \ge 0$, we obtain (2.10).

Consider a sequence (t_k), $t_k \to +\infty$ and define the functions U_k by

$$U_k(s,x) = u(t_k + s, x), \quad s \in (-a,b), \quad x \in \Omega$$

with some fixed $a > 0$, $b > 0$.

Remark 7. By Theorems 6 and 8 (U_k) is bounded in

$$L^p(-a,b;V) \quad \text{and} \quad L^2(-a,b;L^2(\Omega))$$

for a weak solution of (1.2) in $(0,\infty)$.

Define the ω limit set associated to u by

$$\omega(u) = \{u_\infty \in V \cap L^2(\Omega) : \exists t_k \to +\infty \text{ such that } u(t_k,\cdot) \to u_\infty \text{ in } L^2(\Omega_0)\}$$

for any bounded $\Omega_0 \subset \Omega$. By using arguments of [7] one can prove

Theorem 9. *Let the assumptions of Theorem 8 be satisfied. On the operator H assume that there exists a finite $\rho > 0$ such that $[H(u)](t,x)$ depends only on the restriction of u to $(t-\rho,t)$, for any t. Further, there exists $F_\infty \in L^2(\Omega)$ such that*

$$\lim_{T\to+\infty} \int_{T-1}^{T+1} \| F(t) - F_\infty \|_{L^2(\Omega)}\, dt = 0.$$

Then for any solution u of (1.2) in $(0,\infty)$ we have $\omega(u) \neq \emptyset$. If $u_\infty \in \omega(u)$ then there is a sequence $t_k \to \infty$ such that

$$U_k \to u_\infty \text{ in } L^p((-1,1)\times\Omega_0) \text{ for any bounded } \Omega_0 \subset \Omega$$
$$\text{and weakly in } L^p(-1,1;V) \cap L^2(-1,1;L^2(\Omega));$$

further, $u_\infty \in V \cap L^2(\Omega)$ is a solution of the stationary problem

$$\sum_{j=1}^n \int_\Omega f_j(x,u_\infty,\nabla u_\infty) D_j w\,dx + \int_\Omega f_0(x,u_\infty,\nabla u_\infty) w\,dx = \langle F_\infty, w\rangle$$

for all $w \in V \cap L^2(\Omega)$.

Remark 8. The operators H, defined in Remark 3 satisfy the assumptions of this theorem if $\beta_0(s,t,x) = 0$ for $s < t-\rho$, $t - \rho \leq \tau(t)$, respectively.

References

[1] D. Arcoya, J.I. Díaz and L. Tello. S-sharped bifurcation branch in a quasilinear multivalued model arising in climatology. *J. Diff. Eq.*, to appear.

[2] M. Badii, J.I. Díaz and M. Tesei. Existence and attractivity results for a class of degenerate functional-parabolic problems. *Rend. Sem. Mat. Univ. Padova*, **78** (1987), 109–124.

[3] J. Berkovits, V. Mustonen. Topological degree for perturbations of linear maximal monotone mappings and applications to a class of parabolic problems. *Rend. Mat. Ser. VII*, **12**, Roma (1992), 597–621.

[4] H. Brézis. *Opérateurs maximaux monotones et semi-groups de contractions dans les espaces de Hilbert.* North-Holland, Amsterdam, 1973.

[5] M. Brokate, J. Sprekels. *Hysteresis and Phase Transitions.* Springer, 1996.

[6] F.E. Browder. Pseudo-monotone operators and nonlinear elliptic boundary value problems on unbounded domains. *Proc. Natl. Acad. Sci. USA* **74** (1977), 2659–2661.

[7] J.I. Díaz, J. Hernández, L. Tello. On the multiplicity of equilibrium solutions to a nonlinear diffusion equation on a manifold arising in climatology. *J. Math. Anal. Appl.* **216** (1997), 593–613.

[8] J.I. Díaz, G. Hetzer. A quasilinear functional reaction-diffusion equation arising in climatology, in: *PDEs and Applications.* Dunod, Paris, 1998.

[9] J.L. Lions. *Quelques méthodes de résolution des problèmes aux limites non linéaires.* Dunod, Gauthier-Villars, Paris, 1969.

[10] G. Mahler. Nonlinear parabolic problems in unbounded domains. *Proc. Roy. Soc. Edinburgh Sect. A* **82** (1978/1979), 201–209.

[11] V. Mustonen. On pseudo-monotone operators and nonlinear parabolic initial-boundary value problems on unbounded domains. *Ann. Acad. Sci. Fenn. A. I. Math.* **6** (1981), 225–232.

[12] M. Renardy, W.J. Hrusa, J.A. Nohel. *Mathematical problems in viscoelasticity.* Monographs and Surveys in Pure and Applied Mathematics **35**, Longman Scientific and Technical – John Wiley and Sons, Inc., New York.

[13] L. Simon. On perturbations of initial-boundary value problems for nonlinear parabolic equations. *Annales Univ. Sci. Budapest, Sect. Comp.* **16** (1996), 319–341.

[14] L. Simon. On different types of nonlinear parabolic functional differential equations. *Pure Math. Appl.* **9** (1998), 181–192.

[15] L. Simon. On the stabilization of solutions of nonlinear parabolic functional differential equations. *Proceedings of the Conference Function Spaces, Differential Operators and Nonlinear Analysis,* Syöte, 1999, 239–250.

[16] W.A. Strauss. The energy method in nonlinear partial differential equations. *Notas Mat.* **47**, 1969.

[17] I.I. Vrabie. *Compactness methods for nonlinear evolutions.* Pitman Monographs and Surveys in Pure and Applied Mathematics **32**, Longman Scientific and Technical, Harlow, 1987.

[18] E. Zeidler. *Nonlinear functional analysis and its applications II A and II B.* Springer, 1990.

[19] F. Wu *Theory and applications of partial functional differential equations.* Springer, 1996.

László Simon
Department of Applied Analysis
L. Eőtvős University of Budapest
Pázmány Péter sétány 1/C
H-1117 Budapest, Hungary
E-mail address: simonl@ludens.elte.hu

D. Haroske, T. Runst, H.-J. Schmeisser (eds.): Function Spaces, Differential Operators and Nonlinear Analysis. The Hans Triebel Anniversary Volume.

Function Spaces in Presence of Symmetries: Compactness of Embeddings, Regularity and Decay of Functions

Leszek Skrzypczak

Dedicated to Prof. Hans Triebel on the occasion of his 65th birthday

Abstract. The article is a survey of results concerning radial subspaces of Besov and Lizorkin-Triebel spaces.

Embeddings are one of the crucial subjects in the theory of function spaces. Their boundedness and compactness have been studied widely and found different applications. Among conditions implying compactness the "presence of symmetries" seems to be that one which has attracted the least attention. But the families of radial and block radial functions appeared useful for nonlinear PDEs, so the spaces of functions invariant with respect to a compact group of rotations seem to be an interesting and important subject.

Variational methods of solution of partial differential equations usually require a certain type of compactness. This is needed to prove the existence of a solution. The compactness argument can be reduced to the question of compactness of embeddings of proper function spaces. The most classical result in that context in a theorem proved by V.I. Kondrachov in 1945. The theorem asserts that the Sobolev space $W^1_p(\Omega)$ defined on a bounded domain $\Omega \subset \mathbb{R}^n$ is compactly embedded into $L_q(\Omega)$

$$\begin{aligned} &\text{for} \quad 1 \le q < \frac{np}{n-p} && \text{if} \quad p < n \\ &\text{and for} \quad 1 \le q < \infty && \text{if} \quad p \ge n, \end{aligned}$$

cf. [16]. The case $p = q = 2$ was proved earlier by F. Rellich. But the Kondrachov theorem becomes useless if one looks for entire solutions of nonlinear PDE on $\mathbb{R}^n$. In that case, subspaces of Sobolev spaces consisting of functions with some symmetry properties are helpful. This is related to the special decay and smoothness properties of rotation invariant functions from Sobolev classes. The decay properties imply compactness of embeddings of radial Sobolev spaces $RW^1_p(\mathbb{R}^n)$ into $L_q(\mathbb{R}^n)$ if $1 \le p < q$ and $\frac{n}{p} - \frac{n}{q} < 1$.

The last phenomenon was noticed by W.A. Strauss for the radial functions from first order Sobolev classes in 1977, cf. [27]. He proved also the compactness of the Sobolev embedding for $p = 2$. The proof of the compactness is based on the following inequality describing the behaviour of a radial function $f \in W_2^1(\mathbb{R}^n)$ at infinity

$$|f(x)| \leq C_n \left\|f|L_2(\mathbb{R}^n)\right\|^{1/2} \left\|\nabla f|L_2(\mathbb{R}^n)\right\|_2^{1/2} |x|^{-(n-1)/2}, \quad n \geq 2, \quad x \neq 0. \tag{0.1}$$

A similar result was proved by S. Coleman, V. Glazer, A. Martin [3] and in a not explicit form by Berestycki and Lions [1].

In 1982 P.L. Lions published the next paper about the subject in which he proved the inequality analogous to the inequality (0.1) for $p \in [1, \infty)$, cf. [18]. In consequence he was able to prove the compactness of the embeddings for $p \neq 2$. Moreover, another type of symmetry, namely a block-radial (cylindrical) symmetry, is regarded there and the corresponding compactness theorem is proved. One other fact observed in this paper is the phenomenon of the higher local regularity of radial functions outside the origin. It was noticed that if $f \in W_p^1(\mathbb{R}^n)$ is radial and $\widetilde{f}(|x|) = f(x)$ then $\widetilde{f}$ is an element of the Hölder-Zygmund space $\mathcal{C}^{\frac{p-1}{p}}((0, \infty))$.

Both, radial and block-radial symmetries, appeared useful for nonlinear PDE. A nice presentation of the applications of the above results to semilinear elliptic equations can be found in the book of Kuzin and Pohozaev [17].

First results concerning the necessity of the conditions were obtained by Ebihara and Schonbek [4] in 1986. They showed that $RW_2^m(\mathbb{R}^n)$ is not compactly embedded in $L^2(\mathbb{R}^n)$ for any positive integer m. They also showed that if $1 \leq m < \frac{n}{2}$ then $RW_2^m(\mathbb{R}^n)$ is not compactly embedded into $L^q(\mathbb{R}^n)$ with $q = \frac{2n}{n-2m}$. The first result means that the integrability conditions are more important in the case of the embeddings of radial functions than in the case of embeddings of function spaces defined on bounded domains. Estimates of entropy numbers of the embeddings described later are quantitative versions of this rough remark.

Lions' result has found an natural counterpart for Sobolev spaces defined on n-dimensional Riemannian manifolds M. In that case one should regard a subspace of functions invariant with respect to an action of a compact group G of Riemannian isometries of the manifold M. In contrast to $\mathbb{R}^n$ such a group may have no fixed point. So one can prove the global version of the phenomenon of the higher regularity. Roughly speaking, if the functions possess enough symmetries, the Sobolev embeddings are valid in higher L_p spaces. In particular it was proved by Hebey and Vaugon that if M is compact and G has no finite orbits, then the first order Sobolev space $R_G W_p^1(M)$ consisting of functions invariant with respect to the action of G can be compactly embedded into $L_q(M)$ if $1 - \frac{n-k}{p} > \frac{k-n}{p}$, where k is the smallest dimension of the orbits of G in M. The authors regard also noncompact manifolds, cf. [14] and also [12] and [13].

This article is a survey of the results proved by the author and collaborators in the just described field. We work with the Besov and Lizorkin-Triebel function spaces. The main tool we used is the atomic decomposition method, which seems to be new in this context. On the one hand the method allows to work with a wide variety of function spaces, on the other hand it supports the construction of certain counterexamples. We do not want to go into technical details here, so we do not describe the decompositions, but we should mention that our method is just the Frazier-Jawerth approach [9] adapted to the presence of symmetries.

To make the picture more complete, two remarks should be made. The radial Besov and Lizorkin-Triebel spaces on $\mathbb{R}^n$ were investigated by Epperson and Frazier, cf. [8]. The authors construct the radial counterpart of the φ-transform but pay no attention for the problems of compactness of embeddings and improved regularities. One can also force the Sobolev embeddings on $\mathbb{R}^n$ to be compact by putting proper weights. This was investigated by many authors. We refer to the book of Edmunds and Triebel [6] and the survey by Haroske [10] for details.

1. Function spaces and groups

In the paper we deal with Besov and Lizorkin-Triebel scales of function spaces. To be concise we omit, probably well known, definitions. Instead of it, we recall the most prominent examples.

Let $s \in \mathbb{R}$, $0 < q \leq \infty$. Besov spaces $B^s_{p,q}(\mathbb{R}^n)$, $0 < p \leq \infty$ and Lizorkin-Triebel spaces $F^s_{p,q}(\mathbb{R}^n)$, $0 < p < \infty$, are natural generalizations of Sobolev and Hölder spaces, e.g. we have

$$
\begin{aligned}
F^0_{p,2}(\mathbb{R}^n) &= L_p(\mathbb{R}^n) && \text{(Lebesgue spaces)}\\
F^m_{p,2}(\mathbb{R}^n) &= W^m_p(\mathbb{R}^n) && \text{(Sobolev spaces)}\\
F^s_{p,2}(\mathbb{R}^n) &= H^s_p(\mathbb{R}^n) && \text{(Potential spaces)}
\end{aligned}
$$

if $1 < p < \infty$ and

$$
B^s_{\infty,\infty}(\mathbb{R}^n) = \mathcal{C}^s(\mathbb{R}^n) \qquad \text{(Hölder-Zygmund spaces)}
$$

if $s > 0$. The spaces are quasi-Banach spaces, Banach spaces if $p \geq 1$ and $q \geq 1$. For details and further information we refer to Frazier and Jawerth [9] or Triebel [30, 31], in particular also for the definition and various equivalent characterizations. Occasionally we also mention functions spaces not included in the above scales: the space of bounded continuous functions $C(\mathbb{R}^n)$, the space of functions with "local" bounded mean oscillation $\mathrm{bmo}(\mathbb{R}^n)$ and the space of functions with "local" vanishing mean oscillation $\mathrm{vmo}(\mathbb{R}^n)$. We refer once more to [30, 31] for definitions.

Let g be an isometry of $\mathbb{R}^n$. For $\varphi \in \mathcal{S}(\mathbb{R}^n)$ we put $\varphi^g(x) = \varphi(gx)$. If $f \in \mathcal{S}'(\mathbb{R}^n)$ then f^g is a tempered distribution defined by

$$
f^g(\varphi) = f(\varphi^{g^{-1}}), \quad \varphi \in \mathcal{S}(\mathbb{R}^n),
$$

where g^{-1} denote the isometry inverse to g.

Let $O(n)$ denote a real orthogonal group of $n \times n$ matrices. The group $O(n)$ acts on $\mathbb{R}^n$ as a group of linear isometries. The group is a compact Lie transformation group. So its action on $\mathbb{R}^n$ is smooth. By $SO(n)$ we denote the subgroup of $O(n)$ consisting of matrices with determinant equal to 1. It is a connected component of identity of $O(n)$.

Definition 1. *Let H be a subgroup of $O(n)$. We say that the tempered distribution f is invariant with respect to H if $f^h = f$ for any $h \in H$. For any possible s, p, q we put*

$$R_H B^s_{p,q}(\mathbb{R}^n) = \{f \in B^s_{p,q}(\mathbb{R}^n) : f \text{ is invariant with respect to } H\},$$
$$R_H F^s_{p,q}(\mathbb{R}^n) = \{f \in F^s_{p,q}(\mathbb{R}^n) : f \text{ is invariant with respect to } H\}.$$

Convention.

1. Only additional restrictions of p, q and s will be given. Lack of restrictions means that the assertion holds for all possible values.
2. We will write $A^s_{p,q}(\mathbb{R}^n)$ and $R_H A^s_{p,q}(\mathbb{R}^n)$ if we speak about the common properties of $B^s_{p,q}$ and $F^s_{p,q}$, or $R_H B^s_{p,q}$, $R_H F^s_{p,q}$.

Remark 1.

1. The space $R_H A^s_{p,q}(\mathbb{R}^n)$ is a closed subspace of $A^s_{p,q}(\mathbb{R}^n)$. This follows immediately from the continuity of the linear operators $T_h f = f^h - f$, $h \in H$, in $A^s_{p,q}(\mathbb{R}^n)$ and the representation $R_H A^s_{p,q}(\mathbb{R}^n) = \bigcap_{h \in H} \ker T_h$. Thus, they are quasi-Banach spaces with the norm induced from $A^s_{p,q}(\mathbb{R}^n)$. One can show that $R_H A^s_{p,q}(\mathbb{R}^n)$ is a complemented subspace of $A^s_{p,q}(\mathbb{R}^n)$ if $p, q \geq 1$, cf. [25].
2. The group $SO(n)$ acts transitively on the unit sphere $S^{n-1} = \{x \in \mathbb{R}^n : |x| = 1\}$ therefore the subspaces $R_{SO(n)} A^s_{p,q}(\mathbb{R}^n)$ consists of the radial distributions. It is the smallest nontrivial rotation invariant subspace of $A^s_{p,q}(\mathbb{R}^n)$. We will denote this subspace by $RA^s_{p,q}(\mathbb{R}^n)$. Thus, $R_H A^s_{p,q}(\mathbb{R}^n) = RA^s_{p,q}(\mathbb{R}^n)$ if $SO(n) \subset H$. These subspaces were investigated in [22].
3. Another example of rotation invariant subspaces are the subspaces which corresponds to block radial (cylindrical) symmetries. Let $n = n_1 + \cdots + n_m$ and
$$H = SO(n_1) \times \cdots \times SO(n_m)$$
with the action of $h = (h_1, \ldots, h_m) \in H$ on $\mathbb{R}^n$ defined by
$$h(x^{(1)}, \ldots, x^{(m)}) = (h_1(x^{(1)}), \ldots, h_m(x^{(m)})), x^{(i)} \in \mathbb{R}^{n_i}.$$
The H invariant subspaces of the first order Sobolev spaces were investigated in [18].
4. We will assume that H is a closed subgroup of $O(n)$. This assumption seems to be not very restrictive. The groups $SO(n)$ as well as $SO(n_1) \times \cdots \times SO(n_m)$ are of course closed subgroups of $O(n)$. Moreover, closed subgroups of real orthogonal groups $0(n)$ cover all compact Lie groups in that sense that any compact Lie group is isomorphic to a closed subgroup of a certain orthogonal group $O(n)$, [2, Chapter 0, Theorem 5.1].
5. We give another example of identification of the compact Lie group with the subgroup of orthogonal transformations that arise naturally in geometry. Let $X = G/K$ be a homogeneous Riemannian manifold with a Riemannian metric tensor g. Then G is the Lie group of isometries of X and K is a stability subgroup of a fixed point $o \in X$. The group K is compact and acts naturally

on the tangent space T_oX by tangent mappings $d(k)_o$, $k \in K$. The action is orthogonal on T_oX since $g_o(d(k)_o v, (dk)_o w) = g_o(v, w)$, $\quad v, w \in T_oX$.

2. Compactness of embeddings

2.1. Sobolev embeddings

From the point of view of the theory of embeddings the most important information about the space $A^s_{p,q}(\mathbb{R}^n)$ is contained in the number $s - \frac{n}{p}$ called a differential dimension. If $0 < p_0 \le p$ and if $s_0 - \frac{n}{p_0} \ge s - \frac{n}{p}$, then

$$F^{s_0}_{p_0,q_0}(\mathbb{R}^n) \hookrightarrow F^s_{p,q}(\mathbb{R}^n), \qquad 0 < q_0,\, q \le \infty \quad \text{(continuous embedding)} \tag{2.1}$$

and

$$B^{s_0}_{p_0,q_0}(\mathbb{R}^n) \hookrightarrow B^s_{p,q}(\mathbb{R}^n), \qquad 0 < q_0 \le q \le \infty \quad \text{(continuous embedding)}, \tag{2.2}$$

cf. Sickel and Triebel [23]. We shall prove that if we restrict our interest to the subspaces $R_H B^s_{p,q}(\mathbb{R}^n)$ and $R_H F^s_{p,q}(\mathbb{R}^n)$ of H-invariant functions (distributions) than most of the above embeddings become compact. In fact we are able to characterize all situations where those embeddings are compact.

Theorem 1 ([25]). *Let $H \subset O(n)$ be a closed subgroup of $O(n)$, $n \ge 2$. Let $R_H A^s_{p,q}(\mathbb{R}^n)$ be an H-invariant subspace of $A^s_{p,q}(\mathbb{R}^n)$. The embedding*

$$R_H A^{s_0}_{p_0,q_0}(\mathbb{R}^n) \hookrightarrow R_H A^{s_1}_{p_1,q_1}(\mathbb{R}^n)$$

is compact if and only if,

(i) *for any $x \in S^{n-1}$ the orbit $H \cdot x$ has infinite many elements,*

(ii) *$p_0 < p_1$ and $s_0 - \frac{n}{p_0} > s_1 - \frac{n}{p_1}$.*

Remark 2.

1. If $n = 1$ then the only invariant subspace is the subspace of even distributions. In that case the embeddings are not compact, cf. [22].
2. If $H = SO(n)$ then we deal with the radial distributions. In that case the embeddings are compact iff $p_0 < p_1$ and $s_0 - \frac{n}{p_0} > s_1 - \frac{n}{p_1}$. This was proved in [22].
3. In the case of the block radial symmetries the assumption (i) is equivalent to $n_i > 1$ for any i.

As direct consequences of the above theorem we get the following corollaries.

Corollary 1. *Let $0 < p, q \le \infty$. The following conditions are equivalent*

(i) *the embedding $R_H A^s_{p,q}(\mathbb{R}^n) \hookrightarrow \mathrm{vmo}(\mathbb{R}^n)$ is compact,*

(ii) *the embedding $R_H A^s_{p,q}(\mathbb{R}^n) \hookrightarrow \mathrm{bmo}(\mathbb{R}^n)$ is compact,*

(iii) *the embedding $R_H A^s_{p,q}(\mathbb{R}^n) \hookrightarrow C(\mathbb{R}^n)$ is compact,*

(iv) *the embedding $R_H A^s_{p,q}(\mathbb{R}^n) \hookrightarrow L_\infty(\mathbb{R}^n)$ is compact,*

(v) *$p < \infty$, $s - n/p > 0$ and for any $x \in S^{n-1}$ the orbit $H \cdot x$ has infinite many elements.*

Corollary 2. *Let $1 \le p_0, p_1 \le \infty$ and let $m \in \mathbb{N}$. Then the embedding*

$$R_H W^m_{p_0}(\mathbb{R}^n) \hookrightarrow L_{p_1}(\mathbb{R}^n) \tag{2.3}$$

is compact if, and only if, $p_0 < p_1$, $m - n/p_0 > p_1$ and for any $x \in S^{n-1}$ the orbit $H \cdot x$ has infinite many elements.

2.2. The Trudinger-Strichartz embeddings

In this section we assume that $1 < p_0, q_0$. The Sobolev embeddings imply that if $p_0 < p_1 < \infty$ and $s \ge \frac{n}{p_0} - \frac{n}{p_1}$ then

$$F^s_{p_0,q_0}(\mathbb{R}^n) \hookrightarrow L_{p_1}(\mathbb{R}^n)\,. \tag{2.4}$$

If $p_1 = \infty$ then the embedding (2.4) holds if $s > \frac{n}{p_0}$ but does not hold if $s = \frac{n}{p_0}$. So one can look for a larger target space in this case. The corresponding limiting embeddings are usually called Trudinger-Strichartz embeddings but Trudinger's work had some forerunners written by Yudovich and Pohozaev. The new target spaces are the "exponential" Orlicz spaces. We first recall the basic notations concerning these spaces.

For $\nu \ge 1$ and $1 < r < \infty$ we define the Orlicz function $\Phi_{\nu,r} : [0,\infty) \to \mathbb{R}$ by

$$\Phi_{\nu,r}(t) = t^\nu \exp(t^r)\,.$$

The Orlicz space $E_{\nu,r}(\mathbb{R}^n)$ generated by the Orlicz function $\Phi_{\nu,p}$ is the set of all measurable functions $f : \mathbb{R}^n \to \mathbb{C}$ (functions equal almost everywhere being identified) such that

$$\varrho_{\nu,r}(\lambda f) \;=\; \int_\Omega \Phi_{\nu,r}(\lambda|f(x)|)dx < \infty, \tag{2.5}$$

for some positive real λ. Furnished with the Luxemburg norm

$$\|f|E_{\nu,p}\| = \inf\left\{\lambda > 0 : \varrho_{\nu,p}(\lambda^{-1} f) \le 1\right\}\,,$$

the Orlicz space is a Banach space. Using standard methods one can easily prove that

$$\begin{aligned} E_{\nu,r}(\mathbb{R}^n) &\hookrightarrow E_{\mu,r}(\mathbb{R}^n)\,, \qquad \text{for } \nu \le \mu\,,\\ E_{\nu,r}(\mathbb{R}^n) &\hookrightarrow E_{\nu,q}(\mathbb{R}^n)\,, \qquad \text{for } r \ge q\,. \end{aligned}$$

and that the expression

$$\sup_{j\in\mathbb{N}} j^{-1/r} \|f|L_{\nu+j}(\mathbb{R}^n)\|$$

is an equivalent norm in $E_{\nu,r}(\mathbb{R}^n)$. This "extrapolating" norm is of great importance for studying limiting embeddings.

Theorem 2. *Let $1 < p < \infty$ and $1 < q < \infty$.*

(1) *The embeddings*

$$B^{n/p}_{p,q}(\mathbb{R}^n) \hookrightarrow E_{\nu,r}(\mathbb{R}^n)$$

are continuous if and only if $\nu \ge p$ and $r \le q'$.

(2) *The embeddings*

$$F^{n/p}_{p,q}(\mathbb{R}^n) \hookrightarrow E_{\nu,r}(\mathbb{R}^n)$$

are continuous if and only if $\nu \geq p$ *and* $r \leq p'$.

Remark 3.

1. Trudinger [32] and Strichartz [28] worked with Sobolev spaces on bounded domains. The $\mathbb{R}^n$ version is due essentially to Peetre [20]. Our presentation is similar to that one given by Edmunds and Triebel in [5] and [29].
2. Usually the spaces $B^{n/p}_{p,p}(\mathbb{R}^n)$ and $H^{n/p}_p(\mathbb{R}^n)$ are used in literature in this context. The generalization to $F^{n/p}_{p,q}(\mathbb{R}^n)$ follows from a result of Netrusov, cf. [19]. The Besov case with $q \neq p$ is also well known for specialists. A proof can be found in [26].

Theorem 3 ([26]). *Let* $H \subset O(n)$ *be a compact subgroup of* $O(n)$. *Let* $1 \leq \nu < \infty$ *and* $1 < p, q < \infty$.

(1) *The embeddings*

$$R_H F^{n/p}_{p,q}(\mathbb{R}^n) \hookrightarrow E_{\nu,r}(\mathbb{R}^n), \tag{2.6}$$

are compact if and only if

(i) *any orbit* $H \cdot x$, $x \in S^{n-1}$ *has infinite many elements,*

(ii) $r < p'$ *and* $p < \nu$.

The embeddings (2.6) *are continuous but not compact if* $p = \nu$ *and* $r = p'$. *If* $p' < r$ *or* $p < \nu$ *then the spaces* $R_H F^{n/p}_{p,q}(\mathbb{R}^n)$ *are not contained in* $E_{\nu,r}(\mathbb{R}^n)$.

(2) *The embeddings*

$$R_H B^{n/p}_{p,q}(\mathbb{R}^n) \hookrightarrow E_{\nu,r}(\mathbb{R}^n), \tag{2.7}$$

are compact if and only if

(i) *any orbit* $H \cdot x$, $x \in S^{n-1}$ *has infinite many elements,*

(ii) $r < q'$ *and* $p < \nu$.

The embeddings (2.7) *are continuous but not compact if* $p = \nu$ *and* $r = q'$. *If* $p' < r$ *or* $p < \nu$ *then the spaces* $R_H B^{n/p}_{p,q}(\mathbb{R}^n)$ *are not contained in* $E_{\nu,q}(\mathbb{R}^n)$.

3. Decay, regularity and traces of radial functions

Symmetries have strong influence to decay properties of the function and its local smoothness (out of the origin). First we formulate a generalized version of the Strauss lemma, cf. [27] or [17, Chap. 2 § 8]. Here the approach via atomic decompositions has the additional advantage that it gives detailed hints about the sharpness. We describe the results for radial functions.

Theorem 4 (Behavior near infinity, cf. [22]). *Let* $0 < p \le \infty$.

(i) *Let either* $s > 1/p$ *and* $0 < q \le \infty$ *or* $s = 1/p$ *and* $0 < q \le 1$. *Then there exists a constant* C *such that*

$$|f(x)| \le C \, \| f \,|B^s_{p,q}(\mathbb{R}^n)\| \; |x|^{\frac{1-n}{p}} \tag{3.1}$$

holds for all $f \in RB^s_{p,q}(\mathbb{R}^n)$ *and all* $|x| \ge 1$.

(ii) *Let* $(n-1)/n < p$. *Further, let either* $s < 1/p$ *and* $0 < q \le \infty$ *or* $s = 1/p$ *and* $1 < q \le \infty$. *Then for all* $|x| \ge 1$ *there exists a sequence* $\{f_N\}_{N=1}^{\infty}$ *of smooth and compactly supported radial functions (depending on* x*) such that* $\| f_N \,|B^s_{p,q}\| = 1$ *and* $\lim_{N\to\infty} |f_N(x)| = \infty$.

(iii) *Let* $(n-1)/n < p$. *Then for all triples* (s,p,q) *there exists a positive constant* C *such that for all* $|x| \ge 1$ *there exists a smooth and compactly supported function* $f \in RB^s_{p,q}(\mathbb{R}^n)$, $\| f \,|B^s_{p,q}(\mathbb{R}^n)\| = 1$ *(depending on* x*) such that*

$$|f(x)| \ge C \, |x|^{\frac{1-n}{p}} \, .$$

Theorem 5 (Behavior near the origin, cf. [22]). *Let* $0 < p \le \infty$.

(i) *Let either* $s > 1/p$ *and* $0 < q \le \infty$ *or* $s = 1/p$ *and* $0 < q \le 1$. *Then there exists a constant* C *such that*

$$|f(x)| \le C \, \| f \,|B^s_{p,q}(\mathbb{R}^n)\| \; |x|^{\frac{1-n}{p}} \tag{3.2}$$

holds for all $f \in RB^s_{p,q}(\mathbb{R}^n)$ *and all* $0 < |x| \le 1$. *Moreover, if* $n \ge 2$ *and if* $p < \infty$ *we even have*

$$\lim_{|x|\to 0} |x|^{\frac{n-1}{p}} f(x) = 0 \, . \tag{3.3}$$

(ii) *Let* $s < 1/p$ *and* $0 < q \le \infty$. *There exist a sequence of points* $\{x_N\}_{N=1}^{\infty}$ *in* $\mathbb{R}^n$ *and a corresponding sequence of smooth and compactly supported radial functions* f_N *such that* $|x_N| \to 0$, $\| f_N \,|B^s_{p,q}\| = 1$ *and*

$$|f_N(x_N)| \ge N \, |x_N|^{\frac{1-n}{p}} \, . \tag{3.4}$$

Remark 4.

1. Singular distributions in $RB^s_{p,q}(\mathbb{R}^n)$. Let either $s > 1/p$ or $s = 1/p$ and $0 < q \le 1$. Let us denote by $\psi \in C_0^\infty(\mathbb{R}^n)$ a radial cut-off function supported around the origin and satisfying $\psi(x) = 1$ if $|x| \le 1$. Then (3.1) tells us that $f \in RB^s_{p,q}(\mathbb{R}^n)$ implies that $(1 - \psi(\lambda x))\, f(x)$ is a regular distribution in $S'(\mathbb{R}^n)$ for all $\lambda > 0$. That is in some contrast to the general theory on $\mathbb{R}^n$, cf. [23].
2. Inequalities similar to (3.1) and (3.2) hold also for any H-invariant function, if H has no finite orbits out of origin. In that case we have the exponent $-\frac{k}{p}$, $k = \min_{x\neq 0} \dim H \cdot x$, on the right-hand side of the inequalities, cf. [25].
3. The p-version of the Strauss lemma given by P.L. Lions [18] follows now from $W^1_p(\mathbb{R}^n) \hookrightarrow B^{1/p}_{p,1}(\mathbb{R}^n)$, if $p > 1$. However, our result here does not cover the case $p = 1$, because of $B^1_{1,1}(\mathbb{R}^n) \hookrightarrow W^1_1(\mathbb{R}^n)$.

The improved local regularity of radial functions may be seen from the following.

Theorem 6 ([22]). *Suppose* $s > 1/p$. *Let* $f \in RB^s_{p,q}(\mathbb{R}^n)$ *and* $f_0(|x|) = f(x)$. *Then* f_0 *belongs to* $B^{s-1/p}_{\infty,\infty}(0,\infty)$. *In particular, if* $0 < s - 1/p < 1$, *then for each* $\varepsilon > 0$ *there exists a constant* C_ε *such that*

$$\sup_{\varepsilon<|x|<|y|} \frac{|f(x) - f(y)|}{|x-y|^{s-1/p}} < C_\varepsilon \,. \tag{3.5}$$

Remark 5. The H-invariant version of higher regularity looks as follows. Let ψ be a radial smooth cut-off function supported around the origin and satisfying $\psi(x) = 1$ if $|x| \leq 1$. If $s > \frac{n-k}{p}$ and $f \in R_H B^s_{p,q}$ then $(1-\psi)f \in B^\sigma_{\infty,\infty}(\mathbb{R}^n)$ with $\sigma = s - \frac{n-k}{p}$, cf. [25].

At the end of this section we describe traces for the radial subspaces of the Besov classes. We define

$$\mathrm{tr}\ : f(x_1, x_2, \ldots, x_n) \longrightarrow f(t, 0, \ldots, 0)\,, \qquad |x_1| = t\,.$$

To describe the traces we need some weighted function spaces with polynomial . weights

$$w_\alpha(x) = (1 + |x|^2)^{\alpha/2}, \qquad x \in \mathbb{R}^n\,, \quad \alpha \in \mathbb{R}\,.$$

For $0 < t < \infty$, $1 \leq p, q \leq \infty$, and s, $\alpha \in \mathbb{R}$ we put

$$\begin{aligned} B^s_{p,q}\big(\mathbb{R}^n, w_\alpha, (t,\infty)\big) &= \big\{ f \in B^s_{p,q}(\mathbb{R}^n, w_\alpha) \,:\, \operatorname{supp} f \subset \{x \in \mathbb{R}^n : |x| \geq t\} \big\}\,, \\ RB^s_{p,q}\big(\mathbb{R}^n, (t,\infty)\big) &= \big\{ f \in RB^s_{p,q}(\mathbb{R}^n) \,:\, \operatorname{supp} f \subset \{x \in \mathbb{R}^n : |x| \geq t\} \big\}\,. \end{aligned}$$

All types of spaces will be equipped with the natural norm.

Theorem 7 ([15]). *Suppose* $n \geq 2$, $0 < t < \infty$ *and* $1 \leq p, q \leq \infty$.

If $s > 0$ *then the operator* tr *is an isomorphism of* $RB^s_{p.q}(\mathbb{R}^n)\big(\mathbb{R}^n, (t,\infty)\big)$ *onto* $B^s_{p,q}\big(\mathbb{R}_+, w_{(n-1)/p}, (t,\infty)\big)$.

Remark 6. The theorem means that there exists a linear continuous extension operator, "ext".

4. Invariant function spaces on manifolds

We mention shortly the invariant function spaces on manifolds. We restrict our attention to compact manifolds. In that case we can define the function spaces on the manifold using any finite atlas. No additional structure (Riemannian metric, group structure) is needed. We refer to [31] for definition and details of Lizorkin-Triebel spaces and Besov spaces on manifolds.

Let M be a compact manifold. We equip M with the Riemannian metric g. Let H be a compact group of isometries of (M, g). Each such group is a Lie transformation group so we can define subspaces $R_H F^s_{p,q}(M)$ and $R_H B^s_{p,q}(M)$ of $F^s_{p,q}(M)$ and $B^s_{p,q}(M)$, respectively in the similar way to the Euclidean case.

All standard embeddings concerning the spaces of $B^s_{p,q}-F^s_{p,q}$ type on bounded domains of $\mathbb{R}^n$ have their counterparts for compact manifolds. In particular we have a monotonicity with respect to p, i.e. if $p_1 < p_0$ then

$$B^s_{p_0,q}(M) \hookrightarrow B^s_{p_1,q}(M) \qquad \text{and} \qquad F^s_{p_0,q_0}(M) \hookrightarrow F^s_{p_1,q}(M)\,.$$

In contrast to the Euclidean case the action of the group H on M may have no fixed point. This can improve the embeddings between function spaces of H-invariant distributions. Namely, the embeddings may hold with a better smoothness. The sufficient and necessary conditions are described in the following theorem.

Theorem 8 ([24]). *Let M be a compact connected Riemannian manifold of dimension $n = \dim M > 1$. Let H be a closed group of isometries of M. We put $h^* = \min_{x\in M} \dim H \cdot x$. Let $s_1 < s_0$.*

(a) *The embedding*

$$R_H B^{s_0}_{p_0,q_0}(M) \hookrightarrow R_H B^{s_1}_{p_1,q_1}(M) \tag{4.1}$$

is continuous if either $s_0 - s_1 > (n-h^*)\left(\frac{1}{p_0}-\frac{1}{p_1}\right)_+$ or $s_0 - s_1 = (n-h^*)\left(\frac{1}{p_0}-\frac{1}{p_1}\right)_+$ and $q_1 \geq q_0$. The embedding

$$R_H F^{s_0}_{p_0,q_0}(M) \hookrightarrow R_H F^{s_1}_{p_1,q_1}(M) \tag{4.2}$$

is continuous if

$$s_0 - s_1 \geq (n-h^*)\left(\frac{1}{p_0}-\frac{1}{p_1}\right)_+ .$$

If

$$s_0 - s_1 < (n-h^*)\left(\frac{1}{p_0}-\frac{1}{p_1}\right)_+$$

there are no continuous embeddings in both cases.

(b) *The embeddings* (4.1) *and* (4.2) *are compact if and only if $s_0 - s_1 > (n-h^*)\left(\frac{1}{p_0}-\frac{1}{p_1}\right)_+$.*

Remark 7. If a paracompact manifold M is not compact then it can be equipped with the Riemannian metric with bounded geometry. Using such a Riemannian structure one can define Lizorkin-Triebel spaces by uniform localization and afterwards Besov spaces by real interpolation, cf. [31]. For any compact group of Riemannian isometries one defines the corresponding scales of H-invariant function spaces. One may guess that in this case both phenomenons, the improved regularity and compactness of embeddings take place. Three simple models explain that this will be not the case in general. Consider the two-dimensional surfaces of resolution in $\mathbb{R}^n$: a cylinder, a paraboloid, a hyperboloid, regarded as a Riemannian manifold with the group of isometries defined by a rotations with respect to the axis of symmetry. In the first case we have improved regularity of embeddings but they are not compact. In the second case the embeddings are compact but no improved regularity holds. In the third case we have both improved regularity of embeddings and their compactness. The details can be found in [24].

5. Entropy numbers

The k-th entropy number of a bounded linear operator $T : X \to Y$ between Banach spaces id defined as

$$e_k(T) = \inf\{\varepsilon > 0 : \exists y_1, \ldots, y_{2^{k-1}} \in Y, \ T(B_X) \subset \bigcup_{j=1}^{2^{k-1}} (y_j + \varepsilon B_Y)\},$$

where B_X and B_Y stand for the closed unit balls in X and Y, respectively. The operator T is compact iff $e_k(T) \to 0$ if $k \to \infty$. So the asymptotic behaviour of the entropy numbers gives us the quantitative knowledge about the compactness of the operator and can be used, for example, in investigations of spectral properties. For basic properties of entropy numbers we refer to [6] and to the literature mentioned there.

Let Ω be a bounded domain in $\mathbb{R}^n$. It was proved by Edmunds and Triebel that if the embedding id : $A^{s_0}_{p_0,q_0}(\Omega) \mapsto A^{s_1}_{p_1,q_1}(\Omega)$ is compact then $e_k(\mathrm{id}) \sim k^{(s_0-s_1)/n}$, cf. [6]. In certain contrast to this estimate we have the following estimates of entropy numbers of embeddings of radial subspaces.

Theorem 9 ([15]). *Suppose* $1 \leq p_0 < p_1 \leq \infty$, $1 \leq q_0, q_1 \leq \infty$, *and* $s_0 - s_1 - n(\frac{1}{p_0} - \frac{1}{p_1}) > 0$. *Then, using the abbreviation*

$$\beta = n(\frac{1}{p_0} - \frac{1}{p_1})$$

we have the estimates

$$c\, k^{-\beta} \leq e_k\big(\mathrm{id} : RA^{s_0}_{p_0,q_0}(\mathbb{R}^n) \mapsto RA^{s_1}_{p_1,q_1}(\mathbb{R}^n)\big) \leq C\, k^{-\beta}, \tag{5.1}$$

for some constants c and C independent of k.

Remark 8.
1. We recall that in case of compact Sobolev embeddings between spaces defined on bounded domains the asymptotic behaviour of the entropy numbers depends on a difference of smoothness. Namely

$$e_k\big(\mathrm{id} : A^{s_0}_{p_0,q_0}(\Omega) \mapsto A^{s_1}_{p_1,q_1}(\Omega)\big) \sim k^{-\frac{s_0-s_1}{n}},$$

So, the radial case is somewhat different.

2. Theorem 7 suggests the strong connection between the spaces of invariant functions and weighted function spaces. We finish this paper with sketching relations between Theorem 9 and estimates of the asymptotic behaviour of entropy numbers of embeddings of weighted spaces. A number of result in this direction has been proved in [11] and [6].

Let $1 \le p_0 < p_1$, $s_0 > s_1$ and $\delta := s_0 - s_1 + \frac{1}{p_1} - \frac{1}{p_0} > \alpha$. We regard the embedding of the weighted function space $B^{s_0}_{p_0,q_0}(\mathbb{R}, w_\alpha)$ into $B^{s_1}_{p_1,q_1}(\mathbb{R})$, $1 \le q_0, q_1 \le \infty$. By standard arguments we can divide our considerations into two parts: near the origin and near infinity. Entropy numbers of the part near the origin can be estimated from above by $k^{-(s_0-s_1)}$. To estimate the part near infinity one can use Theorem 9 and Theorem 7. If $s_1 > 0$ then the following diagram is commutative

$$\begin{array}{ccc} B^{s_0}_{p_0,q_0}(\mathbb{R}_+, w_{(n-1)/p_0}, (t,\infty)) & \xrightarrow{\text{id}} & B^{s_1}_{p_1,q_1}(\mathbb{R}_+, w_{(n-1)/p_1}, (t,\infty)) \\ \text{ext} \downarrow & & \text{tr} \uparrow \\ RB^{s_0}_{p_0,q_0}(\mathbb{R}^n, (t,\infty)) & \xrightarrow{\text{id}} & RB^{s_1}_{p_1,q_1}(\mathbb{R}^n, (t,\infty)) \, . \end{array}$$

So, if

$$s_0 - s_1 + \frac{1}{p_1} - \frac{1}{p_0} > \frac{n-1}{p_0} - \frac{n-1}{p_1}$$

then

$$e_k\big(\text{id}: B^{s_0}_{p_0,q_0}(\mathbb{R}, w_{(n-1)/p_0}, (t,\infty)) \mapsto B^{s_1}_{p_1,q_1}(\mathbb{R}, w_{(n-1)/p_1}, (t,\infty))\big) \le c k^{-n(\frac{1}{p_0} - \frac{1}{p_1})}.$$

The part near infinity is dominating. Therefore, using standard relations between weighted function spaces we get

$$e_k\big(\text{id} : B^{s_0}_{p_0,q_0}(\mathbb{R}, w_\alpha) \mapsto B^{s_1}_{p_1,q_1}(\mathbb{R})\big) \le C\, k^{-\alpha + \frac{1}{p_1} - \frac{1}{p_0}}, \qquad \alpha = \frac{n-1}{p_0} - \frac{n-1}{p_1}.$$

Using the isomorphic properties of the scales we can remove the restriction $s_1 > 0$. The lower estimates for the entropy numbers are known, cf. [6], so we get

$$e_k\big(\text{id} : B^{s_0}_{p_0,q_0}(\mathbb{R}, w_\alpha) \mapsto B^{s_1}_{p_1,q_1}(\mathbb{R})\big) \sim k^{-\alpha + \frac{1}{p_1} - \frac{1}{p_0}}, \quad \alpha = (n-1)\Big(\frac{1}{p_0} - \frac{1}{p_1}\Big),$$

$n = 2, 3, \ldots$. This improves somewhat the estimates from above given in [6], where an additional term of logarithmic order appeared.

References

[1] B. Berestycki and P.L. Lions, *Existence of a ground state in nonlinear equations of the type Klein-Gordon.* In: Variational Inequalities and complementary theory and applications, Wiley, New York 1979.

[2] G.L. Bredon, *Introduction to compact transformation groups*, Academic Press, New York-London, 1972.

[3] S. Coleman, V. Glazer, and A. Martin, *Action minima among solutions to a class of euclidean scalar field equations,* Comm. in Math. Physics **58** (1978), 211–221.

[4] Y. Ebihara and T.P. Schonbek, *On the (non)compactness of the radial Sobolev spaces,* Hiroshima Math. J. **16** (1986), 665–669.

[5] D.E. Edmunds, R.M. Edmunds and H. Triebel, *Entropy numbers of embeddings of fractional Besov-Sobolev spaces in Orlicz spaces,* J. London Math. Soc. **35** (1987), 121–134.

[6] D.E. Edmunds and H. Triebel, *Function spaces, entropy numbers and differential operators.* Cambridge Univ. Press, Cambridge 1996.

[7] D.E. Edmunds and H. Triebel, *Sharp Sobolev embeddings and related Hardy inequalities: The critical case,* Math. Nachr. **207** (1999), 79–92.

[8] J. Epperson and M. Frazier, *An almost orthogonal radial wavelet expansion for radial distributions,* J. Fourier Analysis Appl. **1** (1995), 311–353.

[9] M. Frazier and B. Jawerth, *A discrete transform and decomposition of distribution spaces,* J. Funct. Anal. **93** (1990), 34–170.

[10] D. Haroske, *Embeddings of some weighted function spaces on* $\mathbb{R}^n$*: entropy and approximation numbers*, An. Univ. Craiova, Ser. Mat. Inform **26** (1997), 1–44.

[11] D. Haroske, H. Triebel, *Entropy numbers in weighted function spaces and eigenvalues distribution of some degenerated pseudodifferential operators, I.* Math. Nachr. **167** (1994), 131–156.

[12] E. Hebey, *Nonlinear analysis on manifolds: Sobolev spaces and inequalities.* AMS, Providence, Rhode Island 1999.

[13] E. Hebey, *Sobolev spaces on Riemannian Manifolds.* LNM 1635, Springer, Berlin 1996.

[14] E. Hebey and M. Vaugon, *Sobolev spaces in the presence of symmetries*, J. Math. Pures Appl. **76** (1997), 859–881.

[15] Th. Kühn, H.-G. Leopold, W. Sickel, L. Skrzypczak, *Entropy numbers of Sobolev embeddings of radial Besov spaces*, (submitted).

[16] V.I. Kondrachov, *Certain properties of functions in the spaces* L^p, Dokl. Akad. Nauk SSSR **48** (1945), 563–566.

[17] I. Kuzin and S. Pohozaev, *Entire solutions of semilinear elliptic equations.* Birkhäuser, Basel, 1997.

[18] P.-L. Lions, *Symétrie et compacité dans les espaces de Sobolev*, J. Funct. Anal. **49** (1982), 315–334.

[19] Y.V. Netrusov, *Spaces of singularities of functions in spaces of Besov and Lizorkin-Triebel type,* Trudy Mat. Ins. Stekl. **187** (1989), 162–177 (Russian) [English translation: Proc. Steklov Inst. Math. **3** (1990), 185–203].

[20] J. Peetre, *Espaces d'Interpolation et Théorème de Soboleff*, Ann. Inst. Fourier **16** (1966), 279–317.

[21] T. Runst, W. Sickel, *Sobolev spaces of Fractional Order, Nemytskij Operators and Nonlinear Partial Differential Equations.* De Gruyter, Berlin, New York, 1996.

[22] W. Sickel and L. Skrzypczak, *Radial subspaces of Besov and Lizorkin-Triebel Classes: Extended Strauss Lemma and Compactness of Embedding*, J. Fourier Anal. App. **6** (2000), 639–662.

[23] W. Sickel and H. Triebel, *Hölder Inequality and Sharp Embeddings in Function Spaces of* B^s_{pq} *and* F^s_{pq} *type,* Z. Anal. Anwendungen **14** (1995), 105–140.

[24] L. Skrzypczak, *Optimal atomic decompositions on manifolds and compactness of Sobolev embeddings,* (to appear).

[25] L. Skrzypczak, *Rotation invariant subspaces of Besov and Triebel-Lizorkin space: compactness of embeddings, smoothness and decay properties*, Revista Mat. Iberoamer. (to appear).

[26] L. Skrzypczak, B. Tomasz, *Compactness of embeddings of the Trudinger-Strichartz type for rotation invariant functions,* Houston J. Math. **27** (2001), 633–647.

[27] W.A. Strauss,*Existence of solitary waves in higher dimensions,* Comm. in Math. Physics **55** (1977), 149–162.

[28] R.S. Strichartz, *A note on Trudinger's extension of Sobolev's inequality,* Indiana Univ. Math. J. **58** (1972), 841–842.

[29] H. Triebel, *Approximation numbers and entropy numbers of embeddings of fractional Besov-Sobolev spaces in Orlicz spaces,* Proc. London Math. Soc. **66** (1993), 589–618.

[30] H. Triebel,*Theory of Function Spaces.* Birkhäuser, Basel, 1983.

[31] H. Triebel, *Theory of Function Spaces II.* Birkhäuser, Basel, 1992.

[32] N. Trudinger, *On embedding into Orlicz spaces and some applications,* J. Math. Mech. **17** (1967), 473–483.

Leszek Skrzypczak
Faculty of Mathematics and Computer Science,
Adam Mickiewicz University
Matejki 48–49
60-769 Poznań, Poland
E-mail address: `lskrzyp@amu.edu.pl`

Participants FSDONA–01

Abdolaziz Abdollahi
Department of Mathematics
College of Sciences, Shiraz University
Shiraz 71454, Iran
abdollahi@math.susc.ac.ir

Arsen G. Bagdasaryan
Yerevan State University
st. A. Manukyan 1
Yerevan, 375049, Armenia
angen@arminco.com

António Bento
Centre for Mathematical Analysis
and its Applications
University of Sussex at Brighton
Falmer
Brighton BN1 9QH, U.K.
a.j.g.bento@sussex.ac.uk

Günter Berger
Mathematisches Institut
Fakultät für Mathematik & Informatik
Augustusplatz 10/11
D-04109 Leipzig, Germany
berger@mathematik.uni-leipzig.de

Marco Paolo Bernardi
Dipartimento di Matematica
Università di Pavia
Via Ferrata, 1
I-27100 Pavia, Italy
bernardi@dragon.ian.pv.cnr.it

Oleg V. Besov
Steklov Mathematical Institute RAN
Gubkina 8
117 966 Moscow GSP-1, Russia
besov@mi.ras.ru

Cristiana Bondioli
Dipartimento di Matematica
Università di Pavia
Via Ferrata, 1
I-27100 Pavia, Italy
bernardi@dragon.ian.pv.cnr.it

Jiří Bouchala
VŠB Technická univerzita Ostrava KAM
17. listopadu
708 33 Ostrava-Poruba, Czech Republic
jiri.bouchala@vsb.cz

Gérard Bourdaud
Institut de Mathématiques de Jussieu
Equipe d'Analyse Fonctionnelle
Université Paris 6
4, place Jussieu
75252 Paris Cedex 05, France
bourdaud@ccr.jussieu.fr

Michele Bricchi
Mathematisches Institut
Fakultät für Mathematik & Informatik
Friedrich-Schiller-Universität Jena
D-07740 Jena, Germany
mbricchi@minet.uni-jena.de

Richard C. Brown
Department of Mathematics
Box 870350
The University of Alabama
Tuscaloosa, Alabama 35487-0350, U.S.A.
dbrown@gp.as.ua.edu

Viktor I. Burenkov
University of Wales
College of Cardiff, School of Mathematics
23 Senghenydd Road
Cardiff CF24 4YH, Wales, U.K.
Burenkov@cardiff.ac.uk

António Caetano
Departamento de Matemática
Universidade de Aveiro
3810-193 Aveiro, Portugal
acaetano@mat.ua.pt

Bernd Carl
Mathematisches Institut
Fakultät für Mathematik & Informatik
Friedrich-Schiller-Universität Jena
D-07740 Jena, Germany
carl@minet.uni-jena.de

Maria J. Carro
Departament de Matemàtica
Aplicada i Anàlisi
Facultat de Matemàtiques
Universitat de Barcelona
Gran Via de les Corts Catalanes 585
08007 Barcelona, Spain
carro@mat.ub.es

Joan Cerdà
Departament de Matemàtica
Aplicada i Anàlisi
Facultat de Matemàtiques
Universitat de Barcelona
Gran Via de les Corts Catalanes 585
08007 Barcelona, Spain
cerda@cerber.mat.ub.es

Andrea Cianchi
Instituto di Matematica
Facoltà di Architettura
Università di Firenze
Via dell'agnolo 14
50122 Firenze, Italy
cianchi@cesit1.unifi.it

Sergey Dachkovski
Mathematisches Institut
Fakultät für Mathematik & Informatik
Friedrich-Schiller-Universität Jena
D-07740 Jena, Germany
dsn@minet.uni-jena.de

Stephan Dahlke
Universität Bremen
Fachbereich 3, ZeTeM
Postfach 330 440
D-28359 Bremen, Germany
dahlke@math.uni-bremen.de

Chiara De Fabritiis
Dipartimento di Matematica
Università di Ancona
Via Brecce Bianche
60131 Ancona, Italy
fabritii@dipmat.unian.it

Djairo G. De Figueiredo
IMECC, UNICAMP
Caixa Postal 6065
13081-970 Campinas, S.P. Brazil
djairo@ime.unicamp.br

Ron De Vore
Department of Mathematics
University of South Carolina
Columbia, SC 29208, U.S.A.
devore@math.sc.edu

Sophie Dispa
Université de Liège
Institut de Mathématique
Grande Traverse, 12, Bt. B37
B-4000 Liège (Sart-Tilman), Belgium
sdispa@ulg.ac.be

Pavel Drábek
Faculty of Applied Sciences
University of West Bohemia
Univerzitni 22, P.O. Box 314
30614 Plzeň, Czech Republic
pdrabek@kma.zcu.cz

David E. Edmunds
Centre for Mathematical Analysis
and its Applications
University of Sussex at Brighton
Falmer
Brighton BN1 9QH, U.K.
D.E.Edmunds@sussex.ac.uk

William Desmond Evans
University of Wales
College of Cardiff, School of Mathematics
Senghenydd Road
Cardiff CF24 4YH, Wales, U.K.
EvansWD@cardiff.ac.uk

Walter Farkas
Mathematisches Institut
Universität München
Theresienstraße 39
D-80333 München, Germany
farkas@rz.mathematik.uni-muenchen.de

Ji Gao
Department of Mathematics
Community College of Philadelphia
1700 Spring Garden Street
Philadelphia, PA 19130-3991, U.S.A.
jgao@ccp.cc.pa.us

Gerardo Emilio García Almeida
40 Plasnewydd Road
Cardiff CF24 3GN, Wales, U.K.
almeidage@cardiff.ac.uk

Petr Girg
Faculty of Applied Sciences
University of West Bohemia
Univerzitni 22, P.O. Box 314
30614 Plzeň, Czech Republic
pgirg@kma.zcu.cz

Mikhail L. Gol'dman
People's Friendship University of Russia
Miklucho-Maklaya street 6
117198 Moscow, Russia
seulydia@glas.apc.org

Amiran Gogatishvili
Mathematical Institute
Academy of Sciences
Žitna 25
11567 Praha 1, Czech Republic
gogatish@math.cas.cz

Karlheinz Gröchenig
Institut für Mathematik, NUHAG
Universität Wien
Strudlhofgasse 4
A-1090 Wien, Austria
groch@tyche.mat.univie.ac.at

Vagif Guliev
Faculty of Mechanics and Mathematics
Baku State University
L. Khalihor str. 23
370 145 Baku, Azerbaijan
guliev@azdata.net

Petr Gurka
Department of Mathematics
Czech University of Agriculture
16521 Praha 6, Czech Republic
gurka@tf.czu.cz

A. Turan Gürkanli
Ondokuzmayis University
Faculty of Art and Sciences
Department of Mathematics
55139, Kurupelit, Samsun, Turkey
gurkanli@samsun.omu.edu.tr

Petteri Harjulehto
Department of Mathematics
P.O. Box 4, Yliopistokatu 5
FIN-00014 University of Helsinki, Finland
petteri.harjulehto@helsinki.fi

Dorothee D. Haroske
Mathematisches Institut
Fakultät für Mathematik & Informatik
Friedrich-Schiller-Universität Jena
D-07740 Jena, Germany
haroske@minet.uni-jena.de

Desmond J. Harris
Pryske Cottage
Lianishen, Chepstown
Gwent NP5 6QD, U.K.

Peter Hästö
University of Helsinki
Kaitalahdenranta 17
02260 Espoo, Finland
peter.hasto@helsinki.fi

Abdolkarim Hedayatian
Department of Mathematics
College of Sciences, Shiraz University
Shiraz 71454, Iran
hedayat@math.susc.ac.ir

Matthias Hieber
TU Darmstadt
FB Mathematik, AG 4
Schloßgartenstr. 7
D-64289 Darmstadt, Germany
hieber@mathematik.tu-darmstadt.de

Reinhard Hochmuth
Institut für Mathematik 1
FU Berlin
Arnimalallee 2–6
D-14195 Berlin, Germany
hochmuth@math.fu-berlin.de

Yavdat Ilyasov
Bashkir State University
Department of Mathematics
Ufa 450074, Russia
IlyasovYS@ic.bashedu.ru

Niels Jacob
Univ. of Wales Swansea
Department of Mathematics
Singleton Park
Swansea SA2 8PP, U.K.
N.Jacob@swansea.ac.uk

Jutta Jäger
Fakultät für Mathematik & Informatik
Friedrich-Schiller-Universität Jena
D-07740 Jena, Germany
jaeger@minet.uni-jena.de

Jon Johnsen
Department of Mathematical Sciences
Aalborg University
Fredrik Bajers Vej 7G
9220 Aalborg East, Denmark
jjohnsen@math.auc.dk

Gennadiy A. Kalyabin
Samara State Aerospace University
Klinicheskaya 24-52
443096 Samara, Russia
kalyabin@mb.ssau.ru

Anna Kamont
Mathematical Institute
Polish Academy of Sciences
ul. Abrahama 18
81-825 Sopot, Poland
a.kamont@impan.gda.pl

Ron Kerman
Department of Mathematics
Brock University
500 Glenridge Avenue
St. Catharines
Ontario, L2S 3A1, Canada
rkerman@spartan.ac.brocku.ca

Vakhtang Kokilashvili
M. Alexidse 1
A. Razmadze Mathematical Institute
380093 Tbilisi, Georgia
kokil@imath.acnet.ge

Victor Kolyada
Department of Mathematics
University Carlos III of Madrid
Avda. Universidad, 30
28911 Leganes (Madrid), Spain
vkolyada@math.uc3m.es

Volodymyr Koshmanenko
Institute of Mathematics
Ukrainian National Academy of Sciences
3, Tereshchenkivska str.
03601, Kyiv, Ukraine
kosh@imath.kiev.ua

Miroslav Krbec
Mathematical Institute
Academy of Sciences
Žitna 25
11567 Praha 1, Czech Republic
krbecm@math.cas.cz

Alois Kufner
Mathematical Institute
Academy of Sciences
Žitna 25
11567 Praha 1, Czech Republic
kufner@math.cas.cz

Thomas Kühn
Mathematisches Institut
Fakultät für Mathematik & Informatik
Augustusplatz 10/11
D-04109 Leipzig, Germany
kuehn@mathematik.uni-leipzig.de

Denis Labutin
Department of Mathematics
ETH Zürich
CH-8092 Zürich, Switzerland
denis@math.ethz.ch

Rüdiger Landes
Department of Mathematics
Physical Sciences Center
601 Elm Street
University of Oklahoma
Norman, Oklahoma 73019, U.S.A.
rlandes@ou.edu

Massimo Lanza de Cristoforis
Università di Padova
Dipartimento di matematica
pura ed applicata
via G. Belzoni, 7
35131 Padova, Italy
mldc@math.unipd.it

Hans-Gerd Leopold
Mathematisches Institut
Fakultät für Mathematik & Informatik
Friedrich-Schiller-Universität Jena
D-07740 Jena, Germany
leopold@minet.uni-jena.de

Uwe Luther
Fakultät für Mathematik
TU Chemnitz
D-09107 Chemnitz, Germany
uwe.luther@mathematik.tu-chemnitz.de

Jan Maly
Department of Math. Analysis
Faculty of Mathematics & Physics
Charles University
Sokolovská 83
186 00 Praha 8, Czech Republic
maly@karlin.mff.cuni.cz

Vladimir Mazya
Department of Mathematics
Linköping University
S-58183 Linköping, Sweden
vlmaz@mai.liu.se

Osvaldo Mendez
University of Texas at El Paso
School of Mathematical Sciences
500W University Ave., 124 Bell Hall
El Paso, TX 79968, U.S.A.
mendez@math.utep.edu

Akihiko Miyachi
Department of Mathematics
Tokyo Woman's Christian University
Zempukuji, Suginami-ku
Tokyo 167-8585, Japan
miyachi@twcu.ac.jp

Mohammad Reza Molaei Taherabadi
Shahid Bahonar University of Kerman
Department of Mathematics
Faculty of Mathematics
& Computer Science
P.O. Box 76135-133, Kerman, Iran
mrmolaei@arg3.uk.ac.ir

Kazem Mosaleheh
Department of Mathematics
College of Sciences, Shiraz University
Shiraz 71454, Iran
mosaleheh@math.susc.ac.ir

Susana Magarida Domingues de Moura
Departamento de Matemática
Universidade de Coimbra
Apartado 3008
3000 Coimbra, Portugal
smpsd@mat.uc.pt

Vesa Mustonen
Department of Mathematics
University of Oulu
SF 90570 Oulu, Finland
Vesa.Mustonen@oulu.fi

Joachim Naumann
Institut für Mathematik
Humboldt-Universität zu Berlin
Unter den Linden 6
D-10099 Berlin, Germany
jnaumann@mathematik.hu-berlin.de

Júlio Severino das Neves
Departamento de Matemática
Universidade de Coimbra
Apartado 3008
3000 Coimbra, Portugal
jsn@mat.uc.pt

Erich Novak
Mathematisches Institut
Fakultät für Mathematik & Informatik
Friedrich-Schiller-Universität Jena
D-07740 Jena, Germany
novak@minet.uni-jena.de

Igor Novikov
Voronezh State University
Department of Mathematics
University Square 1
394693 Voronezh, Russia
igorno@icmail.ru

Petri Ola
Department of Mathematics
University of Oulu
SF 90570 Oulu, Finland
petri.ola@oulu.fi

David Opěla
Department of Mathematics
Campus Box 1146
One Brookings Drive
St. Louis, Missouri 63130-4899, U.S.A.
opela@karlin.mff.cuni.cz

Bohumír Opic
Mathematical Institute
Academy of Sciences
Žitna 25
11567 Praha 1, Czech Republic
opic@math.cas.cz

Peter Oswald
Bell Labs, Lucent Technologies
Room 2C-403
600 Montain Av.
Murrey Hill, NJ 07974-0636, U.S.A.
poswald@research.bell-labs.com

Serap Oztop
Istanbul University
Faculty of Sciences
Department of Mathematics
34459 Vezneciler-Ýstanbul, Turkey
serapoztop@hotmail.com

Lassi Päivärinta
Department of Mathematics
University of Oulu
SF 90570 Oulu, Finland
lassi@rieska.oulu.fi

Stanislav I. Pohožaev
Steklov Mathematical Institute RAN
ul. Gubkina, 8
117 966 Moscow GSP-1, Russia
pohozaev@mi.ras.ru

Evgeniy Pustylnik
Department of Mathematics
Technion – Israel Institute of Technology
Haifa 32000, Israel
evg@techunix.technion.ac.il

Jiří Rakosník
Mathematical Institute
Academy of Sciences
Žitna 25
11567 Praha 1, Czech Republic
rakosnik@math.cas.cz

Thomas Runst
Mathematisches Institut
Fakultät für Mathematik & Informatik
Friedrich-Schiller-Universität Jena
D-07740 Jena, Germany
runst@minet.uni-jena.de

René L. Schilling
Centre for Mathematical Analysis
and its Applications
University of Sussex at Brighton
Falmer
Brighton BN1 9QH, U.K.
r.schilling@sussex.ac.uk

Hans-Jürgen Schmeißer
Mathematisches Institut
Fakultät für Mathematik & Informatik
Friedrich-Schiller-Universität Jena
D-07740 Jena, Germany
mhj@minet.uni-jena.de

Veli Shakhmurov
University of Istanbul
Department of Electronic Engineering
Avcilar, 34850 Istanbul, Turkey
sahmurov@istanbul.edu.tr

Tatyana Shaposhnikova
Department of Mathematics
Linköping University
S-58183 Linköping, Sweden
tasha@mai.liu.se

Winfried Sickel
Mathematisches Institut
Fakultät für Mathematik & Informatik
Friedrich-Schiller-Universität Jena
D-07740 Jena, Germany
sickel@minet.uni-jena.de

László Simon
Department of Applied Analysis
Eötvös Loránd University
Múzeum krt 6-8.
H-1088 Budapest, Hungary
simonl@ludens.elte.hu

Leszek Skrzypczak
Faculty of Mathematics and
Computer Science
Adam Mickiewicz University
Matejki 48/49
PL-60-769 Poznań, Poland
lskrzyp@math.amu.edu.pl

Michael Solomyak
Department of Theoretical Mathematics
The Weizmann Institute of Science
Rehovot 76100, Israel
solom@wisdom.weizmann.ac.il

Vladimir Stepanov
Computer Centre of
Russian Academy of Sciences
Shelest 118-205
680 042 Khabarovsk, Russia
stepanov@as.fe.ru

Kazuya Tachizawa
Mathematical Institute
Tohoku University
Sendai 980-8578, Japan
tachizaw@math.tohoku.ac.jp

Hans Triebel
Mathematisches Institut
Fakultät für Mathematik & Informatik
Friedrich-Schiller-Universität Jena
D-07740 Jena, Germany
triebel@minet.uni-jena.de

Gunther Uhlmann
University of Washington
Department of Mathematics
Box 354350
Seattle, WA 98195-4350, U.S.A.
gunther@math.washington.edu

Mansour Vaezpour
Department of Mathematics
College of Sciences
Yazd University
Yazd, Iran
vaez@yazduni.net

Jan Vybíral
Department of Math. Analysis
Faculty of Mathematics & Physics
Charles University
Sokolovská 83
186 00 Praha 8, Czech Republic
vybiral@karlin.mff.cuni.cz

Tobias Werther
Institut für Mathematik, NUHAG
Universität Wien
Strudlhofgasse 4
A-1090 Wien, Austria
a9301647@unet.univie.ac.at

Anja Westerhoff
Mathematisches Institut
Fakultät für Mathematik & Informatik
Friedrich-Schiller-Universität Jena
D-07740 Jena, Germany
anja1@minet.uni-jena.de

Dachun Yang
Mathematisches Institut
Fakultät für Mathematik & Informatik
Friedrich-Schiller-Universität Jena
D-07740 Jena, Germany
dcyang@minet.uni-jena.de

Abdellah Youssfi
Université de Marne-la-Vallée
Cité Descartes, 5, boulevard Descartes
Champs sur Marne
77454 Marne-la-Vallée Cedex 2, France
youssfi@math.univ-mlv.fr